AF327151

MANGANESE ORES OF SUPERGENE ZONE: GEOCHEMISTRY OF FORMATION

Solid Earth Sciences Library

Volume 8

Manganese Ores of Supergene Zone: Geochemistry of Formation

by

IGOR M. VARENTSOV

Geological Institute,
Russian Academy of Sciences,
Moscow, Russia

KLUWER ACADEMIC PUBLISHERS

DORDRECHT / BOSTON / LONDON

A C.I.P. Catalogue record for this book is available from the Library of Congress

ISBN 0-7923-3906-1

Published by Kluwer Academic Publishers,
P.O. Box 17, 3300 AA Dordrecht, The Netherlands.

Kluwer Academic Publishers incorporates
the publishing programmes of
D. Reidel, Martinus Nijhoff, Dr W. Junk and MTP Press.

Sold and distributed in the U.S.A. and Canada
by Kluwer Academic Publishers,
101 Philip Drive, Norwell, MA 02061, U.S.A.

In all other countries, sold and distributed
by Kluwer Academic Publishers Group,
P.O. Box 322, 3300 AH Dordrecht, The Netherlands.

Printed on acid-free paper

Printed in the Netherlands

TABLE OF CONTENTS

PREFACE											xi

1. INTRODUCTION										1

1.1. Studies before 1990s								1
1.2. Investigations from 1930s–1990s							6
1.3. Methodical Aspects								15

2. MANGANESE ORES IN WEATHERING CRUSTS						20

2.1. Introduction									20
   2.1.1. Formulation of the Problem. Definitions					20
2.2. Tropical Weathering Crusts							23
   2.2.1. Model Test Area: Groote Eylandt Deposit, North Australia			23
      2.2.1.1. Introduction								23
      2.2.1.2. Geological Setting							25
      2.2.1.3. Stratigraphy								25
      2.2.1.4. Structure								27
      2.2.1.5. Manganese Oxyhydroxide Ores						30
         2.2.1.5.1. Structure of the ore bed					32
         2.2.1.5.2. Ore types							32
         2.2.1.5.3. Mineral and chemical composition of the ores			38
      2.2.1.6. Clay Components of Enclosing Sediments				59
      2.2.1.7. Geochemistry: Distribution of the Major Components
           and Trace Elements							70
      2.2.1.8. Paragenetic Assemblages of Components				94
      2.2.1.9. K–Ar Age of the Mn Oxyhydroxide (Cryptomelane) Ores		102
      2.2.1.10. Problem of Genesis							105
      2.2.1.11. Conclusion								110
   2.2.2. Model Test Area: Moanda Deposit, Gabon, West Africa			110
      2.2.2.1. Introduction								110

2.2.2.2. Geological Setting: Structure, Lithology, Mineral
Composition, Geochemistry of the Ore-Bearing Horizon ... 112
2.2.2.3. Genesis: Ore Formation as a Result of Deep
Supergene Alterations (Lateritization) ... 115
2.2.3. Some West African Deposits (Nsuta, Ghana; Mokta,
Côte d'Ivoire; Tambao, Burkina Faso (Upper Volta)) ... 118
2.2.3.1. Geological Setting ... 118
2.2.3.2. Mineralogy and Geochemistry of the Initial Stages
of the Protore Alteration ... 120
2.2.3.2.1. Nsuta ... 121
2.2.3.2.2. Mokta ... 123
2.2.3.2.3. Tambao ... 124
2.2.3.3. Mineralogy and Geochemistry of the Late Stages of
Oxyhydroxide Ore Formation (Manganese Cap (Cuirass)) ... 124
2.2.3.3.1. General characteristics ... 124
2.2.3.3.2. Mokta ... 126
2.2.3.3.3. Tambao ... 126
2.2.3.3.4. Nsuta ... 127
2.2.3.4. Features of the Protore Alteration
(Formation of the Primary Residual Ores) ... 127
2.2.3.5. Formation History of the Major Manganese
Ore Minerals ... 128

3. MANGANESE ORES IN KARST ... 132

3.1. Introduction ... 132
3.1.1. Manganese Ore Karst in the General System of Ground
Water Activity ... 132
3.1.2. Formulation of the Problem ... 134
3.2. Model Test Site: Zone of Oxidation and Accumulation
of the Karst Ores, the Úrkút Manganese Deposit, Hungary
(Mineralogy, Geochemistry, and Genesis) ... 134
3.2.1. Introduction ... 134
3.2.2. Geological Setting ... 136
3.2.3. Mineralogy of the Oxidized Ores ... 139
3.2.4. Geochemistry ... 148
3.2.4.1. The Major Components ... 148
3.2.4.2. Heavy Metals ... 150
3.2.4.3. Rare Earth Elements (REE) ... 152
3.2.5. On Genesis ... 167
3.3. The Giant Manganese Ore Accumulations in Paleokarst: Deposits of
the Postmasburg Region, Griqualand West, Cape Province, South Africa ... 170
3.3.1. Geological Setting ... 170
3.3.2. Genesis ... 173

3.3.2.1. The General Features   173

3.3.2.2. The Main Manganese Deposits: Bishop, Lohatlha, Glosam,
and Kapstewel, Postmasburg Region, Griqualand West   176

3.3.2.3. Processes of the Manganese Ore Formation   180

3.4. Manganese Ore Deposits of North Africa: Imini, Morocco   188

3.4.1. Geological Setting   188

3.4.2. Alterations with the Retaining of the Relatively
Stable Rock Structure   193

3.4.3. Dissolution, Formation of Cavities, and Their Filling   194

3.4.4. Genesis of the Manganese Ore Accumulations   194

3.5. The Other Mn Ore Accumulations in the Oxidation Zone
and Paleokarst   196

3.5.1. Zone of Oxidation and Paleokarst Accumulation of the
Manganese Ores in the Deposits of the Western Ukraine   196

3.5.2. Zone of Oxidation and Karst Ore Accumulation of the
Enisei Ridge Mn Deposits (Central Siberia)   197

3.5.2.1. Geological Setting   197

3.5.2.2. Mode of Occurrence, Lithology, Mineralogy, and
Genesis of Primary (Hydrothermal-Metasomatic) Ores   198

3.5.2.3. Zone of Oxidation and Karst Formation   201

3.6. Manganese Mineralization and Ores in the Regions of the
Intensive Karst Development   203

3.6.1. Manganese Mineralization in Modern Karst Caves   205

3.6.2. Manganese Ore Occurrences in the Karst Regions   206

3.7. On the Geochemical Model of the Karst Mn Ore Formation   210

4. MODEL OF MANGANESE ORE FORMATION IN THE
SUPERGENE ZONE   213

4.1. Formulation of the Problem   213

4.2. Data of Natural Observations   214

4.2.1. Alteration of the Silicate Protore   214

4.2.1.1. Early Stages of Alterations (Transformation of Tephroite)   215

4.2.1.1.1. Stages of Mn(II)-smectite formation   215

4.2.1.1.2. Stage of manganite (γ-MnOOH) formation   215

4.2.1.2. Major States of Weathering (Alteration of Mn-Garnet)   216

4.2.2. Alteration of Carbonate Protore   216

4.2.3. Formation of Manganese Cap (Cuirass)   217

4.3. Experimental Data   219

4.3.1. Introduction   219

4.3.2. Sorption Synthesis of the Oxyhydroxide Phases of
Mn, Fe, Ni, and Co on Iron Oxyhydroxides   219

4.3.2.1. Main Results   219

4.3.2.2. Discussion   226

4.3.3. Sorption Synthesis of the Oxyhydroxide Phase of
Mn, Fe, Ni, Co, and Cu on Manganese Oxyhydroxides 227
4.3.3.1. Experimental 229
4.3.3.2. Results 230
4.3.3.3. Discussion 237
4.3.3.4. Geochemical Interpretation 240
4.3.3.5. Sorption Synthesis of the Oxyhydroxide Phase of
Mn, Fe, Ni, Co, and Cu on 7 Å MnO_2 (Birnessite) 240
4.3.4. The Effect of the Major Ions of the Background Electrolyte
Solution on Sorption of the Heavy Metals 242
4.3.4.1. Formulation of the Problem 242
4.3.4.2. Sorption of Cu(II) from Seawater and NaCl Solutions 243
4.3.4.3. The Role of the Major Ions of the Background Electrolyte
Solution 250
4.3.4.4. Structural Aspects of Electrolyte Aqueous Solutions,
Ion Hydration, Salting out: Structural Properties
of Water near the Solution/Solids Phase Interphase 253
4.3.4.5. Problems of Salting Out 255
4.3.4.6. Structural Properties of Water near the Solution/Solid
Phase Interface 255
4.3.4.7. Features of Uptake of Cu(II) and Other Heavy Metals:
The Relationships between Various Kinds of Sorption 258
4.3.4.8. On the Effect of Major Ions of Electrolytes on
Cu(II) Sorption by Manganese Oxyhydroxides 261
4.3.5. On the Role of Organic Matter in the Sorption Process
of Transition Metals 264
4.3.5.1. The Interaction between Organic Matter and Metals 267
4.3.5.2. Role of Dissolved Organic Matter in the Process
of Fe(II) Sorption by Birnessite 271
4.3.5.3. Role of Dissolved Organic Matter (Citrate-Ion) in
the Process of Cu(II) Sorption by Birnessite 272
4.3.5.4. Experimental Data on Cu(II) Sorption by Birnessite
at Different Concentrations of the Complexing
Component (EDTA) 273
4.3.6. Geochemistry of Formation Processes of Mn Oxyhydroxide
Minerals 274
4.3.6.1. Early Stages: Initial Oxidation of Mn(II) to Mn(III) 274
4.3.6.2. Major Stages: Oxidation of Mn(II) to Mn(III) and Mn(IV) 276
4.4. Conclusion: On the Geochemical Model 279

5. EVOLUTION OF MANGANESE ORE FORMATION IN
THE EARTH'S HISTORY 282

5.1. General Classification of Manganese Deposits 282

5.2. On the Relationship of Formation of the World Ocean
Metalliferous Sediments and Oxyhydroxide Nodules and Crusts   284
    5.2.1. Metalliferous Sediments   284
    5.2.2. Mn–Fe Crusts and Nodules   286
5.3. Geochemical Evolution of Mn and Mn–Fe Ore Formation
in Geological History   288

BIBLIOGRAPHY   303

INDEX   333

PREFACE

The significance of manganese ores is very well known in cast iron and steel production, as well as in various types of chemical raw material and agricultural fertilizers. The world industry development in recent years requires their increased production in the vicinity of the metallurgical centers in different regions of the world; high grade manganese and associated metal ores are needed. Analysis of the world production and consumption of manganese ores by industrial countries indicates convincingly that the highest commercial value belongs to the ores associated with the supergene zone (National Minerals Advisory Board, 1981; Coffman and Palencia, 1984; Doncoisne, 1985; Jones, 1990, 1991; Manganese, 1990; McMichael, 1989).

The remarkable property of manganese, in contrast to many other types of mineral resources, is that the ore accumulations of this metal are distributed in the wide geochronological interval from the Archean to the present time; these ores are deposited in basins and supergene environments of different types from lakes, internal seas to pelagic and abyssal regions of the World ocean, as well as different types of weathering crusts and karst. At the same time the manganese accumulations and features of their mineral and chemical compositions are relatively sensitive indicators, reflecting facies and geodynamic conditions of their formation. These properties aid the investigation of the Earth's evolution processes.

The aim of the present work is to develop the general geochemical model of the supergene manganese ore formation, that would reflect the evolution of these processes in the Earth's history. This model may be designed as a synthetic generalization of partial geochemical models, describing the processes of manganese ore formation in the main types of the supergene environment.

Some 30 years ago the author attempted to generalize the known world data on manganese ores on the basis of detection of manganese-bearing formations. It was emphasized (Varentsov, 1964, p. 4) that "Such an attempt to make apparent the paragenetic associations of manganese deposits with their country rocks can only be regarded as the first step in unravelling the factors controlling their distribution and formation at different stages of the Earth's geological history. The next step is research on the lithological and geochemical associations between the manganese ores and their host rocks in which the problems of the factors controlling these associations can be solved. Among these factors are: location in a definite climatic zone, particulars of the physico-chemical environment of the basin of sedimentation, the nature of the source of the ore material, the determination of the part taken by biological processes in the ore formation, etc."

On the basis of the roughly generalized main regularities of the manganese ore occurrence in the Earth's crust (Varentsov, 1964) the author tries to solve the following problems in the present work:

1. To present the results of detailed investigations of carefully chosen types of ore formation settings or some study areas for the creation of definite geochemical models of the process, which would describe all the major factors controlling the ore formation.

2. Generalization of the available information on the given genetic type of manganese ore formation on the basis of the definite study sites or settings of this type.

3. Experimental modeling of the main links of manganese ore formation processes under laboratory conditions, with regard to the data on physico-chemical and biochemical properties of these phenomena to determine the reaction mechanism and to create relatively general geochemical models of the main types of the ore formation settings.

4. Study of ancient supergene manganese ore deposits of the basic genetic types with relatively simple and distinct relationships of the major facies elements to develop particular geochemical models. The data on modern manganese ore formation under conditions of relatively similar genetic types were used for this purpose too.

5. Generalization of the available information on the basic types of modern and ancient manganese-bearing accumulations, allowing the understanding of the manganese ore formation evolution in the Earth's history. Untangling the geochemical essence of the main historical stages of the manganese ore formation as the result of the geodynamic regime and facies evolution.

6. Preparation of the general geochemical model, enabling to characterize the processes of supergene manganese ore formation in the Earth's history without any contradictions.

These investigations were carried out within the frames of N.S. Shatskii's program (before 1960) on the study of the regularities of mineral resource distribution. In 1965 the author formed a group of physico-chemical experimental modelers, investigating the mineral-ore formation under the auspices of the Laboratory of Lithology of Volcanogenic-Sedimentary Formations of GIN Acad. Sci. U.S.S.R. The parameters of the observed natural events were accepted as the fundamentals of the conducted experimental investigations, and the interpretation of the obtained results was performed on the basis of modern physico-chemical and biochemical concepts and data on processes in modern basins. Thus, experimental investigations represented an integral part of the study, conducted in conjunction with the studies of modern and ancient accumulations of manganese and associated metals.

Data on the supergene manganese deposits of the world, compiled and investigated during these years, was synthesized in the light of the known information into a general geochemical model of manganese ore formation.

This work was initiated by N.S. Shatskii (1956–1960) and later was headed by N.M. Strakhov (1960–1965) at the laboratories of the Geological Institute of the Acad. Sci. U.S.S.R. It is necessary to emphasize that many problems which were considered in the work were subjected to thorough and productive consideration by laboratory personnel, they were repeatedly discussed at the sessions of both the Lithology Sector of GIN Acad. Sci. U.S.S.R. and the Institute Scientific Council.

It should be pointed out that since 1968 the author has been a national representative of Russia and a Secretary of the Commission on Manganese of the International Association on the Genesis of Ore Deposits (IAGOD), and since 1976 he has been a scientific coordinator of a Project on the Manganese Geology and Geochemistry of the International Geologic Correlation Program of the UNESCO-IUGS. It promoted the direct acquaintance with the latest international information on deposits and allowed the discussion of problems of manganese ore formation with well-known specialists: K. Boström, B.R. Bolton, G.P. Glasby, Gy. Grasselly, R. Giovanoli, D.S. Cronan, K. Lalu, F. Manheim, S. Roy, R.K. Sorem, L. Frakes, J.R. Heine, and P. Halbach *et al.*

The author is grateful to these above mentioned persons. The author expresses a deep gratitude to the workers of the Physico-Chemical Experimental Group, many of whom he has worked with since 1965 (Z.M. Rivina, Z.A. Petrakova, N.V. Bakova, V.S. Putilina, L.V. Zaitseva, N.I. Kartoshkina, and N.Yu. Vlasova).

My sincere thanks are due to my late colleague and friend Dr. Alexander M. Leites whose assistance was helpful.

Chapter 1

INTRODUCTION

MODERN STATE OF THE ART OF THE PROBLEM*

> "... The History is not in what we wore, but it is in that, how they let us go naked." (B.L. Pasternak)

> "We observe continuously in the history of science, that one or another thought, one or another phenomenon passes unnoted for more or less long time, but then under new external conditions they disclose before us their unlimited influence over scientific outlook." (V.I. Vernadskii,* p. 230)

1.1. Studies before 1930s

Manganese ores were known to mankind since the ancient times irrespective of the fact that they were often confused with iron and magnesium ores. Particularly widely known were ore minerals such as pyrolusite, that was described for the first time by Plinium as the variety of the iron magnetic oxide (magnetic iron ore or Lapis magnes). The reputation of pyrolusite can be associated with that the fact that remarkably it removes the green tint, imparted by iron oxides to even light varieties of glass, that are the usual mixtures of the quartz sands. Plinium explained the origin of the word magnes, attributing it to the name of a legendary herdsman Magnes, who paid attention to the fact that the iron nails of his boots and the tip of his stick were pulled to the ground in those places where magnetic iron ore deposits were located (Sully, 1955), but there are many other versions in the literature. In manuscripts of the XIVth century, kept in the British Museum, such 'new' terms are used to designate different matters such as magnesia ferrea and magnesia, and the latter was used as an additive for the production of violet glass. Probably, this terminological ambuigity led to the fact that in manuscripts of the medieval investigator Albertus Magnus (1193–1280) there was the term manganesis, that approximates closely to the modern word by its spelling and semantic meaning. But the particular term 'manganese' was used in the work by V. Biringuccio (Biringuccio, 1540, cited by Sully, 1955), who indicated in 1540,

---

*V.I. Vernadskii, 'About Scientific Outlook', in A.Ya. Yanshin (Ed.), *Scientific Thought as a Planetary Phenomenon*, Nauka, Moscow (1991), pp. 191–234.

that the mineral of this name occurred in Germany near the settlement Witerbo, Tusconia, as ferrous 'cinder', from which unlike the iron oxides it was impossible to produce metal during smelting, and that the sublimation was observed during the heating process, and this mineral turned to cinder after the vapor separation.

The discovery of manganese as a chemical element is usually attributed to K. Sheele (1774). In that year I.G. Gun got the first sample of the metal manganese. The general progress of the natural sciences, particularly, of chemistry and geology at the end of XVIIIth and the first half of XIXth century led to the strong scientific foundation for the development of the geology and geochemistry of manganese; this process was actively stimulated by the general industrial development, in particular by metallurgical production. Approximately in the middle of the XIXth century the method of mass production of cheap cast steel were discovered and applied in industry (Bessemer, 1856 and Siemens-Martin, 1864), in which manganese, introduced into the cast iron composition in the form of metal or ferromanganes, was used as a steel deoxidizer and desulfurizer, as well as for production of steels alloyed with manganese. In addition let us emphasize, that considerable quantities of manganese are used in catalytic processes and in organic synthesis.

Thus, in this period of time the cycle was completed, that was repeated many times during the later rather complicated multistage reversible processes of the evolution of geological science, which attributed the discovery of large deposits of manganese and associated metals to the development stages of fundamental scientific and methodological investigations (Varentsov, 1980). Such changes of cycles occur when the accumulated geologic information, including data on accidentally discovered deposits, achieves a critical volume and the available facts cannot be explained by the existing theoretical concepts, the necessity for new ideas and for an unambiguous explanation of all available sums of knowledge is therefore created. The appearance of such ideas raises unavoidably the problem of the development of new investigation methods. First results, obtained during the testing and developing of these ideas with the application of the latest experimental methods, accelerate significantly the development of science and practice along the closed circuit: idea – method (experiment) – practical realization and accumulation of information — new ideas ...

Description of the manganese geology and geochemistry development from the beginning of the XIXth century to the mid-1930s can be found in works by Vernadskii ('Geochemistry Outlines. ... History of Manganese', 1954a; 'Geochemistry of Manganese in Connection with the Doctrine on Mineral Resources', 1954b). The vigorous process of information accumulation on manganese and associated elements in different geologic conditions (rocks, minerals, natural waters, different organisms, and, in general, in media) during the XIXth and in the beginning of the XXth centuries led to the formulated cardinal problems of manganese geology and geochemistry, that in general may be represented by the following theses.

1. Deep contrast of manganese distribution (dispersion and accumulation) in the Earth's crust, that is controlled by endogenous, exogenous, and biochemical processes. "Manganese is present in each organism, in each microbe in quantities completely untraceable for us and nevertheless it gives birth to accumulations of billions of tons" (Vernadskii, 1954a, p. 74).

2. Absence of any significant manganese concentrations in the magmatism zone and in its products. "... The most stable and characteristic form of manganese in juvenile minerals is a solid isomorphic admixture of compounds in the bivalent state. One may be confident, that the pure manganese compound will be unstable in the Earth's shell (magmatosphere), and if it incidentally occurs, then it will unavoidably be subjected to new molecular regrouping and will enter new compounds that are less rich in manganese" (Vernadskii, 1954a, p. 75).

3. Formation of large manganese ore accumulations is associated with processes in the zone of free oxygen existence, in particular, in the hydrosphere: "Vadose manganese minerals represent a completely different picture. Compounds, rich in manganese, both exist and represent the most stable forms of its occurrence in the biosphere and stratosphere. We know of several dozens such minerals. The most common and stable is manganese dioxide – MnO_2. ... Vadose minerals – wads, psilomelanes, and pyrolusites – form the largest manganese accumulations, known to us" (Vernadskii, 1954a, pp. 76, 79).

4. At the end of the XIXth century it was shown by the studies of the World ocean, seas, lakes, and swamp basins that a tremendous ore formation process takes place at the Earth's surface in the Quaternary time. Evaluating the results of these works, Vernadskii wrote: "At the present epoch the wads are formed by large masses in different places on the bottom of the World ocean and neighboring seas. Wad concretions, invariably containing iron and manganese, are deposited in quantities of billions of tons on the bottom of the Indian, Atlantic, Pacific, and Antarctic oceans" (1954a, pp. 77–78); and further "... Similar aggregates, sometimes of other forms, are deposited in thousands of fresh water lakes and swamps, associated with the history of ancient glaciers, and the glaciation epoch of the northern countries – in Scotland, in Scandinavia, in Finland, in the northeastern part of Russia, in states of Canada and the Northern America – they are found here in rivers, where, however, their accumulations in bulk are rather rare. At present wads are formed or they were formed recently in historical and Pleistocene time. Annually hundreds of thousand of tons of manganese, are accumulated by this method."

5. The most important and remarkable point in the manganese geology and geochemistry study may be attributed to the first description of the Chiatura manganese deposit by Komarov in 1854. This deposit was characterized more thoroughly by G.V. Abich in 1858, who described for the first time the main features of the geologic structure, ore mineral composition, occurrence of the ore deposit in the bottom of the Tertiary sediments, reserve scales, and commercial significance of the manganese ores. In a relatively short time these works resulted in the beginning of the industrial exploitation of the Chiatura deposit in 1879, initiated by the Georgian poet A. Tsereteli (Betekhtin, 1964a).

Discovery of the Nikopol' manganese deposit in 1883 by mining engineer V.A. Domter led to the beginning of the systematic study and industrial exploitation of the largest Oligocene South Ukrainian manganese basin in Europe. In the monograph by N.A. Sokolov, *Manganese Ores of the Tertiary Sediments of the Ekaterinoslavskaya Region and the Vicinities of the Krivoi Rog*, published in 1901, the systematic data on manganese ores of the Nikopol' basin was presented for the first time (Betekhtin, 1964b).

Considering the scientific significance of these works, Vernadskii emphasized (1954a, p. 79), that these deposits were the Oligocene equivalents of relatively shallow sea sediments, subjected to relatively insignificant postdepositional alterations, and they may be correlated with the manganese accumulations of the modern basins. Furthermore, the first achievements of the manganese ore investigations at the south of Russia and Caucasus influenced greatly the general level of the manganese geology study in the world. With regard to the genesis of the Archean–Early Proterozoic manganese deposits in India, altered to the granulite stage, Fermor (1909) noted, that such gondite (i.e. spessartine) associations of the Dharwar Supergroup were equivalents of the above mentioned Tertiary deposits by the composition of the primary sediments.

6. In 1774 the Swedish chemist K. Scheele discovered the presence of manganese in plants, this was soon confirmed and accepted as the proved scientific fact by Alexander von Humboldt (Humboldt, 1793, see Vernadskii, 1954a). According to the following reports, the investigations demonstrated a definite high biologic, or wider, biogeochemical role of manganese, evaluating which Vernadskii stated: "We can now say, that billions of tons of manganese atoms, existing in the continuous biogeochemical motion, enter and abandon living matter" (1954, p. 82). Hence, in the beginning of the XXth century the extreme significance of biogeochemical processes in the formation of manganese ore accumulations was accepted by most researchers in spite of the fact, that existing methods of study allowed neither the essence nor mechanism of these processes to be understood, they were therefore acknowledged only as an empirical generalization, i.e. as a phenomenon. Giving the general evaluation to the results of these works, Vernadskii wrote: "In many cases with confidence, in other cases on the basis of quite valid hypothesis the wad formation may be considered as the result of biogeochemical processes, for example, the concretion formation on the bottom of the water basin and at the wet places of the Earth's surface. Observations and experiences of biologists and chemists G. Molish, Shorler, D. Jackson, M. Beierfink, B. Perfil'ev, B. Butkevich, and other mineralogists and oceanographers – D. Bukanan, N. Sokolov, J. Murrey and E. Phylippi, Ya. Samoilov and A. Titov – prove it with complete confidence" (1954a, p. 82). But recognizing the significant importance of the biochemical reactions, associated with bacterial activity, Vernadskii (1954a, p. 83) noted, that "... deep oxidation of the manganese compounds always takes place in association with living matter, to a greater or lesser extent." Developing this concept on the basis of limited number of rather approximated data available to the beginning of the 1930s, Vernadskii suggested the following scheme (1954, p. 86):

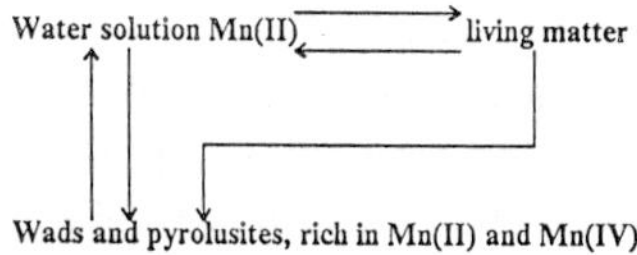

It is particularly remarkable that the real scientific content and biogeochemical processes, shown in general in this scheme, were only recently described in detail with the application of the newest instrumental analytical methods (Skinner and Fitzpatrick, 1992; Bruland, 1983; Landing and Bruland, 1987). Moreover, this scheme may be considered as a forerunner of the presently discussed model of the hydrogenetic accumulation of Mn(II) in nodules and crusts of the modern basins. The concepts are interesting because up to the mid-1970s most investigators were sharply divided between 'biological' or 'chemical' ways of the iron-manganese ore formation. This picture mirrored the state that existed in 1870s in biological chemistry and in natural science as a whole, when there was a vigorous discussion between Liebig and Pasteur about the nature of enzymes (ferments). Discussions stimulated the conduction of active investigations, which the achievements of chemistry and chemical kinetics of that time showed, that fermentative processes (biocatalysis) were a variety of chemical catalytic reactions, and they did not drastically differ from those reactions (Varentsov, 1980, 1989).

7. Investigations of manganese ore accumulations on the bottom of modern basins and in ancient sequences were accompanied by long and slightly productive discussions of 'neptunists' and 'plutonists' from the beginning of XIXth to the 1970s. The arguments, on which the ideas of supporters of one or another concept were based, were determined by the nature of investigated geologic objects. And so long as manganese ore formation under supergene conditions, in weathering crusts, swamp and lake basins, and internal seas was a relatively well-established scientific fact, then the supporters of 'plutonism' occupied a somewhat subordinate position, because they had insufficient direct observations. Having established a wide regional distribution of the Mn–Fe nodules and crusts in the World ocean by the results of the pioneer investigations at the scientific-research ship 'Challenger', Murray (Murray, 1876; Murray and Irvine, 1895; Murray and Renard, 1884, 1891) calculated, that annually 7.8×10^6 t of Mn is supplied by rivers into oceans (Murray, 1887, see Vernadskii, 1954a) and this quantity, in the author's opinion, was sufficient for the manganese accumulation in nodules and sediments. At the same time, the authors recognized the low temperature halmyrolitic process of leaching of this metal from basalt materials of the sea bottom as an additional manganese source. But at about this time K.W. von Gumbel (1881) proposed a concept of hydrothermal-sedimentary origin of these ores by the results of detailed geological and mineralogical investigations of Mn–Fe deposits and concretions in the Jurassic sediments of Bavaria, Alpine region and neighboring areas. Later, on the basis of investigations of Mn–Fe nodules and ore-bearing sediments of the World ocean, carried out during the first expeditions of the research vessels 'Challenger' and 'Gazelle', as well as of mineralization in the modern thermal springs, the reliability of the genetic concepts, proposed by von Gumbel, was confirmed, but it did not get a wide recognition among contemporaries (Gumbel, 1878, 1888a, 1888b, 1889, 1891). The concept of hydrothermal-sedimentary Mn ore formation was accepted with difficulties during the following century, that was illustrated by strong disputes in 1960s–1970s between N.M. Strakhov, who did not attribute any significance to endogenic influence on the sedimentation and ore accumulation in the ocean, and the 'plutonism' supporters (Strakhov, 1960a, 1962a, 1962b, 1965, 1968, 1974a, 1974b).

At present, when the hydrothermal-sedimentary processes of Mn ore formation, often distinguished with gigantic scales both in modern and ancient basins, are studied closely, and the main stages of these processes are experimentally simulated and theoretically calculated, one has to acknowledge that von Gumbel's ideas represent something more than the curiosity in the science history, and they retain considerable interest for the modern researchers.

1.2. Investigations from 1930s–1990s

From 1930s to 1970s the investigations of manganese geology and geochemistry were conducted both in Russia and in foreign countries under rather isolated conditions, with the extremely limited exchange of the scientific information and ideas appearing as publications. Such a state was determined by the general cultural-political situation, existing as a result of all-penetrating totalitarianism of those years in the U.S.S.R. Nevertheless the progress in this field of the Earth sciences in the U.S.S.R. was significant. This was associated with works of the scientific collectives, headed by A.G. Betekhtin, N.S. Shatskii, N.P. Kheraskov, N.M. Strakhov, D.G. Sapozhnikov, and others. Betekhtin – a known researcher in the geology of ore deposits, mineralogy, and geochemistry – made a significant contribution to the investigation of the Chiaturskoe and other manganese deposits. A particular place among his works occupies the monograph *Commercial Manganese Ores of the U.S.S.R.* (1946), summarizing the results of long-term investigations. Betekhtin's conclusions on the association of the Chiaturskoe deposit with the shallow-water coastal areas of the large European Oligocene basin is of long-standing scientific significance. The determined regularity on the facies control of the mineralogical zoning of Mn ores was confirmed with the examples of the South Ukrainian and North Ural deposits. Betekhtin's works on mineralogy and genesis of volcanogenic-sedimentary Mn deposits of the Southern Urals (1940, see References Betekhtin, 1946) were equally significant.

In the beginning of 1950s N.P. Kheraskov (1951) and N.S. Shatskii (1954, 1960, 1965) developed a doctrine on geologic formations as natural paragenetic associations of the rocks, in which minerals, in particular, manganese ores, play a role of the obligatory member of the paragenesis. The theory about the formations was especially fruitful concerning the volcanogenic-sedimentary associations, their natural vertical and horizontal series, for example, formations, associated with basic ('greenstone series') or acidic ('porphyry series') volcanism (Shatskii, 1954). The most important component of the doctrine on formations may be considered that it was the first to enable the verification of inconclusive discussions on the origin of ore matter, enclosed in the host rocks, and to apply some objectivity to the paragenesis of rocks and ores. The most important criterion of ore formations is their repetition in geological history and a rather high characteristic feature for definite development stages of structural regions and basins, as well as confinement to definite intervals of the Earth's crust evolution (for example, iron quartzite formations). At the same time both Kheraskov (1951) and Shatskii (1954, 1960, 1965) emphasized repeatedly, that the formational method should be considered as the initial or preliminary stage of investigations, preceding the detailed structural, lithological facies, and geochemical study of the geological object. Productivity of the formational method was manifested in that it stimulated an extraordinarily great number of works, including works on manganese ores, in which authors often only named different associations of the same rocks in order to emphasize the originality of their investigations (see review, Kosygin and Kulish, 1984; Roy, 1969), but they did not disclose the mineralization essence. Works by Strakhov and the collective headed by him introduced a considerable contribution into the study of the Mn ore formation geochemistry (Kholodov, 1983; Strakhov, 1947, 1960a, 1960b, 1962a, 1962b, 1964, 1965a, 1965b, 1968, 1974a, 1974b, 1976; Strakhov *et al.*, 1967, 1968).

Strakhov's interest in geochemistry was formed in the prewar years (1939–1941), when a series of articles on this problem was published, and the final results were presented as the monograph *Iron Ore Facies and their Analogs in the Earth's History* (1947). In this work on the basis of distribution analysis of bauxites, iron, and manganese ores at the idealized facies profile of some hypothetical basin, Strakhov represented the distribution relatior of ore accumulations of a triad: Al–Fe–Mn. In such an ideal basin under the conditi' of a humid zone with the supply of ore components from exogene sources the bauxites are accumulated within continental locations, the iron ore sediments are concentrated in continental and shallow parts of the basin, and the manganese ores, as a rule, are deposited in relatively deep areas of a sea. Here the accumulation of all components of this ore triad takes place in the lower parts of the transgressive series. The response to this work among geologists and geochemists, was extremely significant. The publication forced the author to consider the Al–Fe–Mn ore formation processes in the general context of the sediment formation process, that later was realized in the three-volume monograph *Fundamentals of the Lithogenesis Theory* (Strakhov, 1960a, 1960b, 1962a) and in the book *Lithogenesis Types and Their Evolution in the Earth's History* (1963). In these monographs there is a detailed presentation of concepts on four types of lithogenesis: humid, arid, glacial, and volcanogenic-sedimentary, existing both in modern conditions and in ancient sedimentation accumulations. But as applied to the mineral resources, this theory needed additional testing on the example of areas with relatively definite relationships of facies zones of the sedimentary basins, modern and of geologic past. To solve this problem, manganese was chosen as a sufficiently indicative informative metal of the ore triad, and the Oligocene manganese deposits of the European south part of the European U.S.S.R. (Nikopol', Bol'shoi-Tokmak, Labinskoe, Chiatura, and Mangyshlak) were chosen as the areas for the sedimentary ore formation investigations. In the period from 1959 to 1968 Strakhov headed the team of researchers, engaged in this problem (Strakhov, 1964, 1968; Strakhov *et al.*, 1967, 1968). The main conclusions may be presented with the following theses:

1. Accumulation of tremendous amounts of Mn ores, the origin of which was related with the exogene events on the source continent took place asynchronously, as Strakhov and others considered, in the Early–Middle Oligocene in separate coastal-shallow water and littoral parts of the South-European basin, mainly, in the environments of the humid climate.

2. The ore formation took place in the tectonically stable zones of the basins at the background of sharply relaxed terrigenous sedimentation, that did not deplete the processes of Mn ore accumulation.

3. At the largest Mn basins (South-Ukrainian and Georgian) the accumulation of Mn ore concentrations was controlled by their position at the facies profile of the basin, that was reflected in the consecutive change with the depth of the Eh regime, dissolved O_2, and directly in the mineral composition of ores: Mn oxyhydroxides (pyrolusite, cryptomelane, 10 Å-manganates, etc.) – manganite + Ca rhodochrosite – Ca rhodochrosite + manganocalcites.

4. Extraordinary role in the ore formation processes belongs to events of redox diagenesis, and this is reflected particularly clearly in the processes of manganite and rhodochrosite

ore accumulations.

These studies allowed to sum up the results of works on the Mn geochemistry, performed in the mid-1960s–beginning of 1970s, and they represented the basis, supporting Strakhov's concept on dominating significance, of four main types of the lithogenesis and related ore formation events. But the 1960s was the period of mature development of geodynamics as an independent branch of the science, and the wide distribution of the concept of plate tectonics drastically influenced the related subjects in the Earth's sciences. Wide-scale research of sedimentation and ore formation in the modern basins and, particularly, achievements in the study of the World ocean, associated with the International Deep Sea Drilling Project of the R/V 'Glomar Challenger', was accompanied by the increased volume of new data on geology and geochemistry of Mn and associated metals.

The first generalizations were presented in the series of Strakhov's publications (1965a, 1965b, 1968, 1974a, 1974b, 1976a, 1976b). These works significantly expanded the above mentioned items on the lithogenesis and sedimentary ore formation. But the increasing volumes of principally new, unknown earlier natural, factual and experimental material and a considerably higher level of the geochemical studies, reflecting progress in physics, chemistry, and instrumentation technology led to the publication of works that gave essentially different and new treatment of the Mn geochemistry problems; it is related mainly to the understanding of the nature and scales of hydrothermal-sedimentary processes, events of the authigenic formation of minerals, and particularly, the relationships of diagenetic and hydrogenetic processes in the formation of the Mn–Fe oxyhydroxide nodules and crusts, and metalliferous sediments, the study of mechanisms of the ore formation processes, the role of biogeochemical events, etc. The existing situation indicates the general progress in the given branch of geology, and, as a whole, that it is not something unusual in the development of natural sciences in the period of the scientific-technological revolution. The importance of Strakhov's works in the Russian lithology and geochemistry is well known. But in a number of western publications there is no adequate evaluation of Strakhov's works. Pointing out this situation, Kholodov (1986, p. 7) wrote in the 'Foreword of the Translation Editor' to the book by S. Roy *Manganese Deposits*, Mir, Moscow (1986):

> ...Attention is drawn to the fact, that the author emphasizes the investigations of 1970, being obviously poorly acquainted with Soviet works of the preceding period. So, for example, in the book there are many references to the papers by I.M. Varentsov, V.P. Rakhmanov, V.K. Chaikovskii and others, though classic monographs by A.G. Betekhtin *Commercial Manganese Ores of the U.S.S.R.*, Moscow, Leningrad: Izd. Acad. Nauk SSSR, 1946; F.V. Chukhrov *Colloids in the Earth's Crust*, Moscow: Izd. Acad. Nauk SSSR, 1936; and N.M. Strakhov *et al. Geochemistry of the Sedimentary Manganese Ore Process*, Moscow: Nauka, 1968 are not even referred to in this work.

How scientifically significant the concepts by Strakhov on lithogenesis and geochemistry of the sedimentary ore formation, and whether they have the potential of self-development, will only be known in the future. Here it is necessary to point out, that the most important feature of the scientific cognition as a collective but not individual process is its polysemantic pluralism of approaches, formulations and realizations of the appearing problems. Obviously the principles of progress of sciences on the Earth, and in particular, geochemistry of Mn ore-formation, is contained in it.

A major contribution to the study of Mn ore formation in the U.S.S.R. belongs to D.G. Sapozhnikov, who is an apprentice and a supporter of Strakhov. Inspite of the results of his first important investigation of Mn ores of the Kardzhal'skoe deposit in the Central Kazakhstan (Sapozhnikov, 1963) the author was compelled later to reconsider, admitting their hydrothermal-sedimentary genesis (Sapozhnikov, 1967), Sapozhnikov's works greatly stimulated the investigations in this region. Heading the investigations of Mn ores in the Institute of Geology of Ore Deposits of Academy of Sciences of the U.S.S.R., Sapozhnikov coordinated successfully the works on this subject in the U.S.S.R., headed the section of Al–Fe–Mn of the Interdepartment Lithological Committee and Council on Ore Formation of the Academy of Sciences of the U.S.S.R., and this was reflected in the success of several All-Union Meetings on Manganese Ores (Sapozhnikov, 1967, 1980, 1984; Sapozhnikov and Vitovskaya, 1987; Vitovskaya, 1985).

The intensive development of geological and geochemical works on study of poly-metallic Mn–Fe ores on the bottom of the World ocean mentioned above, that began in 1960s, was dictated by the assumption, that in future most part of these mineral resources will be recovered from the ocean floor due to the depletion of the deposits on the land. The results of long-term studies were published in a series of monographs and papers (Andreev, 1984; Anikeeva, 1984; Baturin, 1986; Bezrukov, 1970a, 1970b, 1976; Lisitsyn and Bogdanov, 1986; Lisitsyn *et al.*, 1976, 1987; Murdmaa and Skornyakova, 1986; Smirnov, 1979; Volokov, 1979, 1980). The investigations were primarily conducted by collectives of the Institute of Oceanology of the Academy of Sciences of the U.S.S.R. and Ministry of Geology (P.L. Bezrukov, A.P. Lisitsyn, Yu.A. Bogdanov, N.S. Skornyakova, G.N. Baturin, I.O. Murdmaa, I.I. Volkov, I.S. Gramberg, B.Kh. Egiazarov, S.I. Andreev, L.I. Anikeeva, V.S. Shcherbakov, V.M. Yubko, M.F. Pilipchuk, I.M. Mirchink, and others). The published works contain the results of studies on ocean Mn–Fe nodules and crusts, and metalliferous sediments, that may be considered as new types of ores of manganese, copper, nickel, cobalt, and other metals. It is vital to note, that the researches were carried on by the Institute of Oceanology of Academy of Sciences of the U.S.S.R. from 1959 to the present time, and since 1979 the Ministry of Geology of the U.S.S.R. took part in them. Abundant factual material, representing a fundamental basis for empirical generalizations, composes the main value of these works. The results of detailed explo-ration works for the Mn–Fe nodules, conducted at special sites – polygons – selected for the purpose of their further exploitation have a special importance. During the last two decades a number of institutes of the Academy of Sciences of the U.S.S.R., Academy of Sciences of the Ukrainian SSR, scientific-research enterprises of the Ministry of Geology of the U.S.S.R. and of other departments took part in these coordinated investigations. The main regional and local regularities of the distribution of Mn–Fe nodules and of the location of high-grade ore fields were described with reliability and completeness in the considered monographs. Described in these publications the regularities of Mn–Fe nodule distribution, in particular, the localization of the ore accumulations richest in Cu and Ni in the sublatitudinal, subequatorial belt (often named as the 'D. Horn's Zone' by some authors) of the east of the Pacific, were confirmed and studied in detail by the successive works (see the review by Teleki *et al.*, 1987).

Considerable contribution to the ore distribution study of the Indian and Atlantic oceans within the frames of these works was made by the team of the Institute of Geological Sciences of the Academy of Sciences of the Ukrainian SSR, headed by E.F. Shnyukov (Shnyukov and Mitropol'skii, 1987; Shnyukov and Orlovskii, 1979; Shnyukov *et al.*, 1974, 1979, 1984).

Significant successes in the studies of the formation processes of the Mn–Fe oxyhydroxide ores in the modern and ancient basins are associated with investigations of the team of the workers of the Institute of Geology of Ore Deposits (IGEM) of the Academy of Sciences of the U.S.S.R. and the Geological Institute of the Academy of Sciences of the U.S.S.R. It is necessary to admit, that the well-substantiated concepts on the structure and composition of supergenic oxyhydroxides of Mn and Fe compose the essence of the genetic models, representing the basic search criteria for the given type of the mineral resources. But the extremely low level of the structural ordering of the Mn and Fe oxyhydroxides and their high dispersion were an unsurmountable barrier until recently for the successful application to these objects of traditional methods of an X-ray structural analysis. Such problems were solved by the works of F.V. Chukhrov, A.I. Gorshkov, V.A. Drits, B.A. Sakharov and others (1975, 1976, 1978, 1980a, 1980b, 1983a, 1983b, 1987a, 1987b), owing to a new methical approach – application of highly local methods of the electron microdiffraction and an X-ray energy dispersion analysis. On the basis of newly obtained results by the mentioned authors the crystallochemical nature of the whole series of earlier unknown or slightly studied Mn and Fe oxyhydroxide minerals was unravelled. The structural studies of asbolanes, detection of the crystallochemical nature of buserites, determining the disordered nature of the mixed-layered minerals of the type of asbolane-buserite I, buserite I-buserite II, buserite-'defective lithiophorite' and so on represents a special interest. The success of the methods used may be shown by the fact that the wide distribution of MnO_2 in the modern basins and weathering crusts 10 Å-phase may be represented by at least one of seven minerals with clearly distinguishing structural parameters: asbolane, buserite I, buserite II, todorokite, mixed-layered phases of asbolne-buserite I, buserite I-buserite II, and buserite I-'defective lithiophorite'.

Except for the methical procedures mentioned in the works of these researchers, successful application of profile modeling of diffraction images for such widespread minerals as vernadite and birnessite, the effective complexing of diffraction methods with X-ray photoelectronic spectroscopy and fine structure of the X-ray spectra of absorption (EXAFS), were demonstrated. These procedures opened the possibility for the gathering of data on coordination and valency of the main cations, their local structure, distribution of oxygen ions, hydroxyl groups, H_2O molecules and so on in the mineral.

On the basis of obtained results and in the general context of the data on geology, sedimentology, geochemistry, the authors consider the problems of the genesis of the determined minerals and phases.

As it was mentioned above, in the western countries in the 1930s to the beginning of the 1960s the studies of Mn geology and geochemistry were conducted separately from the similar works in the U.S.S.R. The scientific problems and the level of these works were determined mainly by the general progress of geological sciences, achievements of chemical and physical methods of investigations and the presence of corresponding geological bodies. Such a situation may be illustrated with the example of the U.S.A. as the country, that like many industrial nations of the West, is virtually dependent on the import of Mn ores from abroad: from West and South Africa, Brasil, and India. Under these conditions the U.S. Geological Survey had to give a scientifically justified evaluation and prognosis of Mn deposits (Coffman and Palencia, 1984; DeYoung *et al.*, 1984; Doncoisne, 1985; Jones, 1991, 1992; Manganese, 1990; McMichael, 1989; National Minerals Advisory Board, 1981; U.S. Congressional Office, 1985). The detailed geological surveying of perspective areas, and the study of mineralogy and geochemistry, uncovered a great number of relatively small Mn ore deposits in volcanogenic-sedimentary sequences and the oxidation zone at the southwest of the U.S.A. The studies were conducted according to special program. The material obtained was thoroughly described and systematized in the works by D.F. Hewett and others (Fleischer, 1960; Hewett, 1964, 1968; Hewett and Fleischer, 1960; Hewett *et al.*, 1963, 1968). The investigations convincingly demonstrated the existence of proper hydrothermal accumulations of Mn and they preceded the discovery of the largest hydrothermal-sedimentary Mn deposits in South Africa (Kalahari, 79% of the Mn World reserves) and in the Central Kazakhstan. Hand in hand with these works L.S. Ramsdell, W.F. Cole, M. Fleischer, W. Richmond, J. Faust, M.D. Crittenden, A.D. Wadsley, R.K. Sorem and others laid the groundwork for the modern mineralogy of Mn oxyhydroxides with the application of X-ray structural and other physical methods of analysis available in those years (Buser *et al.*, 1954; Cole *et al.*, 1947; Crittenden *et al.*, 1962; Fleischer, 1960, 1964; Fleischer and Faust, 1963; Fleischer and Richmond, 1943; Fleischer *et al.*, 1962; Ramsdell, 1932; Richmond and Fleischer, 1942; Richmond *et al.*, 1969; Sorem and Cameron, 1960; Wadsley, 1950; Wadsley and Walkley, 1951; Zwicker *et al.*, 1962). After these studies the use of X-ray diffraction allowed the unambiguous use of such widespread mineral names as cryptomelane, ramsdellite, lithiophorite, rancieite, nsutite and others, and that produced significant progress in this relatively unstudied part of mineralogy (Rode, 1952). The results obtained as well as the data of the European mineralogists were systematized in Ramdohr and Frenzel's work (Ramdohr and Frenzel, 1956). A series of initiating researches on experimental synthesis of Mn oxyhydroxides were carried out by a team of Swiss scientists (Glemser *et al.*, 1961).

A symposium on the Mn deposits, that took place in 1956 in the frame of the XXth Session of the International Geological Congress, Mexico, and the publication of a five-volume proceedings of this forum were the significant events in geology. In this publication the data and study results on the most important deposits and ore occurrences of the world were for the first time systematically represented, and the main problems of manganese mineralogy and geochemistry were clarified, that reflected the level of the science development in those years (Reyns, 1956a–1956e). But under the regime of totalitarianism, geologists of Russia and former Republics of the U.S.S.R. could not take part in the symposium work and the data on Mn deposits of these regions were left practically inaccessible for the international geological community. In the Mn symposium proceedings there is no information on the Mn–Fe nodules and metalliferous sediments of the modern basins, because the second half of 1950s was the very beginning of wide-spread geological research of the World ocean and the geodynamic initiation, that was based on the concepts of the plate tectonics. Only later the investigations of specialists of many countries on Mn–Fe nodules, performed in the beginning of 1960s, were systematized and generalized in works by J. Mero (1962, 1965), who was the first to determine the general regularities of distribution of their chemical composition and, in particular, who distinguished the most prospective regions of the World ocean with high concentrations of Mn, Ni, Cu, Co and founded their commercial validity as tremendous deposits of high-grade natural-alloyed ores. Thus, generalizations of the international community from the 1950s–mid 1960s supported Vernadskii's statement that the important problems of geology and geochemistry had essentially global character and that their successful resolution was feasible only with close international cooperation.

Acknowledgment of the necessity of international scientific cooperation of geological and geochemical researches was developing and it became the urgent imperative of that time. In 1967 at the Second Symposium of the International Association on Genesis of Ore Deposits (IAGOD) in St. Andrews, Scotland, by initiative of Gy. Grasselly (University of Szeged, Hungary) the Working Group (W.G.) on Manganese (Commission since 1972) was formed, the aim of which was the initiation and coordination of international investigations on problems of manganese geology and geochemistry. I.M. Varentsov was elected its scientific secretary and coordinator in 1968 (Varentsov and Grasselly, 1969).

In 1970 at the first scientific and organizational session of the W.G. in the frames of the joint IMA–IAGOD Symposium, that was held in Tokyo–Kyoto, the main aims of the international researches were determined (Varentsov and Grasselly, 1970). At the session it was decided to hold the Second International Symposium on manganese geology and geochemistry in the frames of the 25th Session of the International Geological Congress, Sidney, Australia, in 1976 and to publish Proceedings in the form of a several volume monograph, reflecting achievements in these areas of the Earth sciences for the last 20 years, i.e. after the First Symposium. Later at a meeting of the IAGOD Commission on Mn, held regularly since 1970: in the frames of the 24th International Congress, Montreal, Canada, 1972 (Dorr, Varentsov, and Grasselly, 1971–1972), 4th Symposium of the IAGOD, Golden Sands, Varna, Bulgaria, 1974 (Graselly and Varentsov, 1974), the problems on the preparation of the 2nd Symposium on Mn and publications of Proceedings, held in close cooperation with other international organizations were discussed and corrected. For example, with the Commission on Marine Geology of the International Council of Geological Sciences (E. Seibold, University of Kiel, F.R.G.) main attention was focused on three problems: (1) mineralogy, geochemistry and methods of study of manganese ores; (2) manganese deposits on continents; (3) manganese on the floor of modern basin (Grasselly and Varentsov, 1975).

It is important to point out, that the time of the 2nd International Symposium on Manganese is related to the period of active development of the scientific-technological revolution, the achievements of which had an effect on the introduction of instrumental, physical, and chemical methods into geological and geochemical practice, allowing to obtain mass and relatively highly precise determinations of chemical composition and structural characteristics of the matter, and the wide-spread application of computers. Such methodical investigations advanced significantly the development of new equipment, that proved to be extremely fruitful in its dir⌐ct application to the geochemistry of manganese and associated metals. International and national programs of study of oceanic Mn–Fe nodules and International Deep Sea Drilling Project (DSDP, IPOD) on D.V. 'Glomar Challenger' verified the validity of main precepts of geodynamics-tectonics of lithospheric plates and revealed the major regularities of genesis of Mn ore accumulations as characteristic assemblages in the history of basins of the World ocean and continents. In a three-volume publication of Symposium Proceedings, united under the general heading *Geology and Geochemistry of Manganese* (Varentsov and Grasselly, 1980a–1980c) works on three main directions were included:

1. Significant reviews, reflecting the achievements in certain areas of geological science (in particular, mineralogy, geochemistry of Mn) of the previous decades and displaying progress in the most prospective directions. The necessity of such works, summing up the results for relatively long intervals of time, is extremely important for the wide community of geologists. Special place in such works occupies the consideration of known scientific ideas, for example, discussions of 'neptunists' and 'plutonists', supporters of the decisive role of microorganisms and inorganic genesis of manganese ores, followers and opponents of mobilism, geodynamic concepts and so on: for example, review by D. Crerar etc. (Crerar *et al.*, pp. 293–334, in Varentsov and Grasselly, 1980a).

2. Original investigations of geochemistry, mineralogy, and structural-tectonical aspects of geology of manganese deposits, in which new ideas were often put forward. These investigations, as a rule, were performed with the application of the newest methods, and have continuous scientific validity. The significance of such works is not only in that they signify the appearance of new scientific directions, but in that the publication of investigations of this type stimulates the concentration of efforts in the main directions and the creation of conditions for the development of new ideas and methods of their introduction into the general scientific practice. The illustration of this statement may be the physico-chemical and structural investigations by R. Giovanoli (Giovanoli, pp. 159–202, in Varentsov and Grasselly, 1980a) of fine dispersed Mn-oxyhydroxide phases, that provoked a vivid discussion on the structure and nature of 10 Å manganates and other minerals (Burns and Burns, 1979). This dispute was developed into an elaboration of the above mentioned original models of Mn oxyhydroxide structures (Bursill and Grzinic, 1980; Potter and Rossman, 1979; Turner and Buseck, 1979). The most fruitful was the application of methods of electron microdiffraction in combination with the X-ray energy dispersion spectroscopy and other physical methods (Chukhrov *et al.*, 1978, 1980a, 1980b, 1983a, 1983b, 1987a, 1987b, 1989).

3. New data on regional and global aspects of manganese deposits on continents and in the World ocean, that are extremely necessary for the creation of the factographical basis of many theoretical concepts and broad hypotheses of the Mn ore formation. For example, a number of propositions on the modern geodynamics at continents (problem of the Tethys and older ocean-like basins), and plate tectonics in the World ocean (problem of metalliferous sediments in axis zones) need to be more thoroughly investigated and in new factographical data for the better understanding of formation processes of minerals, in particular, manganese deposits. Materials of this series compose the greater part of the second and third volumes (Varentsov and Grasselly, 1980b, 1980c). The significant

event of the 2nd International Symposium on Manganese Deposits, Sidney, 1976, was the organization and beginning of activity of Project no. III 'Genesis of Manganese Ore Formations' of the International Program of Geological Correlation IUGS–UNESCO (leaders: Gy. Grasselly, Hungary, S. Roy, India, scientific coordinator I.M. Varentsov, U.S.S.R.). The aim of the project was the broadest coordination of scientific and practical investigations on the three above mentioned problems and practical introduction (in the form of help to developing countries) of obtained achievements by holding international meetings and sessions of the Working Groups on fundamental problems, publication of collected papers and monographs, and special publications of magazines. In the period of 1976–1986 the meetings of the Project were held each 1–2 years and the short reports were published in the journal *Geological Correlation*, UNESCO, Paris. For example, in the Soviet Union the united sessions of the Commission on Mn (Project IGCP No. III) were held during IAGOD Symposium, Tbilisi, 1982, with the visit to the Chiatura deposit and in the frames of the XXVIIth International Geological Congress, Moscow, 1984, with geological excursions to deposits of the Ukraine and Georgia. From 1986 to 1990 the International Cooperation was continued under the aegis of IGCP Project No. 226 'Correlation of manganese sedimentation with paleoenvironments' (leaders: S. Roy, India, B.R. Bolton, Australia), and since 1991 to the present time Project No. 318: 'Genesis and Correlation of Marine Polymetallic Oxides' (leaders: J.R. Hein, (B.R. Bolton, 1991–1993), S. Dasgupta). The distinguishing feature of these projects is that their workshops are held both in frames of Symposia, Congresses and during geological excursions to significant manganese deposits: 1986, International Sedimentological Congress, Canberra, Australia; 1987, Session at deposits Imini and others, Maroc; 1988, International Symposium 'Sedimentology in relation with mineral deposits', Beijing, with excursions to deposits of China; 1989, XXVIIIth Session IGC, Washington, U.S.A., with excursions to Urkut (Hungary) and Varna (Bulgaria) deposits; 1992, XXIVth Session of the IGC, Kyoto, Japan, with excursions to deposits of Hikkaido Island (the Hokkaido Field Workshop); 1993, 16th Colloquium of African Geology, Mbabane, Swaziland with excursions to iron and manganese deposits of the Transvaal Supergroup in Griqualand West, South Africa.

However, it is necessary to stress that with the increasing flow of information and accelerating rates of research of mineral resources, the Commission on Mn of IAGOD and corresponding Projects IGCP could coordinate only part of the investigations on Mn geology and geochemistry. A great number of such works is carried out in the frames of well-subsidized investigations of national programs of the U.S.A., Japan, F.R.G., France and other industrial countries, equipped with the latest instrumentation for land, marine and laboratory projects.

These studies are characterized by two marked tendencies: on one hand, they are directed, first of all, to the acquisition of high quality empirical data with by application of the newest technological achievements, and on the other hand, in some of these works there are attempts of conceptual synthesis of available information (Cannon and Force, 1983; Force and Cannon, 1986). The tendency to global correlation or probably broader consideration of studied events is an inherent feature for significant monographs of the last 20 years (Cronan, 1980; Glasby, 1977; Halbach *et al.*, 1988; Maynard, 1983; Rona *et al.*, 1893; Roy, 1981; Sorem and Fewkes, 1979; Teleki *et al.*, 1987).

1.3. Methodical Aspects

> "The task, ... which is to be solved by natural sciences, is to try to find some regularities in the unlimited variety of events, of the surroundings, in other words, to understand these heterogenous events, reducing them to simple concepts. It is necessary to try to derive a particular from general, to understand definite phenomenon as the result of simple and general laws." W. Heisenberg (1987b, p. 211)

"Geochemistry is a history of atoms in our planet body" – V.I. Vernadskii pointed out many times (1960, p. 272), referring to ideas of M. Faraday and Ch. Schonbein, outlined in their notebooks of 1830–1840s. This concept is wholly justified and for the modern geochemistry of the Mn ore formation, that may be considered as the history of formation of relatively high concentrations, in this case, of manganese and associated elements in the geological bodies. The distinguishing feature of the considered processes is that the Mn ore formation is the characteristic intrinsic unit in the general sequence of sediment accumulation, rock formation at the definite stage of the basin development or geological evolution of the region, corresponding to a certain paleobasin. Data known at the present time and results of investigations, conducted by the author, focused the main attention on the following major items of the work:

I.   Geochemistry of the processes of formation of Mn ores in relatively young modern settings.

II. Experimental research of the processes of Mn ore formation.

III. Geochemistry of Mn ore formation in paleobasins and the supergene settings of the geologic past.

IV. Geochemistry of Mn ore formation in the history of evolution of the Earth's crust.

V.   Geochemical model of Mn supergene ore formation.

The structure of the work is determined by the general logic of our knowledge of Mn geochemistry. Unlike many commercially valuable metals, the geochemical distinguishing feature of manganese is that its ore accumulations are concentrated in a broad range of environments, from weathering crust-lakes to vast abyssal regions of the World ocean. This feature determined the procedure of study of Mn ore formation.

The modern settings of supergene ore formation are considered as the systems, in which there may be distinguished three main structural categories: (a) source of ore forming components; (b) environment of their migration and accumulation; (c) proper ore accumulations. However the duration of ore formation during sedimentary and diagenetic processes may vary from several hours and days to many hundred million years. Thus, similar events are referred to as rarely observed or unobserved events. Along with extremely low rates of the formation of Mn mineral phases, that are so characteristic for settings with relatively low temperatures and pressures, a considerable obstacle for the study is inaccessibility of objects for direct observation and quantitative evaluation of the most important parameters, controlling these processes.

Therefore it is essential that in the geochemical literature the problem is discussed to what degree the chemical equilibrium is attained in natural sedimentary systems. Elucidation of this problem may be possible with a thorough experimental investigation. It is pertinent to note, that in geochemistry the increased experimentation is an extremely justifiable stage, reflecting the most important development trends of modern natural science, characterized by both wide application of technological achievements of the past and the development of complex interdisciplinary investigations. Attempts of synthesis of major theoretical concepts of the geochemistry of sedimentary Mn ore formation and experiment are expedient to consider as the problem of the interrelation of theoretical and experimental empirical knowledge. In other words, the experiment is the definite result of generalization of empirical knowledge, interpreted, in the light of modern theoretical concepts. The model, created on this basis, serves both for the verification of preliminary hypothetical concepts and for acquisition of fundamentally new information on the investigated phenomenon. To evaluate the quality of experimentally obtained information it is necessary to imagine precisely all those possibilities and restrictions, that allow experimental data to be related with real characteristics of natural events.

A distinguishing characteristic for the natural geochemical settings is that the more complicated the sediment ore forming system, the greater the role of the internal relationships, the more variable the influence of the external environments, and the more diverse and complicated the relations between different factors. Further there are fewer possibilities to learn with isolated investigation of component parts, and the greater is the significance of the approach to its integral investigation, i.e. the system. The modern system approach is necessary in order to avoid drawbacks, inherent to analytical methods of description of such complicated developing dynamic systems as Mn ore forming environments.

The above interrelations determine those conditions, which the natural objects should satisfy (lake, sea region, structural zone, deposit on land or test area), that was chosen as a model for detailed and possibly broadest investigation. The fundamental role of primary observation is the basis for the creation of particular models. Successes in adjacent fields of the natural sciences of the last few decades testify that significant progress may be obtained mainly with the deep study of key objects, creation of preliminary models as a result of their investigations, their successive verification, and determination of partial and general interrelations. The application of similar methods in this work is the detailed description of chosen model objects is given, and on this basis an attempt of broad typology of the investigated event is made, that may result in the preparation of the general model. Discussing the efficiency of such an approach in natural sciences W. Heisenberg (1987, p. 108) wrote:

> Attempting to understand events, we notice that any understanding begins from the perception of their similar features and regular relations. Some regularities are perceived then as peculiar cases which are general for different events and that therefore

may be called a fundamental concept. Thus, any attempt to understand the variable multiplicity of events leads with necessity to the search for the fundamental concept.

Rapid growth in the application of formalized models for geochemical study of ore forming processes has recently been observed. Similar models, in their turn, found wide application in rather complicated experimental investigations, that should both serve as the basis for the verification of those models (including computer ones) and represents a significantly new information basis for quantitative calculations and comparison of different versions of genetic models.

Broad application of thermodynamic methods for understanding of natural sedimentation geochemical systems benefitted significantly from the publication of the monographs by Garrels (1960) and Garrels and Christ (1965). Models, based on the equilibrium thermodynamic approach, were used for the investigations of low temperature geochemical processes. But with the application of these models (often in the form of computer programs) especially among geochemists (Potter, 1979), the contradictions became obvious. The contradictions were revealed between observed natural data and calculated results, obtained on the basis of accepted equilibria. The following main reasons for these discrepancies may be distinguished:
1. Inconsistency of the thermodynamic data, used as the model basis.
2. Modeling reactions are controlled by kinetic parameters.
3. Incomplete and inadequate understanding of the natural process mechanisms.
4. Simulated reactions are irreversible.

The attempts to use the thermodynamic approach have considerable difficulties, if one takes into consideration the fields of low temperature processes and biosphere distribution, i.e. the influence of life and its products. Moreover, Garrels and coworkers (Garrels, 1982; MacKenzie and Wallast, 1977) in their later works pay close attention to the fact that in the zone of supergene sedimentation–postsedimentation alterations, the geochemical processes are substantially subordinate to the kinetic control and, in a rather limited measure, they are determined by equilibrium thermodynamics. At the same time in the supergene zone of sedimentation a special role, which is difficult to overemphasize, belongs to microorganisms and organic matter. Presence of organic components in such geochemical systems catalyze the progress of many, in particular, redox reactions, that would not develop without the influence of life or its products (Putilina and Varentsov, 1980).

Thus the most important condition of the application of models of low temperature geochemical processes for the study of natural ore sedimentation processes is the necessity to obtain adequate data, the relative completeness of which should surpass that information, which was used until recent times. The aforesaid is related to the absence of reliable geochemical parameters, precise characteristics of a number of solid phases (their features of chemistry and structure), forms of component occurrence in the solution, and to the shortage of precise, consistent thermodynamic data on these phases and components. Moreover, the understanding of reaction mechanisms was not possible without the latest instrumentation methods of the study of structure, composition, and surface of mineral phases, determination of intermediate or stage products, and evaluation of diffusion processes.

The further development of the kinetic factor study in the last decade was the deep study of natural processes of authigenic phase formation (including Mn–Fe oxyhydroxides): deposition, redeposition, crystallization, and sorption processes. The growth of recent kinetic investigations and understanding of reaction mechanisms of processes, obviously favor the initiation of a new approach to the evaluation of kinetic control of the variation of composition of interacting solutions, and intermediate products in the general dynamics of phase formation.

Noting the difficulties of the application of calculated thermodynamical methods to the study of geochemistry of sedimentation-diagenetic ore-forming processes in the range of weathering crusts, accumulations of their products in fresh water basins – delta-estuaries-internal seas to abyssal regions of oceans – it is important to emphasize that the achievements in investigations of Mn ore formation in recent years were obtained owing to the application of models, in which considerable emphasis was devoted to kinetic, catalytic, and sorption processes (Arrhenius and Tsai, 1981). In a number of our works (Varentsov, 1972a, 1972b; Varentsov and Pronina, 1973; Varentsov *et al.*, 1979) emphasis was placed upon the fact that in particular, generalized models of these processes, the disequilibrium features of such interactions were reflected and controlled by kinetic and sorption factors, as well as the sufficiently complete concepts on process mechanisms, including the nature of irreversible reactions. The definite examples of the application of the above general statements can be described in corresponding parts of Mn ore formation.

In transition from modern basins to ancient ones the precision of relations of the major structural parts of ore forming systems is significantly decreased. As a whole, taking into account the known features of the Earth's history and the development of geodynamic processes, we may assume that in different intervals of the Phanerozoic there were mainly quantitative alterations in relations of major geological factors (structural, endogenic, climatic, biological, etc.), that became apparent in close interrelation. But for the Precambrian in particular, approximately from the bourdar of the Early Proterozoic (1.8–2.2 Ga) there are grounds to ascertain the presence of essential qualitative differences (Schopf, 1983). Such evolutional features are observed in the geochemistry and sedimentological characteristics of Mn ore deposits. As it was shown earlier (Varentsov, 1964; Varentsov and Rakhmanov, 1974; Varentsov *et al.*, 1984) with extremely limited information and variation of primary data on Mn ore formation in ancient supergene settings and basins, the reconstruction of geochemical characteristics of processes is conducted in the frames of known theoretical and experimental data, including the ore formation in modern settings. In these studies the formational method, or the study of paragenetic associations, are close to an initial factographical basis.

Thus, we can say with reasonable confidence that in the geological science, in particular in the geochemistry of Mn ore formation, relatively complete geochemical models are created as the result of close interaction of three methods:

1. Observations: study of modern and ancient natural objects, accumulation, understanding of empirical data, distinguishing of essential information, factors that control the investigated event. To select the essential information the theoretical concepts are necessary, corresponding to the level of modern achievements of the Earth sciences and related disciplines. The result of this stage is the creation of hypotheses as prelim-

inary models, based on these data. Here it is unavoidable to reduce the complicated natural system to some relatively elementary system, in which the main structural relationships of simulated systems are preserved. The most important aspect is the reduction during construction of the model of Mn ore formation in the frames of the considered method from the particular to the general.

2. Modeling: experimental simulation of some parts, process stages with unavoidable, simplification of systems. It is necessary to emphasize, that the understanding of natural events, in particular the geochemistry of Mn ore formation, is shown by how adequately we can model or experimentally reproduce the investigated process. Thus, observations of natural events, being an essential part in our understanding, do not ensure the necessary understanding of the process.

3. The historical method is extremely essential, because the geochemical evolution of the Mn ore formation processes represents the change of settings and conditions in the general development of the basin of sedimentation, region, the planet during geological time.

Chapter 2

MANGANESE ORES IN WEATHERING CRUSTS

2.1. Introduction

2.1.1. FORMULATION OF THE PROBLEM. DEFINITIONS

Among various types of manganese ores the largest accumulations of rich varieties occur in the supergene settings, in particular in laterite weathering crusts. This established empirical fact underlines the justice of the statement, made by V.I. Vernadskii in the beginning of the century that the highest concentrations of manganese were observed in the zone of free oxygen (Vernadskii, 1954a, 1954b). However, many classifications of manganese deposits do not sufficiently reflect the real diversity of genetic types of ores in the supergene zone (Berner, 1981; Betekhtin, 1946; Park, 1956; Roy, 1968, 1969, 1981; Strakhov *et al.*, 1968; Varentsov, 1962; Varentsov and Rakhmanov, 1974). The general scheme by Patterson (1971), used by Banerdji (1981) for a number of minerals, associated with weathering, is devoid of similar drawbacks. It should be stressed that in our work this classification is somewhat modified in order to reflect more adequately the distinguishing features of manganese ores, formed as the result of supergene processes:
1. Deposits associated with laterites and saprolites. For example, deposits of manganese ores in humid tropical zones.
2. Manganese ores of the oxidation zone or secondary enriched ore deposits. This type includes the oxidized manganese associations, widely developed in the largest sedimentary-diagenetic, hydrothermal-sedimentary ore accumulations. For example, in the South Ukrainian and Georgian Oligocene basins, and the Late Devonian oxidized-carbonate ores of the Central Kazakhstan, etc.
3. Manganese deposits associated with supergene processes of leaching, infiltration, and redeposition. The scales of their ore content may vary sharply. For example, the largest accumulations in karst caves of the Postmasburg area, South Africa (reserves: hundreds mln t of Mn) and relatively large ledges in deposits of Porozhinskoe (Eniseiskii ridge), Urkut (Bakoni Mountains, Hungary), Burshtynskoe (Western Ukraine) etc.

If manganese deposits of laterite crusts are typical formations of humid tropical regions, then the second and third types may be formed under the conditions of temperate or, less commonly, semiarid climates.

The basis of the mineral formation, in particular, of manganese ores in weathering zones are the processes of interaction of natural meteoric waters with the rocks under optimum combinations of the main factors: climatic, geomorphological and structural, lithologic-petrographical, and favorable history of the geological development of the area. It should be borne in mind that none of these factors functions in isolation. The critical role belongs to their optimum combination, corresponding to each definite type of ore. The ore forming supergene processes in the regions of humid tropical climate are distinguished by their intensity and extensity. Along with such important parameters as relatively high average annual temperatures and the volume of precipitation, a considerable role belongs to the chemical features of the latter. For example (Bugel'skii *et al.*, 1983), in the regions of dense vegetation the pH values of meteoric waters, passing through dense leaf litter and foliage, is decreased from 6.0–5.5 to 4.5–3.7, and for rain waters, flowing down tree trunks, this decrease is intensified to 2.6 at relative high concentrations of dissolved organic matter (DOM). In processes of supergene formation of manganese ores the cardinal role belongs to the organic matter in ore formation basins (Putilina and Varentsov, 1980). During interaction with rocks of the weathering crust profile, similar waters act not only as rather aggressive solutions, but additionally the presence of some type of DOM, in particular humic and fulvic acids, determines their action as selective solvents in relation to definite components of the rocks of the crust profile.

The experience of the ore formation study in weathering crusts suggests that the most considerable deposits of manganese and a number of other metals of the weathering crusts are associated mainly with large relatively uplifted and generally geomorphologically differentiated, tectonically stable areas of the Earth's crust (Banerdji, 1981, 1984; Bugel'skii, 1979; Bugel'skii *et al.*, 1983; Chowdhury *et al.*, 1965; Holtrop, 1965; Jones, 1992; Lopez-Eyzaguirre and Bisque, 1975; Ostwald, 1992; Parc *et al.*, 1989; Progress of IGCP Project no. 129, 1984; Taylor *et al.*, 1964; Vascocelos *et al.*, 1952).

A combination of tropical humid environments with optimum structural-geomorphological settings determine high drainage and washing up of rocks (water exchange intensity), that significantly increases the rates of crust forming processes: accumulation of ore components and removing of main rock forming components. Most researchers assume that the seasonal, or at least irregular rainfall, in excess of 2000 mm/year, an average-annual temperature of about 30°C with insignificant seasonal differences, a pH of ground waters between 4–3, means that oxidizing conditions are favorable for active development of laterite weathering crusts (Banerdji, 1984; Bugel'skii *et al.*, 1983; and others). The investigation data currently available proves that the chemical and mineral composition of ores is determined by the type of primary rocks to a considerable degree or, to be more precise, by the range of their enrichment in accumulated components. If the formation of bauxite lodes may be developed within a wide range of the bedrocks (excluding quartzites, evaporites, and other rocks, in which the Al content is significantly lower than clark) then for the manganese ores of the considered type the role of substrate (protore), preliminarily enriched in Mn and associated metals, is the key factor. This inference is displayed with great clarity in many regions of Australia, India, Africa, and South America, where laterite manganese ores are developed in Mn-protores of different composition (carbonate, silicate, oxide). Ultrabasites may arbitrarily be categorized as 'protore'. Oxyhydroxide minerals of manganese, having high concentrations of nickel, cobalt, and other heavy metals (Chukhrov *et al.*, 1980a, 1980b; Elias *et al.*, 1981; Schellmann, 1978; Vitovskaya *et al.*, 1984) are accumulated in the weathering crusts in these rocks. However, unlike nickel and cobalt, the ore accumulations of manganese rarely reach large commercial scales. In comparison with other ore-bearing weathering crusts, the geochemistry of ore forming processes in manganese crusts is relatively poorly studied at the present time. As it was emphasized above in the discussion of the methodical aspects (see 'Introduction'), the method of 'key test area' study was accepted in the work. On the basis of the thorough study of geology, mode of occurrence, mineralogy, and geochemistry of the ore and enclosing sediments of genetically simple and rather distinct sites, some preliminary models of supergenic manganese ore formation can be devised. The relative characteristics and correlation of deposits of this type, with genetically allied accumulations during geological history, may serve as the basis for wider generalization of such particular models. This approach is reflected in the general structure of this work.

To prepare the preliminary version of the geochemical model of supergenic manganese accumulations, a number propositions may be accepted:

— Large laterite manganese deposits are associated with areal weathering crusts and, to a lesser degree, with linear crusts.

— Such crusts are usually distinguished by zonal structure. In this case each zone is characterized by typomorphous mineral paragenesis – definite mineral and chemical composition.

— Variation of typomorphous mineral paragenesis, along with the features of the sub-strate rocks, reflects the chemistry of seeping, infiltrating ground solutions with the regime of pH and Eh. As a rule, in the vertical section the pH of the crust varies from acidic to slightly alkaline (from 3–5 to 9), and Eh changes from slightly reducing, and moderately oxidizing to intensely oxidizing (zone of free O_2).

— As in any ore-forming system, in the weathering crust there may be distinguished three main parts, often spatially slightly separated: (a) source; (b) transportation medium; (c) ore deposition setting. Due to the action of solutions, containing carbonic acid, humic and other acids, and probably with the participation of biochemical and bacterial processes, the following processes operate: (i) dissolution and hydrolysis of initial minerals; (ii) migration; (iii) autochthonous or with some transportation, new formation, and accumulation of ore components. In this way the process of manganese ore formation in the supergene zone in the weathering environments is controlled by the mechanism of dissolution-deposition of elements.

— In processes of authigenic formation of oxyhydroxide minerals of manganese ores and associated metals, the significant role belongs to sorption processes, often ac-companied by autocatalytic accumulation of transitional metals. The bacterial ore accumulations fill a special place.

— Inasmuch, the mineral-ore formation in the modern weathering crusts is characterized by extremely low rates, they may be referred to as unobserved processes. In this case the data of experimental investigations, simulating natural processes, is of special importance. In the manganese deposits in the laterite crusts of North Australia and West Africa, the autochthonous and redeposited nature (which is in many cases sufficiently proved such that it is no longer subject for debate) are accepted as a natural model. It is of value too that the mineralogy and geochemistry of these crusts are relatively well studied (Bouladon *et al.*, 1965; Grandin and Perseil, 1977, 1983; Leclerc and Weber, 1980; Ostwald, 1992; Perseil and Grandin, 1978, 1985; Pracejus and Bolton, 1992; Weber *et al.*, 1979).

2.2. Tropical Weathering Crusts

2.2.1. MODEL TEST AREA: GROOTE EYLANDT DEPOSIT, NORTH AUSTRALIA

2.2.1.1. Introduction

The Groote Eylandt manganese deposit is the largest in Australia. The explored reserves (more that 400 mln t) of high-grade manganese ores are sufficient not only for the demands of the Australian metallurgy, but also considerable amounts of enriched ores are exported.

The deposit is situated in Groote Eylandt Island, located at the western part of the Gulf of Carpentaria (Figure 1) 50 km from the Arnhem Land shore and 640 km from Darwin city, which is the closest city and administrative center of the North territory. Despite the fact that Groote Eylandt was found by Dutch seamen in 1624, its development was initiated only after the deposit's discovery in 1962 (Smith and Gebert, 1969).

The problems of the geology of the deposit, its ore mineral composition, enclosing sediments, their geochemistry, and genesis are represented in a limited number of works (Berry and Bergin, 1968; Bolton and Frakes, 1982, 1984; Bolton *et al.*, 1984, 1988, 1990; Frakes and Bolton, 1982, 1984a, 1984b; Frazer and Belsher, 1975; McIntosh *et al.*, 1975; Ostwald, 1975, 1980, 1981, 1982, 1988; Slee, 1980; Smith and Gebert, 1969; Pracejus and Bolton, 1992; Pracejus *et al.*, 1988, 1990; Varentsov, 1982). The authors expressed rather contradictory concepts relative to the nature of oxide ores; this indicates that these works reflect the initial stage of investigations of this deposit, which is unique in scale and ore quality.

In 1976 in Sidney, Australia, the Symposium on Geology and Geochemistry of Manganese was held in the frames of the XXVth International Geologic Congress. After the XXVth International Geologic Congress, the Groote Eylandt Company Ltd. (branch of the Broken Hill Proprietary Ltd.) organized an excursion for participants of the Symposium on manganese to the deposit, the author being a member of this group. They were provided with the opportunity to study the main ore sections and sample the ores and enclosing rocks.*

---

*The author is grateful to W.C. Smith and K.J. Slee, geologists of Company Groote Eylandt Mining for the help provided at the deposit, for discussions during the section description and sampling; the author also wishes to thank I.V. Reid, Research Department Manager, for the discussion of geological problems and the permission to publish the results of the sample investigations.

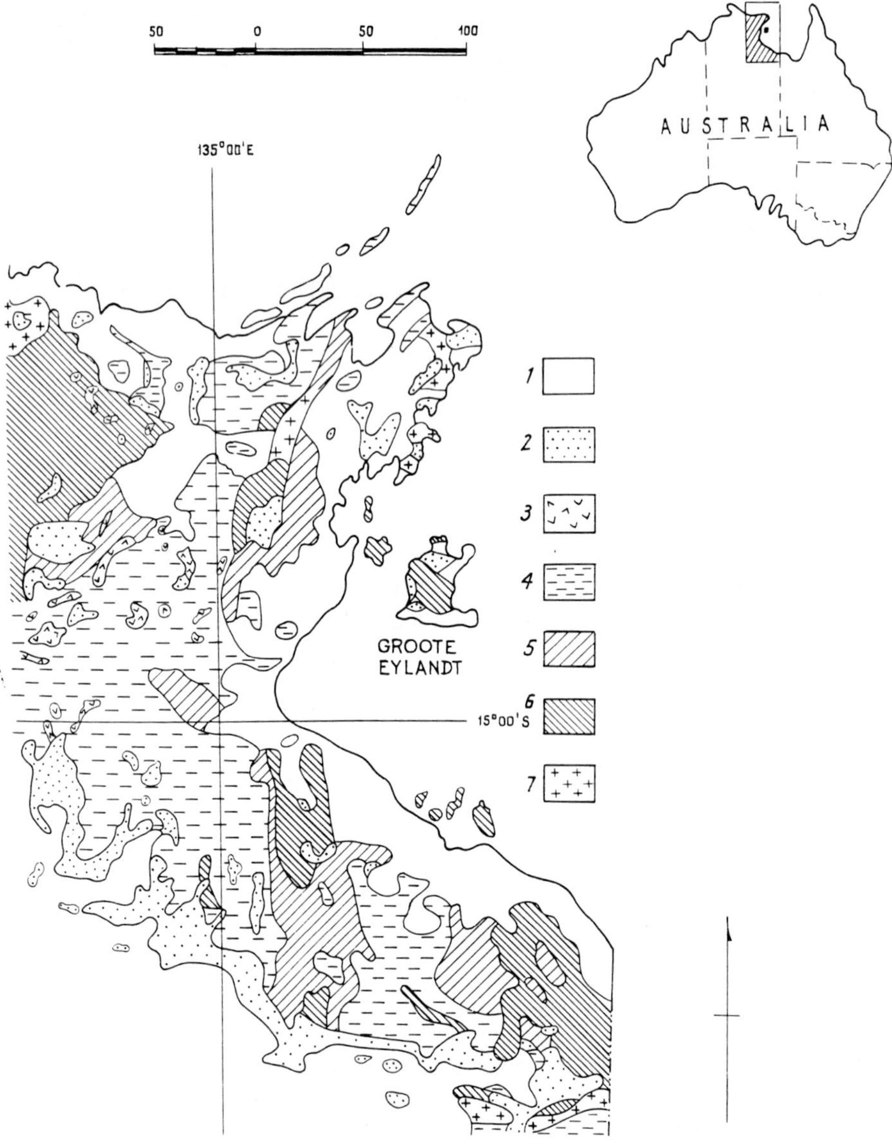

Fig. 1.   Geological map of Arnhem Land and the western part of the Gulf of Carpentaria region (scale in miles). 1. Tertiary and Quaternary sediments (alluvium, sands and continental sediments); 2. Lower Cretaceous deposits (Mullaman Beds); 3. Upper Proterozoic (?) (dolerites); 4. Upper Proterozoic to Middle Cambrian (rocks of the Roper Group, Wessel Group and Lower to Middle Cambrian rocks; 5. Middle Proterozoic (rocks of the McArthur Group and equivalents); 6. Middle Proterozoic (rocks of the Katherine River Group, Parsons Range Group, Groote Eylandt bed and equivalents); 7. Archean to Lower Proterozoic (granites and metamorphic basement rocks). After Smith and Gebert (1969).

2.2.1.2. Geological Setting

Precambrian and Lower–Middle Cambrian rocks, with relicts of thin sediments of the Lower Cretaceous and younger sediments, are widely developed in the region of Arnhem Land and the Gulf of Carpenteria. Sediments related to the interval from the Middle Cambrian to the Lower Cretaceous were not found in this region.

The basement rocks are represented by the Archean (Figures 1–2) and Late Proterozoic gneisses, quartzites, granites, and volcanites. They are overlain by thick sequences of the Middle and Upper Proterozoic and Cambrian sediments (Bolton *et al.*, 1984, 1988, 1990; Gebert and Slee, 1969; Slee, 1980). The Lower Cretaceous sediments occur as thin monadnocks, elevated more than 150 m above sea level. These sediments fill the valleys and they are developed at the coastal plain areas. The intensive development of the laterite weathering profile, in particular in the Lower Cretaceous rocks, is typical for the area.

2.2.1.3. Stratigraphy

Precambrian. Groote Eylandt layers are developed across most parts of the island territory and can be correlated with rocks of Parsons Range Group and Tawallah Group, composing the sequences of the Middle Proterozoic of Arnhem Land. Groote Eylandt layers (thickness about 600 m) are composed of quartz-feldspar sandstones with intercalations of shingles, and in the lower part of conglomerates. A clay matrix is observed in sandstones. It is remarkable for the development of cross-bedding stratification (Plumb *et al.*, 1980; Slee, 1980; Smith and Gebert, 1969).

Lower Cretaceous. The Mullaman beds are related to the Lower–Middle Cretaceous. They were deposited at the land surface, represented by a dissected plateau, composed of rocks of the Groote Eylandt Group. Sediment monadnocks are elevated to 159 m over sea level. These sediments occur in the plateau valleys and in small depressions of the western, southwestern, and probably northeastern margins of the plateau.

Judging from the drilling data in the western and southwestern parts of the island the Mullaman beds (95 Ma, Late Albian–Early Cenomanian (Bolton *et al.*, 1984, 1990)) are composed by quartz sands, clays, manganese-bearing glauconitic marl, and sandy clays, enclosing the layer of manganese oxides. Lately these sediments have been shown to have an absolute age of 95 Ma and are related to the Cenomanian age (Ostwald and Bolton, 1992). In relatively depressed parts (southern depression) the sandy clays and clays are underlain by grey pyritic manganese-bearing marl, the top of which occurs at a depth of less than 6 m below modern sea level (Smith and Gebert, 1969).

As a whole the Cenomanian sediments of the southern basin which are 65 m thick are subdivided into two sequences: (a) lower (clay facies), corresponding to a transgressive cycle, it is represented by sand siltstone (6 m), and calcareous siltstone marl (above 28 m), (b) upper, corresponding to regressive cycle (pisolite facies), composed of calcareous siltstone (marl, 6 m) and sandy siltstone (22 m). If in the calcareous siltstone (marl) the fine dispersed components are represented by glauconitic clays with distinct pellets of glauconite, in sandy siltstone they are represented by ferruginous smectites (Ostwald and Bolton, 1992).

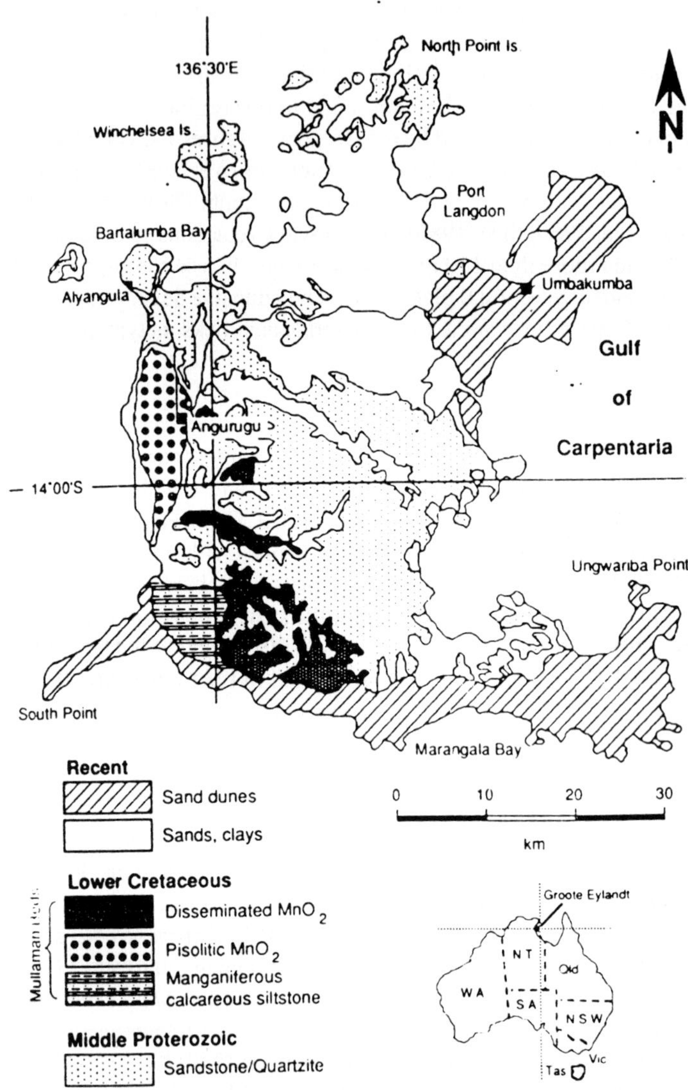

Fig. 2.   Geological map of Groote Eylandt, Northern Territory, Australia. After Pracejus and Bolton (1992a).

The 'Marl' consists of grey laminated clay with glauconite and pyrite, containing small grey, white, and cream-colored concretions of carbonate and several thin intercalations of green chert. According to X-ray determinations (Ostwald, 1987; Smith and Gebert, 1969) the carbonate concretions are composed of minerals of the calcite-rhodochrosite series.

The Mn contents in these carbonates range from 3.3 to 46.8% without any significant amounts of Fe, determined by EPMA (Ostwald, 1987; Ostwald and Bolton, 1992). It is significant, that in those cases, when the boreholes penetrate the layer of oxide manganese ores and the 'marl', one may observe that the main commercial oxide ores of the deposit occur above the 'marl'. At the same time in recent works by Frakes and Bolton (Bolton and Frakes, 1982; Bolton *et al.*, 1988, 1990; Frakes and Bolton, 1984b), the presence of facies with a relatively restricted distribution of oxide manganese ores is emphasized. These ores substitute laterally the Lower Cretaceous carbonate manganese-bearing rocks mentioned above.

Tertiary–Modern sediments. The western part of the Groote Eylandt the Mullaman layers are unconformably overlain by conglomerates that are presumably related to the Late Cenozoic. The sediments are represented by pebbles of limonite sand argillites, cemented by limonite sandy clay of laterites, and by laterite pebbles and pisolites, composed of manganese oxyhydroxides, but, if the conglomerates occur on the quartzite-like sandstones of the Groote Eylandt Group, or on the layer of oxide manganese ores, the boulders and pebbles composing them, reflect the composition of the underlying rocks.

In the western and southern part of the island the sand soils contain oxyhydroxide ferruginous and slightly manganese rich ferric pisolites. These sediments are considered (Pracejus and Bolton, 1992; Slee, 1980; Smith and Gebert, 1969) to be the equivalents of Late Tertiary conglomerates, but it is quite probable that they may be younger. It is supposed that they were formed at the place of occurrence in soils as the result of lateritization processes. The manganese oxyhydroxides of pisolites are represented mainly by lithiophorite. This mineral, in a certain degree, is distributed in the main ore layer too. In soil alluvial accumulations the lithiophorite is considered as an indicator of a laterite weathering.

2.2.1.4. Structure

In some published works the structure of the crystalline basement is extremely schematically described. On the basis of indirect data it is supposed that the basement is intensively dislocated, metamorphosed, intruded, and eroded. Before the initiation of the Groote Eylandt Beds sedimentation, it represented a stable shelf area. The sandstones of the Groote Eylandt Beds occur almost horizontally, displaying in some places the signs of block, fault dislocations, but the fold or fault deformations at the island were not observed (Smith and Gebert, 1969).

Significant oxyhydroxide manganese mineralization is confined to the Mullaman layers in the western, southwestern parts of the island (Figures 1 and 2). In relatively depressed areas of the western coastal plains, gently dipping to the south, two sedimentation basins are easily distinguished: northern and southern (Figure 3, Pracejus and Bolton, 1992). These basins are filled in by the sediments of the Cenomanian epicontinental marine basin. In the southern, relatively deep-water basin the discussed sediments are composed of sand claystones and manganese carbonates, and in the northern basin their equivalents are represented, as Frakes and Bolton (1984) considered, by pisolite manganese oxyhydroxides, which represent more shallow-water facies. In works by the Australian

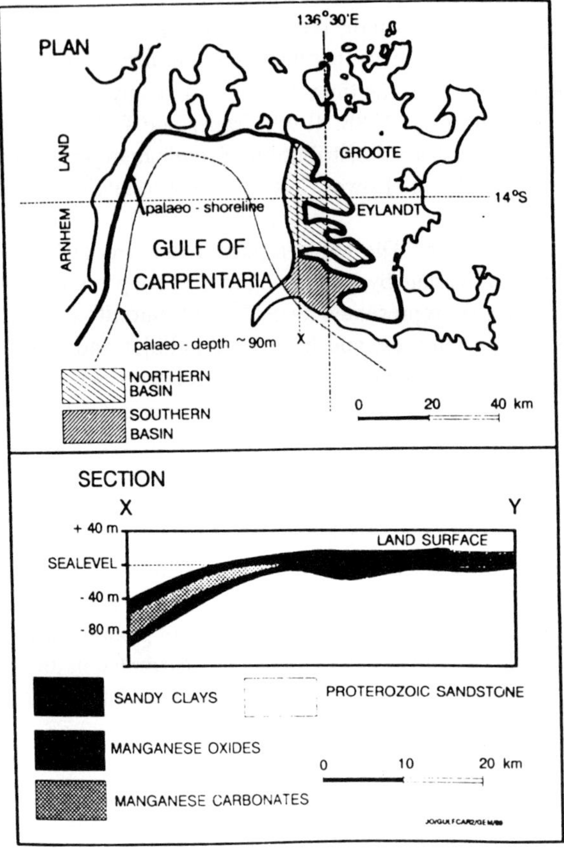

Fig. 3.   Distribution of manganese oxides and manganese carbonates within the Groote Eylandt and section through the Northern and Southern basins. After Ostwald and Bolton (1992a).

geologists (Bolton and Frakes, 1982, 1984, 1988; Bolton *et al.*, 1984; Frakes and Bolton, 1984, 1992; Ostwald, 1980, 1981, 1987, 1988; Pracejus and Bolton, 1988, 1992; Pracejus *et al.*, 1990; Smith and Gebert, 1969), the sedimentary-diagenetic nature of carbonate manganese accumulations in the southern basin and their relicts in the northern basin was convincingly shown. However, the problem is to what degree it is possible to consider the widely supergenically altered sediments and oxyhydroxide ores of the northern basin as shallow water sedimentation equivalent facies of the southern basin; that is the subject of discussions in recent years beginning after the publications of our works (Varentsov, 1982a, 1982b; Varentsov and Golovin, 1987). It is well known that these ores constitute the main commercial value of the deposit. This discussion stimulated the publication of a large volume of factual data, of value for adequate understanding of the nature of

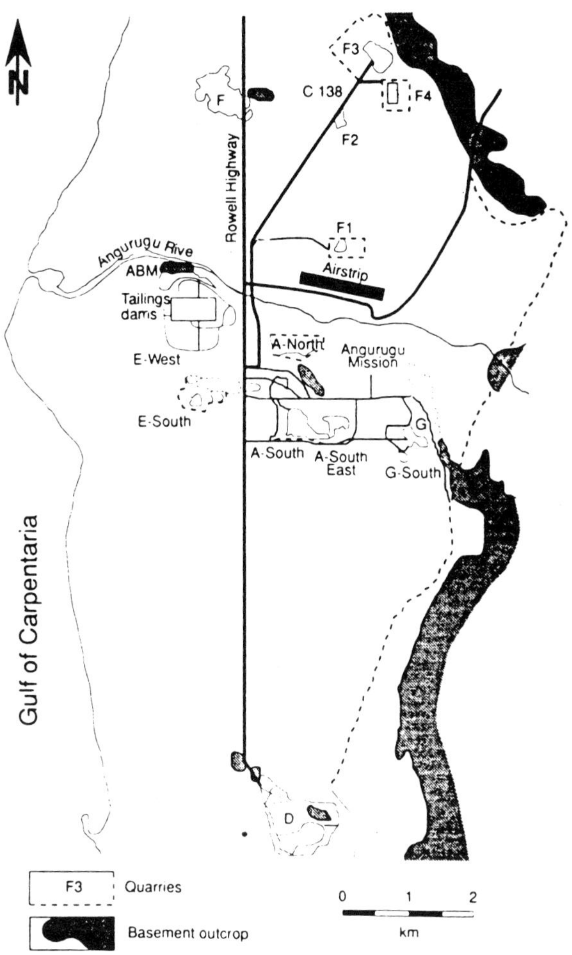

Fig. 4.   Location map of mining areas at Groote Eylandt. After Pracejus and Bolton (1992).

manganese ore accumulations, irrespective of the conceptual trends of the authors.

In the light of the comments above it is vital to note that in the relatively depressed areas of the western coastal plain, the bed of manganese oxyhydroxides reaches the maximum development, forming an almost continuous ore body, elongated to 22 km with a width of 6 km (Figures 2 and 3).

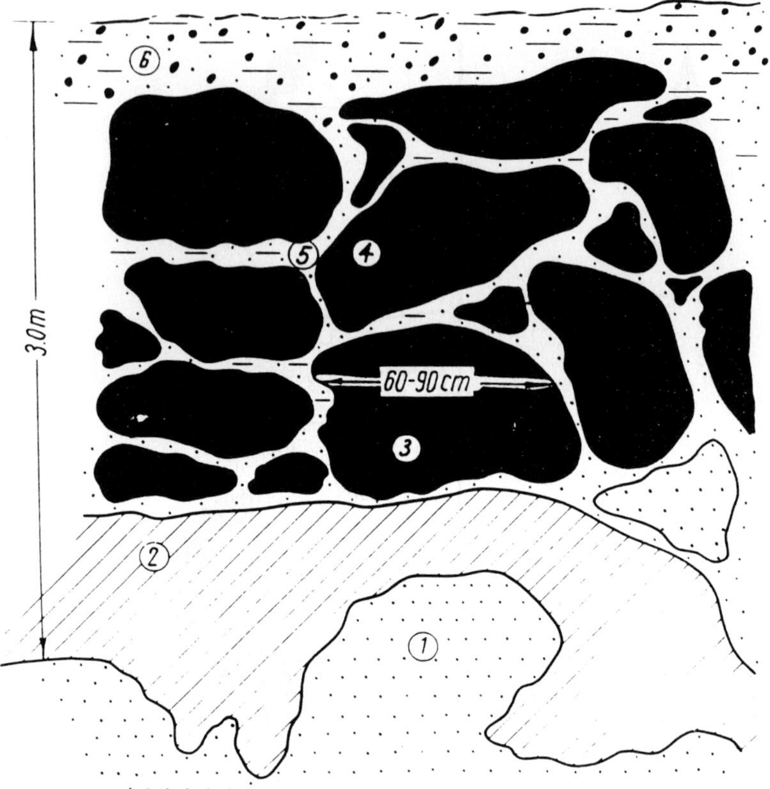

Fig. 5.   Schematic cross-section of an ore stratum, Groote Eylandt deposits, the southern part of the D-quarry (see Figure 4), face near the southeastern outcrop of the sandstones (Precambrian Groote Eylandt Beds). 1. Sand, a product of decomposition of the sandstone (Groote Eylandt Beds, Precambrian); 2. Brown laterite (after Mullaman Beds, Lower–Middle Cretaceous); 3 and 4. Boulder-lumpy oxyhydroxide manganese ores (3. sample no. 2; 4. sample no. 3); 5. Yellowish-brown clay sandy matrix; 6. Brown essentiallly ferrugenous lateritic soil (Varentsov, 1982).

2.2.1.5. Manganese Oxyhydroxide Ores

Cenomanian manganese-bearing marls and clays of the Mullaman Beds occur unconformably on the extremely eroded surface of the Proterozoic sandstones (Groote Eylandt Group). In the northern basin, the observations in the quarry faces and investigations of boreholes indicate that in the basal part of these sediments there are concretions of manganese oxyhydroxides (Figures 3 and 4). The basal complex of sediments is overlain by laterite limonitized quartz sand, above which occur white, red, and yellow predominantly kaolinite clays, and sand clays, including the layer of oxyhydroxide manganese ores. Extremely significant for the understanding of the deposit genesis is the relationship between manganese-bearing marl and the main ore bed, that are divided by a layer of

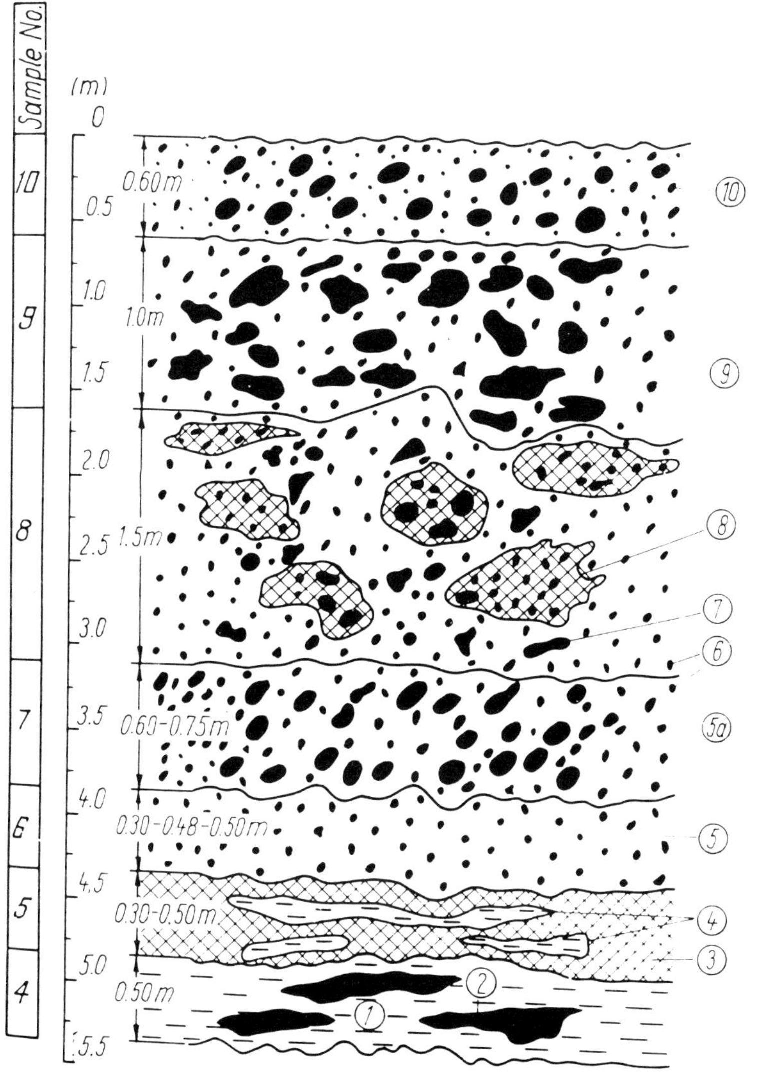

Fig. 6.   Schematic cross-section of an ore stratum, Groote Eylandt deposits, G-quarry, southwestern face (see Figure 4). 1. Ochreous laterite; 2. Iron and manganese oxyhydroxides in the form of irregular lenses; 3. fine-globular Mn oxyhydroxides (1–2 mm), earthy; 4. Lenses and irregular mottled patches of brownish-red kaolinitic and lateritic clay; 5/5a. Oolites and pisolites of Mn oxyhdyroxides, small (< 2–4 mm); 6. Pisolites of Mn oxyhydroxides, their size in the lower part being 3–7 mm, in the upper part a mixture of small (< 2–3 mm) and large (up to 10 mm); 7. A mixture of small oolites (< 2–3 mm); 8. Mn oxyhydroxides, represented by irregularly lenticular patches of earthy mass and ore (fragments); 9. Laterite: Fe, Mn oxyhydroxides, pisolites, and fragments of irregularly tabular shape; 10. Soil composed of red-brown ferrugenous laterite, pisolites, fragments, and clay matrix.

clays (8 m) and pebble sand (2 m). At the flanks of the basement uplift, where the ore bed pinchout is observed, the ores occur directly on the eroded sandstones of the Groote Eylandt Beds (Figures 3, 5 to 9).

2.2.1.5.1. Structure of the ore bed.   Most geologists studying the deposit (Ostwald, 1975, 1980, 1981, 1982; Ostwald and Bolton, 1992; Slee, 1980; Smith and Gebert, 1969) consider that the manganese ores are enclosed in the sequence of the Lower–Middle Cretaceous sands and clays, unconformably overlying the Middle Proterozoic sandstones of the Groote Eylandt Beds. Only in recent works (Bolton and Frakes, 1982; Bolton *et al.*, 1984, 1988, 1990; Frakes and Bolton, 1984b, 1992) is the presence of both relatively limited, laterally distributed oxyhydroxide manganese ores (probably Early–Middle Cretaceous facies) with the significant predominance of carbonate varieties and widely developed oxyhydroxide accumulations – products of relatively late supergene alteration and reworking of sedimentary-diagenetic ores. Recognition by the Australian geologists (Bolton *et al.*, 1984; Pracejus and Bolton, 1992; Pracejus *et al.*, 1988, 1989) of the significant role of the supergene processes in the formation of the oxyhydroxide ores of the Groote Eylandt deposit may be considered to be the result of active discussion of problems of their genesis in recent years (Varentsov, 1982). Our observation in the quarry faces of the deposit (see Figures 4 to 10, the quarry locations are given in the works by Slee (1980) and Pracejus and Bolton (1992)) allow us to consider, that the ore bed is represented by a rather complicated combination of different mineralogic-textural types of ores, intercalating with red, cream-colored, and white clays of predominant kaolinite composition. The thickness of the ore bed varies from 2 to 20 m, averaging about 3 m.

In most cases in the foot of the ore bed there are 'sand and siliceous ores', which vary greatly in thickness up to 4 m, represented essentially by pisolites, small nodular aggregates, composed of loose slightly cemented masses, bonded by sand, clay or manganese oxyhydroxides. Limonite pisolites of 2–15 mm in diameter occur often in the section base.

Lens-like interbeds with a fairly regular strike, composed of oolites and pisolites, often cemented by clayey oxyhydroxide material, occur above. The maximum thickness of the oolitic part of the section is to 4 m.

Slightly textured, massive, sometimes lumpy boulder ores (to 30–40 cm) with the shape of irregular rounded concretion aggregates, bonded by light, reddish clay-sand masses (quartz, kaolinite, goethite) usually overlie oolites usually in the marginal parts of the deposit. A layer of cream-colored reddish sandy kaolinitic clays (to 6 m) occurs in the top of the ore sandstone. These sandstones are overlain by a bed of pedogenous lateritic aggregates (to 4.5 m) and lateritic soils (to 1 m).

It is necessary to emphasize the irregularity of the noted parts of the ore layer and the variable character of interrelations between them.

2.2.1.5.2. Ore types.   Two main types of ores may be distinguished at the Groote Eylandt deposit: oolite (< 1 mm) and pisolite, and massive concretion. Oolite and pisolite ores are

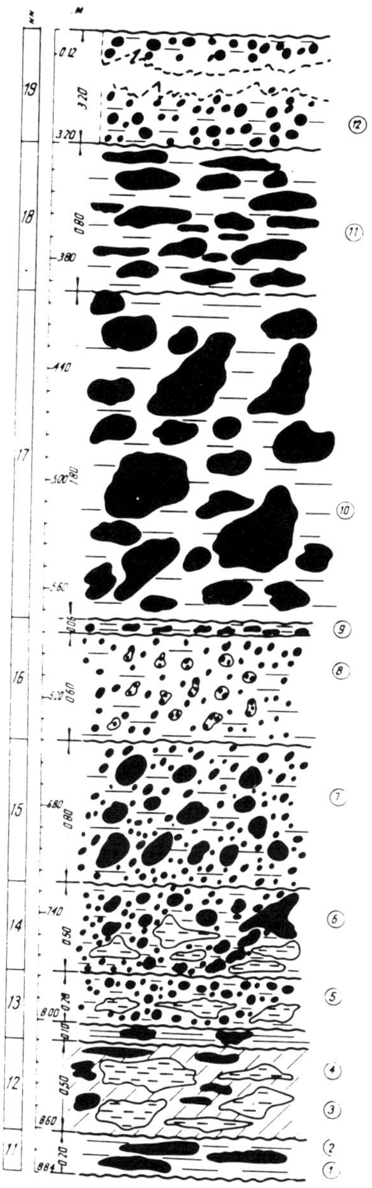

Fig. 7.   Schematic cross-section of an ore stratum, Groote Eylandt deposits, the A-quarry, south part, south face (see Figure 4). 1. Red-brownish kaolinitic clay; 2. Lenticular Mn, Fe oxyhydroxide nodules and hematite patches; 3. Small-globular, earthy-sooty Mn oxyhydroxides, containing small lenticular patches of goethite, hematite, and ochreous material; 4. Lenses and pockets of kaolinite; 5. Oolites of Mn oxyhydroxides incorporated in a kaolinitic matrix containing lenses of bleached kaolinite; 6. Pisolites (3–7 mm), small concretions of Mn oxyhydroxides and lenses of bleached kaolinite incorporated in a red-brownish kaolinitic matrix; 7. Pisolites of Mn oxyhydroxides and their aggregates incorporated in a kaolinitic matrix (60%); 8. Oolites (1–3 mm) of Mn oxyhydroxides in a kaolinitic matrix (60%); 9. Pisolites (up to 7 mm) in a kaolinitic matrix (up to 75–80%); 10. Concretions of Mn oxyhydroxides in kaolinitic clay; 11. Fragments of tabular shape composed of manganese oxyhydroxides in a kaolinitic matrix; 12. Laterite composed of kaolinitic clays with concretionary-pisolitic patches of Mn oxyhydroxides.

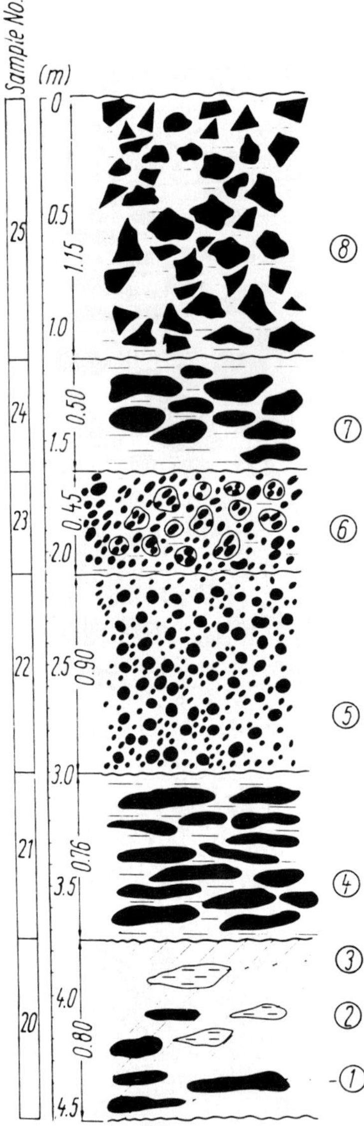

Fig. 8.   Schematic cross-section of an ore stratum, Groote Eylandt deposits, A-quarry, southeastern face (see Figure 4). 1. Flattened nodules and lenses of Mn oxyhydroxides; 2. Red clayey ferrugenous matrix; 3. White lens-like patches of kaolinite; 4. Platy fragments, lenses, and lens-like interbeds composed of compact Mn oxyhydroxides; 5. Small (< 2–3 mm), cemented oolites of Mn oxyhydroxides; 6. Pisolites of Mn oxyhydroxides, cemented to a compact mass; 7. Large (5–15 cm) nodules of Mn oxyhydroxides and their intergrowths and aggregates; 8. Laterite composed of angular gravel fragments of Fe oxyhydroxides (hematite) in a ferruginized kaolinitic matrix.

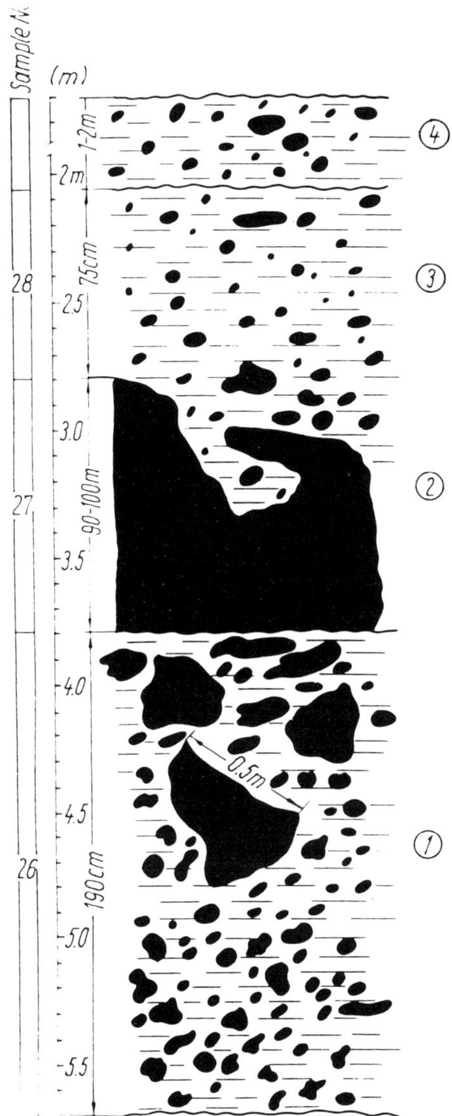

Fig. 9.   Schematic cross-section of an ore stratum, Groote Etlandt deposits, E-quarry, northern part, northwestern face (see Figure 4). 1. The lower part abounds in ferruginous laterite, the upper part contains boulder fragments and Mn oxyhydroxide nodules, the matrix being represented by kaolinitic clay; 2. Boulder-lumpy accumulations of Mn oxyhydroxides (cryptomelane, pyrolusite), higher up; 3. Mottled ferruginous laterite; 4. Brown lateritic soil.

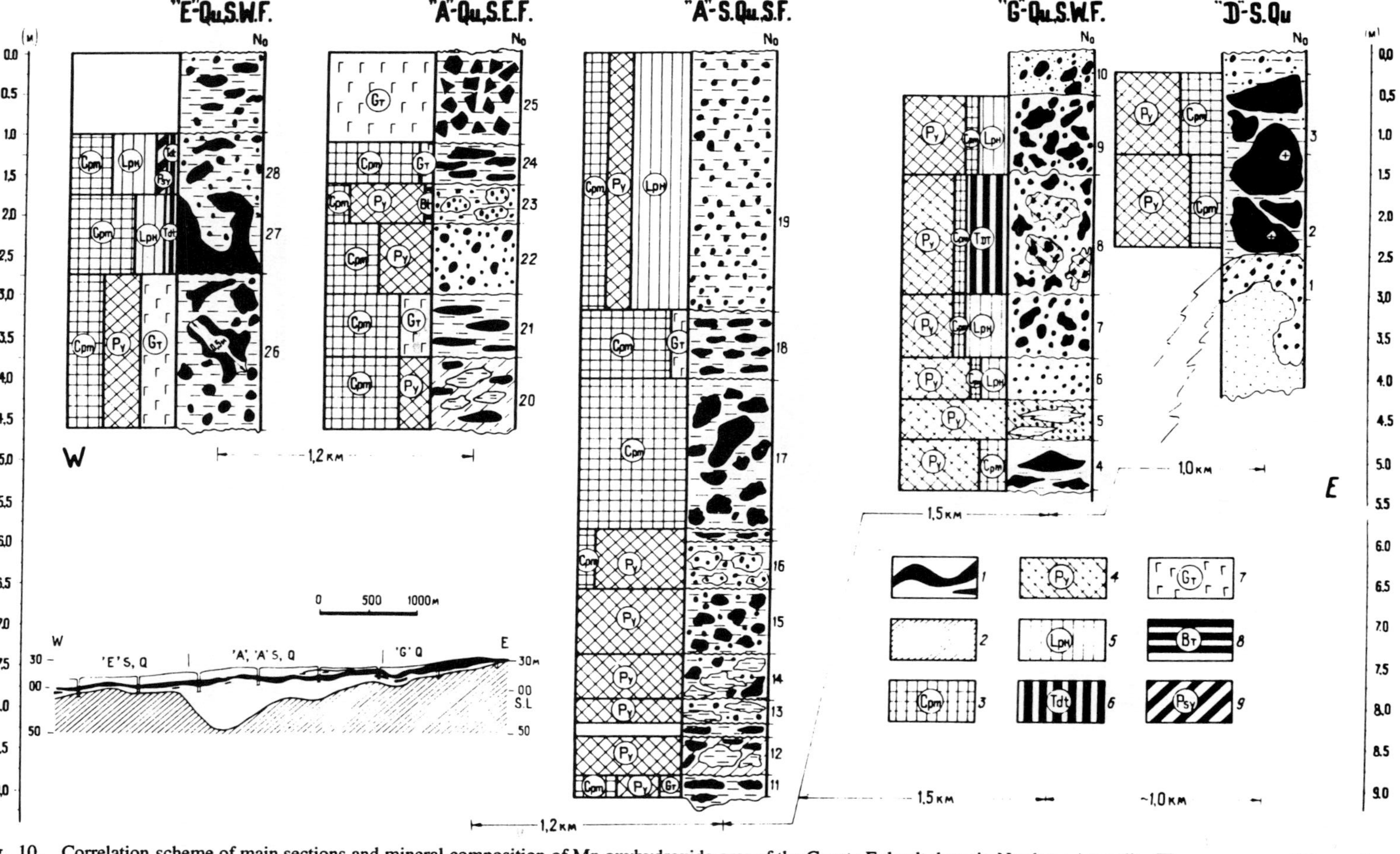

Fig. 10.   Correlation scheme of main sections and mineral composition of Mn oxyhydroxide ores of the Groote Eylandt deposit, Northern Australia. The occurrence of the ore layer and the location of quarries at the deposit are shown in the inset. Description of ore composition is given in the notes to Table 1. 1. Mn oxyhydroxide ores; 2. Precambrian bedrocks; 3. Cryptomelane; 4. Pyrolusite; 5. 10 Å-manganate (lithiophorite); 6. 10 Å-manganate (todorokite); 7. Goethite; 8. Birnessite; 9. Romanechite (psilomelane).

observed in two varieties: loose oolites, pisolites and cemented ones. The dimensions of oolites are < 1 mm, and of pisolite aggregates 1–15 mm (average 2–6 mm). Concretion ores are distinguished by a variability in shapes with dimensions from relatively regular rounded aggregates (5–10 cm) to large lumpy, boulder associations without distinct structural features.

Next to these types, Fraser and Belcher (1975) recognize sandy ores (20–30% of SiO_2) in the form of sand admixtures of quartz, siliceous (more than 30% of SiO_2), ferruginous (more than 10% of Fe_2O_3), dispersed patches in the enclosing clays, and sinter accumulations – wads.

In a recently published work by Pracejus and Bolton (1992) there is a general description of practically the same key sections of the Groote Eylandt deposit. However, relying on the formation model of Mn ores in the stagnant oxygen-free sea basin, proposed by Frakes and Bolton (1984), these authors (Pracejus and Bolton, 1992) distinguish among Mn oxyhydroxide ores, the varieties forming in the sedimentary conditions of the Cretaceous sea basin and extremely similar accumulations, formed in the supergene conditions of tropical weathering crusts. The authors distinguish the following groups:

1. Oolite: ore, composed of loose ovoid to spherical grains of concentric laminated structure. Pracejus and Bolton (1992) refer them to the primary deposits of sedimentary marine environments.
2. Pisolite: ore, represented by ovoid to spherical grains of concentric laminated structure. Pracejus and Bolton (1992) refer them to products of sedimentary origin, accumulated in the sea basin.
3. Relatively small subordinate amounts of cemented infiltrated crust accumulations and nodules, to which are also assigned the supergenous oolites, pisolites, spherulites, and mangcrete.

It should be pointed out that while only slightly differing in their mode of occurrence, mineralogy, and chemical composition from oolite and pisolite ores of the first two groups, the ores of the third group, including oolite and pisolite varieties, are considered (Pracejus and Bolton, 1992) as characteristic supergene accumulations or, according to the terminology of these authors, as 'secondary ores'. According to Bolton *et al.* (1988) the most essential arguments, indicating that the formation of manganese oxyhydroxide pisolites took place in a basinal environment, may be referred to some normal and inverse gradation cycles, distinguished in the section. Here they usually occur in the lower parts of the ore bed. Gradational changes of pisolite dimensions do not usually exceed the ranges from 2 to 25 mm in those intervals with thickness from 1 to 2 m (Bolton *et al.*, 1988; Pracejus and Bolton, 1992). This is very important but its interpretation will be given below in the general context of the data on mineralogy and geochemistry.

It is important to emphasize that the manganese ore accumulations are developed in the western part of the Groote Eylandt, where they compose two isolated basins (northern and southern), located at different hypsometrical levels and separated by the bench of the Precambrian sandstone (see Figures 2, 3, and 10) (Ostwald and Bolton, 1992; Pracejus and Bolton, 1992; Slee, 1980; Smith and Gebert, 1969). Manganese oxyhydroxide ores, described in those works and being discussed in the present contribution, are developed only in the northern basin, where at present their commercial exploitation is conducted.

The manganiferous calcareous siltstone with rhodochrosite and manganocalcite as the major Mn bearing components dominate the southern basin. Frakes and Bolton (1988) assume that the Mn oxyhydroxide ores of the northern basin were formed in an oxidizing environment at relatively shallow water depths (15 to 50 m). At the same time the formation of carbonate manganese-bearing sediments of the southern basin took place in a slightly reducing environment at greater depths (< 90 m). There are no significant disagreements among most investigators on the sedimentary-diagenetic nature of the Mn carbonate ores. This conclusion was emphasized by the essential difference in the composition of the enclosing rocks: nontronite-smectite-hydromica-glauconite clay components, terrigenous quartz, carbonates of Mn and Ca in the southern depression. Whereas the ore-free sediments, enclosing Mn oxyhydroxide ores of the northern depression are composed essentially of kaolinite, terrigenous residual quartz with admixtures of dickite, halloysite and rare relicts of hydromica, and smectite.

In Table 1, borrowed from the work by Pracejus and Bolton (1992), the main characteristics, criteria were summed up, in which the authors distinguished 'primary' and 'secondary' oxide, Mn ores as the products of deep weathering of the northern basin.

2.2.1.5.3. Mineral and chemical composition of the ores. The study of the mineral composition of the ores was carried on by the Australian geologists Bolton and Frakes (1982, 1984, 1985), Bolton *et al.* (1984, 1988, 1990), Frazer and Belcher (1975), Frakes and Bolton (1982, 1984a, 1984b), Ostwald (1975, 1980, 1981a, 1981b, 1982, 1984, 1987, 1988, 1992), Ostwald and Bolton (1990, 1992), Pracejus and Bolton (1992a, 1992b), and Pracejus *et al.* (1986, 1988a, 1988b, 1990a, 1990b). These works made fundamental contributions, that had high factographical significance. However, as it was mentioned above, the results of researches, performed at the Geological Institute of the Russian Academy of Sciences, indicated that commercially valuable Mn oxyhydroxide ores of the Groote Eylandt were represented by the product of deep alteration of initially relatively low-grade Mn-bearing sediments of the Northern basin. The significance of this genetic problem (model) is that it determines the efficiency of search-exploration works for high-grade Mn oxyhydroxide ores in the tropical belts, where the processes of supergenic mineralization dominate at corresponding intervals of the geological history of the Earth.

In the Geological Institute of the Russian Academy of Sciences the ores were studied in thin section, and by the methods of X-ray diffractometry (V.A. Aleksandrova), infrared spectroscopy, DTA, DTG and under transmission and scanning electron microscope. The study of the ore mineral composition and features of its alteration in the ore bed, that was conducted together with Dr. E.A. Perseil, the National Museum of Natural History, Paris, France, for the control samples was especially useful and greatly appreciated.

Chemical composition of ores and their components was determined by the methods of wet chemistry (M.I. Stepanets) and X-ray fluorescence spectroscopy (E.M. Margolin). The presence of six main manganese minerals was distinguished in the ores: pyrolusite and cryptomelane predominate; lithiophorite, romanechite, nsutite, 10 Å-manganate (todorokite) are present in significant volumes. A number of other oxyhydroxide minerals of manganese are observed as the admixtures. The chemical composition of ores is given in Tables 2 and 3. The mineral composition derived as a result of microscope work and the

Table 1.

A. Primary manganese ores, North Basin, Groote Eylandt deposits (after Pracejust and Bolton, 1992).

Name	Structure and geometry	Mineralogy	Secondary overprint
Oolite (uncemented)	Normal or inverse grading, subhorizontal, following the relief of the bedrock, small- to medium scale cross-bedding	Pyrolusite, todorokite, cryptomelane, birnessite, nsutite, romanechite, lithiophorite, vernadite, (manganocacite); nuclei contain silt, sand grains and reworked Mn ores	Replacements: Fe, Mn clays; cementation: Mn compaction
Pisolite (uncemented)	Normal to inverse grading, subhorizontal, following the relief of the bedrock, small- to medium-scale cross-bedding	Pyrolusite, todorokite, cryptomelane, nsutite, romanechite, lithiophorite, vernadite; nuclei contain silt, sand grains, and reworked Mn ores	Replacements: Fe, Mn clays; cementation: Mn, compaction pressure solution along fractures
Siliceous ore	Massive cementation of sand grains; sand grains in matrix of oolitic-pisolitic ores	Pyrolusite, cryptomelane, todorokite, quartz	
Hard ground	Dense, thin layers (cm) occasionally erosional features on surface, laterally continuos ores tens of meters; incorporation of reworked oolitic, pisolitic and massive ores	Cryptomelane, pyrolusite	Filling of cracks by secondary Mn–Fe phases; replacement by Fe oxides and clays; replacement of primary structures and textures by Mn oxides

B. Secondary manganese ores, North Basin, Groote Eylandt deposits (after Pracejus and Bolton, 1992).

Name	Structure and geometry	Mineralogy
Cemented mangcrete (oolitic-pisolitic)	Oolitic-pisolitic ores, massive, very hard, generally subhorizontal	Pyrolysite, cryptomelane, todorokite, nsutite
Massive mangcrete	Massive, hard to very hard cementations in oolites, pisolites and in other ore types, often relic structures of primary features, horizontal to subhorizontal	Cryptomelane, pyrolusite, sometimes pisolites are weathered out
Envelope grain	Laminated envelope containing loose or cemented ooliths or pisoliths within oolite/pisolite units, round angular shapes	Cryptomelane, pyrolusite
Laminar mangcrete	Finely laminated on pipe walls, following vertical and horizontal pipes, in outcrops as Mn karst	Cryptomelane, quartz
Mn cave pearls	Loose, rounded and elongated shapes with concentric layering comparable to laminar mangcrete but with much lower density	Cryptomelane, pyrolusite quartz
Spherulite (manganiferous)	Cementation of different materials (e.g. sand, pebbles in ovoid masses in soil, laterally continuous horizons)	Todorokite, cryptomelane, goethite, hematite, quartz
Radial pyrolusite	Laterally continuous over tens of meters, strongly undulating, erosional upper boundary and flat base, following the top of the ore body (G-south) and within clay units, overprinting oolitic-pisolitic textures . . .	Pyrolusite

Table 1.   (Continued)

Name	Structure and geometry	Mineralogy
Siliceous ore-manganiferous sand	Massive to recrystallized often horizontal, partly cemented sands	Pyrolusite, todorokite, cryptomelane, quartz
Concretions (near surface)	Varying in size, sometimes dehydration features, laterally continuous horizons, upward oriented growth	Cryptomelane, goethite
Digital concretions	Digital or coral-like growth, increasing upward in thickness, dehydration features, below main ore body	Cryptomelane
Chocolade ore	Strongly leached, irregular, brittle matrix with Fe oxyhydroxides and clays, rare pisoliths-ooliths or cemented fragments	Pyrolusite, cryptomelane todorokite, kaolinite, Fe oxyhydroxides

C. Rocks and sediments associated with manganese ores (after Pracejus and Bolton, 1992).

Name	Structure and geometry	Lithology and mineralogy	Secondary overprint
Soil	Irregular, laterally extensive concentrated in low topography, vertical in pipes and channels	Quartz sand, organics, conglomerates, clays	Fe–Mn spherulites, fining up to the top, Mn concretions lateritic alterations
Sand	Laterally extensive, occasionally in pipes, small-medium scale cross-bedding and massive	Mainly quartz; along coast: coral and shell	Fe–Mn impregnations and cementation rare salification
Laterite (Mn)	Laterally extensive underneath soils, vugular	Sands, cementations of oxyhydroxides, clays	Pipes and sink holes with spherulites
Laterite (Fe)	Lateral, underneath manganiferous laterite, vugular	Sands, Fe oxydes-oxyhydroxides, clays	Horizons with Fe enrichment and pipes, replacements of oolites-pisolites
Clay	Planar, regionally up to 50 m thick, sometimes layered, mostly irregular in laterites and pipes, rarely layers in vugs as impregnation in the matrix of ores	Kaolinite (nontronite, montmorillonite) as clays silts, claystones, siltstones and clayey sands	Impregnations of Fe oxides-oxyhydroxides Mn concretions and digital concretions
Fe oolite-pisolite	Soft to hard cementation cross-cutting manganiferous units	Goethite, occasionally hematite	
Massive ferricrete	Dense, hard, lateral, cementations	Goethite	Occasionally patchy manganese oxide minerals, rarely quartz
Fe pipe wall	Dense cementation of relic rocks and minerals, following outer boundary of pipes	Hematite, goethite, kaolinite, quartz	
Conglomerate	Irregular pockets and, lenses discontinuous	Sand, quartzite pebbles, reworked Mn ores, Fe oxides clays	
Quartzite-sandstone	Dense, bedrock with erosion features	Quartz	(Mn) impregnations

Table 2. Chemical composition of manganese ores, Groote Eylandt deposits, Northern Australia.

no.	no. of sample*	SiO_2	TiO_2	Al_2O_3	Fe_2O_3	FeO	CaO	MgO	MnO	MnO_2	Na_2O	K_2O	H_2O^+	H_2O^-	CO_2	$C_{org.}$	P_2O_5	BaO	Total
1	2	6.82	0.18	4.46	2.91	no	0.47	no	0.75	78.49	0.40	1.65	3.06	0.36	no	no	no	0.29	99.84
2	3	2.39	0.21	2.64	2.23	no	0.31	no	0.58	85.30	0.40	1.67	2.48	0.49	0.15	no	no	0.86	99.71
3	4	14.76	0.35	3.69	3.34	no	0.31	no	0.69	71.07	0.28	0.51	3.07	0.82	no	no	no	1.04	99.93
4	4-1	47.35	0.43	5.98	4.69	no	0.27	no	0.49	35.68	0.10	0.61	3.34	0.34	no	no	0.004	0.24	99.52
5	5	20.39	0.30	5.64	1.31	no	0.23	no	0.86	62.58	0.62	2.44	3.60	0.43	no	no	no	1.08	99.48
6	6	12.69	0.11	1.56	1.78	no	0.31	no	0.59	80.67	0.28	0.41	1.00	0.22	no	no	no	0.24	99.86
7	7	1.00	0.12	2.13	2.45	no	0.63	0.11	0.65	82.74	0.34	0.77	4.44	1.39	no	no	no	2.97	99.74
8	8	7.96	0.35	4.63	3.13	no	0.31	0.23	0.60	73.80	0.28	0.82	5.39	1.33	no	no	no	0.95	99.72
9	9	9.36	0.23	3.26	3.18	no	0.08	no	0.51	77.78	0.28	0.61	3.16	0.64	no	no	no	0.60	99.69
10	11	30.41	0.40	6.57	17.85	no	0.16	no	2.43	35.08	0.34	0.87	4.88	0.65	no	no	no	0.04	99.68
11	12	22.39	0.38	5.45	3.56	no	0.21	no	0.27	63.80	0.10	0.33	3.05	0.47	no	no	0.004	0.12	100.13
12	13	8.55	0.35	6.18	4.23	no	0.16	no	0.61	75.05	0.34	0.70	3.13	0.41	no	no	no	0.04	99.75
13	14	1.63	0.11	0.86	2.89	no	0.47	no	0.57	91.13	0.28	0.27	1.31	0.23	no	no	no	0.09	99.84
14	15	2.89	0.16	0.98	2.45	no	0.47	no	0.60	90.09	0.28	0.27	1.11	0.23	no	no	no	0.23	99.76
15	16	6.51	0.30	4.44	3.12	no	0.16	no	0.49	81.38	0.23	0.33	2.24	0.45	no	no	no	0.10	99.75
16	17	50.93	0.94	18.74	8.71	no	0.42	no	0.44	9.56	0.10	0.61	7.67	1.13	no	no	0.02	0.45	99.72
17	18	16.84	0.58	9.58	7.94	no	0.31	no	0.49	54.66	0.51	1.85	5.33	1.03	no	no	no	0.85	99.97
18	19	28.79	1.07	19.42	16.98	no	0.28	0.05	0.32	21.24	0.10	0.72	8.96	1.41	no	no	0.02	0.21	99.57
19	20	77.25	0.54	10.53	2.15	no	0.27	no	0.15	3.69	0.10	0.28	3.85	0.46	no	no	0.04	0.21	99.52
20	21	10.93	0.34	1.81	14.65	no	0.27	no	0.50	64.37	0.29	2.40	3.39	0.51	no	no	0.004	0.45	99.91
21	22	8.64	0.27	2.87	1.98	no	0.21	no	0.38	81.49	0.10	0.78	1.82	0.36	no	no	0.004	0.61	99.51
22	23	5.50	0.32	2.06	3.99	no	0.39	0.11	0.84	82.14	0.28	1.60	1.87	0.26	no	no	no	0.30	99.66
23	24	0.77	0.19	1.78	1.98	no	0.28	no	1.70	86.05	0.29	3.51	2.66	0.63	no	no	0.004	0.70	99.91
24	25	16.12	0.94	12.92	56.42	no	0.28	0.07	0.16	1.00	0.10	0.28	9.77	1.67	no	no	0.09	0.12	99.94
25	26	41.36	0.86	14.67	7.92	no	0.27	no	0.51	25.66	0.10	0.83	6.21	0.91	no	no	0.004	0.40	99.73
26	27	4.33	0.22	4.67	3.17	no	0.27	no	0.43	78.88	0.19	2.44	4.05	1.08	no	no	0,002	0.16	99.89
27	28	16.06	0.59	10.75	9.04	no	0.27	0.07	0.51	52.15	0.19	1.22	6.66	1.30	no	no	0.004	1.21	100.02

*Numbers of the analyzed samples correspond to those on schemes of the ore bed structure (see Figures 5–10), analyst M.M. Stepanets, GIN, Russian Academy of Sciences.

Notes to Table 2

1. Sample 2: black with bluish tint material, composing boulder-lumpy ores (see Figures 5, 10, and 11); composition: pyrolusite with notable admixture of cryptomelane, quartz is present in trace amounts.

2. Sample 3: black with bluish tint fine-crystalline material, composing relatively homogenous boulder-lumpy ore (see Figures 5, 10, and 11); composition: pyrolusite with significant admixture of cryptomelane and slight amounts of lithiophorite.

3. Sample 4: small globules, composed of manganese oxyhydroxides with admixtures of iron oxyhydroxides, and kaolinite material, form lens-like accumulations in ferruginous laterite-like mass (see Figures 6, 10, and 12); composition: pyrolusite with relatively slight admixture of cryptomelane, traces of manganete and considerable admixture of quartz and kaolinite.

4. Sample 4-1: dense, hard, fine-crystalline matter of manganese oxyhydroxides, impregnating laterite quartz accumulations (see Figures 6, 10, and 12); composition: quartz, pyrolusite with subordinate admixture of cryptomelane.

5. Sample 5: finest oolites-globules (1–3 mm) (see Figures 6, 10 and 12); composition: pyrolusite, quartz with admixture of kaolinite.

6. Sample 6: fine-pisolite ore, represented by cemented aggregates of pisolites (see Figures 6, 10, and 12); composed of essentially pyrolusite with slight admixture of lithiophorite, traces of cryptomelane.

7. Sample 7: black cryptocrystalline matter of manganese oxyhydroxides, composing internal part of nodules, external part is represented by iron oxyhydroxides (see Figures 6, 10, and 12); pyrolusite and lithiophorite predominates with traces of cryptomelane and goethite.

8. Sample 8: fine oolites and globules (1–3 mm) of manganese oxyhydroxides, separated from loosely cemented aggregate mass (see Figures 6, 10, and 12); composition: pyrolusite, 10 Å-manganate (todorokite), traces of cryptomelane, probably hollandite, and kaolinite in considerable amounts.

9. Sample 9: black cryptocrystalline microglobular matter of manganese oxyhydroxides, composing nodules, fragments of irregular form, occurring in laterite mass (see Figures 6, 10, and 12); composed essentially of pyrolusite with subordinate lithiophorite, traces of cryptomelane.

10. Sample 11: black matter of manganese oxyhydroxides with admixture of iron oxyhydroxides, composing lens-like ore nodules (see Figures 7, 10, and 13); composition: pyrolusite, cryptomelane, and subordinate goethite and quartz.

11. Sample 12: black matter, composing small globules, bonded into lens-like aggregates (see Figures 7, 10, and 13); represented by almost pure pyrolusite with insignificant admixture of quartz.

12. Sample 13: fine ore pisolites, and oolites (see Figures 7, 10, and 13), composed of almost pure pyrolusite.

13. Sample 14: ore leguminous nodules and pisolites (6–7 mm), composing cemented aggregates (see Figures 7, 10, and 13); composed of almost pure pyrolusite.

14. Sample 15: fine ore pisolites and leguminous nodules (3–7 mm), separated from kaolinite matrix (see Figures 7, 10, and 13); consist of almost pure pyrolusite.

Notes to Table 2 (continued)

15. Sample 16: ore pisolites and oolites (1–3 mm), separated from kaolinite matrix (see Figures 7, 10, and 13); composed of pyrolusite with traces of cryptomelane and kaolinite.

16. Sample 17: white kaolinite matter of the matrix with admixture of manganese and iron oxyhydroxides (see Figures 7, 10, and 13), in which ore concretions occur; composition: quartz, kaolinite, iron oxyhydroxides; Mn ore components are represented mainly by cryptomelane.

17. Sample 18: black with bluish tint fine-crystalline ore matter, composing the laminated fragments in kaolinite matrix (see Figures 7, 10, and 13); composition: cryptomelane with admixture of kaolinite, traces of quartz and goethite.

18. Sample 19: small ore pisolites and fabiform nodules from earthy laterite (see Figures 7, 10, and 13); composition: lithiophorite with considerable admixture of cryptomelane, pyrolusite, kaolinite.

19. Sample 20: black fine-crystalline matter, composing flat nodules and ore lenses; represented by essentially cryptomelane with slight admixture of pyrolusite, considerable amount of quartz, subordinate clayey (essentially kaolinite with traces of hydromica) material (see Figures 8, 10, and 14).

20. Sample 21: black with bluish tint fine-crystalline matter, composing flattened fragments, lenses, and lens-like intercalations of ore, composed of cryptomelane with subordinate goethite, and with rather slight admixture of kaolinite-hydromica clay material (see Figures 8, 10, and 14).

21. Sample 22: black fine-crystalline matter, composing small (2–3 mm) ore pisolites, oolites, cemented; composition: essentially pyrolusite,cryptomelane with admixture of quartz, kaolinite clay material (see Figures 8, 10, and 14).

22. Sample 23: black matter, composing ore fabiform nodules and pisolites (5–7 mm), cemented into dense, hard mass (see Figures 8, 10, and 14); composition: pyrolusite with subordinate cryptomelane, traces of birnessite, slight admixture of clay material (kaolinite, smectite).

23. Sample 24: black fine-crystalline matter, composing large (5–15 cm) ore concretions, their intergrowths and aggregates (see Figures 8, 10, and 14); composition: cryptomelane with rather slight admixture of goethite, quartz, clay (hydromica, smectite) matter, traces of manganite.

24. Sample 25: iron oxyhydroxides in mixture with clay-siliceous material, composing ore angular gravel fragments (see Figures 8, 10, and 14); composition: in ferrugenous kaolinite mass of laterite, goethite with essential admixture of kaolinite.

25. Sample 26: black cryptocrystalline matter of manganese oxyhydroxides with admixture of brown iron oxyhydroxides and light kaolinite material, composing the laterite-like mass (see Figures 9, 10, and 15); represented by cryptomelane, pyrolusite and goethite with essential admixture of quartz, kaolinite clay material.

26. Sample 27: black with bluish tint crystalline-grained fine-porous matter, composing boulder-lumpy ore fragments (see Figures 9, 10, and 15); composition: cryptomelane with admixture of lithiphorite, traces of 10 Å-manganate (todorokite), essential quantities of quartz, traces of kaolinite.

27. Sample 28: black with bluish tint fine-crystalline matter, composing the ore nodules, pisolites and fragments in the mottled ferruginous laterite mass; consisting essentially of cryptomelane, lithiophorite with admixture of quartz, romanechite (psilomelane), 10 Å-manganate (todorokite), traces of kaolinite, pyrolusite (see Figures 9, 10, and 15).

Table 3.   Contents of trace elements in manganese ores, Groote Eyland deposit, Northern Australia (ppm).

no.	no. of sample*	Cu	Rb	Sr	Y	Zr	La	Ce
1	2	93.0	56.0	482.0	58.0	49.0	12.0	50.0
2	3	117.0	48.0	567.0	28.0	80.0	23.0	19.0
3	4	65.0	58.0	850.0	56.0	104.0	29.0	52.0
4	4-1	–	–	–	–	–	–	–
5	5	65.0	90.0	721.0	48.0	23.0	12.0	8.0
6	6	70.0	55.0	1348.0	20.0	65.0	12.0	46.0
7	7	91.0	65.0	2445.0	80.0	87.0	12.0	35.0
8	8	94.0	77.0	4350.0	59.0	96.0	18.0	22.0
9	9	56.0	20.0	1549.0	36.0	47.0	21.0	35.0
10	11	46.0	59.0	174.0	35.0	58.0	28.0	8.0
11	12	66.0	41.0	133.0	112.0	76.0	101.0	33.0
12	13	87.0	60.0	244.0	78.0	102.0	31.0	50.0
13	14	55.0	57.0	192.0	84.0	44.0	13.0	45.0
14	15	96.0	46.0	367.0	91.0	71.0	40.0	20.0
15	16	68.0	48.0	218.0	72.0	69.0	44.0	70.0
16	17	76.0	61.0	432.0	63.0	110.0	12.0	110.0
17	18	47.0	30.0	517.0	99.0	89.0	27.0	167.0
18	19	59.0	61.0	208.0	46.0	409.0	53.0	177.0
19	20	36.0	42.0	448.0	48.0	53.0	12.0	55.0
20	21	78.0	42.0	701.0	55.0	53.0	17.0	46.0
21	22	79.0	41.0	629.0	42.0	53.0	12.0	61.0
22	23	90.0	20.0	283.0	89.0	44.0	17.0	51.0
23	24	83.0	104.0	688.0	74.0	67.0	26.0	210.0
24	25	45.0	20.0	21.0	56.0	150.0	15.0	23.0
25	26	71.0	20.0	207.0	53.0	104.0	32.0	183.0
26	27	107.0	42.0	514.0	157.0	40.0	101.0	223.0
27	28	72.0	58.0	286.0	88.0	124.0	43.0	433.0

*Description of samples, see Table 2, Figures 5 to 10. Determinations were done from separated portions by the method of X-ray fluorescence spectroscopy, analyst E.M. Margolin, GIN, Russian Academy of Sciences.

data of the X-ray diffractometry studies is given in notes to Tables 2, 3 and Figures 11 to 26.

Distribution and transformation of the manganese oxyhydroxide minerals in the ore bed. The section of quarry-D (see Figures 5, 10, and 11): Ores are represented by two varieties: boulder-lump and pisolite.

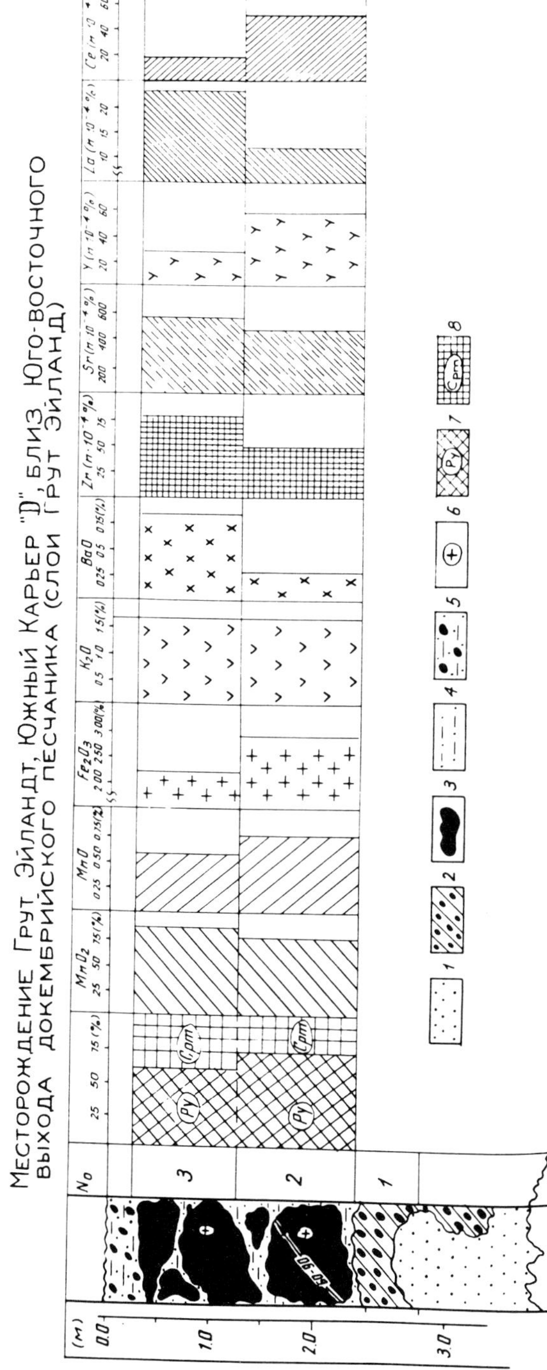

Fig. 11.   Structure of ore bed and distribution of mineral and chemical components in manganese oxyhydroxide ores of the Groote Eylandt deposit, southern quarry-D (see Figures 4 and 10), near the southeastern outcrop of the Precambrian sandstone (see Figure 5) (Groote Eylandt Beds). 1. Precambrian sandstone (Groote Eylandt Beds); 2. Brown laterite (after Mullaman Beds, Middle Cretaceous); 3. Boulder-lumpy oxyhydroxide manganese ores; 4. Yellowish-brown clay-sandy matrix; 5. Brown essentially ferruginous soil; 6. Places of sampling; 7. Pyrolusite; 8. Cryptomelane.

The boulder-lump ores of massive structure are made of pyrolusite with essential amounts of cryptomelane and quartz admixture.

The pisolites up to several mm in size are bonded by manganese cement, composed of cryptomelane and nsutite. The most common component of pisolites is the pyrolusite, often substituted by cryptomelane (II). Numerous cracks are filled in with lithiophorite. As a whole cryptomelane predominates. In small pisolites (sample 2, see Figures 5, 10, and 11) the cores consist of clastic components. The alternation of the concentric layers of pyrolusite and cryptomelane is observed in pisolites; sometimes they are composed by a fine mineral mixture, i.e. intergrowth of pyrolusite and cryptomelane. The fragments of the clay composition (kaolinite) are incorporated into the matrix of manganese oxyhydroxides (Figure 16). Boulder-lump ores of the upper part of the section (sample 3, see Figures 5, 10, and 11) contain black with a bluish tint fine-crystalline material, represented by pyrolusite with a rather significant admixture of cryptomelane and small amounts of lithiophorite.

The section of quarry-G (see Figures 6, 10, and 12): At the base of the section the ochreous ferruginous laterite occurs, incorporating small globules of manganese oxyhydroxide with admixtures of iron oxyhydroxides and kaolinite material in the form of irregular lenses (sample 4, see Figures 6, 10, and 12). Globules are composed of pyrolusite with relatively slight admixtures of cryptomelane, traces of manganite and significant amounts of quartz and kaolinite.

Pisolites predominate in the upper section. Two types of these accumulations are distinguished:

(a) Fine, soft pisolites with a diameter of less than 4 mm. The main components are pyrolusite and cryptomelane; kaolinite is present both in peripheral and in the internal parts of pisolites. This type of pisolite composes the main structural components of the ore bed (interlayers, lenses and so on), that are depicted by numbers 4, 5, 6, 8, and 9 in Figures 6, 10, 12 (samples 5 to 9). Small concretions of pyrolusite are substituted by cryptomelane. The development of two almost simultaneous processes are observed, as a rule, at one and the same level of the bed: (1) pisolites with cryptomelane core are enriched in pyrolusite to the periphery; (2) the concretions, the central part of which is composed of pyrolusite, are substituted by cryptomelane.

Variation of the structure and mineral composition in the lower part of the ore bed is of interest too (samples 5 to 7, see Figures 6, 10, and 12).

Impersistent intercalation composed of small globular oolithic (1–3 mm) earthy manganese oxyhydroxides, incorporating lenses and irregular-mottled aggregates of brown kaolinitic lateritic clay, overlie the laterite-like mass of the section base (see above sample 4, Figures 6, 10, and 12). These oolite-globules are composed of pyrolusite and quartz with an admixture of kaolinite (sample 5, see Figures 6, 10, and 12).

Small (< 2–4 mm) oolites and pisolites of manganese oxyhydroxide (sample 6, see Figures 6, 10, and 12) cemented, as a rule, to aggregates, predominate above. They contain essentially pyrolusite with a slight admixture of lithiophirite and traces of cryptomelane.

Pisolites of manganese oxyhydroxides (sample 7, see Figures 6, 10, 12) have developed further up the section. In the lower part of the intercalation their size is 3–7 mm, in the upper part there is a mixture of small (< 2–3 mm) and large (to 10 mm) pisolites. The internal part of pisolites is composed by black cryptocrystalline matter of manganese oxyhydroxide,

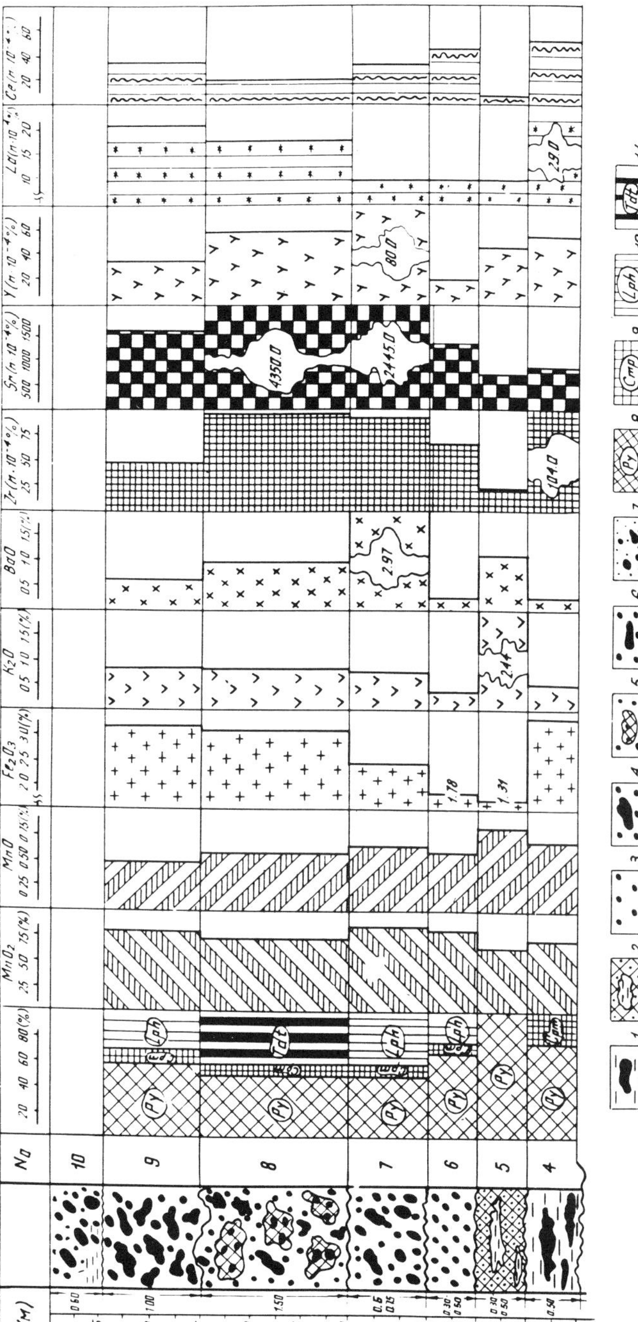

Fig. 12. Structure of ore bed and distribution of mineral and chemical components in manganese oxyhydroxide ores of the Groote Eylandt deposits, quarry-G, southwestern face (see Figures 4, 6, and 10). 1. Ochreous laterite with iron and manganese oxyhydroxides in the form of irregular lenses; 2. Fine-globular Mn oxyhydroxides (1–2 mm), earthy with lenses and irregular mottled patches of brownish-red kaolinitic and lateritic clay; 3. Oolites and pisolites of Mn oxyhydroxides, small (< 2–4 mm); 4. Pisolites of Mn oxyhydroxides, their size in the lower part being 3–7 mm, in the upper part being a mixture of small (2–3 mm) and large (up to 10 mm) pisolites, a mixture of small oolites (< 2–3 mm) also occurred; 5. Mn oxyhydroxides represented by irregularly lenticular patches of the earthy mass and ore fragments; 6. Laterite: Fe, Mn oxyhydroxide pisolites and fragments of irregularly tabular shape; 7. Soil composed of red-brown ferruginous laterite: pisolite, fragments and clay matrix; 8. Pyrolusite; 9. Cryptomelane; 10. Lithiophorite; 11. Todorokite.

its external part by iron oxyhydroxides. Pyrolusite and lithiophorite amounts to traces of cryptomelane and goethite predominate.

In pisolites of sampling levels 8 and 9 (see Figures 6, 10, and 12) lithiophorite becomes the essential component, whereas at lower levels (samples 4 to 6, see Figures 6, 10, and 12) this mineral occurs in trace amounts. In some cases the lithiophorite forms the cores of pisolites (samples 8 and 9, Figures 6, 10, and 12), around which the cryptomelane is developed; sometimes it is found in the form of small nodules. A point that should be mentioned is that at the level of sample 8 (see Figures 6, 10, and 12) ores are represented both by mixtures of oolites (< 2–3 mm) and accumulations of manganese oxyhydroxides in the form of irregular lens-like uncompacted earthy patches and ore fragments. They consist of pyrolusite, todorokite with traces of cryptomelane, and probably hollandite. Kaolinite is observed in considerable quantites (Figures 25, 26). In ores of the upper interval of the bed, represented by the nodules and fragments of irregular shape, enclosed in the laterite mass (sample 9, see Figures 6, 10, and 12), the main component is black cryptocrystalline and microglobular matter of managanese oxyhydroxides. It is represented essentially by pyrolusite with subordinate lithiophorite and traces of cryptomelane.

It is interesting to note the presence of nsutite at levels of samples 8 and 9 in the nodules, composed of cryptomelane. The reflectivity and anisotropy of nsutite is considerably higher than those of cryptomelane.

(b) Pisolites are relatively harder, their diameter is >10 mm (see sample 8, component 7, Figures 6, 10, and 12), they are characterized by the predominance of cryptomelane; lithiophorite occurs at the extreme periphery of pyrolusite aggregates and it is associated with kaolinite.

The section of the southern quarry-A (see Figures 7, 10, and 13): At the section base (sample 11, see Figures 7, 10, and 13) there are lens-like ore nodules, composed of pisolites up to several cm in size; their cement is represented by the ferruginous–manganese matter, that contains fragments of ferruginous and clayey composition, quartz grains, and fragments of concretions. Iron and manganese oxyhydroxides form an ultrafine mixture (Figure 17, photos A, B). Concretions and pisolites are composed of amorphous iron oxy-hydroxides partially altered into goethite and substituted by cryptomelane and pyrolusite within the same nodules. These accumulations show signs of slight redeposition. At the higher level (sample 12, see Figures 7, 10, and 13) pisolites and small globules, bonded into lens-like aggregates, are developed. Fine crystalline pyrolusite, closely associated with cryptomelane, predominates in their composition. The black matter, often composing the whole of these pisolites, is represented by almost pure pyrolusite with an insignificant admixture of quartz. This association is also observed in the pisolites of the higher level too (sample 13, see Figures 7, 10, and 13). The pisolites of the level considered (sample 13) are composed of almost pure pyrolusite like those of the lower level (sample 12). The remarkable feature of these pisolites is the presence of romanechite (psilomelane) relicts in the form of residual relict islands in the cryptomelane mass. At higher horizons (samples 14, 15, and 16, see Figures 7, 10, and 13) an increase of pisolite sizes is observed. The ubiquitous development of cryptomelane from the periphery to the center is of special interest. It is significant that to the level of sample 14 (see Figures 7, 10, and 13) the pyrolusite remains the dominant component.

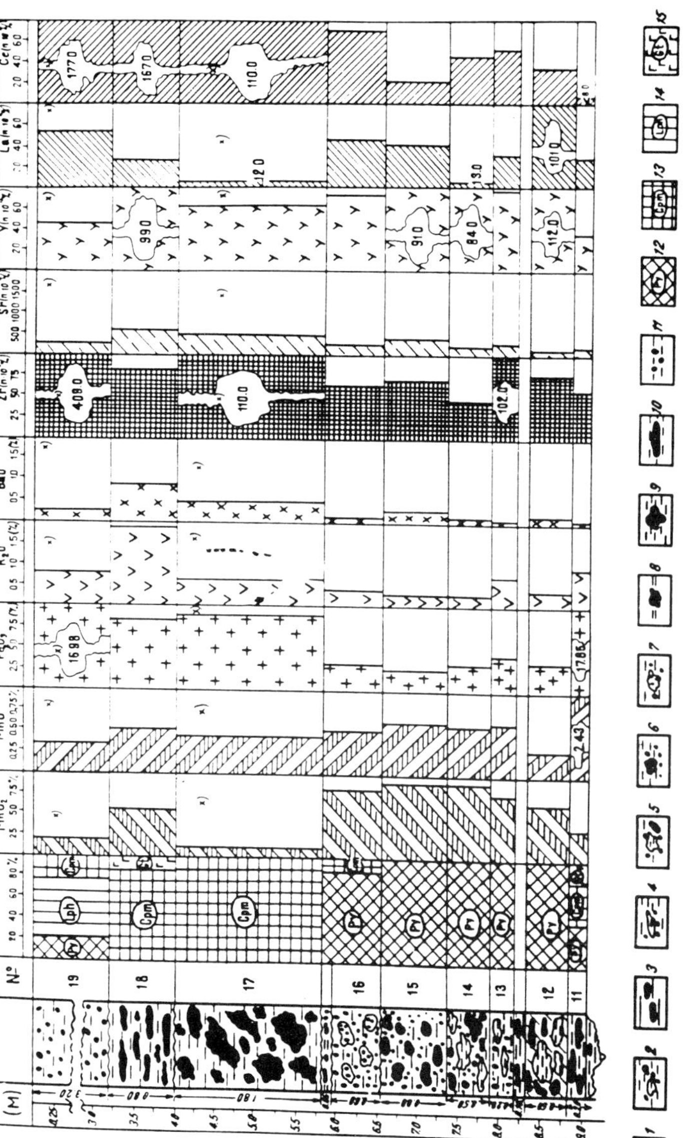

Fig. 13. Structure of the ore bed and distribution of mineral and chemical components in manganese oxyhydroxide ores of the Groote Eylandt deposits, southern part of the quarry-A, southern face (see Figures 4, 7, and 10). 1. Lens-like nodules, of Mn and Fe oxyhydroxides in brownish kaolinite mass; 2. Mn oxyhydroxides fine-globular earthy-sooty, containing small lens-like aggregates of hematite, and ochreous material; 3. Concretions of Mn oxyhydroxide; 4. Lenses, pockets of kaolinite; 5. Oolites of Mn oxyhydroxides in kaolinite ground mass, comprising lenses of light kaolinite; 6. Pisolites (3–7 mm), small concretions of Mn oxyhydroxides, lenses of light kaolinite in brownish kaolinite matrix; 7. Pisolites of Mn oxyhydroxides, their aggregates in kaolinite matrix (60%); 8. Concretions of Mn oxyhydroxides in kaolinite clay; 9. Concretions, their intergrowths, aggregates of Mn oxyhydroxides in kaolinite matrix; 10. Fragments of tabular form, composed of Mn oxyhydroxides in kaolinite matrix; 11. Laterite, composed of kaolinite clays with concretion-pisolite aggregates of Mn oxyhydroxides; 12. Pyrolusite; 13. Cryptomelane; 14. Lithiophorite; 15. Goethite.

However, beginning at the level of sample 15 (see Figures 7, 10, and 13) change of this mineral's occurrence takes place; the dominating pyrolusite becomes porous; in polished section it is yellow-brown, sometimes its progressive substitution by cryptomelane is observed (Figure 17, photo D). Romanechite appears at the same level. The predominance of pyrolusite with traces of cryptomelane and kaolinite is observed in the ore pisolites and oolites (1–3 mm), enclosed in kaolinite matrix of the level of sample 16 (see Figures 7, 10, and 13).

From the level of sample 17 (see Figures 7, 10, and 13) the development of massive concretion ore is distinguished. Cryptomelane sometimes with romanechite (psilomelane) predominates in its composition, and lithiophorite and nsutite are present as admixtures. Numerous ferruginous fragments, enclosed in kaolinite clay with halloysite (Figures 25 and 26) are observed in the cryptomelane mass, along with clastic quartz grains with obvious signs of corrosion and dissolution. In many cases these fragments are represented by goethite.

At the level of the sample 18 (see Figures 7, 10, and 13) with the marked predominance of cryptomelane there is the same paragenesis of minerals as at the level of sample 17. Furthermore, it is emphasized that cryptomelane and/or nsutite of the concretion structure is developed at the place of quartz dissolution (Figure 17). Similar relationships of the minerals are described in the work by Grandin and Perseil (1977) for African manganese deposits in the weathering crusts. The concretion–pisolite aggregates of manganese oxyhydroxides are enclosed in the pedogenic laterite (sample 19, see Figures 7, 10, and 13), composed of kaolinite clays at the top of the section. These aggregates consist of lithiophorite with significant admixtures of cryptomelane, pyrolusite, and kaolinite.

The section of quarry-A, southeastern face (see Figures 8, 10 and 14): At the base of the section flat bundles and lenses occur (sample 20, Figures 8, 10, and 14), composed of aggregates of pisolites, enclosed in a reddish clay-ferruginous matrix. Fragments and grains of quartz, being in the stage of dissolution, are observed in the pisolite core. It is necessary to emphasize the geochemical significance of these quartz clastic particles. The black microcrystalline matter, composing nodules and pisolites, is represented predominantly by cryptomelane with a slight admixture of pyrolusite, significant amounts of quartz and subordinate clayey (mainly kaolinite with traces of relict hydromica) material. Cryptomelane predominates in the pisolites and cement. Ramsdellite is usually in the center of pisolites or appears in the form of thin streaks. At this level the alteration and transformation of pisolites are observed: pyrolusite and ramsdellite are developed after cryptomelane.

Upper ores (level of sample 21, see Figures 8, 10, and 14) are represented by tabular fragments, lenses, and lens-like intercalations, composed of dense manganese oxyhydroxides. They are distinguished by massive structure and relatively high hardness. Cryptomelane with subordinate goethite and with a rather small admixture of kaolinite and traces of hydromica material predominates in their composition. The presence of romanechite (psilomelane), well crystalline and with numerous clastic quartz grains with distinguished signs of corrosion and dissolution (see Figure 18), is characteristic. Alteration of the ore (predominantly cryptomelane) matter, observed below, is continued at this level too. With a general predominance of cryptomelane, some alteration of this matter

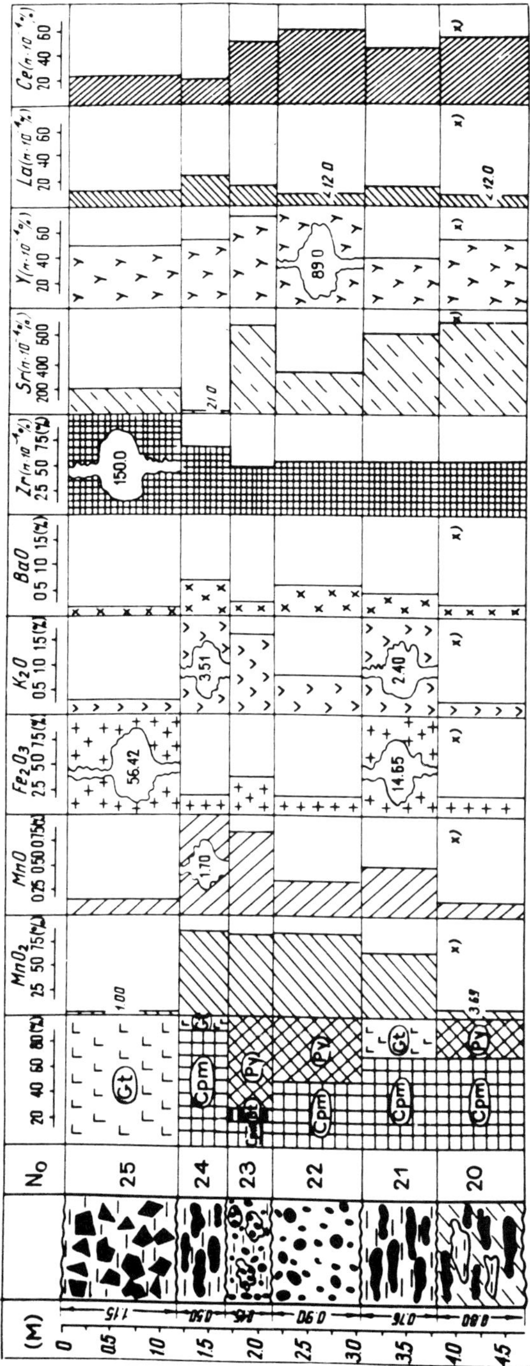

Fig. 14.   Structure of the ore bed and distribution of mineral and chemical components in manganese ores, the Groote Eylandt deposits, quarry-A, southeastern face (see Figures 4, 8, and 10). 1. Dense, hard nodules and lenses of Mn oxyhydroxides; 2. Tabular fragments, lenses, and lens-like intercalations composed of dense, hard Mn oxyhydroxides; 3. Small (2–3 mm), cemented oolites of Mn oxyhydroxides; 4. Pisolites (5–7 mm) of Mn oxyhydroxides, cemented into dense, hard mass; 5. Large concretions (5–15 cm) of Mn oxyhydroxides, their intergrowths and aggregates; 6. Laterite composed of angular gravel fragments of Fe oxyhydroxides (hematite) in ferruginous kaolinite mass; 7. Pyrolusite; 8. Cryptomelane; 9. Birnessite; 10. Goethite.

into nsutite and the appearance of lithiophorite is observed (Figure 18).

Higher up the level of sample 22 (see Figures 8, 10, and 14) the small (< 2–3 mm) oolites of manganese oxyhydroxides, often cemented into aggregates, are developed. These are represented by pyrolusite and cryptomelane with admixture of quartz and kaolinite. Usually, pyrolusite predominates both in the concentric layers of oolites and in the cement. However, pyrolusite is substituted by cryptomelane, developed in the finest cross-cutting fractures from the periphery to the internal parts of oolites. In a number of cases cryptomelane is the predominating component often this mineral forms fine-grained intergrowths with pyrolusite.

At the level of sample 23 (see Figures 8, 10, and 14) pisolites from 5 to 7 mm in size are widely distributed. They are cemented into dense aggregates, lumps, and lenses; pyrolusite with subordinate cryptomelane predominates in their composition, there are traces of birnessite, admixture of kaolinite, and relics of smectite. Kaolinite and quartz are common in the composition of the pisolite cement of the samples 22 and 23.

Rising up the section to the level of the sample 24 (see Figures 8, 10, and 14) the cryptomelane becomes the predominating mineral, composing large (5 to 15 cm) ore concretions, their aggregates, and intergrowths. Insignificant admixtures of goethite, quartz, relics of hydromica, smectite, and traces of manganite are characteristic. The ground mass of cryptomelane in some parts is cross-cut by numerous streaks of romanechite. Even if the differences in reflectivity of these two components is difficult to differentiate, the pleochroism, and effects observed in crossed nicols, of romanechite distinguish it from cryptomelane. The section top (sample 25, see Figures 8, 10, and 14) is represented by pedogenic laterite, composed of angular gravel fragments of iron oxyhydroxides (goethite) without any significant traces of manganese in the ferruginous mass, containing mainly kaolinite with an admixture of halloysite (Figures 25, and 26).

The section of quarry-E, northern part, northwestern face (see Figures 9, 10, and 15): The samples of manganese oxyhydroxide ores and enclosing sediments, sampled from the section of the ore bed of the quarry-E, northern part, northwestern face (see Figures 9, 10, and 15), mainly comprise components of the laterite weathering crust. The presence of kaolinite in the whole section indicates the stability of the acidic regime of pH during mineralization and the presence of essential reserves of free alkaline earth elements.

It is necessary to emphasize the following features of this section:

Dissolution of quartz grains is a general phenomenon. It is followed by the substitution of romanechite, observed at all levels of the section.

From the base to the top of the section the ores become harder and more massive and the predominance of cryptomelane and romanechite is observed. According to Perseil (personal communication), the manganese ores from the laterite weathering crust of the Tambao deposit, Burkina Faso (Upper Volta), are of the same nature.

The romanechite (Figure 19), occurring in the form of thin acicular crystals, is formed at the place of dissolution of quartz clastic grains; this can be seen in the example of relations with quartz relics (A). Such a type of alteration of the mineral paragenesis is rather typical for both this section and for general laterite manganese-bearing weathering crusts.

It is important to point out that at the lower part of the section ferruginous laterite

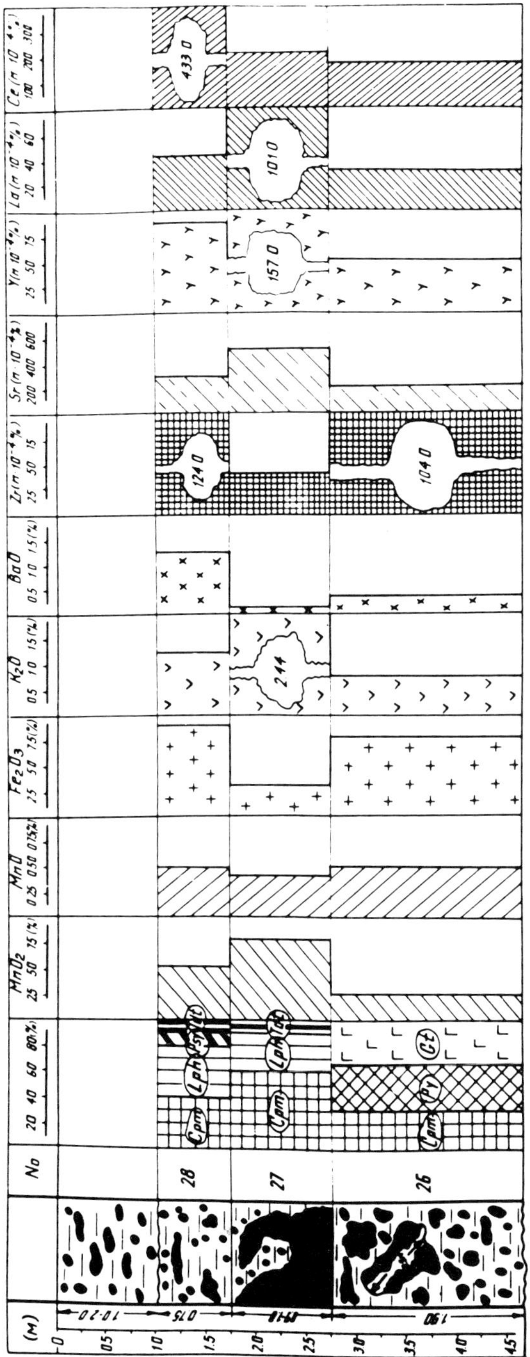

Fig. 15. Structure of the ore bed and distribution of mineral and chemical components in manganese ores, the Groote Eylandt deposits, northern part of the quarry-E, northwestern face (see Figures 4, 9, and 10). 1. Boulder fragments, nodules of Mn oxyhydroxides, the ground mass is kaolinitic clay; 2. Boulder-lumpy accumulations of Mn oxyhydroxides; 3. Mottled ferruginous laterite with tabular fragments, and pisolites of Mn oxyhydroxides; 4. Mottled ferruginous pedogenic soils; 5. Cryptomelane; 6. Pyrolusite; 7. Lithiophorite; 8. 10 Å-manganate (todorokite?); 9. Romanechite (psilomelane); 10. Goethite.

predominates. Upwards it is replaced by boulder fragments of manganese oxyhydroxides (see Figures 9, 10, and 15). The sample 26 (laterite-like mass) is represented by a black cryptocrystalline matter of manganese oxyhydroxides with admixtures of brown iron oxyhydroxides and light kaolinite material; its mineral composition is: cryptomelane, pyrolusite with significant admixtures of quartz and kaolinite.

The boulder-lump ore fragments (sample 27, see Figures 9, 10, and 15) are composed of black with bluish tint crystalline, earthy, fine-porous matter, represented by cryptomelane with an admixture of lithiophorite, traces of todorokite, significant quantities of quartz, and subordinate kaolinite.

Manganese ore nodules, leguminous concretions, and fragments, enclosed in mottled ferruginous laterite masses, are developed at the section top (sample 28, see Figures 9, 10, and 15). The spotting is caused by the irregular distribution of iron oxyhydroxides (mainly goethite, less frequently manganese oxyhydroxides) in the kaolinite matrix. The ores consist mainly of cryptomelane, lithiophorite with admixture of romanechite, todorokite, quartz, traces of pyrolusite and kaolinite (Figure 19). The microscopic study of samples from the lower and middle parts of the section (samples 26 and 27, Figures 9, 10, and 15) indicates that cryptomelane predominates among them. Initial progressive substitution of fragmentary quartz grains by cryptomelane is observed in fissures and corrosion contacts. This type of alteration is well distinguished in samples 26 and 27. The formation of concentric zones, composed of romanechite (see Figure 19), is observed at the final stages of such dissolution.

The essential features of the mineral composition of the ores.   Cryptomelane is related to oxyhydroxide minerals of manganese (arbitrarily α-MnO_2), comprising channels (tunnels) in their structure, in particular, to the group of hollandite type (Burns and Burns, 1979; Chukhrov *et al.*, 1989). In general, the crystallochemical formula of minerals of this group may be represented in the following form:

$$R_x\ (Mn^{4+}, Mn^{3+}, Fe, Al, \square)_8\ (O,OH)_8\ nH_2O,$$

where R = K, Ba, Pb or Na, and $\square$ are vacancies in octahedral bands. The value of x depends on the valency of tunnel cations and the charge of the octahedral bands, forming due to isomorphic substitutions and vacancies, and from the relations of O and OH groups in the anion framework. Large channels of the structure are usually occupied by cations with a relatively large ionic radius: K, Ba, Pb or Na. The following minerals are dependent on the dominant cation nature: hollandite, cryptomelane, coronadite, and monjiroite, in which Ba, K, Pb, or Na correspondingly predominate.

The most common forms of cryptomelane occurrence in the ores of the Groote Eylandt deposits are represented as distinct textural and microscopic varieties, in which, as may be seen under the optical microscope, the cryptocrystalline oxyhydroxide material contains abundant inclusions of clastic quartz (Figures 16, 17, and 19). Fine globular structure is distinctly observed under the scanning electron microscope (Figures 20 to 22). Moreover, for early stages of recrystallization of the finest globules the development of the thinnest, elongated along c-axis laminated crystals (see Figure 22) or acicular crystals elongated in

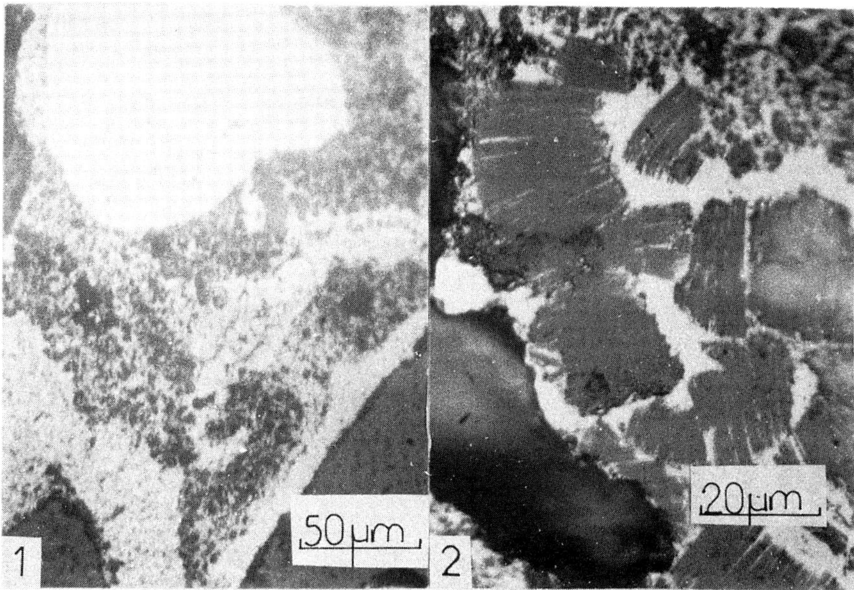

Fig. 16. Photomicrographs of a part of the boulder-lumpy manganese ore; polished section, plain light, sample 2, southern part of the quarry-D, (see Figures 4, 5, and 10), the Groote Eylandt deposits (photo by Dr. E.A. Perseil). Mineral composition: pyrolusite with essential admixture of cryptomelane. 1. Clastic quartz grains in Mn oxyhydroxide cement. 2. Pockets of kaolinite, impregnating, substituting the grains of Mn oxyhydroxides.

the c-(or b-)axis is characteristic. Similar morphological features are explained (Chukhrov *et al.*, 1989) by the maximum rate of the growth attained for individual octahedral bands, and the margins of which were characterized with maximum valence non-saturation. Ostwald (1980) points out that similar varieties (i.e. earliest generation) compose the main mass of oolite, pisolite, and concretion ores. Streaks of several younger generations of cryptomelane are developed, and often cross-cut this variety of cryptomelane; the younger generations have a different microcrystalline structure, usually represented by elongated prismatic crystals. Sometimes such crystals are as large as 3 mm. They are distinguished by relatively small quantities of admixtures. Thin notched and mamillate acicular aggregates, with the typical form of frost dendrites, are recognized among relatively late generations, and are products of crystallization. It is particularly remarkable that aggregates of acicular crystals of cryptomelane fill the voids in ores (Figures 20 (2, 3), 23, and 24). The following forms of cryptomelane occur (Ostwald, 1975, 1980, 1982): (a) concentric layers in pyrolusite-cryptomelane pisolites; (b) zones and layers in the cemented pisolites, ore pebbles, boulders, and large concretions; and (c) colloidal.

The history of cryptomelane alteration begins with its early generation – the grey cryptocrystalline component – as a slightly anisotropic modification of $\alpha\text{-MnO}_2$. Here the appearance of the patches (i.e. the small segregations) of ramsdellite is observed.

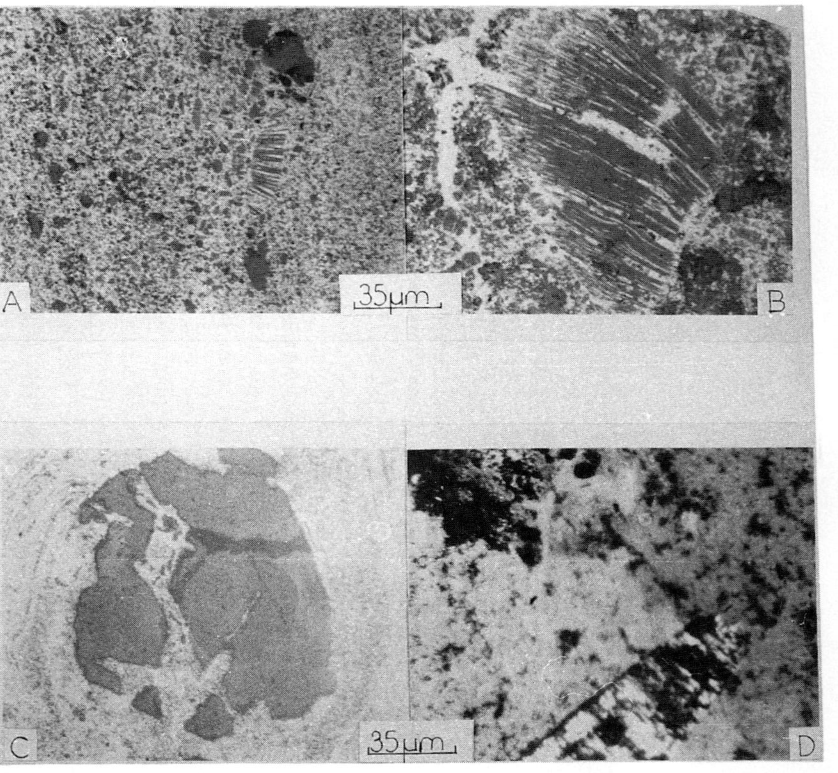

Fig. 17.   Photomicropgraphs of Mn oxyhydroxide ores (lens-like nodules and cemented pisolites). Polished section, plain light, southern part of the quarry-A (see Figures 4 and 10), the Groote Eylandt deposits (photo by Dr. E.A. Perseil). Mineral composition: pyrolusite and cryptomelane, in subordinate amounts: goethite and quartz. A and B. Fragments and grains of ferruginous and clay composition in cement, bonding the Mn oxyhydroxide pisolites. Sample 11. Base of the ore bed (see Figure 7). C. Relicts of clastic quartz grains, preserved after dissolution (sample 18). Top of the ore bed (see Figure 7). D. Decomposition of the pyrolusite patch and its substitution by cryptomelane. Sample 15 (see Figure 7).

Enrichment in manganese takes place due to the removal of silica. A similar process was noted during the deep weathering of the series of quartzites with Mn garnets in Western Africa (Perseil and Grandin, 1978).

The morphology of prismatic crystals of romanechite, filling geodes in ores, the typical representative of which is sample 28 (see Figures 10 and 19), is identical, as Perseil considers (written communication), to similar varieties of the romanechite from the African deposits. In these minerals there is distinct crystallographical completion and development of the tops of prismatic crystals.

The patches and segregations in concretion ores are of special interest for the understanding of formation stages (see Figures 17 to 19). The clay admixture and quartz grains (light-grey cross-cutting streaks of the fine-grained structure, filling in the cracks of syneresis in concretions), and grey cryptocrystalline varieties, often composing interlayers and banded isotropic pisolites, are characteristic for this form.

The distribution of cryptomelane is shown in Figures 10 to 15. The following features have engaged our attention: (a) in sublatitudinal profile of the main sections of the deposit the relative amounts of cryptomelane generally increase from the east to the west according to the dip of the ore bed; (b) the richest, practically pure cryptomelane ores are observed in the central part of the deposit (quarry-A, southern face, see Figures 10 and 13), where the ore bed has the maximum thickness (9 m). Here the proper cryptomelane varieties (samples 17 and 18) are observed in the middle part of the ore body, composed of large concretions and lumpy ores, slightly altered by later events of pedogenic lateritization; (c) the flank has rather thin sections (quarry-G and D in the east and E in the west, see Figure 10). Cryptomelane has, as a rule, a residual nature after substitution by pyrolusite, romanechite (psilomelane), birnessite, todorokite, lithiophorite, and goethite.

Pyrolusite occurs in the form of granular masses to fine-grained intergrowths of prismatic fine crystals. The following forms of its occurrence are recognized: (a) concentric layers, zones, and patches of fine-laminated pyrolusite and cryptomelane pisolites; (b) patches of cryptomelane substituted by pyrolusite in pisolites; (c) cementing material of pisolites; (d) growth crusts on the surface of pisolites and, to a lesser degree, the larger fragments; (e) the finest crystals of pyrolusite and fragments of shells of this mineral in the form of soft aggregates; the aggregates are the result of pisolite substitution. In most cases with distinct interrelations it is seen that pyrolusite is developed after early generation cryptomelane and it is a relatively late supergenous product. But as it was shown above, the inverse interrelations, i.e. development of cryptomelane (more recent generations) after pyrolusite, was noted. Pyrolusite is distributed in the lower half of the ore bed essentially in the central and western parts of the deposits (see Figures 10 to 15, sections of quarries-A, southern, G and D). This distinct localization of relatively pure pisolite varieties, in different structure types of the ores, allows the consideration of these accumulations as the products of relatively early supergenous alteration in the general sequence of the primary carbonate material alteration. In the upper parts of the ore bed the pyrolusite is developed, in most cases, after cryptomelane and is subjected, in its turn, to relatively late substitution by romanechite, todorokite, birnessite, lithiophorite, and goethite.

Manganite occurs mainly in quartz clastic rich manganese ores essentially as relict skeleton forms, preserved during alteration of this mineral into pyrolusite. Ostwald

(1980) described elongated prisms of manganite up to 5 mm long. As a whole, however, manganite is an intermediate residual product in the sequence of alteration of manganese ore matter to forms which are stable in the environment of deep supergenous weathering.

Lithiophorite is observed in the ores of the deposit in two main associations: as an admixture in places of lateritization of the main ore horizons, and the material formed in sand soils of the western and southern parts of the island due to lateritization (Ostwald, 1975, 1980, 1982; Smith and Gebert, 1969). Within the ore bed the most typical forms of lithiophorite occurrence are the patches of both cryptomelane and pyrolusite substitution, and the filling in of the voids, formed in eroded and leached pisolites or massive ores. In these cases there is a relation of lithiophorite with silica, indicating their formation during deep weathering and lateritization. The late generations of lithiophorite together with cryptomelane are present in the form of streaks, cross-cutting lens-like accumulations, dispersed in dense yellowish-white kaolinitic clay (see Figures 25 and 26), underlying the laterite surface in the southwestern part of the island (Ostwald, 1975, 1980, 1982). Lithiophorite is developed, as a rule, in the upper half of the ore bed in most parts of the deposits (see Figures 10 to 15). Nevertheless, in some parts of the quarry (quarry-G, southwestern face) lithiophorite occurs in most of the ore bed; this may point to the depth of distribution of manifestations of relatively late supergenous alterations of the laterite nature.

Romanechite (psilomelane) is distinguished in ores of the deposits by the study of polished sections under microscope, X-ray structural investigations, and by the data, obtained with the microprobe (Ostwald, 1975, 1980, 1982). This made possible the presence of microzones, composed of barium-bearing manganese oxides, to be established. Inspite of visible enrichment of BaO ores (see Table 2), it is necessary to distinguish the barium cryptomelane and romanechite (psilomelane) properly, using the X-ray diffraction data. In those cases, when the romanechite amounts are relatively large (see Figures 9, 10, 15, 18, and 19, section of the quarry-E), this mineral occurs mainly in the upper part of the ore bed in association with cryptomelane, todorokite, and lithiophorite. Ostwald (1980) pointed out that the widest distribution of romanechite is observed in the form of rims in the pisolite varieties, composed of pyrolusite and cryptomelane. Romanechite fills in voids of leaching in primary (Albian–Cenomanian) manganese clays. Hence, the supergenous nature of this mineral, irrespective of differences of its generations, is sufficiently determined.

Nsutite is found in discernable amounts in cemented pisolite ores. The main forms of nsutite occurrence are banded patches and parts of substitution in pisolites and the cementing matrix (Ostwald, 1975, 1980, 1982).

Todorokite is distinguished at the deposit in subordinate quantities in association with lateritized concretion ores as well as the products of substitution of cryptomelane pisolitic varieties (see Figures 10 to 15). It is noticed that todorokite forms colloform banded patches, often aggregates of radial crystals (Ostwald, 1975, 1980, 1982). The development of this mineral is revealed most distinctly in the upper parts of the ore bed as one of the products in the general sequence of relatively late supergenous alteration. It is interesting to point out that Ostwald (1982) emphasized the importance of the understanding of oxide ore genesis: in the quarry-G of the Groote Eylandt deposits the high concentration of minerals

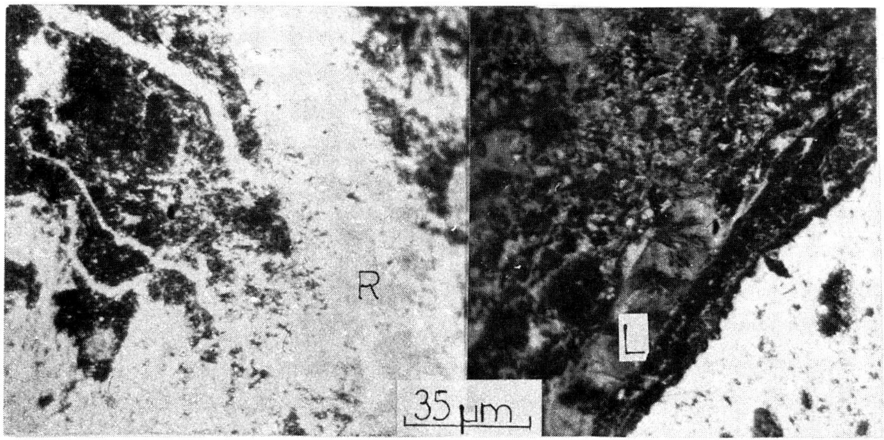

Fig. 18. Photomicrographs of Mn oxyhydroxide ores. Polished section. Plain light. Sample 21. Quarry-A (see Figures 4 and 10). The Groote Eylandt deposits (Photo by Dr. E.A. Perseil). Relationships and development features of romanechite (R) and lithiophorite (L) in the massive Mn oxyhydroxide ores, representesd by tabular fragments, lens-like intercalations, composed essentially of cryptomelane, subordinate goethite, quartz, admixture of kaolinite with relicts of hydromica and clay material (see Figures 8 and 14).

with relatively disordered structure is observed. These minerals are represented by the association: todorokite – birnessite – vernadite – nsutite. The author carried out detailed investigations with electron microprobe, X-ray diffractometry and microdiffraction. On the whole, Ostwald's conclusions coincide with our data (see Figure 10), according to which the section of the quarry-G is characterized by wide distribution of the products of deep (laterite) alteration (lithiophorite, todorokite), developed predominantly after cryptomelane and pyrolusite. Meanwhile, it should be pointed out that the last two minerals reflect relatively early stages of deep supergenous alteration of manganese carbonates.

2.2.1.6. Clay Components of Enclosing Sediments

The recent studies of manganese ores of different nature definitely showed that clay components of ore-enclosing sediments, features of their composition, structural characteristics and associations, reflected rather reliably the conditions of the main stages of the sedimentation history. They may be sensitive indicators of the early stages of formation of sedimentary material: both settings are areas of decomposition of parent rocks, weathering crusts and authigenic postsedimentary and supergenous alterations. Using the clay components of the Groote Eylandt deposits as an example, we try to consider them as an intrinsic part of the whole set of sedimentation (including ore) materials in the general process of the ore formation, as a result of which the deposits formed.

Materials and investigation methods. Clay components were studied in two size fractions: (a) < 0.01 mm and (b) < 0.001 mm. The X-ray diffraction was carried out in the GIN of the Russian Academy of Sciences using the DRON-2 diffractometer (Russia)

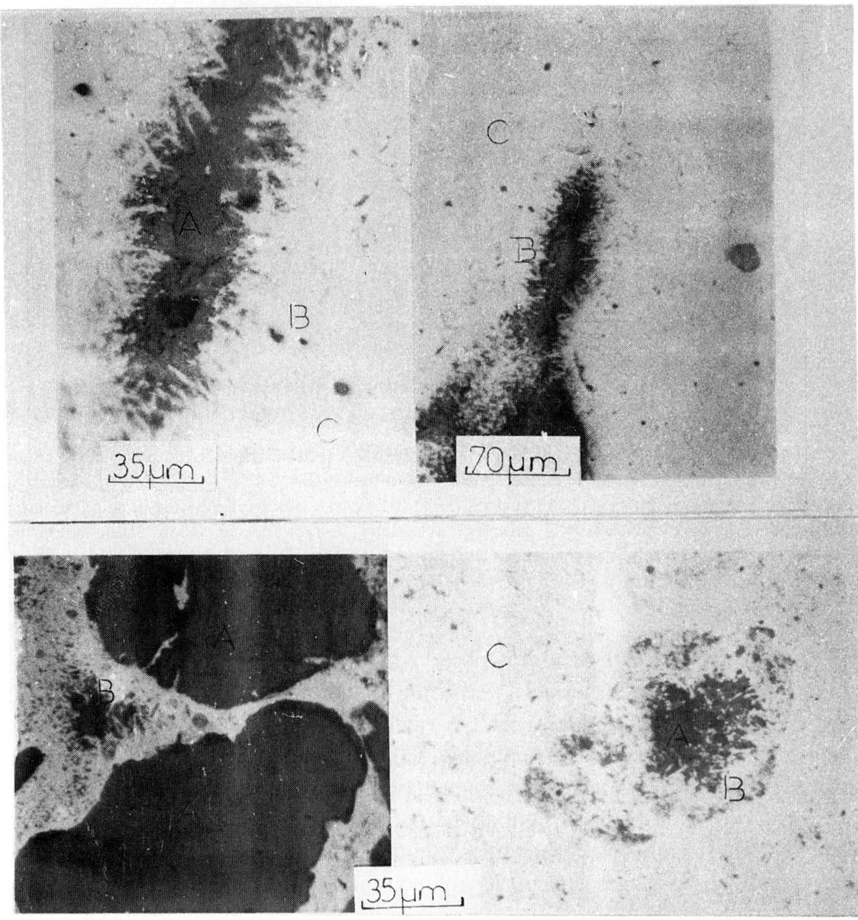

Fig. 19.   Photomicrographs of Mn oxyhydroxide ores. Polished section. Plain light. Sample 28. Quarry-E (see Figures 4, 10), the Groote Eylandt deposit (Photo by Dr. E.A. Perseil). Alteration features of Mn oxyhydroxides in the top of the ore bed (see Figure 9). Development of concretion romanechite (B), substituting in the form of finest acicular crystals, dissolving quartz (A). Relict grains of quartz (black), enclosed in the cryptomelane mass (C), are well visible in the photo.

with radiation Cu, K_{α}, 35 kV and 20 mA. The rate of scanning was 2°/min, and for detailed investigation -1°/min. We obtained diffraction patterns for: (a) air-dried samples, (b) some of them were treated with glycerol (in control cases with ethylene glycol), (c) and heated to 550°C, and if necessary to other temperatures. T.G. Eliseeva took part in the work; interpretation of the obtained data was conducted together with B.A. Sakharov. To determine the structural ordering of the samples we studied oblique features of the representative samples using an ER-100 electron diffractometer, whichh was very efficient for investigations of layered silicates (Zvyagin, 1964). The investigations were

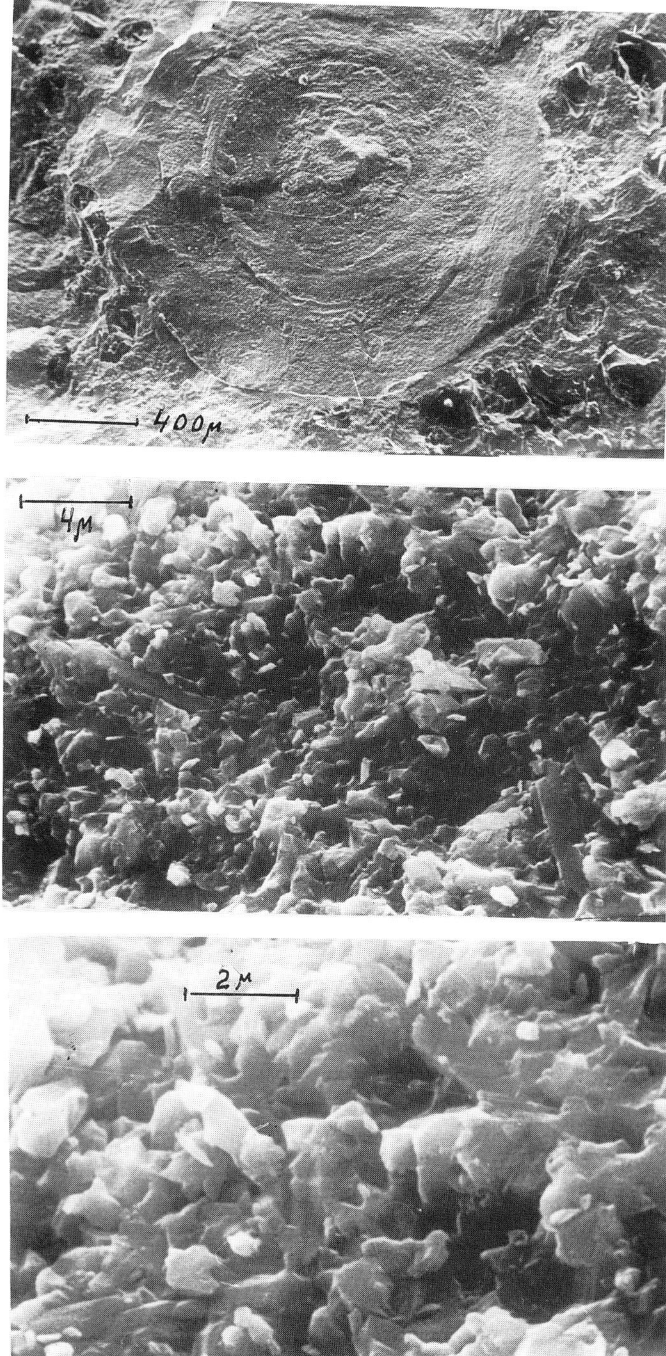

Fig. 20.   Scanning electron microscope photomicrographs of Mn ores. Micro-olite with indistinct concentric structure in black with bluish tint fine crystalline matter composing the tabular fragments and lens-like ore interbeds, consisting of cryptomelane with subordinate amounts of goethite. Sample 21, quarry-A, the Groote Eylandt deposits (see Figures 4, 8, 10, and 14). 1. General view of micro-olite. 2. Central part of the micro-olite (see Photo 1), composed by globular slightly recrystallized matter. 3. Close-up of (1) and (2). The distinct recrystallization of Mn oxyhydroxide matter into aggregates of finest crystals of cryptomelane is shown.

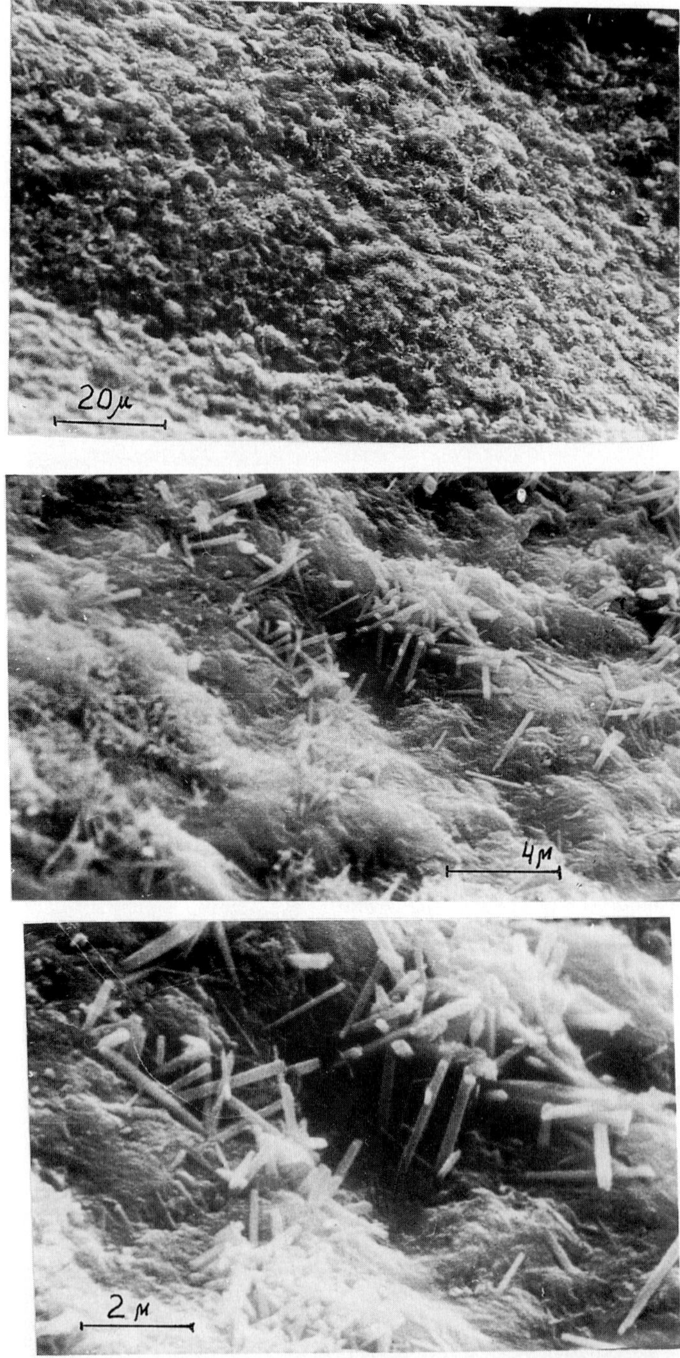

Fig. 21.   Scanning electron microscope photomicrographs of Mn ores. The ground mass, composed of black with bluish tint fine crystalline matter, composing the tabular fragments and lens-like ore interbeds. The ore consists of cryptomelane with subordinate amounts of goethite. Sample 21, quarry-A, the Groot Eylandt deposit (see Figures 4, 8, 10, and 14). 1. General view of the ground mass of the indistinct globular structure. 2. Close-up of (1). The distinct recrystallization of the ground mass into fibrous-fine laminated aggregate of finest crystals of cryptomelane and dispersed rod-shaped crystals of goethite are visible. 3. Close-up of (2).

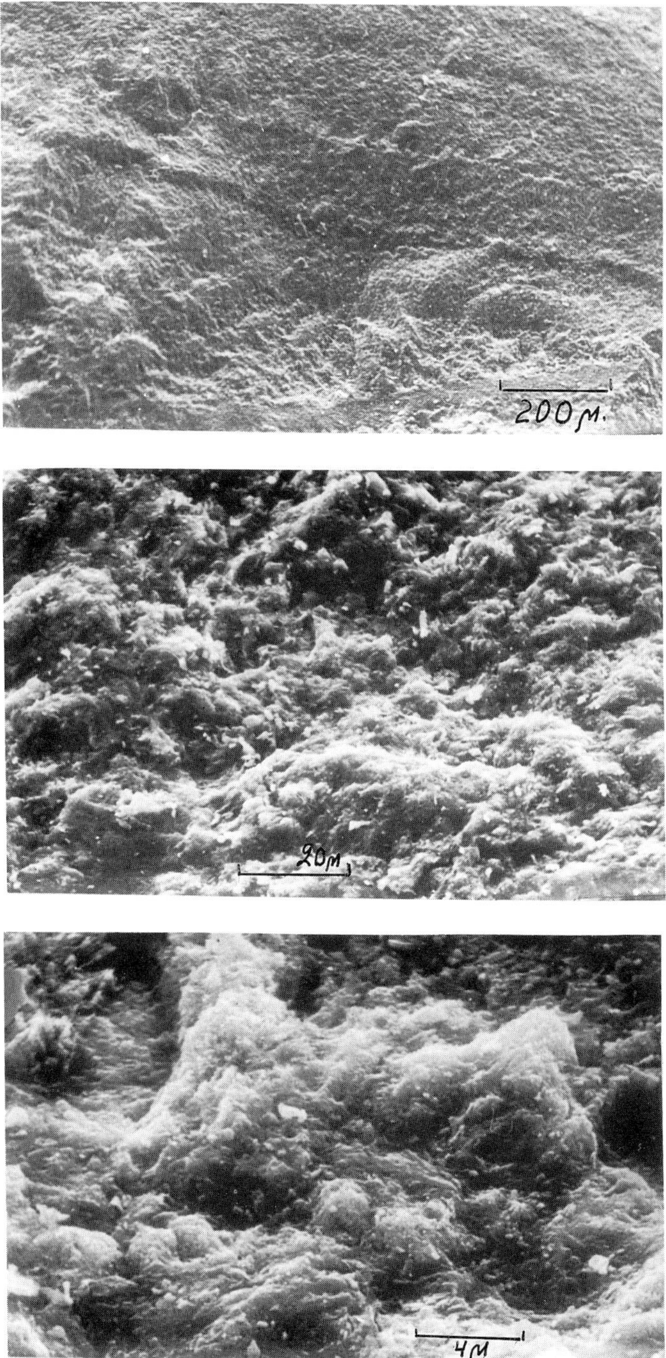

Fig. 22.   SEM photomicrographs of Mn ores. The ground mass, composed of black fine-crystalline matter, composing large (5–15 cm) ore concretions, their intergrowths and aggregates; composition: cryptomelane with a rather small admixture of goethite, quartz, clay matter and traces of manganite. Sample 24. Quarry-A, the Groote Eylandt deposits (see Figures 4, 8, 10, and 14). 1. General view of the microglobular ground mass. 2. Close-up of (1). A distinct microglobular character of the ground mass, composed of cryptomelane, rare dispersed rod-like finest crystals of goethite are shown. 3. Close-up of (2). The microlayered fine-laminated structure of separate globular aggregates of cryptomelane is distinctly visible.

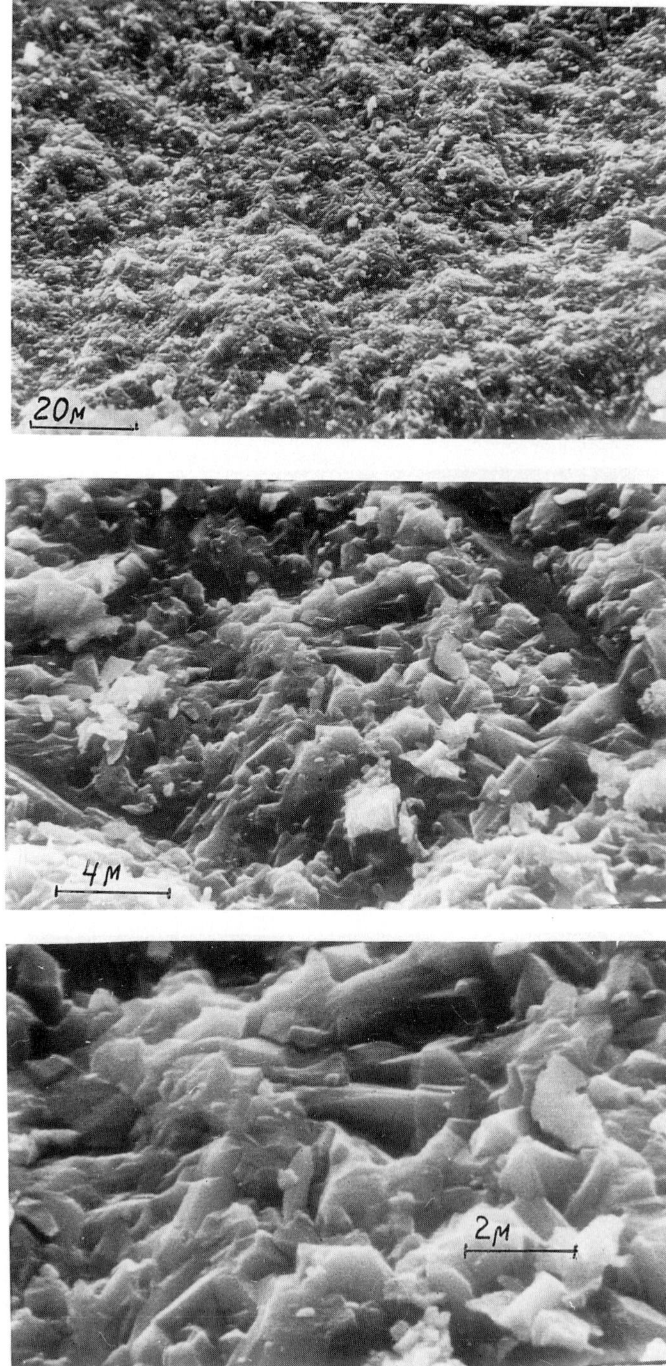

Fig. 23.   SEM photomicrographs of Mn ores. Light-grey with high reflectivity recrystallized variety of cryptomelane (second generation), represented by fine streakes and interlayers in the ground mass of the ore (see Figure 22), composed of concretions, their intergrowths and aggregates. Sample 24, quarry-A, the Groote Eylandt deposits (see Figures 4, 8, 10, and 14). 1. General view of light-grey with high reflectivity recrystallized matter (cryptomelane). 2. Close-up of (1). Aggregates of cryptomelane (second generation) crystals in the form of layered intergrowths are distinctly displayed. 3. Close-up of (2). Irregular intergrowth of laminated prismatic crystals of cryptomelane (second generation).

Fig. 24.   SEM photomicrographs of Mn ores. Cryptomelane of the second generation, light-grey with high reflectivity, filling in streaks and interlayers in the ground mass of the ore (see Figure 22), represented by concretions and/or their intergrowths and aggregates. Sample 24, quarry-A, the Groote Eylandt deposits (see Figures 4, 8, 10, and 14). 1. Contact of cryptocrystalline cryptomelane (I) of the ground mass of the ore, with streakes, aggregates of cryptomelane (II), composed of distinctly crystalline aggregates. 2. Close-up of (2). The distinct contact of crystalline-grained cryptomelane (II), and cryptocrystalline ground mass of the ore. 3. Aggregates, intergrowths of prismatic crystals of cryptomelane (II), composing streakes, interlayers in the ground mass of the ore (cryptomelane I).

Table 4.   Mineral composition of clay components, ore-enclosing sediments, the Groote Eylandt deposits, the data of the X-ray diffractometry.

No.	Quarry section	No. of sample	Mineral composition (fractions: (a) < 0.01 mm; (b) < 0.001 mm)	Notes
1	G	4(1)-a	Kaolinite, admixture of hydromica, pyrolusite, goethite	See Fig. 12
2	G	4(1)-b	Kaolinite, admixture of hydromica, pyrolusite, goethite	See Fig. 12
3	G	4(2)-a	Kaolinite, hydromica, traces of goethite	See Fig. 12
4	G	5-a	Kaolinite, admixture of pyrolusite and dispersed goethite	See Fig. 12
5	G	5-b	Kaolinite, admixture of pyrolusite	See Fig. 12
6	G	8-a	Kaolinite, pyrolusite, admixture of todorokite	See Fig. 12
7	G	8-b	Kaolinite, admixture of todorokite	See Fig. 12
8	A-south	11-a	Kaolinite (90%), hydromica (10%)	See Fig. 13
9	A-south	12-a	Kaolinite, admixture of mixed-layered phase of mica-smectite with slight content of expanded layers (hydromica)	See Fig. 13
10	A-south	13-a	Kaolinite with admixture of hydromica	See Fig. 13
11	A-south	16-a	Kaolinite with admixture of hydromica, and quartz	See Fig. 13
12	A-south	17-a	Kaolinite, admixture of quartz, and traces of hydromica	See Fig. 13
13	A-south	17-b	Kaolinite, admixture of quartz, and traces of hydromica	See Fig. 13
14	A-south	18-a	Kaolinite, admixture of quartz, and traces of hydromica	See Fig. 13
15	A-south	19-a	Kaolinite, traces of hydromica, quartz and goethite	See Fig. 13
16	A	21-a	Kaolinite, traces of hydromica, and goethite	See Fig. 14
17	A	25-a	Kaolinite, admixture of goethite	See Fig. 14
18	A	25-b	Kaolinite, admixture of goethite	See Fig. 14
19	E	26(1)-b	Kaolinite, traces of quartz and hydromica	See Fig. 15
20	E	26(2)-a	Kaolinite, admixture of quartz, and traces of goethite	See Fig. 15
21	E	26(2)-b	Kaolinite, traces of hydromica and goethite	See Fig. 15

performed by S.I. Tsipurskii, GIN, Russian Academy of Sciences. Infrared spectra in the range of 400–4000 cm^{-1} were taken for all studied samples. The most promising samples were studied under a transmission electron microscope. Clay minerals were determined in accordance with recommendations of the International Commission on nomenclature of clay minerals (Bailey *et al.*, 1979).

Features of distribution in the main sections.   The results of the X-ray diffraction study of clay components are summarized in Table 4. The main mineral (composing more than 80–90% of all ore-free fine-dispersed components) is kaolinite, occurring both as light-colored essentially white kaolines and as varieties pigmented to different degrees by oxyhydroxides of iron (essentially goethite) and manganese (pyrolusite, rarer cryptomelane, todorokite). Hexagonal plates (see Figure 25), their intergrowths and more rarely their particles, fragments of irregular form, size in cross-section to 1.5–2.0 micrometer are as a rule characteristic for kaolinite.

The forms of reflexes of the second ellipse (reflection of type 201 131) are visible in electron diffraction patterns of oblique textures of samples 8, 17, and 25 (Figure 26, location of samples is shown in Figures 12, 13, and 14). It characterizes the triclinity of the kaolinites under investigation. Weak resolution of reflexes with $K \neq 3K'$ and greatly diffused backgrounds should indicate the low structural ordering and crystallographic perfectness of minerals, that contradicts observations with the transmission electron microscope (see Figure 25). Moreover, the electron diffraction patterns of the oblique textures of sample 25 (see Figure 26) are only slightly informative due to essential diffusion background between reflections of the given arcs of reflexes.

The observed images of electron diffraction may be explained by samples having a visible admixture of halloysite; this forms textures with imperfect orientation and therefore electron diffraction patterns have reflexes in the form of the elongated arcs. As the data of the X-ray diffraction and distinctly crystallographically shaped hexagonal plates of kaolinite suggest (see Figure 25, nos. 1–3), the supposition of a low degree of its structural ordering, and the absence of the strict period along the axis 1 or of the mechanical destruction of the sample during the preparation of fractions of < 0.001 mm, are improbable for this mineral in these cases. In a number of sections in the sediments, composing the lower part of the ore bed (quarries-G and A southern, see Figures 12 and 13) the visible amounts (10%) of hydromica are recorded. The minerals represent the mixed-layer phase mica-smectite with a small proportion of expanded layers (samples 4 and 11). Up the section the hydromica content is decreased significantly to trace amounts or complete absence. This tendency is distinctly observed in the section of relatively thick ore beds in the south of quarry-A (sequence of samples from the bottom: 11 to 19, see Table 4, Figure 13).

In the western part of the deposits among the general mass of the enclosing sediments of kaolinite composition, according to data of the X-ray diffraction and directly under transmission electron microscope, the essential admixture of halloysite is distinctly observed (section of the quarry-E, sample 26, Figure 25, no. 4, see Figure 15). The length of halloysite crystals, represented in suspension electron microscope mounts by tubular particles is up to 3 micrometers with cross-section of 0.05–0.5 micrometers; in a number of cases the splitting of some crystals at their end is observed.

Thus, the essentially kaolinitic composition of ore enclosing sediments, high crystallographic shape of its particles, sharp decrease of the admixture of hydromica as relatively unstable primary components up the section and the presence of halloysite in separate sections (especially in upper parts) offer a clearer view of how the constituents have arisen. We believe that these are the products of deep aerial chemical weathering, not subjected to significant transportation. It is shown in works by Chukhrov *et al.*, (1965, 1984), that

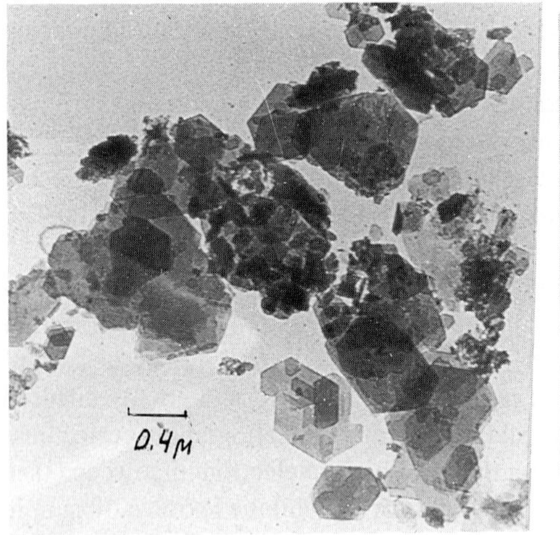
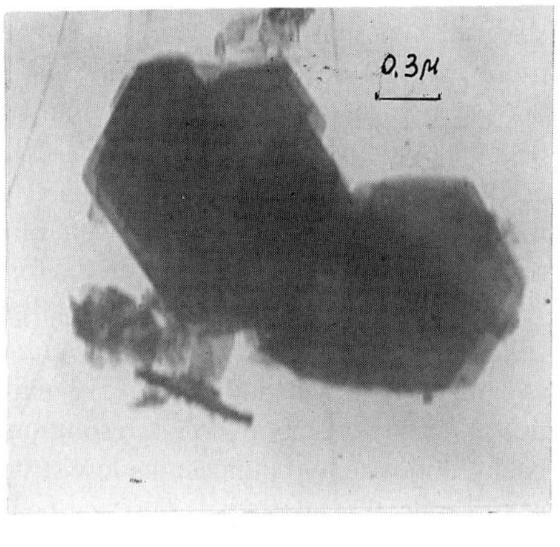
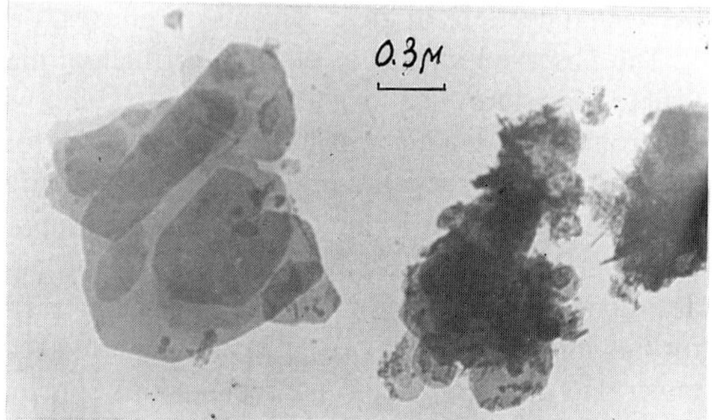
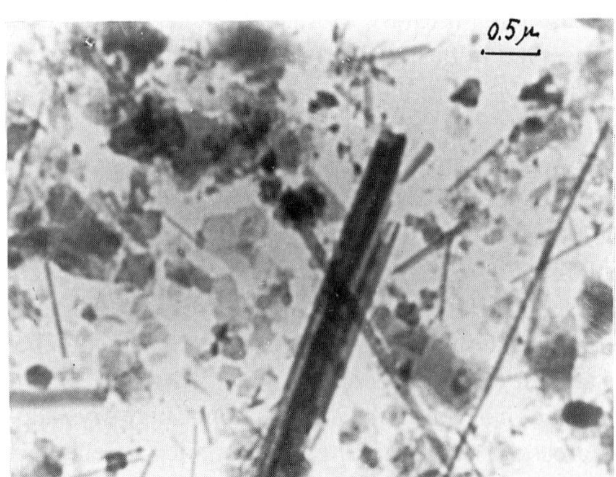

Fig. 25.   Photomicrographs of components of clay fraction (< 0.001 mm) taken by the transmission electron microscope. Ore enclosing and interore sediments, the Groote Eylandt deposits. 1. Aggregates and intergrowths of kaolinite plates with characteristic hexagonal habits and the admixture of the dispersed Mn oxyhydroxides. Sample 8, quarry G. The ground mass, matrix, bonding oolites, pisolites, microglobules, composed essentially of pyrolusite and todorokite (see Figure 12). 2. Intergrowths of kaolinite crystals. In the lower part of the photo a rod-like crystal of cryptomelane is shown. Sample 17, quarry-A, southern face. The main ore-free ground mass, bonding concretions of cryptomelane (see Figure 13). 3. Intergrowths, aggregates of kaolinite plates, to the right there is a fibrous aggregate of the finest crystals of cryptomelane. Sample 25, quarry-A, southeastern face (see Figure 14). 4. Mixture of kaolinite particles, plate-like, hexagonal habits and rod-like to tube-like particles of halloysite (products of substitution of kaolinite), admixture of dispersed Mn oxyhydroxides is visible. Sample 26, quarry-E, northwestern face (see Figure 15).

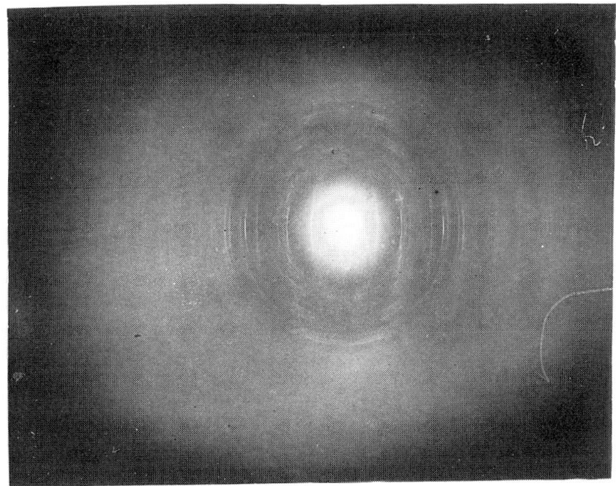

Fig. 26.   Electron diffraction patterns of oblique texture of kaolinite clays. Mn ore enclosing sediments. Groote Eylandt Mn oxide deposits. 1. Kaolinite clay, ground mass, and matrix, bonding oolites, pisolites, and microglobules, composed of pyrolusite, todorokite. Sample 8, quarry-G (see Figures 4, 6, 10, 12, and 25). 2. Kaolinite clay of the ground mass and matrix enclosing concretions of cryptomelane. Sample 17, quarry-A, southern face (see Figures 4, 7, 10, 13, and 25). 3. Kaolinite clay. Sample 25, quarry-A, southeastern face (see Figures 4, 8, 10, 14, and 25).

similar associations of halloysite with kaolinite may indicate the formation of these minerals as a result of the general processes of the development of laterite weathering crusts. If kaolinite, for which the relatively regular crystalline structure is characteristic, is formed, as a rule, due to slowly developing supergenous processes of alumosilicate decomposition, then halloysite, as a mineral with a slightly ordered structure, is synthesized owing to faster reactions. It is known that the products of such rapid alterations are structurally poorly ordered substances. In other words, halloysite is a product of more active supergenous alterations of primary alumosilicate matter than kaolinite; in the example of the Groote Eylandt deposits this is especially clear in the upper part of the ore bed or in the whole of the bodies with relatively small thicknesses (quarry-E, see Figure 15).

The conclusions on the composition and distribution of clay components in the section of the ore bed are valid for the Northern basin of the Groote Eylandt deposits that contains the main reserves of the commercial oxyhydroxide Mn ores, stripped by open mines. In the southern basin, occurring considerably hypsometrically lower, the correlated equivalents of the ore bed, represented by Mn carbonates, are sufficiently well isolated from the supergenous weathering by the covering sediments. According to Ostwald and Bolton (1992) the components of ore-bearing Cenomanian sediments of the Southern basin are represented by glauconite (in the form of pellets enclosed in the manganese marls), ferruginous smectite and mixed-layered phases of 'glauconite-smectite', distinguished in the form of pellets and substitutions of clastic components of the rocks. It is believed that the authors described the widely known processes of postsedimentation glauconitization or development of Fe-mica after Fe-smectites. The statement (Ostwald and Bolton, 1992, p. 1340) that ferruginous smectite and glauconite are widely distributed clay components of sediments of the Northern basin, synchronous with accumulations of manganese oxyhydroxides, contradict the data given above and the results of investigations with participation of one of the coauthors of the above article (Pracejus and Bolton, 1992a, 1992b). Elsewhere in the same work Ostwald and Bolton (1992, p. 1340) write that: "Recent studies have shown that this clay is commonly mixed with Tertiary weathering products such as kaolinite, lithiophorite and anataze, resulting from lateritization (Ostwald, unpublished data)."

As a whole, the analysis of the available information suggests that clay components of the commercial ore bed of the Groote Eylandt deposits are the products of deep laterite weathering of the Cenomanian Fe mica-smectite clays and terrigenous orthoquartzite ore-enclosing sediments. The relics of such primary clay components are registered in trace amounts, essentially in the lower part of the ore bed (see Table 4). The general sequence of supergenous alterations corresponds to the development trend of the laterite weathering crust: (glauconite + Fe mixed-layered mineral mica/smectite + Fe smectite) – kaolinite – halloysite. In works by Pracejus and Bolton (1992a, 1992b) normative (calculated) gibbsite as a typical product of the laterite nature was used.

2.2.1.7. *Geochemistry: Distribution of the Major Components and Trace Elements*

Distribution of chemical elements in ores is shown in Tables 2 and 3, Figures 11 to 15. It is necessary to emphasize the essential features of the chemical and mineral composition of the ores.

1. In virtually stripped sections of the ore bed (see Figures 11 to 15) the lower part is represented by sand-siliceous varieties of the ores, composed of pisolites, small nodules, concretions, consisting of pyrolusite, developing after cryptomelane often with rather essential admixture of goethite (see Tables 2 and 3, samples 11, 21, and 26, Figures 13 to 15). These ores are characterized by relatively slight oxidation (Mn(II)/Mn to 7.82×10^2, see Figures 13 to 15), increased ferruginity (Mn/Fe $\ll 4.0$), presence of essential admixture of free SiO_2 ($SiO_2/Al_2O_3 \gg 1.20$), increased amounts of K (K/Zr > 40), and relatively low concentrations of Ba, Cu, Rb, Sr, Y, Ce (see Tables 2 and 3, Figures 13 to 15).

2. In overlying discontinuous lens-like layers, the ores are composed of oolites and pisolites. In these ores the main minerals are pyrolusite with essentially subordinate amounts of cryptomelane of a residual relict nature (see Tables 2 and 3, samples 13 to 16 and 22 to 23, Figures 13 and 14). In those cases, when the influence of the laterite alteration has a pronounced character, the ores contain an admixture of todorokite (10 Å-manganate), lithiophorite, and birnessite (section of quarry-G, southwestern face). Geochemical features of these ores are rather remarkable: the highest oxidation (Mn(II)/Mn $\ll 0.8 \times 10^{-2}$), slight ferruginity (Mn/Fe $\gg 20$), small silicity ($SiO_2/Al_2O_3 \ll 3.0$), and notable enrichment in Ba, Cu, Rb, Y, and Ce.

3. In the layer above the ore bed, composed of slightly structurized massive, lumpy ore concretions, the main mineral is cryptomelane often with rather large amounts of pyrolusite. In those sections, where the influence of pedogenous laterite weathering is considerable, there is significant amounts of lithiophorite, todorokite often with an admixture of goethite. Such ores are characterized by relatively low oxidation (Mn(II)/Mn $\gg 1.00 \times 10^{-2}$), high ferruginity (Mn/Fe < 2.5), enrichment in quartz ($SiO_2/Al_2O_3 \gg 1.75$) and K_2O, BaO, Cu, and rare earth elements. Special attention must be given to the maximum concentrations of trace elements in the upper part of the ore bed, bordering with the proper laterite soil accumulations. Such ores are represented by large-lumpy and boulder-concretion varieties (Figures 11 to 15).

4. In the proper laterite pedogenous accumulations, occurring over the manganese-bearing bed, the oxyhydroxide compounds of manganese are developed too; they are the products of deep supergenous alteration of underlying accumulations. Lithiophorite, pyrolusite, and cryptomelane usually with the admixture of birnessite and todorokite (or the varieties of 10 Å-manganate) predominate among observed oxyhydroxide minerals. A rather typical component is goethite, in some cases almost completely composing (sample 25, see Figure 14) concretion-lump varieties.

Geochemical features of the considered accumulations are distinguished for consistency with variable mineral composition (see Figures 11 to 15); the sharp depletion in trace elements, relatively weak oxidation of the manganese phases (Mn(II)/Mn to 1.5×10^{-2}), enrichment in Zr, and high ferruginity (see Tables 2 and 3) are characteristic.

Thus, the presence of some symmetry engages the attention during consideration of the features of distribution of structural ore types, and their mineral and chemical composition in the section. The lower (soil) and top parts of the ore bed are represented by relatively large-lumpy concretion-nodular varieties, composed essentially of cryptomelane and, to a lesser degree, pyrolusite often with an admixture of goethite, developing after cryptomelane. In those cases, when laterite pedogenous weathering is considerably developed, lithiophorite, 10 Å-manganate (todorokite), birnessite, and romanechite (psilomelane) are developed. The main features of ore chemistry of these intervals are considered above (see Tables 2 and 3, Figures 11 to 15). It is important to emphasize the presence of some (as a rule, admixture and trace) amounts of manganite (samples 4 and 24, see Figures 12 and 14, Tables 2 and 3) in the lower and top parts of the ore bed. In these cases the manganite is a preserved relict in the general sequence of alteration of manganese-ore matter: Mn carbonate → manganite → cryptomelane → pyrolusite ... This fact is essential for understanding the oxyhydroxide ore formation process of the deposits. The middle part of the ore bed is composed of essentially oolite-pisolite varieties, represented mainly by pyrolusite with a notable admixture of cryptomelane of the early generation displaying a residual and relict character.

The above data on the mode of occurrence, structure of the ore bed, mineral composition of the manganese ores and enclosing sediments, distribution of main components, trace elements and REE in them reinforce the conclusions, made in our earlier works (Varentsov, 1982a, 1982b; Varentsov and Golovin, 1987). It was shown in these works that commercial oxyhydroxide manganese ores of the Groote Eylandt deposits are redeposited and/or autochthonous products of tropical (laterite) weathering crusts, developed within the Northern basin after the initially low-grade carbonate and probably oxide (?) sedimentary-diagenetic ores, which were nearly eroded by the present time. Correlation equivalents of these ores, protected from supergenous alterations by relatively thick sequences of overlying sediments and penetrated by drilling holes, occur hypsometrically considerably lower in the southern basin.

Similar concepts significantly differed from the views of some Australian geologists on the genesis of the Groote Eylandt deposits (Frazer, 1975; McIntosh *et al.*, 1975; Ostwald, 1975, 1980; Slee, 1980; Smith and Gebert, 1969).

In this state of affairs, interest was rekindled in the study of problems of the deposit's genesis, reflected in works of the Australian specialists (Bolton and Frakes, 1982, 1984; Bolton *et al.*, 1984, 1988, 1990; Frakes and Bolton, 1984a, 1984b; Ostwald, 1980, 1981, 1987, 1988, 1992; Ostwald and Bolton, 1992; Pracejus *et al.*, 1988, 1990; Pracejus and Bolton, 1992a, 1992b). These investigations were distinguished by the variety of problems raised and the large volume of data on the factual evidence. In particular the essential role of supergenous processes in the formation of the commercial manganese ores of these deposits was recognized. At the same time the authors distinguished deeply supergenous altered Mn oxyhydroxide ores in the section and primary sedimentary-diagenetic oxyhydroxide types of ores, enclosed in a quartz-kaolinite mass.

With the high definition of the data, such subdivision of the Mn ores was made in the work by Pracejus and Bolton (1992a), that was reflected in the cited Table 1. The authors subdivide the ores into two main categories: (a) primary manganese ores, and (b) secondary manganese ores (see Table 1).

Oolites (uncemented), pisolites (uncemented), siliceous ores and hardgrounds (dense and hard, thin layers of Mn oxyhydroxides several cm thick often with sign of erosion at the surface laterally elongated to dozens of m, incorporating altered oolites, pisolites, and massive ores are assigned to the 'primary', that is sedimentary, oxyhydroxide Mn ores.

Pracejus and Bolton (1992a) attribute cemented varieties of pisolites and oolites (cemented mangcrete), massive mangcrete, laminated mangcrete, and different types of concretions (see Table 1) to the secondary (that is supergenous) products. The distinguished types were thoroughly characterized by the statistical data on normative mineral composition and distribution of the main components, trace elements and rare earth elements (Tables 5 to 7).

Emphasizing the distinctions, noted above, between concepts on the origin of Mn ores, the problem is raised inevitably as to how significant the differences are in the given data on the normative mineral and chemical composition of the distinguished (Pracejus *et al.*, 1990; Pracejus and Bolton, 1992a) 'primary' and 'secondary' ores. In particular, how well constrained is the assignment of uncemented oolites and pisolites to sedimentary diagenetic products, and the cemented varieties of Mn oxyhydroxides to accumulations of the supergenous nature? To solve this problem we calculated the characteristic ratios (Table 8) on the basis of the statistical data on the distribution of the main components, trace elements and rare earth elements (Pracejus *et al.*, 1990; Pracejus and Bolton, 1992a) for the most representative types of Mn ores (Figures 27 to 33).

Distribution of the normative-mineral composition can be seen in Table 8 and Figure 27. The recalculation procedures of the chemical analysis data to the normative minerals of Mn oxyhydroxides is given in works by Pracejus (Pracejus, 1990; Pracejus *et al.*, 1988). In spite of the fact that the data of recalculation of the chemical analysis to normative minerals corresponds rather formally and approximately to the results of instrumentation determination of the virtual mineral composition of rocks or Mn ores, this computational approach allows us to carry on the quantitative comparison on the qualitatively uniform basis at the known, arbitrarily accepted, assumptions.

There is no definite trend (see Table 8, Figure 27) observed from the comparison of the normative mineral composition of uncemented and cemented oolite and pisolite varieties in interrelations between the two main ore-forming minerals: pyrolusite (Pr) and cryptomelane (Cm). For example, if the pyrolusite content is increased from uncemented oolites (U.O.), 43.78% to cemented varieties (C.O.), 61.55%, with the decrease of the cryptomelane amounts, Cm, 12.09% (U.O.) and 7.71% (C.O.), then, for the corresponding varieties of pisolites, the inverse relationships are characteristic: Pr, 68.93% (U.P) and 41.02% (C.P.); Cm, 4.47 (U.P.) and 18.62 (C.P.). The given data may be interpreted as evidence that the method of the normative recalculations for these types of ores either do not determine any definite trend of the supergenous alterations, or that the virtual mineral composition of cemented and uncemented varieties may be varied in rather wide ranges, and as a whole it is sufficiently similar. In particular the closeness of the normative mineral compositions of cemented oolites and uncemented pisolites supports at this assumption: Pr, 61.55% (C.O.) and 68.93% (U.P.); Cm, 7.71 (C.O.) and 4.77% (U.P.) (see Table 8, Figure 27).

At the same time the normative recalculations reflect a significant decrease in the pyrolusite content and a relative increase of cryptomelane of the second generation in concretion ores, developed in the upper half of the ore body. In other words, as it was emphasized above during the description of the distribution of Mn minerals in the section, the definite evolution of the supergenous alterations is observed. In this succession

Table 5. Statistics for major and minor elements (%), Groote Eylandt, Australia (Pracejus and Bolton, 1992).

	Al	Ba	Ca	Fe	K	Mg	Mn	Na	P	Sr	Sc	Ti
Cemented oolites* (cm 18.62; rm 5.86; to 6.22; pr 41.02; w 11.48)												
Avg	3.4	1.88	0.07	1.4	0.49	0.41	52.3	0.12	0.07	4.5	0.34	0.14
Max	4.7	2.93	0.12	2.4	0.67	0.61	55.6	0.19	0.09	13.1	0.53	0.23
Min	2.0	0.65	0.04	0.9	0.24	0.25	46.3	0.07	0.05	1.9	0.20	0.10
n	8	8	8	8	8	8	8	8	8	8	8	8
Std. dev.	1.0	0.72	0.03	0.5	0.16	0.15	3.0	0.05	0.01	3.5	0.13	0.04
Cemented pisolites (cm 7.71; rm 6.74; to 4.11; pr 61.55; w 7.55)												
Avg	4.7	1.62	0.07	2.7	1.59	0.33	49.2	0.23	0.08	5.0	0.22	0.16
Max	8.1	5.84	0.22	23.2	2.66	0.67	56.0	0.57	0.42	14.2	0.40	0.29
Min	1.8	0.01	0.03	0.6	0.71	0.01	31.8	0.12	0.03	1.2	0.03	0.07
n	31	31	31	31	31	31	31	31	31	31	31	31
Std. dev.	1.7	1.60	0.04	4.8	0.49	0.22	5.2	0.08	0.09	3.1	0.12	0.07
Chocolate ores (cm 1.90 rm 0.80; to 1.20; pr 11.10; w 0.40)												
Avg	24.0	0.21	0.09	7.5	0.36	0.04	9.8	0.04	0.02	38.8	0.02	1.41
Max	27.3	0.34	0.08	9.2	0.76	0.09	47.9	0.17	0.06	49.2	0.03	1.62
Min	5.5	0.05	0.01	2.6	0.22	0.03	1.9	0.01	0.01	10.1	0.01	0.36
n	10	10	10	10	10	10	10	10	10	10	10	10
Std. dev.	6.6	0.09	0.02	1.9	0.17	0.02	13.7	0.05	0.02	11.5	0.01	0.37
Kaolinitic clays												
Avg	21.5	0.02	0.01	6.0	0.14	0.02	0.6	0.01	0.02	57.7	0.02	0.96
Max	33.7	0.03	0.01	16.9	0.26	0.04	3.2	0.02	0.14	78.7	0.03	1.71
Min	6.5	0.01	0.01	0.7	0.06	0.01	0.1	0.01	0.00	45.5	0.01	0.30
n	8	3	5	8	8	8	8	8	8	8	3	8
Std. dev.	11.9	0.01	0.00	6.8	0.06	0.01	1.1	0.00	0.05	11.4	0.01	0.55
Concretions (cm 44.40; rm 4.40; to 6.50; pr 14.30; w 13.50												
Avg	3.2	1.09	0.09	5.1	3.72	0.07	49.1	0.19	0.16	3.1	0.09	0.16
Max	4.2	2.12	0.20	35.4	5.35	0.20	54.9	0.26	0.24	7.6	0.18	0.20
Min	2.1	0.23	0.04	0.9	0.61	0.03	20.5	0.07	0.06	1.2	0.05	0.11
n	10	10	10	10	10	10	10	10	10	10	10	10
Std. dev.	0.9	0.66	0.05	10.7	1.33	0.05	10.2	0.06	0.06	2.1	0.05	0.03
Digital concretions (cm 26.60; rm 19.60; to 2.60; pr 13.50; w 10.70)												
Avg	4.2	4.65	0.04	4.8	3.28	0.03	41.8	0.08	0.19	11.4	0.08	0.23
Max	6.0	8.91	0.06	10.2	4.33	0.03	52.4	0.14	0.28	35.6	0.28	0.32
Min	3.5	0.15	0.02	1.2	0.80	0.02	31.3	0.04	0.09	1.7	0.01	0.20
n	6	6	6	6	6	6	6	6	6	6	6	6
Std. dev.	0.9	4.17	0.02	3.4	1.36	0.01	7.1	0.04	0.08	15.0	0.10	0.05

Table 5.   (Continued)

	Al	Ba	Ca	Fe	K	Mg	Mn	Na	P	Sr	Sc	Ti
Fe pipe walls (cm 0.50; rm 0.50; to 0.40; pr 1.60; w 0.20)												
Avg	19.6	0.23	0.01	13.4	0.18	0.01	1.8	0.01	0.02	47.8	0.01	0.97
Max	22.7	0.63	0.01	20.2	0.29	0.02	6.3	0.03	0.04	60.4	0.01	1.15
Min	14.1	0.01	0.01	7.1	0.12	0.01	0.2	0.01	0.01	36.5	0.01	0.74
n	5	3	1	5	5	5	5	5	5	5	1	5
Std. dev.	3.4	0.35	4.9	0.07	0.00	2.5	0.01	0.01	11.2	0.16		
Fe pisolites (cm 2.50; rm 0.30; to 0.60; pr 5.20; w 1.80)												
Avg	5.7	0.09	0.01	46.6	0.21	0.03	4.8	0.02	0.39	9.7	0.03	0.25
Max	8.4	0.33	0.02	59.9	0.94	0.14	30.8	0.11	0.53	48.7	0.13	0.41
Min	1.4	0.01	0.01	22.7	0.02	0.01	0.3	0.01	0.05	1.3	0.01	0.04
n	18	14	7	18	18	18	18	18	18	18	12	18
Std. dev.	2.3	0.10	0.00	10.8	0.22	0.04	7.4	0.02	0.14	12.1	0.04	0.11
Hardgrounds (cm 8.40; rm 3.60; to 4.10; pr 66.10; w 7.70)												
Avg	2.7	1.02	0.05	0.9	0.67	0.54	54.1	0.15	0.09	4.2	0.18	0.13
Max	4.5	1,79	0.08	1.3	0.70	0.90	55.8	0.26	0.10	4.4	0.30	0.19
Min	1.7	0.47	0.03	0.7	0.64	0.31	50.8	0.08	0.08	3.8	0.10	0.08
n	3	3	3	3	3	3	3	3	3	3	3	3
Std. dev.	1.6	0.69	0.03	0.3	0.03	0.31	2.80	0.10	0.01	0.3	0.11	0.06
Laminar mangcretes (cm 18.50; rm 7.90; to 7.30; pr 30.70; w 10.40)												
Avg	6.3	2.16	0.09	3.0	1.74	0.07	44.1	0.31	0.07	11.2	0.06	0.29
Max	8.9	4.21	0.19	4.3	3.07	0.17	47.4	0.50	0.32	28.8	0.12	0.40
Min	2.5	0.10	0.03	1.6	1.29	0.01	37.7	0.12	0.02	5.8	0.04	0.11
n	7	7	3	7	7	7	7	7	7	7	7	7
Std. dev.	2.0	1.82	0.09	1.1	0.65	0.05	3.5	0.12	0.11	8.1	0.03	0.09
Laterites (cm 1.40; rm 0.50; to 0.70; pr 4.90; w 0.60)												
Avg	23.4	0.14	0.01	12.5	0.30	0.03	5.0	0.02	0.02	38.8	0.02	1.20
Max	28.3	0.46	0.06	27.0	0.78	0.05	16.3	0.10	0.07	66.3	0.05	1.56
Min	16.5	0.01	0.01	0.9	0.08	0.01	0.1	0.01	0.01	26.9	0.01	0.81
n	26	24	24	26	26	26	26	26	26	26	23	26
Std. dev.	3.3	0.10	0.01	5.6	0.16	0.01	4.5	0.02	0.02	9.4	0.01	0.21
Massive mangcretes (cm 34.90; rm 2.50; to 5.60; pr 20.70; w 13.80)												
Avg	4.9	0.66	0.09	6.6	3.01	0.04	45.5	0.16	0.16	5.3	0.12	0.26
Max	12.6	3.16	0.20	30.5	4.36	0.12	54.6	0.25	0.42	14.4	0.53	0.67
Min	2.0	0.01	0.01	1.4	1.83	0.01	26.0	0.03	0.04	1.6	0.02	0.10
n	13	13	13	13	13	13	13	13	13	13	13	13
Std. dev.	3.6	0.85	0.06	9.1	0.89	0.03	8.6	0.07	0.11	4.1	0.13	0.18

Table 5. (Continued)

	Al	Ba	Ca	Fe	K	Mg	Mn	Na	P	Sr	Sc	Ti
Radial pyrolusites (cm 3.80; rm 0.60; to 1.90; pr 80.90; w 1.20)												
Avg	3.0	0.17	0.03	2.7	0.35	0.07	54.8	0.07	0.06	3.4	0.03	0.19
Max	3.9	0.50	0.07	4.0	2.09	0.13	56.6	0.28	0.18	4.7	0.11	0.25
Min	2.1	0.06	0.01	1.8	0.06	0.01	52.2	0.01	0.03	2.1	0.01	0.13
n	9	9	8	9	9	9	9	9	9	9	7	9
Std. dev.	0.6	0.14	0.02	0.8	0.66	0.03	1.4	0.10	0.05	0.9	0.04	0.04
Mn-impregnated sands (cm 3.19; rm 0.97; to 0.61; pr 1.92; w 1.85)												
Avg	2.7	0.26	0.01	0.9	0.30	0.01	4.5	0.02	0.02	85.2	0.01	0.15
Max	10.6	2.70	0.07	5.3	1.67	0.03	24.5	0.06	0.11	99.4	0.04	0.54
Min	0.6	0.01	0.01	0.1	0.03	0.01	0.1	0.01	0	52.7	0.01	0.04
n	18	18	15	18	18	18	18	18	18	18	18	18
Std. dev.	2.7	0.65	0.02	1.3	0.52	0.01	8.7	0.01	0.03	15.4	0.01	0.15
Siliceous ores (cm 17.79; rm 1.59; to 4.48; pr 39.06; w 7.24)												
Avg	2.7	0.43	0.06	1.1	1.59	0.06	42.5	0.15	0.07	24.0	0.09	0.10
Max	4.5	0.84	0.13	3.6	2.28	0.14	54.7	0.25	0.15	48.9	0.16	0.17
Min	1.8	0.02	0.02	0.4	0.91	0.01	27.3	0.10	0.01	5.9	0.03	0.08
n	10	10	10	10	10	10	10	10	10	10	10	10
Std. dev.	0.8	0.30	0.03	0.9	0.42	0.05	11.8	0.05	0.04	19.0	0.05	0.03
Uncemented oolites (cm 4.77; rm 2.99; to 2.45; pr 68.93; w 1.90)												
Avg	4.9	1.53	0.10	2.3	0.74	0.57	46.4	0.18	0.07	9.2	0.55	0.22
Max	14.7	2.47	0.16	10.1	1.48	0.86	55.5	0.30	0.53	30.0	1.07	0.64
Min	2.4	0.41	0.01	0.7	0.28	0.09	21.9	0.03	0.04	2.0	0.11	0.10
n	39	39	39	39	39	39	39	39	38	39	39	39
Std. dev.	3.2	0.43	0.03	1.6	0.25	0.20	8.0	0.07	0.08	7.4	0.24	0.13
Uncemented pisolites (cm 12.09; rm 5.28; to 5.66; pr 43.78; w 13.45)												
Avg	5.4	0.81	0.04	2.1	0.45	0.10	50.1	0.07	0.06	7.7	0.04	0.24
Max	13.0	4.09	0.12	5.6	1.01	0.40	61.1	0.19	0.10	19.7	0.19	0.56
Min	0.8	0.03	0.01	0.1	0.02	0.04	36.4	0.01	0.03	0.9	0.02	0.04
n	19	19	19	19	19	19	19	19	19	19	19	19
Std. dev.	3.0	1.34	0.03	1.2	0.31	0.08	7.1	0.07	0.02	5.8	0.05	0.13
Spherulites (cm 1.80; rm 1.50; to 0.80; pr 9.20; w 1.70)												
Avg	16.6	0.46	0.04	16.7	0.28	0.07	11.0	0.02	0.04	31.8	0.04	0.76
Max	23.9	1.76	0.10	37.5	0.59	0.25	46.3	0.07	0.15	55.0	0.20	1.17
Min	4.7	0.06	0.01	2.4	0.10	0.02	0.7	0.01	0.01	13.1	0.01	0.23
n	15	15	9	15	15	15	15	15	15	15	12	15
Std. dev.	5.2	0.51	0.03	10.8	0.15	0.06	13.9	0.02	0.03	16.3	0.06	0.26

*Average content of normative Mn oxide minerals in ores (%), pr – pyrolusite, rm – romanechite, cm – cryptomelane; to – todorokite; w – wad, vernadite, and other poorly identified Mn oxides (Pracejus and Bolton, 1992).

Table 6.   Statistics for trace elements (ppm), Groote Eylandt, Australia (Pracejus and Bolton, 1992).

	As	B	Co	Cr	Cu	Mo	Ni	Pb	Sc	Th	U	V	Zn	Zr
Cemented oolites														
Avg	61	17.32	77.39	14.95	138.80	73	342	57	8	6	12	330.80	115.30	43.17
Max	72	22.89	134.2	17.11	184.70	77	427	58	10	6	13	493.00	161.90	61.88
Min	50	12.83	35.88	13.93	98.84	69	258	56	5	5	11	256.80	99.24	31.48
n	2	8	8	8	8	2	8	2	2	2	2	8	8	8
Std. dev.	16	2.91	35.58	1.07	29.64	6	63	1	4	1	1	76.99	22.20	12.15
Cemented pisolites														
Avg	39	14.13	95.06	14.50	89.50	43	132	41	9	4	10	200.50	97.53	38.27
Max	55	31.24	171.80	18.40	191.20	56	344	72	15	8	13	516.00	185.70	72.75
Min	26	0.00	35.78	13.40	42.90	34	28	23	4	0	6	37.33	65.29	14.79
n	14	31	31	31	31	14	31	14	14	14	14	31	31	31
Std. dev.	8	7.28	28.15	1.25	36.77	6	71	15	4	2	2	101.20	28.49	15.75
Chocolate ores														
Avg	n.d.	17.14	40.05	23.65	50.42	n.d.	22.0	n.d.	n.d.	n.d.	n.d.	261.0	49.70	257.5
Max	n.d.	26.36	63.66	26.33	83.09	n.d.	100.0	n.d.	n.d.	n.d.	n.d.	396.9	81.04	274.8
Min	n.d.	6.41	24.25	17.13	38.48	n.d.	5.3	n.d	n.d.	n.d.	n.d.	159.7	44.95	142.3
n	n.d.	10	10	10	10	n.d.	10	n.d.	n.d.	n.d.	n.d.	10	10	10
Std. dev.	n.d.	7.28	12.70	2.51	12.83	n.d.	29.0	n.d.	n.d.	n.d.	n.d.	72.8	11.05	40.7
Kaolinitic clays														
Avg	2	65.02	17.87	40.16	25.26	2	2.7	10	14	14	3	136.60	42.76	193.00
Max	2	511.60	22.33	76.38	40.70	3	7.7	13	15	15	4	923.70	49.48	279.50
Min	1	0.00	13.50	7.09	14.12	2	0	6	13	12	1	5.58	36.23	63.34
n	2	8	8	8	8	2	8	2	2	2	2	8	8	8
Std. dev.	1	180.50	2.87	24.99	9.07	0	2.9	5	1	2	2	318.80	4.81	78.15
Concretions														
Avg	n.d.	13.83	118.70	15.16	158.70	n.d.	112	n.d	n.d.	n.d.	n.d.	122.60	120.70	33.26
Max	n.d.	43.57	156.30	18.82	287.80	n.d.	314	n.d.	n.d.	n.d.	n.d.	296.00	205.40	54.37
Min	n.d.	1.73	89.22	13.26	53.28	n.d.	24	n.d.	n.d.	n.d.	n.d.	53.82	32.28	22.86
n	n.d.	10	10	10	10	n.d.	10	n.d.	n.d.	n.d.	n.d.	10	10	10
Std. dev.	n.d.	11.00	22.06	1.53	70.50	n.d.	96	n.d.	n.d.	n.d.	n.d.	83.22	51.89	8.01
Digital concretions														
Avg	n.d.	13.67	131.00	16.33	131.20	n.d.	59	n.d	n.d.	n.d.	n.d.	1154.0	122.30	46.89
Max	n.d.	23.67	174.10	19.98	162.40	n.d.	103	n.d.	n.d.	n.d.	n.d.	1522.0	141.20	74.44
Min	n.d.	6.80	85.47	13.98	72.92	n.d.	24	n.d.	n.d.	n.d.	n.d.	344.0	76.02	28.97
n	n.d.	6	6	6	6	n.d.	6	n.d.	n.d.	n.d.	n.d.	6	6	6
Std. dev.	n.d.	5.63	30.67	2.07	34.47	n.d.	26	n.d.	n.d.	n.d.	n.d.	423.9	25.16	20.42

Table 6. (Continued)

	As	B	Co	Cr	Cu	Mo	Ni	Pb	Sc	Th	U	V	Zn	Zr
Fe pipe walls														
Avg	50	0.55	20.05	37.10	32.55	32	1.8	57	14	9	5	286.70	44.93	209.10
Max	50	2.61	28.87	50.38	40.98	32	2.9	57	14	9	5	398.10	54.87	256.20
Min	50	0.00	14.27	27.19	26.92	32	1.2	57	14	9	5	83.05	32.13	155.60
n	1	5	5	5	5	1	5	1	1	1	1	5	5	5
Std. dev.	0	1.15	6.08	9.00	5.33	0	0.7	0	0	0	0	131.90	8.17	37.54
Fe pisolites														
Avg	333	17.41	50.19	23.04	190.30	120	324	112	16	5	20	864.46	191.10	118.80
Max	436	133.10	97.37	79.13	418.70	203	938	163	24	9	26	1535.00	281.50	180.30
Min	228	–	23.86	12.85	52.13	67	96	62	13	1	15	70.89	123.50	50.75
n	5	18	18	18	18	5	18	5	5	5	5	18	18	18
Std. dev.	94	34.60	22.27	16.26	113.80	57	195	47	5	3	5	426.00	49.21	41.32
Hardgrounds														
Avg	n.d.	61.58	75.27	14.14	91.63	n.d.	198.0	n.d.	n.d.	n.d.	n.d.	171.40	86.47	58.97
Max	n.d.	97.92	82.87	14.36	118.50	n.d.	207.0	n.d.	n.d.	n.d.	n.d.	185.60	89.77	85.56
Min	n.d.	9.48	62.72	13.13	55.71	n.d.	191.0	n.d.	n.d.	n.d.	n.d.	159.70	82.20	37.29
n	n.d.	3	3	3	3	n.d.	3	n.d.	n.d.	n.d.	n.d.	3	3	3
Std. dev.	n.d.	46.28	10.95	0.22	32.37	n.d.	8.1	n.d.	n.d.	n.d.	n.d.	13.12	3.88	24.51
Laminar mangcretes														
Avg	n.d.	13.16	73.54	15.36	112.7	n.d.	62	n.d.	n.d.	n.d.	n.d.	509.4	79.65	62.94
Max	n.d.	31.73	173.70	19.61	144.0	n.d.	186	n.d.	n.d.	n.d.	n.d.	1067.0	97.19	82.58
Min	n.d.	6.80	27.23	15.59	100.04	n.d.	18	n.d.	n.d.	n.d.	n.d.	161.50	63.86	43.81
n	n.d.	7	7	7	7	n.d.	7	n.d.	n.d.	n.d.	n.d.	6	7	7
Std. dev.	n.d.	8.50	46.34	1.45	14.9	n.d.	67	n.d.	n.d.	n.d.	n.d.	303.4	14.41	11.45
Laterites														
Avg	52	4.61	41.70	40.70	43.52	5	13	51	24	17	6	301.7	50.95	243.60
Max	94	44.97	92.26	105.70	68.58	18	32	82	33	21	9	738.4	72.83	268.60
Min	31	0.00	14.67	18.79	20.43	1	0.3	25	19	13	4	9.2	39.37	180.60
n	7	26	26	26	26	7	26	7	7	7	7	26	26	26
Std. dev.	23	8.77	21.71	21.62	13.55	6	10	19	6	3	2	168.0	8.26	19.85
Massive mangcretes														
Avg	12	10.47	169.20	15.94	168.50	17	90	56	15	8	10	360.3	136.20	62.58
Max	12	21.41	294.90	18.79	415.96	17	311	56	15	8	10	1189.0	248.0	166.30
Min	12	1.52	90.98	14.33	51.39	17	19	56	15	8	10	66.2	61.22	21.67
n	1	13	13	13	13	1	13	1	1	1	1	13	13	13
Std. dev.	0	4.98	69.44	1.38	119.2	0	95	0	0	0	0	385.0	61.12	45.93

Table 6.   (Continued)

	As	B	Co	Cr	Cu	Mo	Ni	Pb	Sc	Th	U	V	Zn	Zr
Radial pyrolusites														
Avg	96	31.51	23.14	16.03	51.66	28	46	122	4	6	7	397.5	77.19	35.74
Max	96	42.72	58.88	20.65	82.69	28	88	122	4	6	7	1030.0	100.80	41.10
Min	96	17.73	8.89	11.62	29.75	28	20	122	4	6	7	127.4	54.36	28.19
n	1	9	9	9	9	1	9	1	1	1	1	9	9	9
Std. dev.	0	6.48	15.62	2.83	20.35	0	25	0	0	0	0	351.0	15.41	4.72
Mn-impregnated sands														
Avg	2	2.06	53.18	16.30	62.92	4	1.2	9	4	3	1	161.40	42.01	55.64
Max	12	29.90	191.40	28.18	341.60	26	8.1	53	17	10	7	1462.40	86.96	121.60
Min	0	0.00	8.71	14.34	11.63	1	0	2	1	0	0	1.09	32.78	32.20
n	15	18	18	18	18	15	18	15	15	15	15	18	18	18
Std. dev.	3	6.99	49.72	3.25	98.16	6	2.3	13	4	2	2	392.70	15.92	22.26
Siliceous ores														
Avg	n.d.	15.53	149.30	14.97	135.20	n.d.	84	n.d.	n.d.	n.d.	n.d.	251.40	118.10	33.41
Max	n.d.	25.60	216.30	17.05	262.80	n.d.	187	n.d.	n.d.	n.d.	n.d.	828.50	195.86	49.92
Min	n.d.	3.87	86.69	14.32	68.45	n.d.	24	n.d.	n.d.	n.d.	n.d.	51.39	74.38	16.60
n	n.d.	10	10	10	10	n.d.	10	n.d.	n.d.	n.d.	n.d.	10	10	10
Std. dev.	n.d.	7.35	46.49	0.81	69.27	n.d.	43	n.d.	n.d.	n.d.	n.d.	230.50	34.60	10.97
Uncemented oolites														
Avg	53	13.48	83.39	14.67	128.70	79	440	97	8	6	13	329.30	137.00	61.39
Max	66	21.13	149.90	24.44	190.20	108	703	124	12	12	18	512.80	224.40	170.90
Min	39	4.07	26.48	12.53	40.36	7	29	38	3	2	8	144.30	67.14	20.99
n	17	39	39	39	39	17	39	17	17	17	17	39	39	39
Std. dev.	9	4.17	26.98	1.83	37.13	26	160	26	2	3	3	94.58	36.78	36.13
Uncemented pisolites														
Avg	70	37.49	55.34	16.16	71.68	70	98	58	6	5	13	245.40	84.78	69.07
Max	90	114.60	102.80	27.88	120.40	125	233	79	12	8	52	532.80	123.50	136.40
Min	48	0.00	22.24	14.11	39.22	9	41	32	3	0	4	91.59	64.10	11.16
n	8	19	19	19	19	8	19	8	8	8	8	19	19	19
Std. dev.	16	32.35	23.62	3.01	26.89	42	53	16	4	2	16	111.40	17.44	35.87
Spherulites														
Avg	28	6.59	51.18	55.51	72.69	7	63.0	49	23	16	7	512.7	66.13	197.80
Max	28	21.85	83.41	119.10	138.90	7	266.0	54	26	18	7	1471.0	123.30	263.90
Min	27	0.00	18.17	17.11	39.30	7	5.5	44	19	13	6	172.4	42.67	61.88
n	2	14	15	15	15	2	15	2	2	2	2	15	15	15
Std. dev.	1	6.85	18.07	29.33	32.20	0	90.0	7	5	4	1	379.0	28.00	59.00

Table 7.   Summary of REE analyses and statistics of ore and rocks (ppm), Groote Eylandt, North Territory, Australia (Pracejus *et al.*, 1990).

Sample	La	Ce	Nd	Sm	Eu	Gd	Yb	REE
Cemented pisolites								
n	31	31	31	31	31	31	31	31
Avg	53	74	54	13.9	3.3	19.2	5.1	222.5
Max	79	164	98	25.8	7.9	64.8	9.4	448.8
Min	33	12	28	5.2	1.3	10.3	2.7	92.5
Std. dev.	12	44	16	5.1	1.4	11.6	1.5	91.3
Cemented oolites								
n	8	8	8	8	8	8	8	8
Avg	52	55	90	22.7	6.5	24.8	6.9	257.9
Max	60	127	114	27.7	7.9	29.3	10.4	376.2
Min	38	32	51	15.6	3.8	21.5	4.2	166.1
Std. dev.	7	31	21	3.6	1.4	2.8	2.3	69.6
Chocolate ores								
n	9	9	9	9	9	9	9	9
Avg	21	227	26	5.0	0.6	23.2	2.4	305.2
Max	30	337	39	10.0	1.4	30.0	3.4	450.9
Min	16	118	17	2.1	0.2	16.9	1.3	171.5
Std. dev.	4	65	7	2.7	0.4	3.9	0.6	83.6
Clays								
n	8	8	8	8	8	8	8	8
Avg	9	28	12	0.7	0.2	7.9	0.7	58.5
Max	14	40	19	3.7	0.6	33.5	2.2	112.9
Min	0	7	0	0	0	0	0	7.0
Std. dev.	5	12	6	1.3	0.2	11.4	0.7	36.9
Concretions								
n	10	10	10	10	10	10	10	10
Avg	37	116	34	13.7	2.2	20.7	2.6	226.2
Max	55	226	56	20.8	3.9	75.9	6.5	377.1
Min	16	11	18	8.7	0.0	8.7	1.1	133.7
Std. dev.	11	68	14	3.4	1.0	19.9	1.8	77.3
Digital concretions								
n	6	6	6	6	6	6	6	6
Avg	29	52	27	8.5	1.7	23.7	5.8	147.7
Max	33	68	34	14.7	2.4	39.0	7.2	170.1
Min	23	44	17	2.4	0.6	5.8	4.0	97.3
Std. dev.	3	9	7	4.3	0.7	11.5	1.1	26.8

Table 7.   (Continued)

Sample	La	Ce	Nd	Sm	Eu	Gd	Yb	REE
Fe-pipe walls								
n	5	5	5	5	5	5	5	5
Avg	11	35	11	1.5	0.0	31.0	1.5	91.0
Max	16	67	16	3.1	0.0	46.4	2.0	134.1
Min	8	12	9	0.2	0.0	16.8	1.0	52.1
Std. dev	3	21	3	1.4	0.0	11.0	0.5	29.6
Fe-pisolites								
n	18	18	18	18	18	18	18	18
Avg	17	64	20	12.4	0.0	93.5	6.5	213.4
Max	35	95	36	17.7	0.5	112,7	11.0	281.4
Min	7	31	0	7.6	0	48.7	2.1	139.7
Std. dev.	8	16	9	2.4	0.1	17.3	2.1	35.7
Hardgrounds								
n	3	3	3	3	3	3	3	3
Avg	40	29	52	14.8	3.5	17.3	4.4	161.0
Max	44	34	61	22.1	4.9	24.9	5.6	179.3
Min	38	18	44	3.3	1.7	9.2	2.9	135.3
Std. dev.	3	9	9	10.1	1.7	7.9	1.4	22.9
Laminar mangcretes								
n	7	7	7	7	7	7	7	7
Avg	36	117	36	9.7	2.0	16.4	2.8	219.9
Max	47	231	68	19.8	4.0	25.1	4.7	314.9
Min	28	43	23	4.9	0.9	11.5	1.6	147.8
Std. dev.	8	73	17	5.1	1.1	4.9	1.2	70.9
Laterites								
n	26	26	26	26	26	26	26	26
Avg	22	158	30	7.3	1.1	29.6	2.8	250.8
Max	52	417	95	17.6	5.4	61.5	7.0	557.3
Min	11	28	12	0.4	0.0	0.5	0.4	81.3
Std. dev.	11	105	22	4.6	1.5	12.5	1.5	133.0
Massive mangcretes								
n	13	13	13	13	13	13	13	13
Avg	38	170	41	12.5	2.3	25.5	4.4	293.7
Max	61	368	64	20.5	4.1	71.9	10.6	506.3
Min	22	36	20	7.0	0.0	11.8	2.0	116.5
Std. dev.	12	116	17	4.4	1.5	18.4	2.6	144.3

Table 7.   (Continued)

Sample	La	Ce	Nd	Sm	Eu	Gd	Yb	REE
Mn-impregnated sands								
n	18	18	18	18	18	18	18	18
Avg	11	33	12	1.1	0.6	1.8	0.5	60.0
Max	19	87	29	8.4	1.3	13.1	4.3	162.1
Min	6	19	5	0.0	0.1	0.0	0.0	30.1
Std. dev.	4	16	5	2.1	0.3	3.5	1.2	31.3
Radial pyrolusite								
n	9	9	9	9	9	9	9	9
Avg	77	292	213	68.1	12.8	49.4	3.6	715.9
Max	112	396	316	98.3	19.1	72.4	5.2	1002.5
Min	46	193	61	20.1	3.9	20.5	2.7	347.0
Std. dev.	23	61	78	24.9	4.8	15.4	0.9	198.2
Siliceous ores								
n	10	10	10	10	10	10	10	10
Avg	35	75	40	12.4	2.5	13.9	3.5	182.3
Max	50	187	57	17.5	3.6	20.3	7.4	342.9
Min	19	38	22	5.1	1.4	7.9	1.8	95.2
Std. dev.	12	48	15	4.6	0.8	4.7	1.9	87.0
Spherulites								
n	14	14	14	14	14	14	14	14
Avg	27	146	37	9.8	1.1	45.3	3.6	269.8
Max	53	205	103	22.0	5.9	82.8	5.6	382.6
Min	14	62	18	3.0	0.0	22.8	1.8	207.6
Std. dev.	11	55	21	5.2	1.6	19.9	1.2	49.7
Uncemented oolites								
n	39	39	39	39	39	39	39	39
Avg	44	51	58	12.1	3.7	20.9	7.1	196.8
Max	58	106	92	24.7	6.5	34.8	13.4	335.3
Min	29	8	39	4.1	1.4	6.9	1.6	89.9
Std. dev.	7	24	14	5.1	1.3	7.4	2.8	61.3
Uncemented pisolites								
n	19	19	19	19	19	19	19	19
Avg	54	80	66	17.7	4.3	21.6	3.8	247.4
Max	92	178	121	30.7	10.1	37.6	8.5	477.9
Min	30	14	26	6.0	1.5	6.9	1.4	85.7
Std. dev.	14	49	24	7.3	2.0	8.7	1.8	106.6

Table 8. Average geochemical coefficients and normative mineral composition of ores, Groote Eylandt, North Territory, Australia (recalculated data, after Pracejus *et al.*, 1990, 1992).

1	2	3	4	5	6	7	8	9	10	11	12	13	14	15	16	17	18	19	Notes
La/Yb	6.2	14.2	7.5	10.4	9.1	10.0	14.2	5.0	2.6	7.3	12.9	8.6	7.9	21.4	7.5	12.49	11.82	9.65	
La/Sm	3.6	3.1	2.3	3.8	2.7	2.8	2.7	3.4	1.4	7.3	3.7	3.0	3.0	1.1	2.8	5.63	4.45	3.78	
Sm/Yb	1.7	4.7	3.3	2.7	3.4	3.5	5.3	1.5	1.9	1.0	3.5	2.8	2.6	18.9	2.7	2.22	2.66	2.55	
Eu/Sm	0.31	0.24	0.29	0.24	0.24	0.20	0.16	0.20	0.0	0.0	0.21	0.18	0.15	0.19	0.11	0.21	0.22	0.19	
Ce/ΣREE $\times 10^2$	25.9	32.3	21.3	33.3	18.0	41.1	51.3	35.2	25.3	38.5	53.2	57.9	63.0	40.8	54.1	35.35	55.43	37.82	
REE (ppm)	196.8	247.4	257.9	222.5	161.0	182.3	226.2	147.7	213.4	91.0	219.9	293.7	250.8	715.9	269.8	230	1326	156	
Mn/Fe	20.2	23.9	37.4	18.2	60.1	38.6	9.9	8.7	0.10	0.13	14.7	6.9	0.4	20.3	0.7	0.02	1.46	0.65	
Al/Ti	22.3	22.5	24.3	29.4	20.8	27.0	20.0	18.3	22.8	22.8	21.7	18.8	19.5	15.8	21.8	17.39	4.03	9.29	
Ti/Zr	35.8	38.0	32.4	41.8	22.0	29.9	48.1	49.0	21.0	12.0	46.1	41.5	4.9	53.2	38.4	28.75	11.96	–	
B/Zr	0.22	0.59	0.40	0.37	1.04	0.47	0.42	0.29	0.15	0.003	0.21	0.17	0.02	0.88	0.03	0.63	0.54	–	
Co/Zr	1.36	1.14	1.79	2.48	1.28	4.51	3.57	2.79	0.42	0.10	1.17	2.70	0.17	0.65	0.26	0.12	4.82	–	
Ni/Cr	7.17	1.55	7.92	3.45	3.40	2.54	3.37	1.26	2.73	0.01	0.98	1.44	0.05	1.29	0.32	0.43	11.79	–	
Cu/Zr	2.10	1.14	3.21	2.34	1.55	4.05	4.77	2.80	1.60	0.16	1.79	2.69	0.18	1.44	0.37	0.28	8.04	–	normative minerals:
Normative minerals*																			
Cm	12.99	4.77	7.71	18.62	8.40	17.79	44.40	26.60	2.50	0.50	18.50	34.90	1.40	3.80	1.80	–	–	–	cryptomelane
Rm	5.28	2.99	6.74	5.86	3.60	1.59	4.40	19.60	0.30	0.50	7.90	2.50	0.50	0.60	1.50	–	–	–	romanechite
To	5.66	2.45	4.11	6.22	4.10	4.48	6.50	2.60	0.60	0.40	7.30	5.60	0.70	1.90	0.80	–	–	–	todorokite
Pr	43.78	68.93	61.55	41.02	66.10	39.06	14.30	13.50	5.20	1.60	30.70	20.70	4.90	80.90	9.20	–	–	–	pyrolusite
W	13.45	1.90	7.55	11.48	7.70	7.24	13.50	10.70	1.80	0.20	10.40	13.80	0.60	1.20	1.70	–	–	–	other Mn oxy-hydroxides: wads
Gt	1.89	1.71	1.16	1.16	0.70	0.97	6.00	5.50	62.50	8.40	2.70	7.40	9.30	2.60	15.60	–	–	–	goethite
Ka	13.17	15.01	8.50	8.50	6.80	8.12	6.90	8.40	15.70	60.20	17.50	12.90	69.40	8.20	53.00	–	–	–	kaolinite
Q	3.47	1.62	1.03	1.03	1.30	20.61	0.70	8.20	5.60	25.10	4.10	0.20	11.20	0.20	13.00	–	–	–	quartz

1. Coefficient, component, ore type; 2. Uncemented oolite; 3. Uncemented pisolite; 4. Cemented oolite; 5. Cemented pisolite; 6. Hardground; 7. Siliceous ore; 8. Concretion; 9. Digital concretion; 10. Fe-pisolite; 11. Fe-pipe wall; 12. Laminar mangcrete; 13. Massive mangcrete; 14. Laterite; 15. Radial pyrolusite; 16. Spherulite; 17. European shales (for REE) (Haskin and Haskin, 1966) and clays (for other elements) (Turekian and Wedepohl, 1961); 18. Deep-sea nodules (average) (Baturin, 1986; Piper, 1974); 19. Manganese nodules, Central part of the Baltic sea (Ehrlich, 1968; Ingri and Ponter, 1987; Varentsov, 1975).

*Normative minerals: Cm – cryptomelane, Rm – romanechite, To – todorokite, Pr – pyrolusite, W – other Mn oxyhydroxides: wads, Gt – goethite, Ka – kaolinite, Q – quartz.

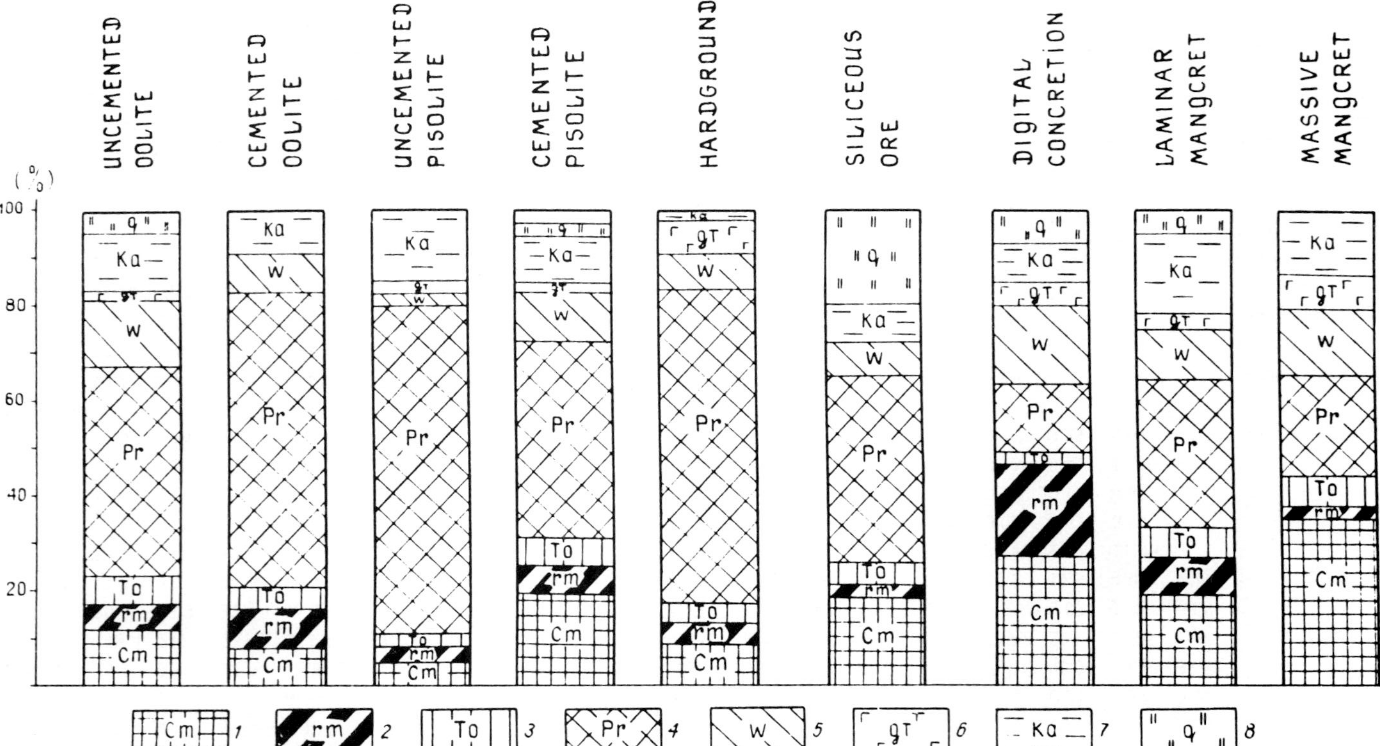

Fig. 27.   Normative mineralogical composition of manganese oxyhydroxide ores, Groote Eylandt deposits, North Territory, Australia, (compiled after the data of Pracejus and Bolton, 1992a). 1. Cm – cryptomelane; 2. Rm – romanechite; 3. To – todorokite; 4. Pr – pyrolusite; 5. W – wad and other poorly identified Mn oxyhydroxides; 6. Gt – goethite; 7. Ka – kaolinite; 8. Q – quartz.

the oolite and pisolite varieties are the early products of the supergenous alterations in comparison with more altered overlying sediments: digital concretion, laminar mangcrete, and massive mangcrete (see Table 8, Figure 27).

The distribution of the ratio Mn/Fe (Table 8, Figure 28). In the distribution of Mn/Fe values there is, almost, a twofold increase of this ratio during the transition from uncemented oolites (20.2) to the cemented varieties (37.4). But for the corresponding types of pisolite ores the values of Mn/Fe differ insignificantly: 23.9 (U.P.) and 18.2 (C.P.). Such relationships cannot be used for the determination of some significant differences between the 'primary' and 'secondary' Mn ores of the deposits. On the whole, with the development of relatively late supergenous cementation of altered, corroded, redeposited oolites, pisolites, and fragments of massive Mn ores resulted in the formation of dense thin interlayers several cm thick (i.e. hardground), the value of Mn/Fe is increased to 60.1 (see Table 8, Figure 28). Similar separation of Mn and Fe clearly distinguish sharply these supergenous ores from Mn–Fe oxyhydroxide concretions of the World ocean (Mn/Fe, 1.46) or internal seas (Central part of the Baltic sea, 0.65) (see Table 8). At the same time, the products of manganese laterite enrichment of the initial sedimentary-diagenetic accumulations (Mn/Fe, 18 to 60) differ significantly from later and essentially ferruginous varieties (digital concretion, laminar mangcrete, massive mangcrete and others), developed in the upper half of the section: Mn/Fe, from 8.7 to 20.3 (see Table 8, Figure 28), most strongly developed in proper ferruginous products (Mn/Fe, 0.10–0.13). In other words, the history of the supergenous alteration development, registered in the section of the ore bed, is sharply subdivided into relatively early manganese lateritic processes and comparatively late ferruginous lateritic processes in the upper part of the bed, and that is reflected in the distribution of the value of Mn/Fe (see Table 8, Figure 28).

The distribution of the Co/Zr ratio (Table 8, Figure 29). Geochemical passivity of the three most inert elements (Al, Ti, and Zr) of zones of sedimentation and supergenesis in the settings of formation of the considered Mn ores is reflected in relatively limited variations of ratios (see Table 8) of Al/Ti (from 15.8 to 29.4) and Ti/Zr (from 12.0 to 53.2). Especially demonstrative is the stability of these ratios for so called (Pracejus and Bolton, 1992 a) 'primary' and 'secondary' ores of similar morphological type. For example, for uncemented–cemented oolite and pisolite varieties the variations of these ratios are rather insignificant: Al/Ti 22.3 (U.O.) – 24.3 (C.O.), 22.5 (U.P.) – 29.4 (C.P.), and Ti/Zr 35.8 (U.O.) – 32.4 (C.O.), 38.0 (U.P.) – 41.8 (C.P.). Hence, the distribution of Co/Zr values may serve as a measure of relative accumulation (depletion) of Co in the Mn ores, being especially clear against the background of this ratio in clays of the Earth's crust (0.63, see Table 8, Figure 29).

In the distribution of Co/Zr in ores of the Groote Eylandt a number of features have engaged our attention:

(a) There is relative accumulation of Co in cemented varieties of oolite and pisolite ores (see Table 8, Figure 29): 1.36 (U.O.) – 1.79 (C.O.), and 1.14 (U.P.) – 2.48 (C.P.).

(b) It is particularly remarkable that the accumulation of Co is observed in concretion ores of relatively late stages of the manganese supergenous (laterite) process. For example, for siliceous ores and concretions the following Co/Zr ratio is characteristic: 4.51 and 3.57 respectively, whereas in later ferruginous laterite products the Co/Zr values are about 0.10

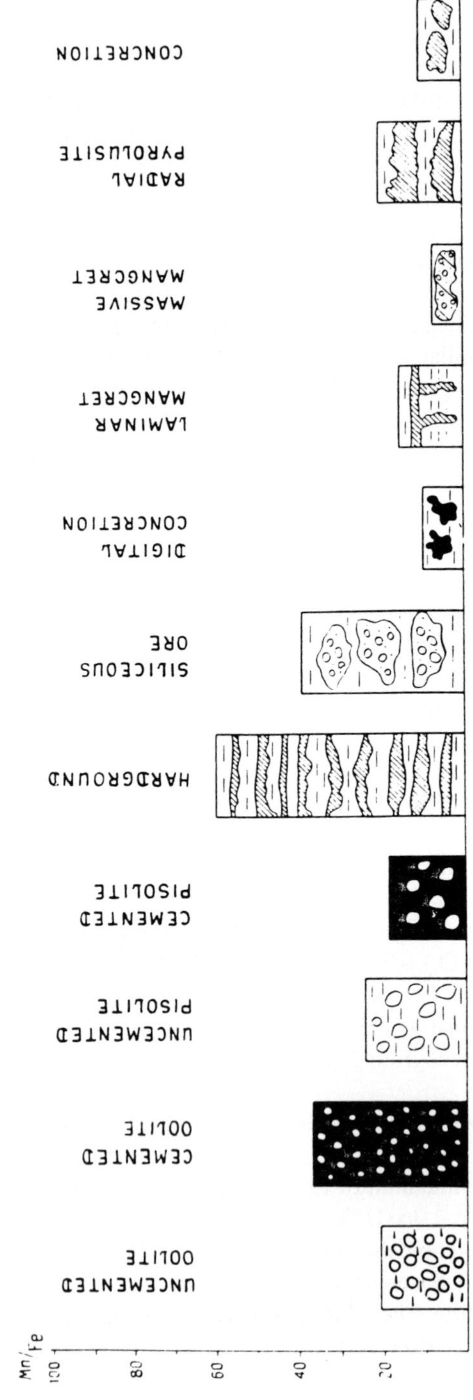

Fig. 28.   Distribution of Mn/Fe in different types of manganese oxyhydroxide ores, Groote Eylandt deposits, North Territory, Australia (compiled after the data of Pracejus and Bolton, 1992a).

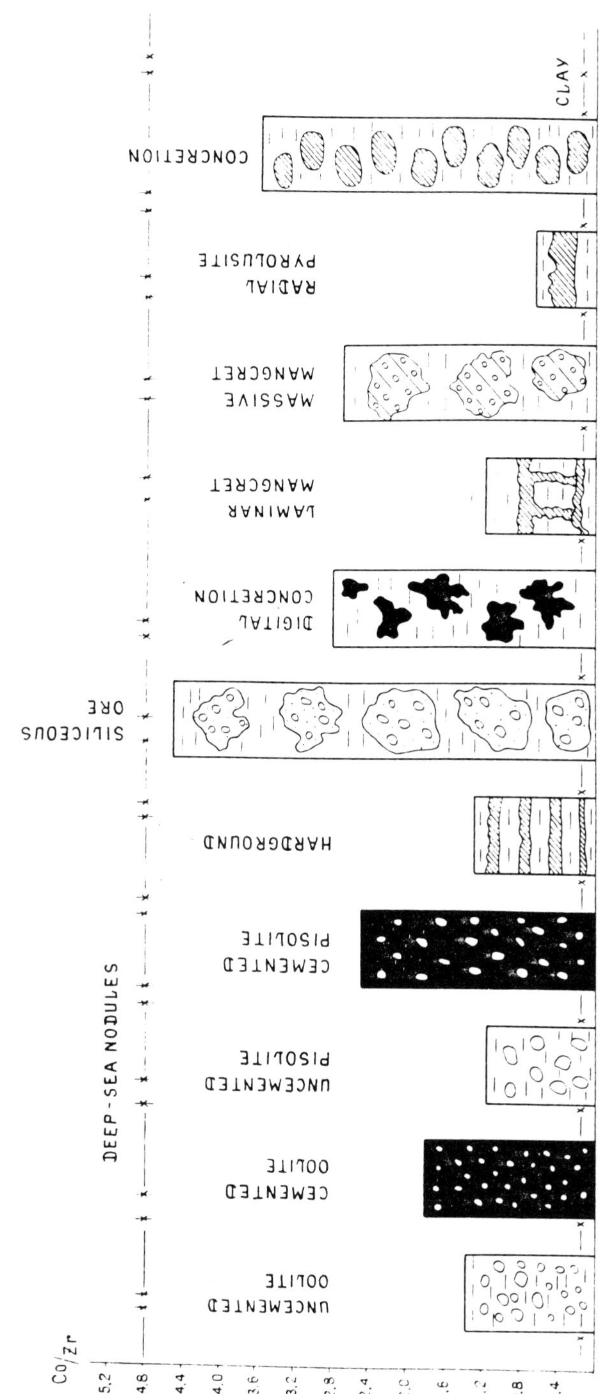

Fig. 29. Distribution of Co/Zr in different types of manganese oxyhydroxide ores, Groote Eylandt deposits, North Territory, Australia (compiled after the data of Pracejus and Bolton, 1992a). Deep-sea nodules, World ocean, average (Baturin, 1986) clay-average (Turekian and Wedepohl, 1961).

to 0.42 (see Table 8, Figure 29). The given data signifies at the geochemical generality of Mn and Co behavior in manganese ore laterite processes. With relatively high amounts of Co in the primary parent rocks, this process was distinguished in the formation of Co 10 Å-manganates (asbolanes). In the section of the Groote Eylandt deposits 10 Å-manganates are developed essentially in the upper half of the ore bed.

The distribution of B/Zr ratio (Table 8, Figure 30). At present it is customary to assume that B, as a rule, is not an element accumulated in the products of humid continental sedimentation and supergenesis. This concept is confirmed by the B/Zr values in Mn oxyhydroxide ores of the Groote Eylandt deposits being considerably lower than that of similar ratios for clays of the Earth's crust (Turkian and Wedepohl, 1961) and Mn deep-sea nodules of the World ocean (Baturin, 1986) (see Table 8, Figure 30). Two types of ores represent an exception: hardground and radial pyrolusite, products of relatively early supergenous alterations, that took place in ore-depositing solutions with relatively low pH values.

At the same time a number of concretion types of ores, products of relatively deep supergenous alteration (digital concretion, laminar mangcrete, massive mangcrete) are characterized by considerably lower values of B/Zr compared with oolite and pisolite ores. Even lower B/Zr ratios (see Table 8, Figure 3) are typical for ferruginous products of the laterite profile (0.003 to 0.15 against 0.59, uncemented pisolite) and correspondingly of the soil laterites (0.02), and spherulites (0.03).

The ratios mentioned in the distribution of B/Zr may be interpreted as an indication of the pronounced stages in the development of the weathering profile: the intensity of supergenous alterations is increased to the top of the ore bed.

The distribution of La/Yb ratios (Table 8, Figure 31). Rare earth elements (REE) are sufficiently sensitive indicators of the formation conditions of Mn ores of a sedimentary and hydrothermal-sedimentary nature (Varentsov, 1993; Varentsov *et al.*, 1991, 1993). In the environments of formation of Mn ores of the supergenous origin the REE reflect both the composition of primary parent rocks (ores) and the conditions, dominating in the formation processes of the weathering crust profile.

The ratio of La/Yb in the Mn oxyhydroxide ores is the measure of separation of heavy and light REE. As a whole in the distribution of La/Yb in different types of ores of the Groote Eylandt there is no distinct trend: as a rule, the values of La/Yb are considerably lower than in shale, Mn–Fe concretions of the World ocean and the Central part of the Baltic sea (see Table 8, Figure 31, La/Yb respectively, 12.49; 11.82; and 9.65). It is interesting to emphasize, that the value of La/Yb of oolite and pisolite (uncemented and cemented) types of Mn ores of the Groote Eylandt deposits differs slightly from concretion or lens-layered varieties of relatively late stages of the Mn laterite process. On the basis of such a homogenous distribution of this ratio the fact that La/Yb in ores is considerably lower than in shales, and similar to the value of the ratio of Mn–Fe concretions of the sedimentation basins, may indicate the considerable removal of La as the most mobile REE in the supergenous settings has occurred. At the same time, high values of La/Yb (21.4; see Table 8, Figure 31) are observed in such ore types as radial pyrolusite, distinguished in the lens-like intercalations at dozens of m with relicts of oolite and pisolite structures and reproducing the contour of the ore bed top. This behavior of La/Yb may indicate the

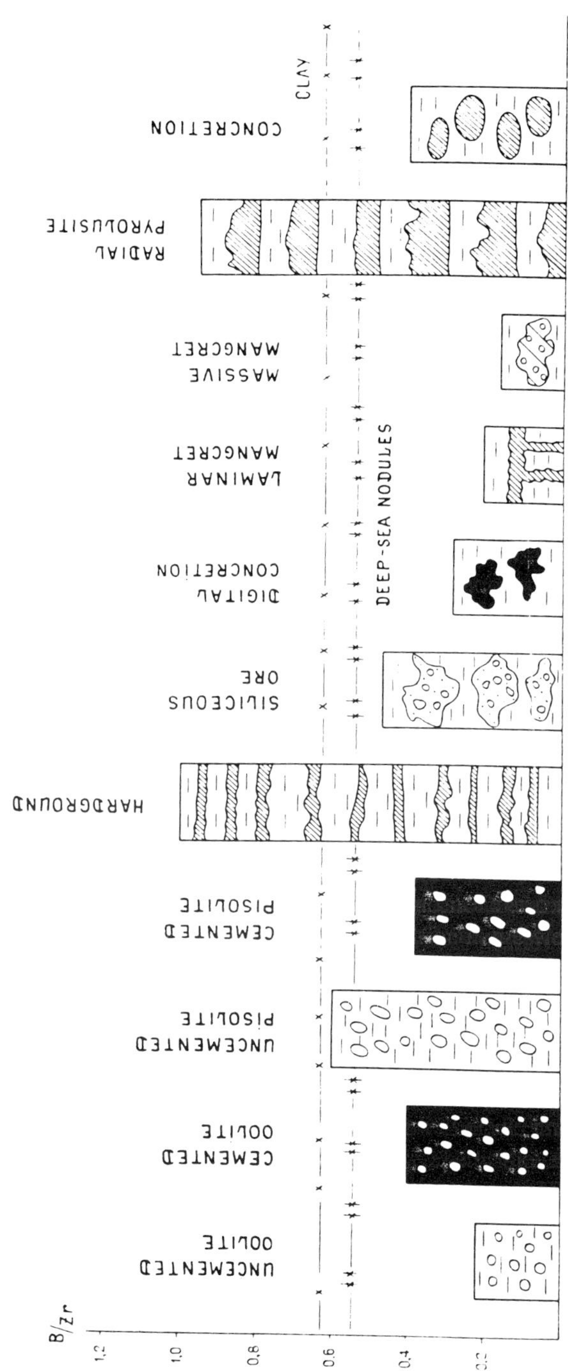

Fig. 30.   Distribution of B/Zr in different types of manganese oxyhydroxide ores, Groote Eylandt deposits, North Territory, Australia (compiled after the data of Pracejus and Bolton, 1992a). Deep-sea nodules, World ocean, average (Baturin, 1986), clay-average (Turekian and Wadepohl, 1961).

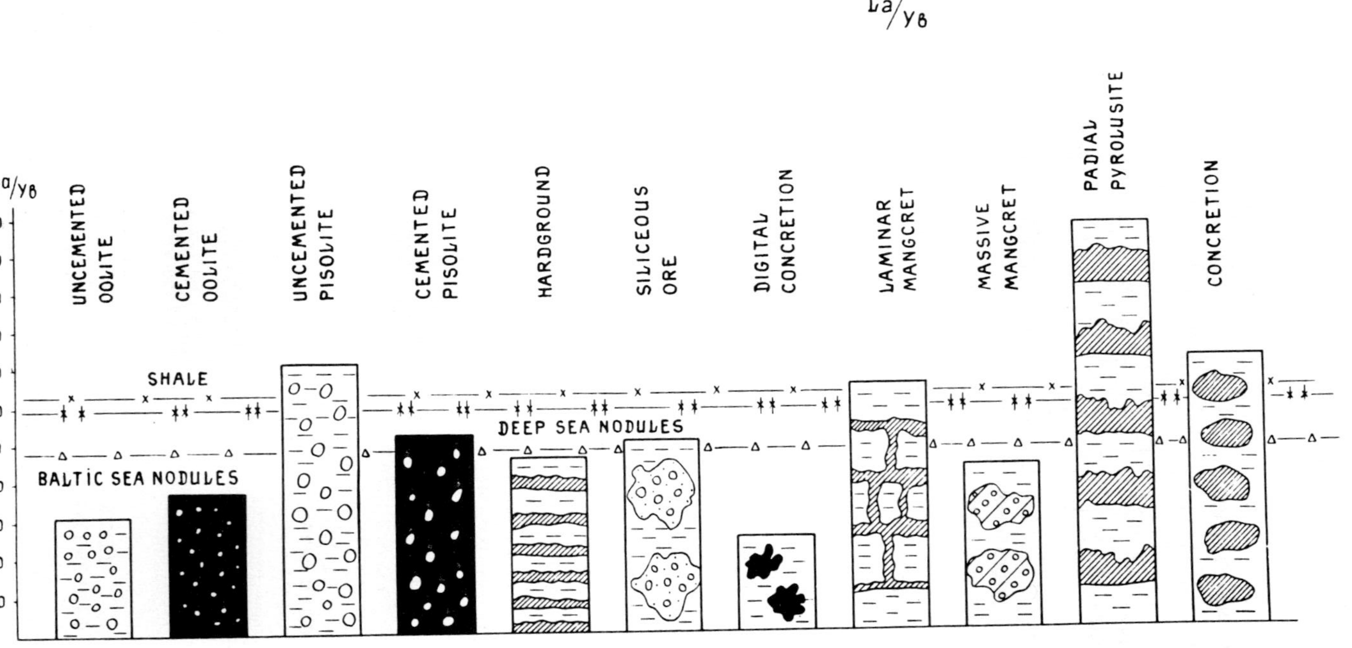

Fig. 31.   Distribution of La/Yb in different types of manganese oxyhydroxide ores, Groote Eylandt deposits, North Territory, Australia (compiled after the data of Pracejus *et al.*, 1990a). Shale – European shales average (Haskin and Haskin, 1966). Deep-sea nodules, World ocean, average, (Piper, 1974), manganese nodules average, central part of the Baltic sea (Ehrlich, 1968; Ingri and Ponter, 1987).

relative accumulation of La under the conditions of formation of the actual supergenous pyrolusite, where low values of pH are characteristic.

Distribution of Ce/ΣREE × 10^2 *(Table 8, Figure 32)*. The value of Ce/ΣREE × 10^2 ratio corresponds to relative percentage of Ce content. The behavior of Ce in the process of Mn–Fe concretion formation in the modern basins is sufficiently well studied (De Baar *et al.*, 1985; Fleet, 1984; Glasby *et al.*, 1987; Varentsov *et al.*, 1991; Varentsov, 1993). The accumulation of this element in Mn–Fe concretions of the World ocean and shelf seas is controlled by processes of chemosorption, autocatalytic accumulation from seawater and pore solutions, accompanied by the interphase oxidation of Ce(II) → Ce(IV), that is characteristic for a number of transitional elements of variable valency (Mn, Fe, Co, and others).

Manganese ores of the Groote Eylandt deposits are subdivided into two groups by the ratio distribution of Ce/ΣREE × 10^2:

(a)   The group of oolite and pisolite (uncemented and cemented) ores as well as the hardground (see Table 1) is distinguished by the values of Ce/ΣREE × 10^2 that are considerably lower than that in shale and Mn–Fe concretions of the Central Baltic (correspondingly 35.5 and 37.82, see Table 8, Figure 32).

(b)   The group of concretion, lens-like ores with the values of Ce/ΣREE × 10^2 higher than in shale and approaching the values of this ratio for Mn–Fe concretions of the World ocean (55.43). It should be recorded that the formation of the latter ores takes place in sufficiently different settings with generality of reaction mechanisms.

The given data may be interpreted as an indication of the accumulation of Ce in relatively stable conditions of concretion ores (concretion, laminar mangcrete, massive mangcrete, see Table 1), and Mn laterite profile of the weathering crust with relatively high values of Eh (zone of free oxygen exchange).

The distribution of Eu/Sm ratio (Table 8, Figure 33). The distribution of Eu/Sm ratio in Mn oxyhydroxide ores of the Groote Eylandt deposits is dependent on the displayed trend. The highest values of Eu/Sm, which considerably exceed the values that are characteristic for shale (Eu/Sm = 0.21), are observed in oolite ores: (U.O.) 0.31 and (C.O.) 0.29; they are slightly lower in pisolite varieties and hardgrounds (0.24). In concretionary lens-like ores of higher horizons of the ore bed, forming deeper stages of the laterite alteration, lower values of Eu/Sm than in shale are observed (to 0.16) (see Table 8, Figure 33).

Such distribution of Eu/Sm reflects at least two main geochemical features of Eu behavior in the process of formation of the weathering crust profile:

(a) It is believed that high values of Eu/Sm in Mn oolite ores of the deposit representing relatively initial products of the supergenous alteration, correspond to the REE composition in the primary parent rocks. It is worthy to note that Eu/Sm value of 0.31 is characteristic for the mean composition of the continental crust as a whole (Taylor and MacLennan, 1981, 1985). As the alteration of the manganese ores and enclosing rocks proceeded, the removal of Eu(III) took place and, correspondingly the values of Eu/Sm were decreased in the products forming with deep alteration of the matter relative to the early generation (concretion, Eu/Sm = 0.16). The practically complete removal of Eu in this process took place during the formation of Fe products of lateritization (Fe-pisolite and Fe-pipe wall, Eu/Sm = 0, see Table 8).

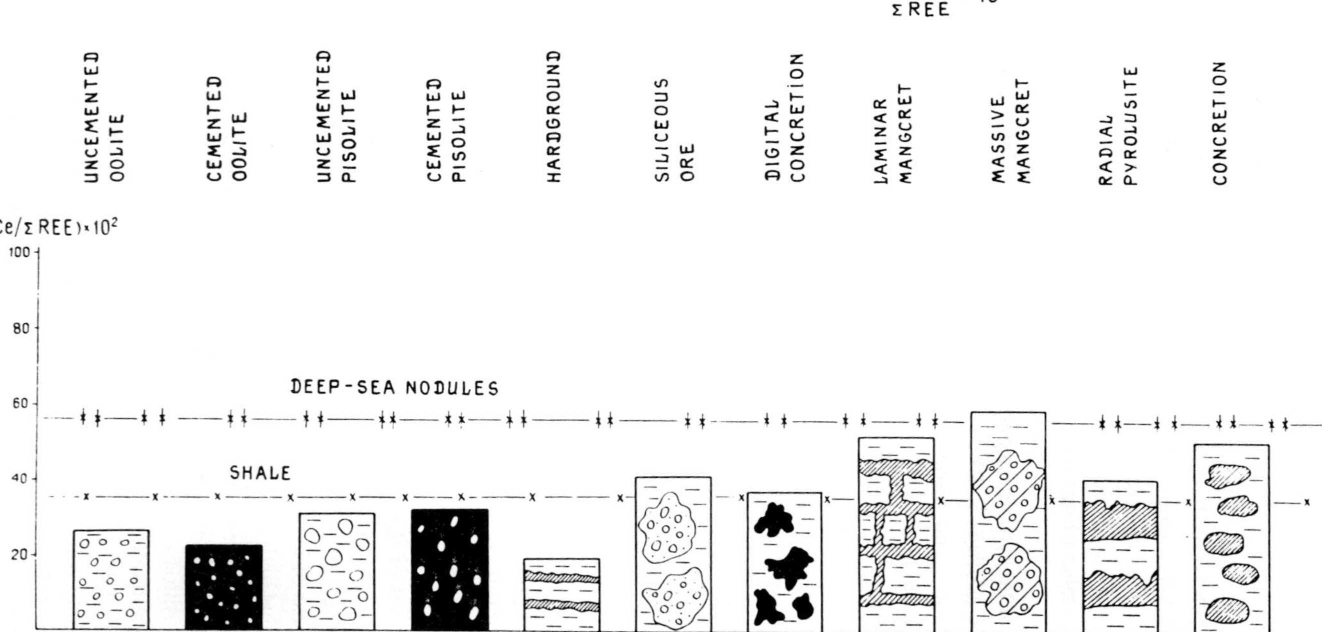

Fig. 32.   Distribution of Ce/ΣREE $\times$ 10^2 in different types of manganese oxyhydroxide ores, Groote Eylandt deposits, North Territory, Australia (compiled after the data of Pracejus *et al.*, 1990). Shale – European shales, average (Haskin and Haskin, 1966). Deep-sea nodules, World ocean, average (Piper, 1974).

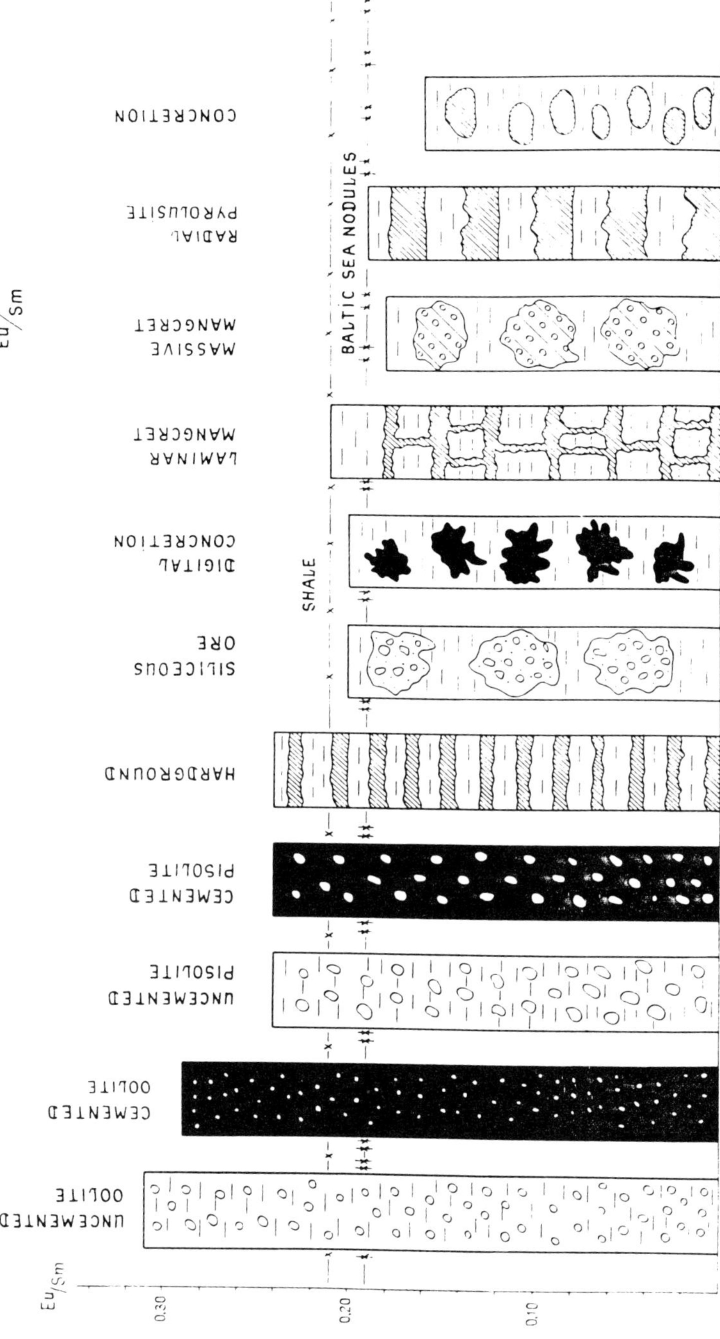

Fig. 33. Distribution of Eu/Sm in different types of manganese oxyhydroxide ores, Groote Eylandt deposits, North Territory, Australia (compiled after the data of Pracejus *et al.*, 1990). Shale – European shales, average (Haskin and Haskin, 1966). Baltic sea nodules, average, central part of the Baltic sea (Ehrlich, 1968; Ingri and Ponter, 1987).

(b) At the same time, the successive series of decreasing values of Eu/Sm from Mn oolite ores to concretion, lens-like varieties (see Table 8, Figure 33) is the inverse trend of the distribution of Ce/REE $\times$ 10 (see Figure 32). Such alterations in these two geochemical indicators testifies to the significant enhancement of the laterite process to the top of the manganese ore bed, accompanied by an increase in oxidation conditions.

2.2.1.8. Paragenetic Assemblages of Components

The features of the structure and mineral and chemical composition of the oxyhydroxide manganese ores of the deposit, find distinct reflections in the isolation of the paragenetic assemblages (associations) of chemical components factor groups (clusters). Such groups are revealed after processing the chemical analysis data by factor analysis; the interpretation of their shades of meaning allows one to determine the geochemical essence of such assemblages: compounds or phases, or specific forms of occurrence of the determined groups of components. The analytical data was processed by computer in the Laboratory of Mathematical Methods GIN, Russian Academy of Sciences using factor analysis (R- and Q-mode, Davis, 1973; Harman, 1967).

The separation of the associations was made by the grouping of components with factor loading greater than or equal to an absolute value of 0.30. Each chemical component is characterized by a value of factor loading (in brackets) which is closely associated with the interrelationships between the components of a specific assemblage. The number of the assemblage corresponds to the number of the factor. It is presumed that characteristic loadings on other factors for each given assemblage are close to 0. These results reveal components of the assemblages and their distribution throughout the section. The data was transformed by rotation methods to obtain more distinct values of factor loadings (Varimax by Kaiser, see Davis, 1973). The experience of the successful application of the factor analysis was described in a number of works (in particular, Varentsov, 1980; Varentsov *et al.*, 1981, 1983; and others). It is worthy to note that the basis for the interpretation of paragenetic combinations of chemical components are the direct data of the study of mineral composition and parameters of those processes, the products of which may, almost certainly, be the considered phases (clusters).

Assemblage IA(+): SiO_2 (0.63),* TiO_2 (0.89), Al_2O_3 (0.87), Fe_2O_3 (0.68), MgO (0.41), H_2O^+ (0.87), H_2O^- (0.77), P_2O_5 (0.69). BaO (0.12), Zr (0.69), La (0.04), Ce (0.19). The set of components and the results of the study of mineral composition of ores definitely indicate that this group, in the most characteristic manifestation (Tables 9 and 10, Figure 34), is represented by the residual, essentially terrigenous components (quartz, kaolinite, halloysite, associated with it oxyhydroxides of iron, and phosphates), being the products of supergenous alterations of Fe-smectites, Fe-hydromicas, manganese carbonates and oxyhydroxide varieties (in particular, manganite, cryptomelane, and others) of relatively early generations. Distinctly pronounced localization of this assemblage is characteristic: it is developed in the western part of the sublatitudinal profile (sections of quarry-A south, A, and southeast, E), gaining the highest development in the lower

---

*The value of the factor loading of the component to the given factor is shown here and below in brackets.

Table 9.   Results of the factor analysis (R-mode) for chemical components (wt. %) of oxyhydroxide manganese ores of the Groote Eylandt deposit, North Australia.

		Factor loadings (after rotation)		
no.	Component	Factor I	Factor II	Factor III
1	SiO_2	0.63	-0.38	-0.26
2	TiO_2	0.89	-0.20	0.14
3	Al_2O_3	0.87	-0.12	0.10
4	Fe_2O_3	0.68	-0.30	0.32
5	CaO	-0.02	0.44	0.02
6	MgO	0.41	0.39	0.15
7	MnO	-0.55	0.28	0.21
8	MnO_2	-0.79	0.37	0.22
9	Na_2O	-0.59	0.34	0.003
10	K_2O	-0.13	0.56	0.29
11	H_2O^+	0.87	0.14	0.23
12	H_2O^-	0.77	0.35	0.30
13	P_2O_5	0.69	-0.08	-0.15
14	BaO	0.12	0.85	-0.23
15	Cu	-0.41	0.56	0.29
16	Rb	-0.23	0.42	-0.02
17	Sr	-0.32	0.72	-0.35
18	Y	-0.04	0.06	0.73
19	Zr	0.69	0.07	0.23
20	Ba	-0.05	0.75	-0.34
21	La	0.04	-0.21	0.78
22	Ce	0.19	-0.03	0.53
Input in dispersion (%)		32.8612	14.4000	10.6801
Total dispersion (%)		32.8612	47.2612	57.9412

and upper parts of the ore bed. As it was noted during the analysis of the geochemical features (see Figures 10 to 15), it is in these parts of the section that there are relatively high amounts of terrigenous components and the residual silicate phases, retained after the alteration of the 'primary' sedimentary-diagenetic Mn-bearing sediments, including early manganese compounds, into the products of deep supergenous alteration.

Assemblage IB(-): MnO (-0.55), MnO_2 (-0.79), Na_2O (0.59), K_2O (-0.13), Cu (-0.41), Rb (-0.23), Sr (-0.32). The given cluster group is an antagonist of the assemblage IA(+). It is relatively widely distributed in the ores of the deposit, with dispersion of 32.86% (see Table 9). The main component is MnO_2, where other components play the role of admixtures. Comparison of the most significant manifestations of this assemblage (Figure 35) with the mineral composition of ores (see Figures 10–15) and the features of

Table 10.   Distribution of factor scores (R-mode) for chemical components (wt. %) of oxyhydroxide manganese ores of the Groote Eylandt deposits, North Australia*).

| | | Factor score (after rotation) | | |
no.	Sample	Factor I	Factor II	Factor III
1	2	-0.78	0.33	-0.04
2	3	-0.47	0.87	-0.63
3	4	-0.23	-0.30	-0.09
4	4(I)	0.30	0.14	-0.30
5	5	-0.62	0.63	-1.28
6	6	-1.14	-0.14	-2.05
7	7	-0.07	2.53	-0.18
8	8	0.41	1.89	-0.49
9	9	-0.39	-0.39	-1.05
10	11	-0.24	-1.58	0.52
11	12	-0.04	-1.37	0.93
12	13	-0.58	-0.86	0.65
13	14	-1.58	-1.04	-0.13
14	15	-1.41	-0.32	0.55
15	16	-0.70	-1.19	0.79
16	17	1.73	0.59	-0.54
17	18	0.53	0.50	0.70
18	19	2.12	0.19	0.79
19	20	1.21	-1.11	-2.30
20	21	-0.14	0.27	-0.12
21	22	-0.33	-0.02	-1.21
22	23	-0.77	-0.002	0.57
23	24	-0.86	1.21	0.99
24	25	2.25	-1.76	-0.09
25	26	1.14	-0.30	0.34
26	27	-0.48	0.13	2.33
27	28	1.16	1.08	1.38

*Chemical, mineral composition of the samples, their position on the area and in section of the ore bed is shown in Tables 2 and 3, Figures 10 to 15, 34 to 37.

its distribution in the sections, provides reason enough to believe that it is represented by the oxyhydroxide phases of pyrolusite type, and to a lesser degree, by cryptomelane with an essential deficit of K^+. It is important to keep in mind that its occurrence in the lower half of the ore bed section is distinguished by the distribution of the given

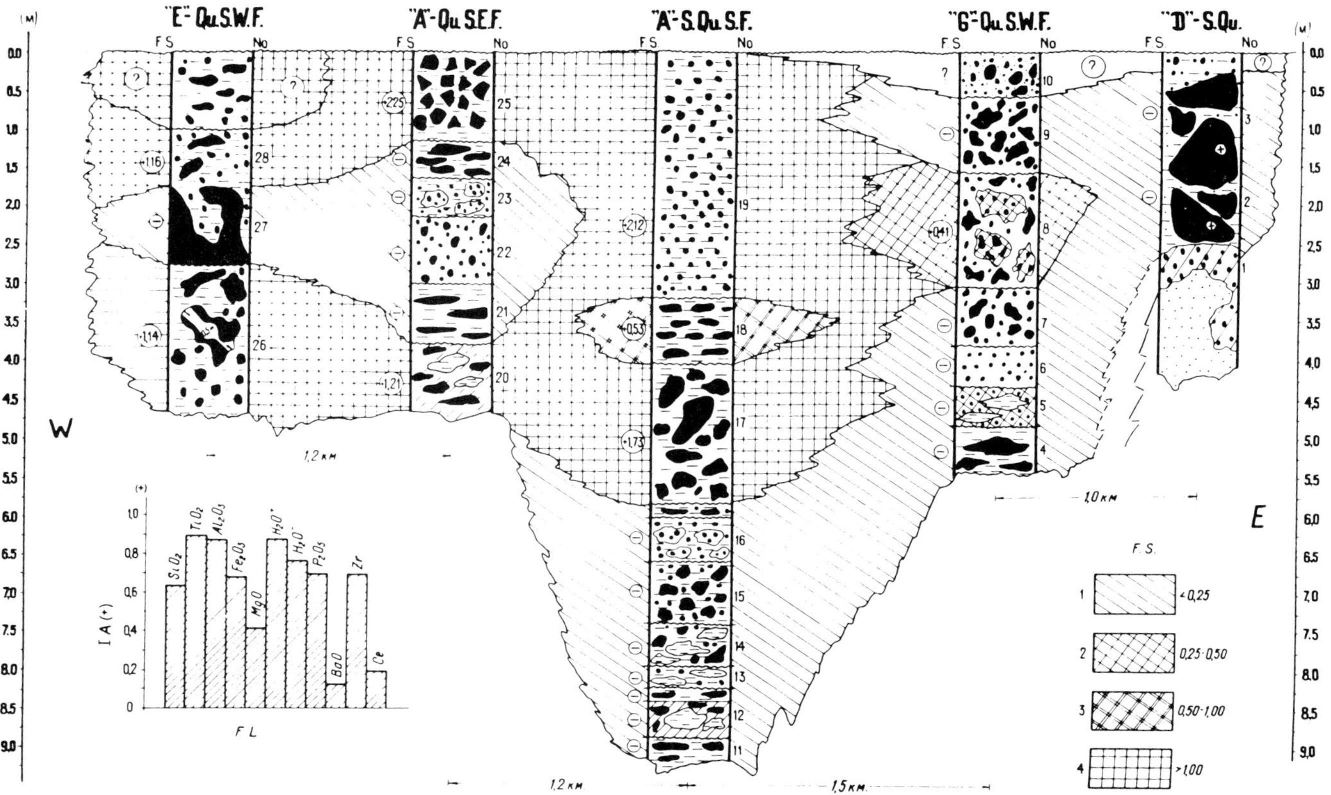

Fig. 34. Distribution of factor assemblage IA(+) in the section of ore-bearing sediments, the northern basin of the Groote Eylandt deposits (see Figures 4 to 10). The assemblage (see Table 9) is represented by a set of chemical components, corresponding to terrigenous and residual, newly formed constituents (kaolinite, halloysite, associated with oxyhydroxides of Fe and phospates). They are the products of supergenous alteration of primary Fe-smectites, Fe-hydromicas, Mn carbonates, as well as Mn oxyhydroxide varieties (manganite and cryptomelane).

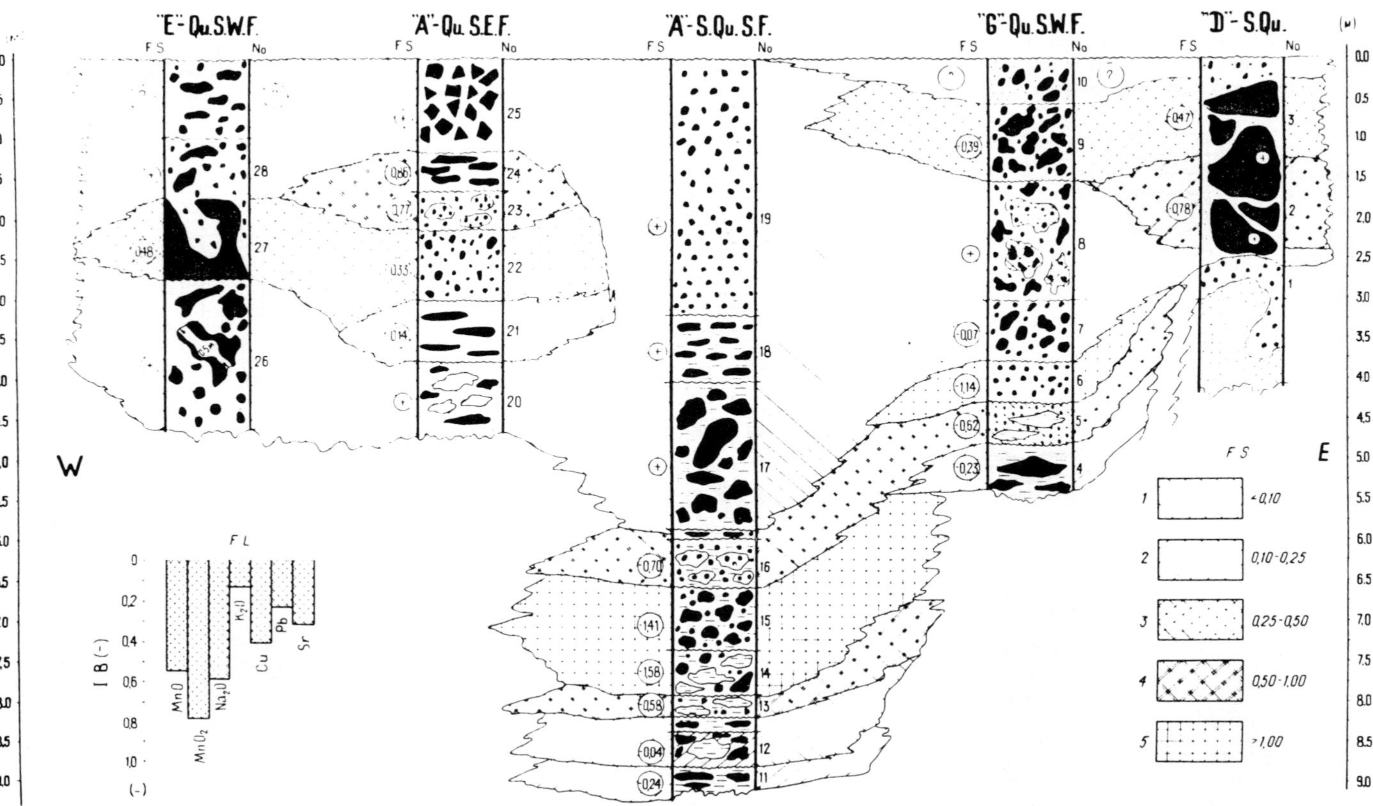

Fig. 35.   Distribution of factor assemblage IB(-) in the section of ore-bearing sediments, the northern basin of the Groote Eylandt deposits (see Figures 4 to 10). The assemblage (see Table 9) is represented mainly by MnO_2. The correlation of the most significant manifestations of this assemblage with the mineral composition of ores (see Figure 10) shows that it is represented essentially by pyrolusite, and to a lesser degree by cryptomelane with significant deficit of K^+. Its occurrence in the lower half of the ore bed is characteristic.

assemblage. This fact may be interpreted as an indication that pyrolusite formation took place at relatively early stages of alteration relative to the early manganese compounds. However, the ores of this generation were preserved under the cover of the overlaying sediments from later supergenous alterations of the pedogenic laterite nature.

Assemblage IIA(+): CaO (0.44), MgO (0.39), MnO (0.28), MnO_2 (0.37), Na_2O (0.34), K_2O (0.56), H_2O^+ (0.14), H_2O^- (0.35), BaO (0.85), Cu (0.56), Rb (0.42), Sr (0.72), Y (0.06), Zr (0.07). This cluster group is represented by oxyhydroxide phases of manganese, in the composition of which the essential role belongs to such large ions as K^+, Ba^{2+}, Sr^{2+}, Cu^{2+} and others. They include mainly tectonomanganates, that is minerals with the tunnel structure (Chukhrov *et al.*, 1989; Giovanoli, 1985), as well as cryptomelane, todorokite and admixtures of birnessite, romanechite (psilomelane) and lithiophorite. Due to these features, the highest factor scores of the given assemblage are observed in those parts of the section where there are combinations of these minerals (see Tables 9 and 10, Figures 10 to 15 and 36): quarries G and E. In these cases, when the relatively pure cryptomelane varieties (middle part of the ore bed in the quarry-A, samples 17 and 18, see Figures 10 and 13) are developed, high factor values are observed (see Tables 5, 10, Figure 36). Thus, this group of components has the dual genetical nature: on the one hand it is represented by different manganates, in which a considerable role belongs to ions with large radius, and these phases are of relatively late character; on the other hand it reflects the presence of cryptomelane varieties, being relatively early or old products of supergenous alterations of the primary sedimentary-diagenetic accumulations of manganese.

In this context the presence of CaO (0.44) and Mg (0.39), united into the same cluster group with MnO (0.28) and MnO_2 (0.37) attracts attention. During the calculation of the pair coefficients of correlation of the relatively large file of analytical data (Pracejus and Bolton, 1992a, p. 1332), it is shown that the considered components are associated by relatively close correlation relations: CaO–Mn ($r = 0.706$); MgO–Mn ($r = 0.719$); L.O.I–Mn ($r = 0.604$). The comparison of the data suggests that there is a generality of the geochemical fate of these components, in the process of supergenous alterations, of the primary manganese-bearing carbonate sediments in the Mn oxyhydroxide accumulations.

Assemblage IIB(-): SiO_2 (-0.38), TiO_2 (-0.20), Al_2O_3 (-0.12), Fe_2O_3 (-0.30), P_2O_5 (-0.08), La (-0.21). The set of components of this group and its antagonistic role relative to the assemblage IIA(+), and its distinguishing features of distribution in sections (see Tables 9 and 10) indicate that the given cluster is represented by the residual clastogenic silicate components. Attention is called to the fact that the largest factor scores of the assemblage are observed in the lower parts of the ore bed (see Tables 9 and 10). This conclusion is confirmed by the data considered above (see Figures 10 to 15) on the enrichment of the ore bed base in the residual clastic redeposited material.

Assemblage IIIA(+): TiO_2 (0.14), Al_2O_3 (0.10), Fe_2O_3 (0.32), MgO (0.15), MnO (0.21), MnO_2 (0.22), K_2O (0.29), H_2O^+ (0.23), H_2O^- (0.30), Cu (0.29), Y (0.73), Zr (0.23), La (0.78), Ce (0.53). The set of components of this cluster and the comparison with the factual mineral composition, and the features of distribution of the highest factor scores in the sections of the ore body (see Tables 9 and 10, Figure 37), allow this conclusion; that this group is represented by the oxyhydroxides of manganese and iron of the residual nature, with which the relatively stable relict phases with TiO_2 and Al_2O_3 are closely

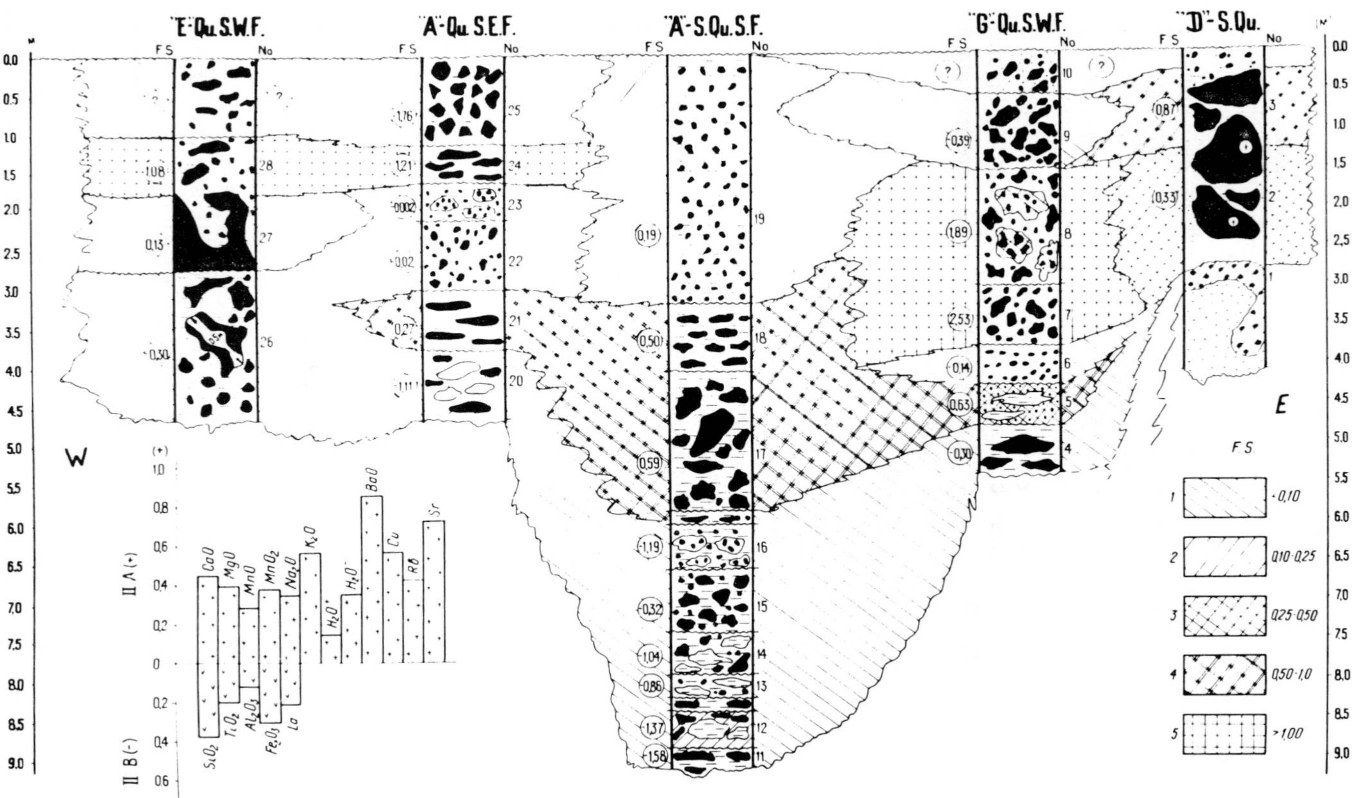

Fig. 36.   Distribution of factor assemblage IIA(+) in the section of ore-bearing sediments, the northern basin of the Groote Eylandt deposit (see Figures 4 to 10). The assemblage (see Table 9) is represented by Mn oxyhydroxides, in the composition of which the essential role belongs to such large cations as K^+, Ba^{2+}, Sr^{2+}, Cu^{2+} and others. They include essentially manganates of the tunnel structures and to a lesser degree Mn oxyhydroxides of the layered structure: cryptomelane, todorokite, birnessite, romanechite (psilomelane), and lithiophorite.

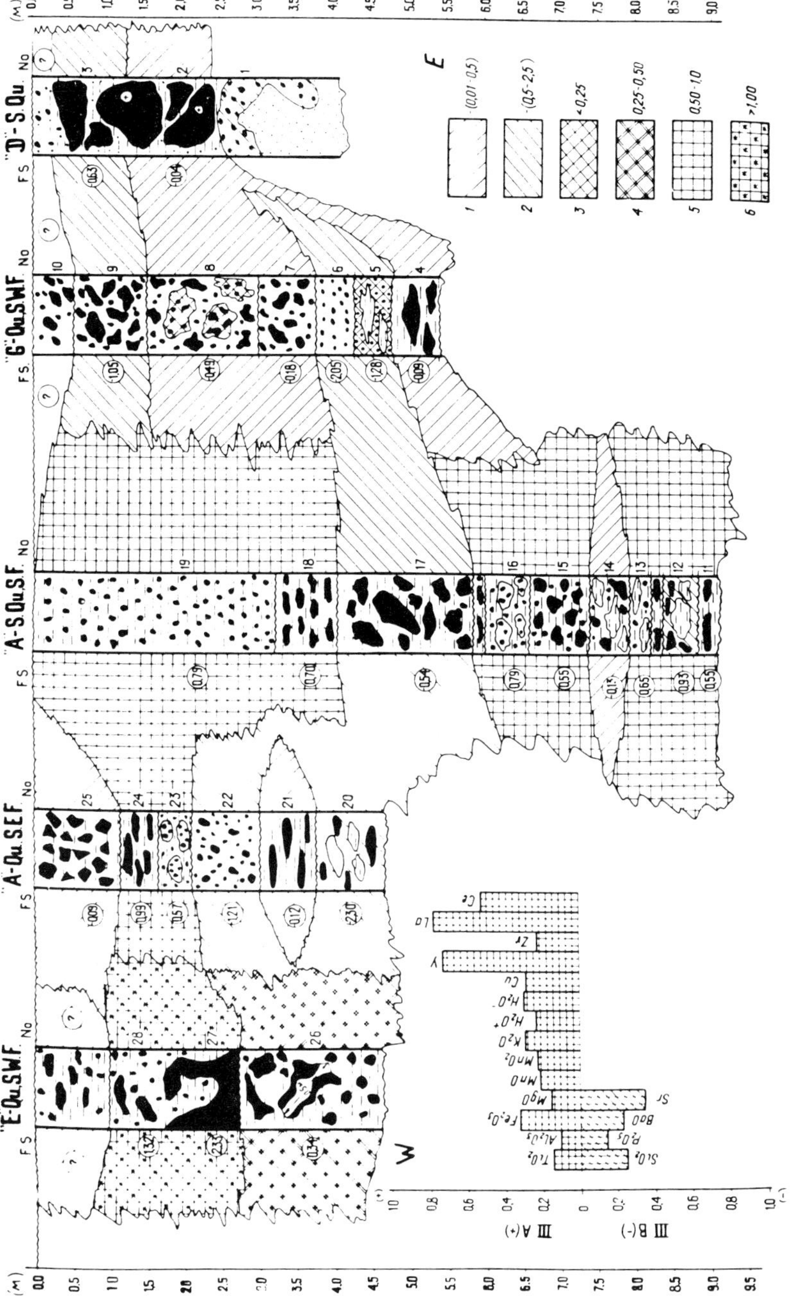

Fig. 37. Distribution of factor assemblage IIIA(+) in the section of ore-bearing sediments, the northern basin of the Groote Eylandt deposits (see Figures 4 to 10). The given cluster group (see Table 9) is represented by a set of chemical components, corresponding to oxyhydroxides of Mn and Fe of the residual supergenous nature, with which the relatively stable relict phases with TiO_2 and Al_2O_3 are associated. Such slightly mobile and accumulating elements as Y, Zr, La, Ce are associated with these compounds.

associated. These compounds are closely related with such slightly mobile elements as Y, Zr, La, and Ce accumulating in supergenous conditions. The conclusions find a reflection in the localization of the considered group: the highest factor scores are observed in the central (quarry-A south and A) and western (quarry-E) sections in the lower and upper parts of the ore bed, that is in ores represented by products of relatively deep supergenous alteration. In the western section (quarry-E, see Figures 10 to 15) the considered forms are the residual products of the alteration relative to early alteration in cryptomelane into such minerals as todorokite, romanechite (psilomelane), and lithiophorite (samples 27 and 28, see Figure 15). In the central sections (quarry-A and A south, see Figures 10, 13, and 14) the given assemblage represents the residual phase after the alteration of cryptomelane: (a) into pyrolusite (samples 11 to 13 and 15 to 16); (b) into pyrolusite, and lithiophorite (sample 19); (c) at the substitution by goethite (samples 18 and 24).

Assemblage IIIB(-): SiO_2 (-0.26), P_2O_5 (-0.15), BaO (-0.23), Sr (-0.35). The considered cluster is an antagonist to the assemblage (IIIA(+), see Figure 37) discussed above; its residual ore-free nature is shown most clearly in cryptomelane ores, that are altered to a limited degree by substitution processes by pyrolusite, romanechite (psilomelane), or other oxyhydroxides of Mn (see Figures 10 to 15), for example, in quarry-A, A south, G, and D.

Hence, the consideration of the distribution features of the paragenetic assemblages of ore components, made with all available geological, mineralogical, and geochemical information, allows the definition of the main characteristics of ore formation and the development of this process within the deposits to be shown in a clear form. The multistage, lateral, heterogeneous character of the supergenous mineralization process is distinctly distinguished.

2.2.1.9 *K–Ar Age of the Mn Oxyhydroxide (Cryptomelane) Ores*

It was emphasized above that among the geologists who studied the manganese ores of the Groote Eylandt deposits, there are contradictions both on the nature of these accumulations and regarding their age. The majority of the Australian authors (Bolton and Frakes, 1982, 1984, 1988, 1990; Bolton *et al.*, 1984, 1989, 1990; Frakes and Bolton, 1982, 1984b; Frazer and Belcher, 1975; McIntosh *et al.*, 1975; Ostwald, 1975, 1980, 1981a, 1981b, 1982, 1984, 1987, 1988, 1992; Ostwald and Bolton, 1990, 1992; Pracejus and Bolton, 1992a, 1992b; Pracejus *et al.*, 1986, 1988a, 1988b, 1990a, 1990b; Slee, 1980) share the concepts of Smith and Gebert (1969) that the oxyhydroxide manganese ores are the sedimentation (or geochemical) facies, accumulated in the Albian–Cenomanian time in the shallow-water intracraton basin: oxyhydroxide ores are laterally substituted by carbonate varieties. Accepting the probability of the existence of relatively low-grade primary (sedimentary-diagenetic) Mn ores, the above data substantiates the validity of conclusions, made in our early works, on the dominating role of the supergenous processes in the formation of the commercial ores (Varentsov, 1982a, 1982b). At the same time it is vital to note that in the recent works of Bolton, Pracejus and others (Bolton *et al.*, 1984, 1988, 1990; Pracejus and Bolton, 1992a, 1992b; Pracejus *et al.*, 1988a, 1988b, 1990a, 1990b) the essential role of supergenous processes in the formation of oxyhydroxide ores of the deposit is discussed.

Therefore, the determination of the absolute age of oxyhydroxide ores has an importance for understanding the processes of the deposit formation. Cryptomelane represents the most important mineral for K–Ar dating (Varentsov and Golovin, 1987). The positive experience of application of this mineral is described in works by Chukhrov *et al.* (1965), Yashvili and Gukasyan (1973), Vasconcelos *et al.* (1992). It should be mentioned that i such tectonomanganates as cryptomelane, K^+ is present in the tunnels of the structi the stability of which is to a considerable degree determined by the relations of sizes of, almost, square sections of the tunnel space (≈ 3.0 Å) and the filling cations with relatively large radius (1.5 Å, for example, Ba^{2+} – hollandite, K^+ – cryptomelane, Pb^{2+} – coronadite, NH_4^+ – synthetic phases and others). The presence of similar relatively large cations is essential for the stabilization of this structure and prevention of its destruction. In the natural manganese oxyhydroxides of the tunnel structure (cryptomelane, hollandite and others) their stability at relatively high temperatures is determined by the presence of such ions as K^+ and Ba^{2+} in their structure (Burns and Burns, 1979; Chukhrov *et al.*, 1989; Giovanoli, 1985; Turner and Buseck, 1979). As a rule, the natural varieties of cryptomelane are distinguished with relatively high stability: some of them at the temperature of 600°C are transformed to bixbyite, and at 825°C to hausmannite, others are altered directly to hausmannite, avoiding bixbyite (Frenzel, 1980). On the substantiation of this type of dating there were unsuccessful attempts to find some information in literature on the behavior of K–Ar isotopic systems of cryptomelane. There is no data on energies of argon removal from the crystalline lattice and its tendency to exchange with K cations, but according to the thermal analysis data (Frenzel, 1980) one may suppose that the beginning of argon removal from the structure is associated with its destruction at temperatures higher than 400°C. The composition of studied sediments, and mineral parageneses of the ores allows for the absence of their visible temperature alteration to be discussed with confidence. Under these conditions the tendency of cryptomelane to low temperature ($< 100°C$) loss of argon is extremely unlikely. Thus, the features of K^+ occurrence in the structure of cryptomelane make the mineral as suitable for the K–Ar dating as mica or feldspar, assuming the same levels of error. In other words, at the preservation of cryptomelane structure, one may believe that the relationships of the corresponding isotopes of K and Ar reflects the age of the mineral formation.

Relatively pure phases of cryptomelane with the most considerable contents of K were selected: samples 21, 24, and 27, and for comparison pyrolusite with relatively high amounts of K^+. The mineral and chemical composition of these ores is given in Table 10a, location in sections of the ore bed is shown in Figures 8 to 10, 14, and 15. The determinations of K–Ar dating were carried out by D.I. Golovin, GIN, Russian Academy of Sciences (Varentsov and Golovin, 1987). To study the isotopic composition of argon the sample weighted portions of 150–200 mg were melted in a quartz tube at a temperature of 1200°C after the six hour ageing at 150°C. The emitted argon was treated with the standard technology on Cu–CuO and Ti–Zr–Al absorbers. The isotopic relations of $^{40}Ar/^{38}Ar$ and $^{38}Ar/^{36}Ar$ were measured by the mass spectrometer MI 1201 in a one beam regime with the registration of ion current intensities by digital voltameter and direct statistical processing of mass spectra with a computer. The number of spectra, determined for each sample, was about 30 with the RMS error (2δ) for the ratio $^{40}Ar/^{38}Ar$ was 0.8%,

Table 10a.   Results of K–Ar absolute age determinations of the manganese oxyhydroxide ores, the Groote Eylandt deposit, Northern Australia (Varentsov and Golovin, 1987).

no. of sample	Mineral composition (see Table 2)	Potassium content (%, weight) $\pm 1\%$	Argon content radioact. (mm^3/g)	Absolute age (Ma)	Geological age (Harland *et al.*, 1982)
13	almost pure pyrolusite	0.52	0.00350	168 ± 20	Middle Jurassic, Callovian age?
21	cryptomelane with subordinate amounts of goethite	1.96	0.00319	$41.5 \pm 0.$	Late/Middle Eocene
24	cryptomelane with rather slight admixture of goethite, traces of manganite	2.74	0.00343	32.0 ± 0.4	Beginning of the Late Oligocene
27	cryptomelane with slight admixture of lithiophorite	1.90	0.00176	24.0 ± 1.0	The end of the Late Oligocene

and for the ratio $^{38}Ar/^{36}Ar$ 1.5%. The measurements were performed by the method of isotopic dilution. The contents of K in the samples was determined by the method of flame photometry with the error less than 1%.

In Table 10a the results of the sample analysis are listed; these were obtained for several independent measurements with the error determined by their convergence (Varentsov and Golovin, 1987).

From the data, given in Table 10a, it is seen that K–Ar dating of samples, represented by cryptomelane, are subordinated to a definite regularity. The sample 21 (see Figures 8, 10, and 14), located in the lower part of the ore bed of quarry-A, has the absolute age of 41.5 ± 0.5 Ma (Late–Early Eocene). This value reflects the time of formation of the relatively early generation of cryptomelane, and the effect of admixtures of potassium-bearing terrigenous, clay components is rather insignificant (Al_2O_3 1.81%). The sample 24, picked up from the top of the ore bed (see Figures 8, 10, and 14) some 2.5–2.75 m above sample 21, has the absolute age of 32.0 ± 0.4 Ma (the beginning of the Late Oligocene). The quantity of quartz and alumosilicate admixtures in this case is relatively lower (SiO_2 0.77, Al_2O_3 1.78%), this allows for the assumption that almost all the potassium and the products of its decomposition are conserved in the lattice of cryptomelane. These determinations do not contradict the K–Ar dating of sample 27 (quarry-G, middle part of the ore bed, see Figures 9, 10, and 15), according to which the age was 24.0 ± 1.0 Ma. In this case the influence of potassium-bearing alumosilicate admixtures was extremely insignificant too (see Table 2). However, at the interpretation of this determination it is necessary to keep in mind that quarry-G section is located in the marginal western part of the deposit, where the eluvial nature of manganese ore accumulations is relatively clearly marked.

The K–Ar age of sample 13 (lower part of the ore bed of quarry-A south, see Figures 7, 10, 13), determined as 168 ± 20 Ma (the Middle Jurassic) differs sharply from the above data. It is explained by the fact that this sample is composed of almost pure pyrolusite (β-MnO_2), the mineral, characterized by the tunnel structure of the rutile type, in which due to the relatively small diameter of channels there is no space for K^+ ions and other cations (Chukhrov *et al.*, 1989). Potassium available in this sample is associated with essential admixtures of alumosilicate micaceous material, whose presence is indicated by the value of Al_2O_3 6.18%. Any reasonable interpretation of this determination is extremely difficult.

The example of the evaluation of the absolute age of the samples, represented on the one hand by cryptomelane and on the other hand by pyrolusite, shows how important it is to account for the structural features of any potassium present in the mineral for its age determination, as well as the sampling of pure minerals with minimum admixtures.

Thus, the K–Ar dating of cryptomelanes of the deposits was determined to the interval of 41.5–24.0 Ma, that is from the Late Eocene to the end of Oligocene–beginning of Miocene. However, taking into account that if there are alterations in the sequence of the primary carbonate-oxyhydroxide manganese ore matter of Albian–Cenomanian age (97.5–95.0 Ma), there are both more early products, for example, manganite patches, and later varieties (pyrolusite, 10 Å-manganate (todorokite), birnessite, romanechite (psilomelane), lithiophorite – this geochronological range (17.5 Ma) should be expanded. Its upper limit may be the Late Tertiary–Quaternary laterite, soil formation processes.

2.2.1.10. Problem of Genesis

The majority of the Australian geologists (Bolton and Frakes, 1982, 1984, 1988, 1990; Bolton *et al.*, 1984, 1988, 1990; Frakes and Bolton, 1982, 1984a, 1984b; Frazer and Belcher, 1975; McIntosh *et al.*, 1975; Ostwald, 1975, 1980, 1981a, 1981b, 1982, 1984, 1987, 1988, 1992; Ostwald and Bolton, 1990, 1992; Pracejus and Bolton, 1992a, 1992b; Pracejus *et al.*, 1986, 1988a, 1988b, 1990a, 1990b; Slee, 1980) that studied the Groote Eylandt deposits, accept the point of view of Smith and Gebert (1969), with different variations on the shallow-water sedimentary genesis of oxyhydroxide manganese ores.

The above mentioned Australian geologists consider that the primary types of oxyhy-droxide ores (see Table 1) of the Groote Eylandt deposit (Pracejus and Bolton, 1992a) were accumulated under shallow-water conditions at the Albian–Cenomanian time, and that the closest analog of the Groote Eylandt deposits is the Chiatura Oligocene deposit in Georgia (Frakes and Bolton, 1984). Considerable interest in the genesis study of manganese ores of the Groote Eylandt deposits is reflected in works by Bolton and Frakes (Bolton and Frakes, 1982, 1984, 1985; Bolton *et al.*, 1984, 1988, 1990). In contrast to earlier investi-gations of the Australian geologists (McIntosh *et al.*, 1975; Ostwald, 1980, 1981, 1982; Slee, 1980; Smith and Gebert, 1969), Bolton and Frakes, as well as Pracejus and Bolton (1992a), assume that only pisolite and oolite ores and some sand ores should be considered as the Albian–Cenomanian primary ores (sedimentary-diagenetic). All other types of ores are, according to their opinion, secondary (that is, later and associated with weathering, dissolution, and redeposition) taking place during the lateritization process. The evidence for the 'primary' origin of pisolites is: (1) features of their structure, different from the pedogenic pisolites; (2) their occurrence in cross-bedded thin complexes; (3) abraded pisolites, forming thin intercalations at definite levels in the ore bed, are observed in relatively thick pisolite layers. The pisolite formation is considered as an event, associated with sedimentation of oxyhydroxides, with a lesser degree of manganese carbonate during the oxygeneration phase, i.e. establishing a normal oxygen regime in the water of the basin, that takes place after the accumulation of dissolved manganese due to long period of widely developed stagnation. This concept is confirmed by regional stratigraphical data: at the basin center the given ore varieties are identified (the late age of the black shales complex) as sediments of stagnated oxygen-free environments, indicated by the paleontological findings, e.g. foraminifers, that were observed below the ore, and the Late Cretaceous remnants of planktonic organisms in the ore zone.

The above results of the studies and the available factual material may be unambiguously interpreted, considering, that the Groote Eylandt deposits represent the locally redeposited products of deep laterite weathering, developed after the Lower–Middle Cretaceous manganese-bearing clay-marl sediments, containing Mn carbonates, and to a lesser degree Mn oxyhydroxides. The determinations of the K–Ar age of cryptomelane ores of the Groote Eylandt deposits discussed earlier (Late/Middle Eocene–Early Miocene/Late Oligocene, 41.5 ± 0.5 to 24.0 ± 1.0 Ma) correspond to the determinations of the absolute age of the laterite weathering crusts of different regions of Australia, made by the paleomagnetic method (see review, Indurm and Schmidt, 1984). In this context the conclusions are of interest, that the Late Paleozoic and Mesozoic rocks of the Pert Basin, Western Australia were subjected to intensive chemical weathering during the Late Tertiary (Schmidt and Embleton, 1976). The most extreme laterite processes were found in the northern part of the Pert Basin. The paleomagnetic determinations performed by the authors show that the period of deep weathering is spanned by the interval from the Late Oligocene to early Miocene. These results correspond very well to other data on the wide regional lateritization in the Northern Australia. The sharp polarity of the natural residual magnetization testifies in the favor of this processes duration.

The wide development of the intensive weathering during the Middle–Late Tertiary was established by the investigations of Schmidt and others (Schmidt *et al.*, 1976) in southern Australia. On the basis of study of the regional material the authors come to an important generalizing conclusion: the Australian continent as a whole was subjected to deep chemical weathering during the Middle–Late Tertiary.

Later the conclusions by Schmidt and others (Schmidt *et al.*, 1976) were refined by Indurm and Senior (Indurm and Senior, 1978), who investigated several regionally separated weathering profiles in southwestern Queensland. It was established that the age of the relatively ancient profile corresponded to the range from Maastrichtian to Early Eocene, and the relatively young one – dated by a wider range – corresponded approximately to Late Oligocene (30–15 Ma). The authors point out that the given datings do correlate with the time intervals of high paleotemperatures, determined by the oxygen isotopes of the marine fauna fossils in the section of sediments of the south part of the Indian ocean. The relatively young age of the deep weathering is related to the

Early Miocene ocean warming up, and that does not contradict the given age range, taking into account the error range of the dating interval (30–15 Ma). The position of the first paleomagnetic pole (later weathering) differs slightly from the pole positions, determined earlier for the deeply weathered rocks of the western, southern, and northern regions of Australia, and this confirms the hypothesis of the existence of wide continental Middle Tertiary weathering on the Australian continent.

It was possible to compare the conclusions of Schmidt *et al.* (1976), Indurm and Senior (1978) with the lateritization intervals of the Indian peninsula. During the study of the paleomagnetic age of the deep weathering profiles at 25 regions of the Indostan peninsula (Schmidt *et al.*, 1983) it was found that laterites, occurring at present at considerable absolute levels, were formed in the Late Cretaceous and in the beginning of the Tertiary, while the weathering crusts, developed on plains and low hills near the ocean coast were formed in the Middle and at the end of the Tertiary period. These age determinations correspond very well with the above data on the laterite age of Australia, with which India composed the unit continental plate 53 Ma ago (Early Eocene).

In accordance with the solution of the problem of the genesis of the Groote Eylandt deposit it is necessary to point out that the study of laterites (in particular Late Cenozoic) developed on different continents, shows the relatively single-type residual profile, composed by the following sequence (from base to top): parent bedrock – transitional zone – saprolite – laterite – residual soil, and this is characteristic for these accumulations. The similar profile is typical for vast plain territories with periodical alternations of dry and humid, or a predomination of humid environments without any visible development of the slope processes.

However in the case of local slope redeposition at the areas of accumulations of the crust material, its residual autochthonous nature turns out to be destroyed and deeply altered with time. Under these conditions the weathering profile reflects the sequences of sediment depositions and accompanied infiltration events.

The essential problem of the genesis of laterite weathering crusts, both residual and allochthonous, is the formation of pisoids (pisolites) as the most important structural component of these accumulations. The data available at present (Alexander and Cady, 1962; Banerdji, 1981; Beadle and Burges, 1953; Chowhury *et al.*, 1965; Chukhrov *et al.*, 1984; Proceeding, 1981; Varadan, 1979) on occurrence conditions, morphology, structure, composition, and chemistry lead to the conclusion that their origin cannot be explained by the mechanisms of genesis alone. This is especially justified for the long formed redeposited products of the weathering crusts, reflecting different zones of the laterite profile and different stages of alteration (maturity), in which there are both development stages of authigenic pisoids, including growths, crusts, plates, concretions, and their clastic derivates. It is necessary to add to the above the great diversity of alteration of the material of the soil laterite and laterite profile, associated with the ground water activity. Both these types of laterites may be visible together.

The results of the study of laterite crusts, illustrated by the example of Western Africa, indicates that pisolites, forming at the profile base, generally incorporate and include the products of relatively early weathering stages as well as unweathered minerals. In the case of the Groote Eylandt deposits such unstable minerals, occurring at the base of the ore bed, near the contact with the parent bedrocks are represented by hydromica (to 10%) with the general predominance of kaolinite. The relict manganite may play a similar role. And finally these minerals may be found at the profile top, where the pisolite accumulation takes place due to the removal of interpisolite material. In the works of a number of authors preference is given to mechanical removing, that is carried out due to climate change and the removal of protecting vegetation; this favors the penetration of rain waters and the washing out of relatively light thin material, leaving the heavy pisolite remnants to accumulate.

Thus, it may be assumed that the influence of the deep laterite weathering on the Early–Middle Cretaceous beds and the accumulation of its products took place in a relatively wide geochronological interval (main stages: Middle–Late Tertiary), accompanied by local redeposition of deeply altered matter in the form of cryptomelane and to a lesser degree pyrolusite material. Cryptomelane is a relatively stable and quickly preserved product of manganese carbonate oxidation, but in a number of cases it is possible to observe, that cryptomelane is developed after manganite and preserved in admixture amounts (Ostwald, 1980). Pyrolusite and in the areas of deep laterite alterations (upper zones of the weathering profile) – lithiophorite, 10 Å-manganite (todorokite) with the admixture of birnessite, nsutite, romanechite (psilomelane) are developed after cryptomelane of the early generation under supergenous conditions. Some distribution symmetry of the ore types, their mineral and chemical composition in the section of the ore bed, may be interpreted as an indication of the local redeposition of manganese crust at the beginning of the relatively late weathering stage. According to Bolton and Frakes (Bolton and Frakes, 1982, 1984) an important argument for the substantiation of the primary-shallow water (sedimentary-diagenetic) nature of the pisolite facies of the Albian–Cenomanian age, represented by manganese oxyhydroxides (pyrolusite, cryptomelane and romanechite) are the textural features of pisolite layers: cross-bedding, inverse and normal character of the gradational sequence of the pisolite dimension distribution, as well as a number of other features.

Nonetheless this data does not contradict the concepts on the probability of the local redeposition of pisolite products of the deep weathering within the same basins of the general sedimentation depression in the western, northern, and southwestern parts of the Groote Eylandt island, where the Early–Middle Cretaceous manganese-bearing sediments up to 100 m thick were accumulated.

Additional light is given to the solution of the genesis problem of the oxyhydroxide manganese pisolites by the above determinations of K–Ar age of cryptomelane ores, underlying the pisolite layer (sample 21, quarry-A, see Figure 14, 41.5 ± 0.5 Ma) and those covering it (sample 24, 32.0 ± 0.4 Ma). These datings seem to indicate the Late Eocene–Early Oligocene age of the accumulations of the pisolites considered and it corresponds to the results of paleomagnetic age determinations (Indurm and Senior, 1978; Schmidt and Embleton, 1976; Schmidt *et al.*, 1976, 1983) of the regional lateritization of the Australian continent in the Middle–Late Tertiary time.

Thus, the available evidence indicates, conclusively two main stages in the deposits formation: (1) Albian–Cenomanian – the deposition of shallow-water marine, essentially carbonate-clay manganese-bearing, sediments took place, the formation of which was satisfactorily explained by the model of the intercraton basin during regressive changes of the stagnation regime to environments of a normally aerated basin (Bolton *et al.*, 1990; Frakes and Bolton, 1982, 1984); (2) Middle–Late Tertiary. During this time the high-grade oxyhydroxide manganese ores were formed. The main role belonged to processes of deep chemical weathering (lateritization). It is quite possible that similar supergenous alterations began in the Maastrichtian, and their development continued to the present time. Deposits of iron-manganese ores in the area of Ripon Hills and manganese ore of Pilbara Province, Western Australia (de la Hunty, 1965; Denholm, 1977) could be close to the Groote Eylandt deposit supergenous in nature. These deposits were formed due to the Cenozoic, and possibly earlier deep laterite weathering of the Proterozoic ferruginous-manganese-bearing shales. As a result of these processes the areal deposits of supergenous Fe–Mn ores were formed at the upper zone at the depth of about 11 m (30 m) from the surface. The concentrations of Mn and Fe are decreased below 11 m. The essential ore accumulation process is the wide but rather irregular substitution of these parent rocks by Fe and Mn oxyhydroxides, during leaching and intensive removal of other components. The ores are represented by fine-grained pyrolusite, occurring in close intergrowth with fine-grained hematite; lithiophorite aggregates are observed as well, and the relicts of deeply altered bixbyite and braunite were found. The enclosing sediments are made up by deeply altered essentially kaolinite material. It is important to point out that the initially Mn and Fe enriched parent rocks (protores) served as the main source of these metals during the formation of supergenous high-grade ores. As long as the depth and intensity of the substitution of the substrate rocks by Fe and Mn oxyhydroxides vary, the considerable variation in Mn and Fe contents is observed in different deposits. If one takes 15% of Mn for the threshold of ore concentrations, then the average contents in such ores are: Mn 19.5%; Fe 26.0%, and the ore reserves of this grade is 56 mln t.

The data given allows us to draw the conclusion, taking into account the similar double-stage formation of initially relatively low-grade carbonate-oxide ores and the successive formation of the products of deep chemical weathering after them, that the Groote Eylandt deposits may be compared to known shallow-water marine manganese ore accumulations, in particular with the Oligocene manganese deposits of Ukraine, Georgia, Bulgaria, and other regions.

By the character of the interrelations of manganese accumulations with the initial rocks, mineralogy, and structural-textural features of the ores, the Groote Eylandt deposits may be compared, to some degree, with the Moanda deposits of the Republic of Gabon considered below. In works by Bouladon and Weber *et al.* (1965), Leclerc and Weber (1980), Perseil and Bouladon (1971) it is shown that the largest accumulations of oxyhydroxide manganese ores (essentially pyrolusite, cryptomelane, lithiophorite) of Western Africa are the products of deep laterite weathering of initially poor manganese-bearing rocks: Precambrian (Francevillian) black carbonaceous shales, containing manganese carbonates (Mn, 13.5%). The manganese deposits, associated with developed weathering crusts, are relatively insufficiently described in literature. It is known that the largest reserves of the high-grade manganese ores in India, West and Central Africa, and Brazil are associated with deposits of such type.

2.2.1.11. Conclusion

The deposit of high-grade oxyhydroxide manganese ores at Groote Eylandt is the largest in Australia. Manganese ores are contained in bed-like body of complicated structure from 2 to 20 m thick, with an average thickness of 3 m. They are developed in the western and southwestern parts of Groote Eylandt in the form of a band 6 km wide and 22 km long. The oxyhydroxide manganese ores occur disconformably on the underlying rocks. The oxyhydroxide ores are represented in the main by oolite, pisolite, large-fragment, lump, and sinter-earthy textural varieties. They are composed essentially of cryptomelane and pyrolusite with the admixtures of lithiophorite, todorokite, nsutite, and romanechite (psilomelane). The enclosing sediments are represented by clays of kaolinitic composition with an admixture of quartz sand.

On the basis of the main section study of the deposit, and analysis of the available material, the manganese oxyhydroxide ores of the Groote Eylandt deposits are considered as the products of the local redeposition of developed weathering crust, formed after the Lower–Middle Cretaceous manganese sediments. The Groote Eylandt deposits belong to the group of manganese deposits, associated with laterite weathering crusts. The results of determinations of K–Ar age of cryptomelane ores point to the interval of the Late/Middle Eocene–Early Miocene/Late Oligocene (from 41.5 ± 0.5 to 24.0 ± 1.0 Ma) that corresponds to a time of wide regional development of the laterite weathering over the Australian continent. Rather clear representatives of deposits of this type are known in Western Africa, for example, Moanda deposits, Republic of Gabon, in Brazil, and in other regions.

2.2.2. MODEL TEST AREA: MOANDA DEPOSIT, GABON, WEST AFRICA

2.2.2.1. Introduction

The Moanda manganese deposit is the largest in West Africa (Figure 38): the reserves of high grade commercial oxyhydroxide ores are evaluated in 200 mln t (Leclerc and Weber, 1980). At present the investigators who studied this deposit, consider that the formation of high grade oxyhydroxide ores is associated with processes of deep supergene alteration – lateritization of the Precambrian manganese-bearing black shales, the Fransvillian Series, attributed to the group of iron-siliceous formations (Bonhomme *et al.*, 1982; Bouladon *et al.*, 1965; Bros *et al.*, 1992; Gauthier-Lafaye and Weber, 1989; Hein and Bolton, 1993; Hein *et al.*, 1989; Nahon *et al.*, 1983; Perseil and Bouladon, 1971; Weber, 1968; Weber *et al.*, 1979). High grade manganese ores are developed, essentially, on a number of plain uplifts, among which two plateaus are well studied: Bangombe and Okouma. Sequential gradations of supergene alterations into high grade oxyhydroxide manganese ores were penetrated by numerous mines and exploration bore holes. The age of the Fransvillian Series of the Lower Proterozoic is 2.095 ± 0.045 Ga (Bonhomme *et al.*, 1982; Bros *et al.*, 1992). Stratiform manganese carbonate accumulations alternate with black shales, dolomites, and sandstones with dolomite and pyrite cement. In the upper part of the patches of Mn mineralization is developed with sandstones, black shales, and to a lesser

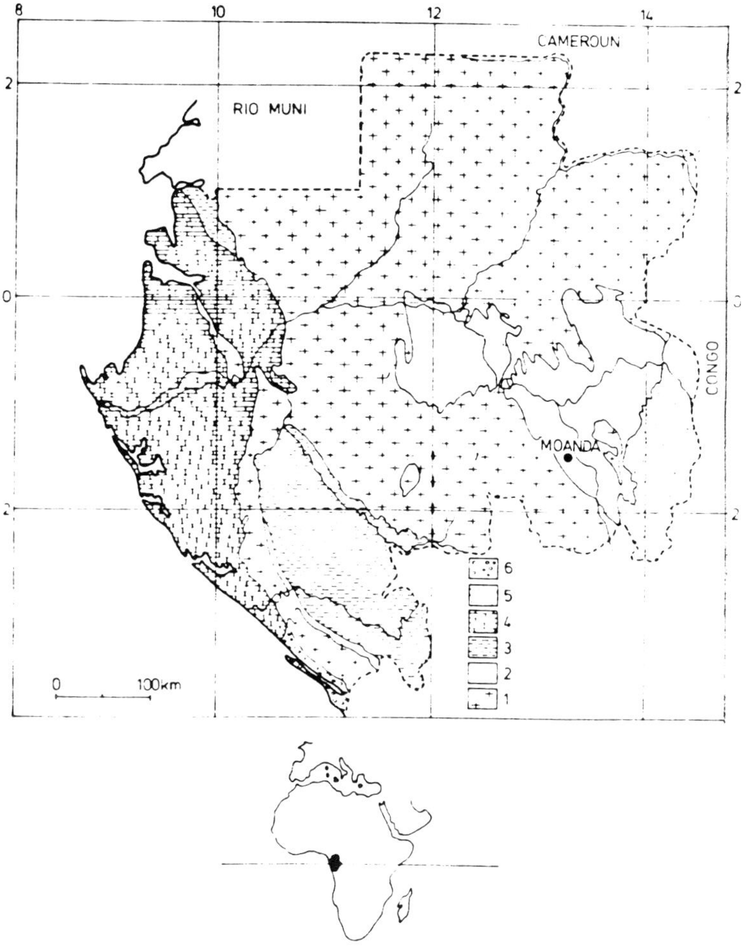

Fig. 38. Geological sketch of Gabon. 1. Lower Precambrian; 2. Middle Precambrian; 3. Upper Precambrian; 4. Mesozoic and Cenozoic; 5. Tertiary of the Bateke Plateau; 6. Quaternary (Leclerc and Weber, 1980).

degree volcanogenic rocks and banded siliceous rocks (cherts and quartzites) associated. According to Weber (Leclerc and Weber, 1980; Weber, 1968) in the representative section of the Bangombe Plateau the following units stand out (from the base): (1) Rocks of the proper Fransvillian Series – black shales with accumulations (stratiform bodies, lenses, and concretions) of rhodochrosite (in some places with dolomite and pyrite); in some areas near the border with covering sediments of the weathering crust rhodochrosite amounts may reach more than 10%; the weathering crust: (2) the pyrolusite interbed (2– 5 cm), separated from the underlying rhodochrosite by a thin manganite interlayer, occurs at the base; (3) discontinuous, intermittent lens-like beds of manganese oxyhydroxides

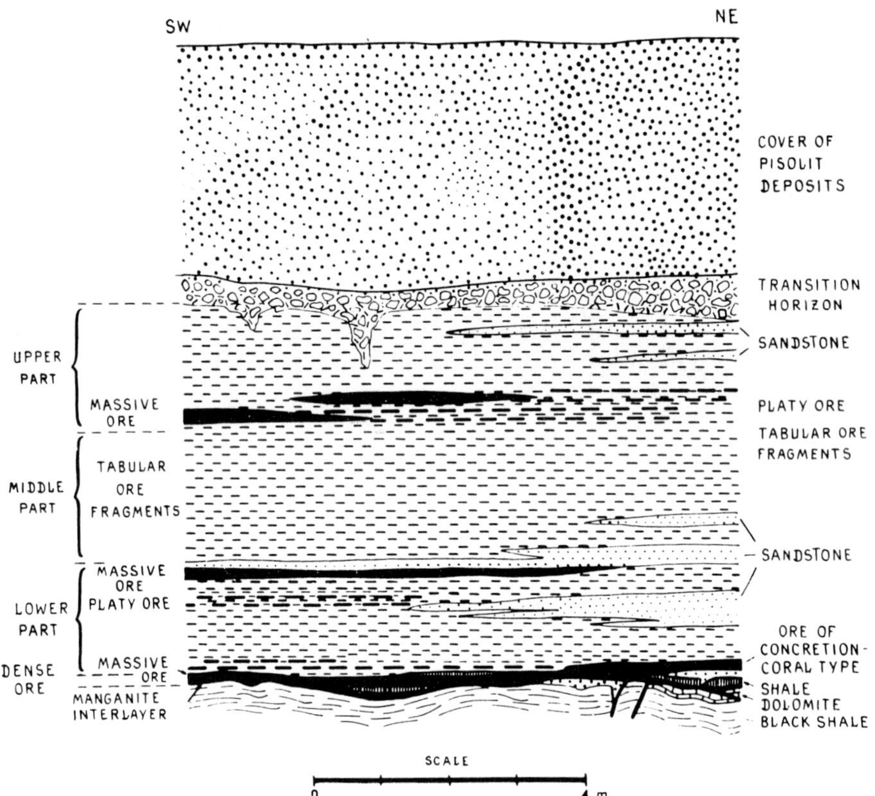

Fig. 39.   Schematic weathering crust profile, Moanda deposit, Gabon (Bouladon *et al.*, 1965).

(cryptomelane, pyrolusite, nsutite, and lithiophorite) 3–9 m thick, alternating with fine-grained ferruginous rocks; (4) ferruginous sandstone (to 60% of Fe_2O_3) (Hein and Bolton, 1993; Leclerc and Weber, 1980; Weber, 1968) and a conglomerate bed (0.5–1.0 m), composed essentially of hematite, goethite and cryptomelane with an admixture of quartz and clay matter (kaolinite); (5) pisolite unit 5–6 m thick; pisolites are represented by gibbsite, goethite, and cryptomelane; the matrix is composed of the same minerals, as well as of quartz, kaolinite, and lithiophorite. The commercial ores are mined below the third and fourth units.

2.2.2.2. *Geological Setting: Structure, Lithology, Mineral Composition, Geochemistry of the Ore-Bearing Horizon*

In the profile of deep chemical weathering of black manganese-bearing shales of the Fransvillian Series (Figure 39) the following units (zones, interbeds, levels and so on) may be distinguished by progressive development of the processes of the primary material alteration (Leclerc and Weber, 1980; Weber, 1968; Weber *et al.*, 1979).

1. Unaltered black shales with dispersed concretion aggregates of multicomponent manganese carbonates, the composition of which may be characterized by the following formulae: $(Mn_{0.64}, Ca_{0.22}, Mg_{0.13}, Fe_{0.01})CO_3$ and $(Mn_{0.45}, Ca_{0.26}, Mg_{0.29})CO_3$ (Boeglin *et al.*, 1980; Nahon *et al.*, 1983). The main components of the Precambrian sediments are illite, manganese carbonate (with average Mn 13.5%), pyrite aggregates, and some amounts of clastic grains of quartz. A significant feature is the enrichment of these rocks in organic matter, on average about 7%, and in some places more than 20% of the total organic carbon (Hein and Bolton, 1993).

2. Slightly altered black shales. The initial stages of alteration are distinguished with depth, for example, at about 10 m in a number of holes in the carbonate nodules a rapid increase of the relative amounts of manganese and the substitution of Ca–Mg–Mn-carbonate by clear crystallized rhodochrosite is noted. Usually rhodochrosite is observed in the form of ovoid microconcretions with sharp outlines. Such microconcretions consist of three or four intergrown crystals, but sometimes they may be represented by rhodochrosite monocrystals. The matrix of microconcretions is composed of illite, goethite, and silty quartz. The average chemical formula of such rhodochrosite (Nahon *et al.*, 1983) is: $(Mn_{0.991}, Ca_{0.007}, Fe_{0.001}, Mg_{0.001})CO_3$. This change takes place both below and above the zone of pyrite oxidation without any alteration of the rock structure. Twofold increase of the manganese content takes place owing to such transformation. In rather large, slightly drained areas, in particular, in the center of the Bangombe Plateau the distinct recrystallization of rhodochrosite keeps pace with partial alteration of the rock microtexture and with the preservation of its microscopic structural features. The recrystallization is accompanied with manganese enrichment. In such settings the thickness of the Mn carbonate sediment, with transformations of such type, is up to several dozens centimetres. They are developed below the compacted basal layer of the proper supergene deposit (see Figure 39).

3. The zone of manganese leaching. The position of the weathering profile horizon is changed depending on the area relief. At the well-drained slopes of the hills manganese is considerably leached, and the black shales are converted into black clay, whereas the higher levels are altered into yellow clay, represented by illite, kaolinite, goethite, and quartz. In these rocks the manganese content does not exceed 1%. At the ares of relatively limited drainage the thickness of the zone of manganese leaching is less than 20 cm.

4. Productive zone.

4a. The compacted basal layer (thickness 90–50 cm). At the contact with the rhodochrosite-bearing black shale, as a rule, the thinnest intercalation (several mm) of manganite occurs, which is overlain by a pyrolusite interbed 2–5 cm thick. The border between the rhodochrosite-bearing rock and manganite intercalation is rather distinct; this is emphasized by a number of investigators (Bouladon *et al.*, 1965; Leclerc and Weber, 1980; Nahon *et al.*, 1983; Perseil and Bouladon, 1971; Weber, 1968; Weber *et al.*, 1979). It is of value to point out that this border cuts disconcordantly the thin-layered black shale. The observed preservation of oval and egg-like outlines of the microconcretions even after their alteration into manganite is of interest. However, in some places the border of the manganite intercalation may also cross the microconcretions. The matrix of the manganite microconcretions consists essentially of kaolinite and goethite. Illite is present at this level as an insignificant admixture, while organic matter is practically absent. The weathering, observed in the profile base and developed around the borders of the manganite microconcretions and particularly of the secondary manganite growths, retaining the perfect optical homogeneity (orientation) with the substrate manganite, is of considerable importance for the understanding of the stage of formation of the manganese oxyhydroxides supergenous accumulations. It is emphasized that similar covering growths are the results of relatively later manganite deposition. The average chemical composition of the manganite, determined with the electronic microprobe, is close to Mn OOH (Nahon *et al.*, 1983): SiO_2 0.20; Al_2O_3 0.15; Fe_2O_3 no; Cr_2O_3 0.05; Mn_2O_3 81.74; FeO no; MgO 0.01; CaO 0.01; Na_2O not determined; K_2O not determined; TiO_2 0.02; H_2O 17.82; total 100.00%.

The pyrolusite interbed is overlain by the massive stratified ore, composed essentially of X-ray amorphous manganese oxyhydroxides with manganite, groutite, polianite (clear crystalline pyrolusite), and nsutite. The observed basal layer replicates, concordantly the stratification features of the shales. But the pyrolusite interbed often crosses the shale stratification. In some places in the pyrolusite interbed there are growths of the concretion aggregates in the form of 'lime nodules' or 'loess dolls' (Leclerc and Weber, 1980), that often penetrate into the underlying substrate rocks. In such cases the initial lamination of the shale is preserved both in the pyrolusite interbed and in these concretion intergrowths. The observed relationships suggest the whole substitution of the shale by almost pure manganite, altered later into pyrolusite. It is emphasized (Leclerc and Weber, 1980) that in these cases the position of the basal ore layer corresponds to the top of the ground water horizon.

In some places of the deposit, where the rocks are represented by decarbonatized shale, the compacted basal layer is also developed. However, in these cases the manganite interbed is not preserved at the base of the pyrolusite layer. The laminated fragments of manganese oxyhydroxides are developed, instead of it, under the basal interbed in the decarbonatized shale. The formation of a new basal layer under its relatively ancient equivalent is sometimes observed that is associated with the change (depression) of the ground water level.

4b. The horizon of laminated fragments (thickness 3–9 m). The ores of this level, located above the basal layers, form the main productive horizon of the deposit. The tabular ore fragments, as a rule, have laminar texture, their thickness varies from one to several cm, such fragments are enclosed into ochreous matrix, containing small ore pisolites as well. The ores are usually developed in the form of horizontal layers. In those places, where karst cones or depressions are developed, such plates are oriented vertically, and they are mixed with the aggregates from the higher levels (in particular, from the transitional (4c) and pyrolusite (4d) levels). It is pointed out that the presence of the tabular ores, as well as the boulder, and large lumpy varieties and the fragments of pyrolusite interbed in the horizon composition may indicate the ancient destruction and redeposition of ores from the compacted basal layer (Leclerc and Weber, 1980). The clearest evidence of such redeposition is the concretion intergrowths–'lime nodules', distinguished at different levels of the ore bed. But as it was mentioned above, they were formed initially in shale below the ore bed base. Within the horizon of the tabular ore fragments at different levels there is an alternation of thin sandstone interbeds, cemented by manganese oxyhydroxides, and the red ferruginous shale. The main components of ores of this horizon are mixtures of amorphous or cryptocrystalline Mn oxyhydroxides, as well as polianite (pyrolusite), lithiophorite, nsutite, and cryptomelane. The ochreous material of the matrix is represented by the mixture of iron oxyhydroxides and alumina.

4c. The transitional horizon (thickness 0.5–1.0 m). The ores of this level differ from the underlying ones by the absence of laminated texture and in the sharp heterogeneity of the composing components. These components include fragments of the tabular ores, aggregates of pisolites, cemented by concretion cryptomelane, fragments of sandstones and crusts, composed of manganese and iron oxyhydroxides. In the areas of the depressed relief and on the plateau slopes these accumulations acquire the nature of compacted hard ores owing to their cementation by the concretion cryptomelane. Such zones of compacted, rather hard ores are sometimes distributed to the underlying horizon, where there is cementation of the tabular fragments and small pisolites. Formed in such a manner the crusts or 'caps' are often cut by water channels, draining the plateau, which have caused cliffs and rocks, composed by the massive ore, and accumulations of ore boulders on slopes to be formed.

4d. The pisolite subhorizon (thickness 5–6 m). These accumulations cover underlying ore levels (4b and 4c). They are represented by spheroidal pisolites ranging from 2–10 mm, enclosed in a yellow ochreous earthy matrix, composed of goethite and gibbsite. The pisolite cores usually include fragments of ore, around which is observed an alternation of concentric layers of gibbsite, goethite, and lithiophorite. The manganese content in this productive subhorizon is about 15% (Leclerc and Weber, 1980).

4e. Clay-sand horizon (thickness 0.1–0.5 m). The specific feature of these supergene sediments is their considerable depletion in manganese due to leaching, but sometimes they contain small amounts of relict pisolites in them.

In the diagram of distribution of rare earth elements (REE), normalized to chondrite (Figure 40, Hein and Bolton, 1993) a distinct single-type feature and extraordinary similarity of the curve configurations for the rhodochrosite accumulations and Mn oxyhydroxides products of the weathering crust of the Moanda deposit is displayed. Such behavior of the REE indicates the inherited nature of their distribution in products of the laterite alteration of parent rocks (Fransvillian Series) that as a whole is rather typical for the weathering crusts (Boiko *et al.*, 1986; Burkov and Podporina, 1967; Duddy, 1981; Geochemistry of elements-hydrolisates, 1980; Nesbitt, 1977; Podporina and Burkov, 1977; Popov *et al.*, 1967).

2.2.2.3. *Genesis: Ore Formation as a Result of Deep Supergene Alterations (Lateritization)*

The analysis of the data of the geologists who studied the Moanda deposit (Bouladon *et al.*, 1965; Hein and Bolton, 1993; Hein *et al.*, 1989; Leclerc and Weber, 1980; Nahon *et al.*, 1983; Perseil and Bouladon, 1971; Weber, 1968; Weber *et al.*, 1979) gives ground to

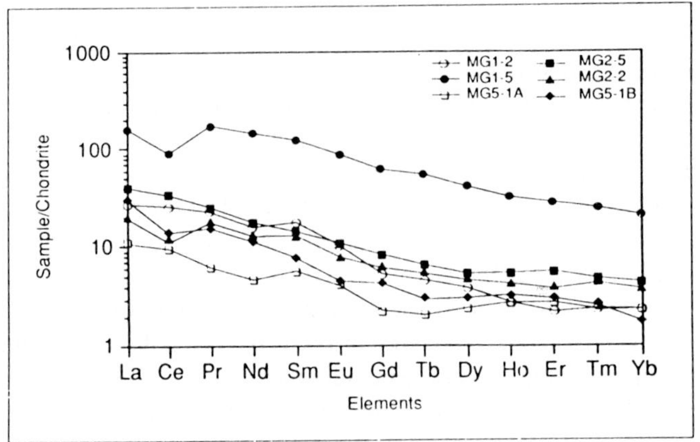

Fig. 40.   REE plots relative to chondrites for Moanda oxide ores, except MG5-1B which is for rhodochrosites (Hein and Bolton, 1993).

distinguish a number of stages in the formation process of the considered laterite ores.

1. The alteration of manganese carbonate. In the lowest horizon of the weathering profile (see above horizon 2), significantly below the ground water level in a relatively reducing environment, the alteration of three component carbonates – (Mn, Ca, Mg)CO$_3$ – into well crystalline rhodochrosite proceeds. The characteristic feature of this stage is in formation of a fairly enriched variety from the carbonate which is relatively depleted in manganese. It is of interest that iron does not take part in these processes under these conditions at the clearly distinguished mobility of manganese ions and complexes. It was emphasized above that iron was present in these slightly altered black shales in the form of pyrite. Thus, manganese at the most early weathering stage is the main ore-forming element, capable of migration in the lower horizon of the weathering profile.

2. Oxidation of pyrite. The further enrichment of the forming carbonates in manganese, without any significant destruction of the shale structure, continues near the upper level of the ground water in the zone of pyrite oxidation.

3. Manganese oxidation. Rather complicated multistage processes of the manganese oxidation go on above the ground water level. The most important place in this process is occupied by the formation of pyrolusite, continuing through a number of intermediate transformations and phases. The relatively complete substitution of the shale by pyrolusite may be considered as the process completion (Leclerc and Weber, 1980; Weber, 1968; Weber *et al.*, 1979).

As it was pointed out (see above the ores of the productive zone (4)), the earliest stage of manganese oxidation is the manganite formation during rhodochrosite (MnCO$_3$) weathering. The significance of this event was emphasized by the geologists who studied the formation of manganese ores in tropical and equatorial zones: the rhodochrosite weathering in deposits Nsuta of West Africa (Perseil and Grandin, 1978), Moanda (Nahon *et al.*, 1983; Perseil and Bouladon, 1971), or in some regions of Brazil, at the Conselheiro Lafaiete deposit (Buttencourt, 1973; Nahon *et al.*, 1983), Serra do Novio, Amapo (Rodrigues *et al.*, 1986), Lucínio de Almeida, Bahia (Basilio and Brondi, 1986), Miguel Congo, Minas Gerais (Barcelos and Büchi, 1986), Urucum, Corumbá, Mato Grosso do Sul (Haralyi and Walde, 1986), Azul, Serra dos Carajas, Pará (Coelho and Rodrigues, 1986), Serra de Buritirama, Pará (De Andrade *et al.*, 1986). Manganite may be the weathering product of different manganese silicates as well.

Such manganese oxyhydroxides as manganite and groutite were obtained as the result of experimental modeling of processes of laterite weathering of rocks, consisting of pyroxmangite (75%), rhodonite (8%), rhodochrosite (6%), Mn-amphibole, and serpentine at the first stage of interaction (Melfi *et al.*, 1973; Melfi and Pedro, 1973).

Manganite is developed not only as the substitution product of some initial manganese-bearing minerals, but as an authigenic new product in the form of acicular crystals, filling in fractures and voids, and this is significant for the understanding of the initial stage of oxidation in the weathering profiles of the Moanda deposit, as well as others, such as Nsuta. It also is important to distinguish relatively late manganite growths around earlier substitution areas by this mineral (Nahon *et al.*, 1983; Perseil and Grandin, 1978). However under the conditions of relatively hard weathering manganite is not encountered as a widely distributed product of the supergene oxidation of the earlier manganese minerals. It is observed essentially in the form of interbeds several mm (to one cm) thick, occurring in the base of the weathering profile (general thickness 10–12 m). Such a limited distribution of manganite is explained by the extremely narrow interval of the geochemical parameters, controlling its formation. The manganite formation may be proceeded after such intermediate phase as hausmannite (or hydrohausmannite). Birnessite, cryptomelane, nsutite, and pyrolusite are usually later products of manganite alteration (Eswaran *et al.*, 1973; Roy, 1968). The general process of the oxidative accumulation of manganese in the weathering profile is determined by significant alteration (decrease) of the hydrostatic level due to a change of the ground water circulation. In other words, after some drop of the hydrostatic level, the intensive leaching and oxidation of the rhodochrosite matter goes on. The parent rocks are usually represented by material (protore) preliminary enriched with manganese, and occurring directly under the basal layer. The richest manganese ore beds were formed as the result of oxidation at the place of rhodochrosite aggregates, and later the growth of these oxide accumulations was controlled by the supply of dissolved manganese, that was leached from around the Fransvillian rocks during the decarbonatization of the black shales. The most important role in this process of accumulation of manganese and other transitional metals belongs to their chemisorption concentration, as a rule, accompanied by autocatalytic oxidation (Varentsov *et al.*, 1979, 1980). The partial washing out of clay components due to the plateau draining by surface and ground waters took place during the later stages. Similar water saturation and the hydration of clay components led to their considerable expanding, general volume increase, and corresponding formation of cracking and breaking textures. Owing to these events the oxyhydroxide manganese ore is represented by disordered accumulations of laminated fragments, enclosed in the clay matrix. Similar character of the drainage system distribution (Leclerc and Weber, 1980), controlling the mechanism of ore formation, may be observed in mines and quarries. Some areas of these drainage systems operate to the present time.

    4. Redistribution of manganese. Redistribution or, as defined by G. Bouladon and others (Bouladon *et al.*, 1965), remobilization of manganese above the basal horizon is distinguished in the development of the wide cementation zone. The example of local redistribution of manganese in the weathering crust may be the formation of coat and cementation growth of the cryptomelane composition, i.e. cap. At the same time it is important to point out that the final result of the development of such a deep process of manganese redistribution is the leaching and depletion of the earlier rich ores.

2.2.3. Some West African Deposits (Nsuta, Ghana; Mokta, Côte d'Ivoire; Tambao, Burkina Faso (Upper Volta))

In spite of some differences in the geological position, and settings, as well as the features of composition of both parent, primary and preliminary enriched source rocks (protore), mineralogy, and structure of ore beds of the West African manganese deposits, they can be characterized by clearly distinguished genetic generality of the ore formation. The distinct geochemical trend of alterations and the history of climate and formation of this region's relief were reflected here. It is pertinent to note that the Mokta and Nsuta deposits are located in the climatic zone of modern tropical forest, and the Tambao deposit in the zone of semidesert steppe sachel (African savanna at the north of Burkina Faso). They differ in the scales of mineralization: if Nsuta (Ghana) is included in a number of large world deposits of high grade ores (reserves more than 10 mln t), then Mokta and Tambao are related to considerably smaller ones (Figure 41). The main aspects of geology, mineralogy, geochemistry, and the genesis of these deposits are reported in a number of works by, essentially, French researchers (Grandin, 1976; Grandin and Perseil, 1977; Perseil and Grandin, 1978, 1985; Sorem and Cameron, 1960; and others).

2.2.3.1. Geological Setting

All three deposits (see Figure 41) are located in the laterite weathering zone, developed in rocks of the crystalline basement of the West African Platform, the formation of which took place about 2.0 Ga ago (Black, 1967; Bonhomme, 1962; Rocci, 1965). The basement is composed by the Birrimian volcanogenic-sedimentary Series, containing intrusive bodies of granites. The remarkable feature of the Birrimian Series is the presence in its composition of the manganese-bearing horizons, traced to more than 100 km. These manganese-bearing rocks are associated with sericite schists of tuffogenous nature and with jasper-like quartzites. But the manganese mineralization is developed discontinuously within these horizons in the form of lens-like complexes, enclosed in ore-free, practically manganese-free sequences. The thickness of these lens-like complexes, as a rule, does not exceed several meters, and their lateral persistence is of several dozens to several hundreds meters.

    If the metamorphism of basement rocks at the areas of the Mokta and Nsuta deposits is manifested relatively insignificantly to the upper facies of micaceous schists, then at the Tambao area the rocks are altered into fine-grained gneisses and amphibolites.

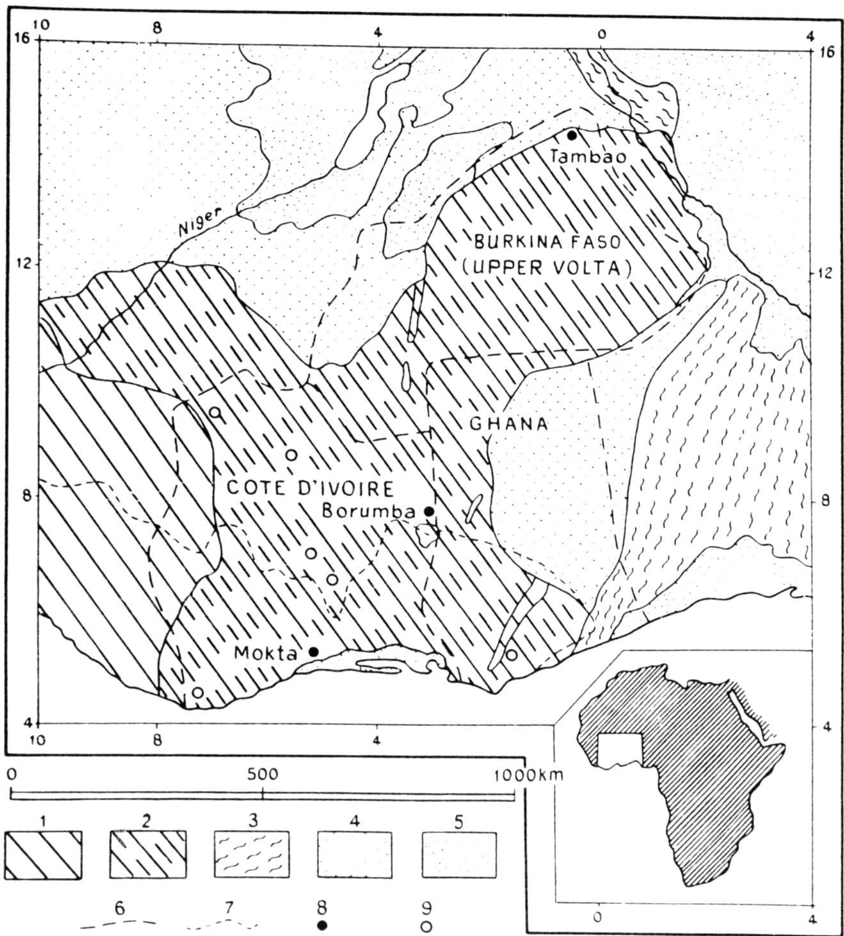

Fig. 41.   Schematic map of main manganese deposits of the weathering crusts, West Africa: Côte d'Ivoire, Burkina Faso, Ghana. 1. Rocks of the Liberian basement, Lower Precambrian; 2. Basement rocks of Côte d'Ivoire (Éburnéen), Middle Precambrian; 3. A chain of Panafrican structures; 4. Sediments of the platform cover; 5. Mesozoic–Cenozoic sediments, essentially continental; 6. Borders; 7. Borders of distribution of dense forests; 8. Studied deposits; 9. Other deposits or occurrences (after Grandin and Perseil, 1977).

The basement rocks of these manganese ore deposits were subjected to transformations, resulted in formation of the rocks, relatively enriched in manganese (protore). The main minerals of the protore are rhodochrosite, spessartine and braunite, sometimes hausmannite, seldom rhodonite, and tephroite. Thus, in protores of Nsuta and Tambao deposits the predominant manganese mineral is rhodochrosite, and at the Mokta deposit spessartine.

All three deposits are located on hill-like uplifts, elevated above gently hilly or flat plains. The characteristic feature of these uplifts is their variations in the depth of slope development as drainage systems. Such uplifts with deposits differ in absolute marks:

(a) for the Mokta deposit the continuous chain of uplifts with the top (140 m) is characteristic;

(b) for the Nsuta deposit a chain of hills with steep slopes and the dome-shaped tops, elevated to 180 m, is characteristic;

(c) the Tambao deposit: elongated hills with narrow ridge-shaped tops and steep slopes, elevated to 350 m, are characteristic.

In spite of the fact that these deposits are separated by almost 9 degrees of latitude, a number of general features are characteristic for them (Grandin, 1976; Perseil and Grandin, 1978, 1985):

1. Relative elevation of the deposits is about 100 m.

2. The material, making up the surface rocks of tops and talus at the uplift slopes, is represented by lumpy and gravel-sandy fragments of ferruginous cap, whose chemical composition is typical for the so called cap of the 'transitional facies'. Such facies of ferruginous cap are distinguished by a number of investigators at the Atlantic coast of West Africa and in the bend of the Niger (Eschenbrenner and Grandin, 1970; Perseil and Grandin, 1978). Its formation proceeded at the end of the Tertiary period and was caused by the wide development of a slope system in the relief of the plain regions. The authors mentioned often determine such forms of the relief as the 'transitional surfaces', with accumulations of litomarges of kaolinite composition.

3. The large outliers and uplifts of different type, in which the relicts of transitional caps are preserved, are located several km (3 to 7) from each of the three considered deposits.

The origin of the hills, at which these three deposits are situated, is determined by the formation of the cut relief on the ancient surface of planation. However, the formation of the relict uplifts took place during a longer interval – to the beginning of the Quaternary time – when relatively high steep slopes were formed. The deposits had a comparable geomorphological and climatic settings up to this stage of the supergene processes. Some features of the differences, associated with the modern climatic zoning, began developing in them later. This is manifested in the known morphological varieties of the hill-like uplifts.

2.2.3.2. *Mineralogy and Geochemistry of the Initial Stages of the Protore Alteration*

Mineral and chemical composition of the primary Precambrian rocks (protore), preliminarily enriched in manganese and represented in some cases by Mn-carbonate, and in other cases by silicate, plays a critical role in the formation of oxyhydroxide supergene ores. The protore significance is displayed especially clearly on the background of mono-type geomorphological development of the deposit regions. The composition of the rocks controlled the relief formation and the trend of supergene alterations, as well as the distribution of the surface material redeposition. All these processes went on during a certain interval of geological time: from the second half to the end of the Tertiary period, when the conditions of humid climate, favoring the development of the kaolinite weathering crust with the distinctly manifested ferralitization, dominated in the boundaries of a rather wide band (more than 10 degrees in latitude).

2.2.3.2.1. Nsuta. Alteration of the carbonate protore. In the carbonate protore of this deposit the manganese contents do not exceed 30–35%. The most characteristic are complexes, and protore layers, preserved at the lowest parts of quarries. They are represented by carbonate schists rich in manganese, and black limestone sandstones, enriched in carboniferous matter and composed essentially of recrystallized rhodochrosite. The beginning of oxidation is determined by appearance of the dispersed mottled patches of manganite (Table 10b, sample GH-15).

Further transition to the completely oxidized ore is specified by an extremely sharp character. In some cases the alteration of manganese carbonate to completely oxyhydroxide ore is observed at a distance of several dozens of cm through the area of intensive vugular porosity, where voids are filled in with manganite crystals (Perseil and Grandin, 1978). The main components of the oxyhydroxide ores are manganite, pyrolusite, and cryptomelane. As a whole the authors determine the following generalized scheme of the mineral transformation:

Mn-carbonate → manganite –› pyrolusite → cryptomelane → nsutite.

But in some samples, with a considerable predominance of pyrolusite, the partial substitution of this mineral by ramsdellite is observed (Perseil and Grandin, 1978). In other cases of ores rich in manganite there are two or three phases of MnO_2, but pyrolusite is completely absent. This data may be interpreted as an indication that in the alteration of Mn-carbonates into oxide phases the process is controlled by at least three main parameters: (1) pH regime; (2) Eh regime (content of dissolved O_2); (3) composition, chemistry, and masses of circulating solutions. In the weathering crust profile of the deposit where the settings differ considerably both in the composition of the altered protores and in the relationships of these three categories of parameters, then the definite deviations from the general scheme with variation of these factors in time may indicate the local specific features of some transformations. This general conclusion finds confirmation in the identification, on the basis of microtextural relations, of two generations of manganite: (1) early as the product of direct alteration of Mn-carbonates; (2) late, a considerably well-crystallized generation, appearing as the last phase in the oxyhydroxide ore formation.

In the final result of the whole sequence of mineralization the tight association is developed: manganite, nsutite, and cryptomelane.

Alteration of the silicate protore. As it was mentioned above, the main manganese-bearing mineral of the protore of this deposit is carbonate, while garnet-containing quartzite plays a secondary role.

In the general sequence of the supergene alteration of garnets, lithiophorite is formed in the beginning, and cryptomelane later. The process development is favored by the presence of fractures and other disturbances in the rocks, in which concretions of lithiophorite are formed. These concretions at subsequent stages are substituted and altered into

Table 10b.   Chemical composition of manganese ores – products of the protore alteration, the Nsuta and Tambao deposits (after Perseil and Grandin, 1978).

Component	NSUTA				TAMBAO	
	GH15[1]	GH3b		GH16[4]	HV14a[5]	HV5[6]
		Mineral insoluble residue[2]	Soluble fraction[3]			
SiO_2	14.16	4.90		61.00	30.00	15.20
Al_2O_3	1.55	19.60	1.02	18.90	17.81	15.87
Fe_2O_3	0.80	1.75	8.85	3.24	3.80	6.92
TiO_2	–	0.55	–	–	–	0.75
MnO_2	2.39	–	–	7.74	27.50	43.50
MnO	47.14	0.06	30.10	–	8.80	4.30
CaO	3.00	0.11	10.70	0.26	1.63	0.05
MgO	2.40	0.59	5.20	0.31	0.54	0.06
Na_2O	0.16	0.16	–	1.42	0.07	0.12
K_2O	0.08	3.33	–	1.74	0.38	0.32
Li_2O	–	0.37	–	–	–	–
BaO	–	–	–	–	–	2.60
NiO	–	0.55	–	–	–	0.22
CoO	–	–	–	–	–	0.19
H_2O	0.70	0.70	–	3.80	8.90	10.9
CO_2	27.87	–	13.20	–	–	–
Total	–	32.67	69.07	98.41	98.63	100.38
		101.74				

The Nsuta deposit:

1. Sample GH-15. Slightly oxidized carbonate manganese ore. The noted alterations: rhodochrosite, and manganite.

2. Sample GH-3b. Slightly oxidized carbonate manganese ore, mineral insoluble residue.

3. Sample GH-3b. Slightly oxidized carbonate manganese ore; of the analysis extract after weak acid treatment.

4. Sample GH-16. Oxidized ore, strongly leached. The noted alterations: garnet – cryptomelane + goethite.

The Tambao deposit:

5. Sample HV14a. Partially oxidized ore. Protore alteration according to scheme:

garnet ⟶ cryptomelane + goethite / lithiophorite

6. Sample HV5. Oxidized ore. Alteration of Mn-silicate according to scheme:

garnet ⟶ cryptomelane / lithiophorite

cryptomelane and nsutite. The process begins with the appearance of the finest aggregates of cryptomelane, dispersed in the mass of lithiophorite, and during their further maturization a close association of lithiophorite and cryptomelane is observed. Often during the intensive progress of the alteration process of Mn-silicate, lithiophorite, as an initial phase, is not preserved or may not be present at all, and the direct substitution according to the scheme is observed: Mn-garnet – cryptomelane + goethite (see Table 10b, sample GH-16).

The formation features of these minerals allow one to consider that at the initial stage of alteration the hydrolithic decomposition of garnet is associated with intensive leaching, that determines the transportation of hydrolysis products to the areas of lithiophorite formation beyond the garnet grains. It is interesting to point to the presence of the lens-like zones, pockets, and patches, composed of white siliceous matter and developed at the periphery of areas, enriched in manganese oxides.

2.2.3.2.2. Mokta. Alteration of the carbonate protore. At the deposit with rather limited distribution of the rhodochrosite protore, the last alteration stages of this mineral into manganese oxyhydroxide phases are usually distinctly manifested:

groutite → ramsdellite

cryptomelane → nsutite → ramsdellite.

Alteration of the silicate protore. At this deposit the high grade ores were mainly formed by the intensive substitution of garnet-containing quartzites by manganese oxyhydroxides. It is significant that rhodochrosite and braunite are not present in this type of protore in any considerable amounts. Two alteration stages are distinguished (Grandin and Perseil, 1977; Perseil and Grandin, 1978, 1985). During the first stage of garnet alteration hydrolysis reactions take place, that are usually accompanied by leaching and dissolution. There is binding of manganese, released from garnet essentially with aluminium as a residual, slightly mobile component of the hydrolithic interaction: as the result of this lithiophorite is formed. Perseil and Grandin (1978, 1985) distinguish two types of garnet substitution by lithiophorite: (a) 'centrifugal' (autochthonous) – a grain of garnet is substituted from the center to periphery, that responds to quick manganese binding to an oxyhydroxide phase without any considerable transportation; (b) 'centripetal' (allochtonous) – substitution from the grain periphery to its center, that is accompanied by transportation of part of manganese due to dissolution of nearby components. In the result of similar alterations the contours of garnet grains, as a rule, are preserved (Table 11, samples 65, 68). The geochemical properties of transformations of the first stage of alteration of the silicate protore are rather well observed in the comparison of samples 65 and 68 (see Table 11): SiO_2 amounts in the transformation process are decreased by a factor of 3.2, and MnO_2 and MnO quantities are increased by a factor of 10.7 and 47.7, respectively.

Within the second stage the formation of cryptomelane initially proceeds along the silicate and quartz streaks in garnets. Later cryptomelane fills in hexagonal areas in the altered garnet grains. The substitution is usually distributed from the periphery to the center along the microfractures or other defect zones. During this stage within the massive cryptomelane accumulations the laminar aggregates of nsutite are formed (see Table 11, samples 66, 67, and 42). The chemical evolution of the ores, formed at this stage of the protore transformation, is distinctly seen from the comparison of contents of samples 66, 67, 42 (see Table 11, Grandin and Perseil, 1977): the content of MnO_2 is increased from 64.80 to 78.73% with some oxidation increase, $MnO:MnO_2 = 0.111 - 0.072 - 0.071$, whereas the amounts of Fe_2O_3, Al_2O_3, and CaO are considerably decreased.

2.2.3.2.3. Tambao. Alteration of the carbonate protore. The highest grade ores of this deposit (more than 40% of Mn) are mainly associated with the protore, represented by rather pure coarse-crystalline rhodochrosite. In the areas of the oxyhydroxide ores the relicts of preserved textures and the residuals of the primary carbonates occur. The following general scheme of alterations is suggested (Perseil and Grandin, 1978):

Mn-carbonate $\rightarrow$ birnessite $\rightarrow$ nsutite and ramsdellite.

In those cases, when pyrolusite is registered, it, as a rule, is substituted by cryptomelane or nsutite and ramsdellite.

Alteration of the silicate protore. A rather insignificant part of the ores of this deposit is genetically associated with deep weathering of the garnet-bearing quartzite. The latter forms sharply subordinated interbeds, interbedding in the mass of sericite schists, among which a significant place belongs to manganese carbonates. A rather limited quantity of manganese silicates in the section and their slight preservation during the history of the deposit relief formation considerably complicate the distinguishing of its alteration stages into oxyhydroxide manganese ores.

At the first stage of the garnet alteration the same features are distinguished that were considered for the Mokta deposit (Perseil and Grandin, 1978, 1985). The following features are characteristic for the second stage: cryptomelane is developed both after lithiophorite and at the rims of partially altered grains of garnet, the central parts of which were not subjected to hydrolytic decomposition. In the latter case two separated, newly formed phases may occur: lithiophorite and cryptomelane (see Table 10b, samples HV14a and HV5). It should be noted that in sharply oxidizing settings of deep protore weathering there may be either fast transition from the first stage to the second one, or destruction and annihilation of products of the first stage owing to the second stage development. The stage is displayed especially clearly near the topographical surface. The stages of the ore formation may be manifested repeatedly at new levels during the formation of the new relief forms and its rejuvenation. In such sharply oxidizing conditions with the relatively high acidity of the draining waters the formation of cryptomelane takes place.

2.2.3.3. Mineralogy and Geochemistry of the Late Stages of Oxyhydroxide Ore Formation (Manganese Cap (Cuirass))

2.2.3.3.1. General characteristics. The most important genetic feature of the above ores is the presence of deeply altered, limitedly distinguishable textural features and components of primary rocks/protores in them: for example, layered textures, relicts of garnet,

Table 11. Chemical composition of manganese ores, Mokta deposit, Côte d'Ivoire (after Grandin and Perseil, 1977).

Components	Residual lens-like-stratified ores (after silicate protore)						Ores of cuirasses			
	65	68	66	67	42	32	73	80	74	62
SiO_2	83.38	26.20	11.50	9.70	3.50	1.80	1.00	1.00	0.90	0.00
MnO_2	4.35	45.76	64.60	72.86	78.73	84.60	80.90	78.73	65.90	87.56
MnO	0.13	6.20	7.17	5.28	5.66	2.49	7.07	8.10	6.67	7.46
Fe_2O_3	1.73	4.66	3.00	2.41	2.16	1.76	0.50	0.90	3.88	0.24
Al_2O_3	5.52	7.08	4.67	3.54	2.60	2.95	3.42	3.60	9.50	0.38
CaO	0.34	0.35	0.80	0.70	0.23	0.14	0.06	0.06	0.06	0.12
MgO	–	–	0.20	0.30	0.07	0.07	tr.	tr.	0.11	0.02
Na_2O	–	0.15	0.20	0.15	0.21	0.13	0.39	0.45	0.39	0.14
K_2O	–	0.44	0.64	0.32	0.42	0.44	1.46	1.66	1.46	0.43
BaO	–	–	–	–	–	–	–	–	0.14	–
NiO	–	–	–	–	–	–	–	–	0.45	–
CoO	–	–	–	–	–	–	–	–	0.20	–
TiO_2	–	–	–	–	–	–	–	–	0.26	–
Li_2O	–	–	0.10	0.10	0.05	0.10	–	–	0.05	–
H_2O	0.96	6.78	6.30	5.20	4.40	5.50	5.00	5.80	8.80	3.00
L.O.I.	3.30									
Total	99.71	99.02	99.18	100.56	98.03	100.00	99.81	100.30	99.27	99.35

Residual lens-like-stratified ores, progressively enriched in oxyhydroxides.

1. Sample 65. Deeply leached low grade silicate ore with relicts of thin lamination. Initial uncompleted alteration. Manganese oxyhydroxides are developed in the altered garnets in the form of cutting streaks.

garnet ↗ kaolinite
↘ lithiophorite → cryptomelane

2. Sample 68. Oxyhydroxide silicate ore. Transitional stage of enrichment.

garnet ↗ lithiophorite
↘ cryptomelane

3. Sample 66. Oxyhydroxide ore. Transitional stage of enrichment (for mineral composition see sample 68).

4. Sample 67. Oxyhydroxide ore. Transitional stage of enrichment (for mineral composition see sample 68).

5. Sample 42. Oxyhydroxide massive ore; relatively complete stage of alteration with the loss of features of the primary protore; composed of thin layers of cryptomelane and nsutite.

Ores of cap (cuirass)

6. Sample 32. Oxide ore with relicts of the protore components, hard, dense with small concretions. Composition: manganite – pyrolusite – cryptomelane – nsutite – ramsdellite.

7. Sample 73. Rich oxide ore of fine fibrous, felt-like structure, composed of cryptomelane.

8. Sample 80. Rich oxide ore.

9. Sample 74. Rich oxide ore. Nsutite concretions with cavities filled by kaolinite and lithiophorite are enclosed in cryptomelane mass.

10. Sample 62. Rich concretion ore. The distinctive feature of it is absence of kaolinite and iron oxide; composed of the close association of cryptomelane and nsutite.

braunite, rhodochrosite, and ore-free minerals. However, in the deposits considered other ores are also developed, the formation of which is not directly associated with the alteration events of the primary material immediately *in situ*. These ores form so called 'secondary' or relatively late ledge among stratified clay accumulations or in Colluvium. Their formation is associated with the lateral migration of manganese-bearing solutions down the slope.

Such ores form layers 3–5 m thick, occurring parallel to the relief surface and covering sediments many m thick, deprived of any considerable concentrations of manganese. Such ore layers form hard armored horizons – caps (cuirasses) – occurring at the base of ancient ferralite soils (Grandin and Perseil, 1977; Perseil and Grandin, 1978).

The characteristic feature of the cap ore is its concretion-bearing nature and high enrichment in manganese: to 50–60% of Mn or to 80–95% of MnO_2. In rare cases, among the masses of the concretion ore there are relicts of the primary structures (remnants of garnet grains, and layered structures), as well as foreign inclusions (quartz pebbles, ferruginous gravel fragments). These relicts and inclusions belong to the primary components of the soils, from which the given manganese caps were formed together with the relict remnants of lens-like ores, fragments of vein quartz and transitional ferruginous cap (Tables 10 to 11).

2.2.3.3.2. Mokta. The manganese caps are the characteristic features of the deposit (see Figure 41). They make up about 15% of the reserves of high grade ores (see Table 11). Both well-preserved varieties and ores, associated with well-developed soil profile, are distinguished among them. Perseil and Grandin (1978) point out that the given caps are formed on the ancient basements of slopes before the beginning of the relief dissection in the Quaternary period during the time interval, when the deposit was emerged in the relief as a large elevated ridge, whose slope steepness did not exceed 15–20 degrees.

The cap ores are represented by concretion accumulations, in which laminated or fine fibrous and diverse columnar textures are distinguished (see Table 11, samples 73, 80, 74, and 62), as well as breccia-like varieties sometimes being observed. Usually, the residual clay matter of the cap is represented by white or red (ferruginous) kaolinite material. In the cap numerous remnants of plant roots occur, relative to which the manganese oxyhydroxides are the later phases. This fact indicates the relatively late secondary origin of caps and their formation within the soil layer (Perseil and Grandin, 1978).

The most distributed association of the cap minerals was formed in the following sequence:

$$\text{cryptomelane} \rightarrow \text{nsutite} \rightarrow \text{ramsdellite} \rightarrow \text{pyrolusite} \rightarrow \text{cryptomelane}.$$

2.2.3.3.3. Tambao. The cap ores of this deposit are represented by the same varieties characteristic of the Mokta deposit: concretion, hard, or breccia-like. But the typical feature of the given ores is the pisolite varieties. Manganese pisolites sized from 6 to 10 mm are usually composed of cryptomelane, nsutite, and ramsdellite, and have cores of clay material.

The same minerals compose most of the concretion ores. However, in the center of some concretions there are tabular grains of garnet or relicts of braunite, altered into lithiophorite. It is of interest that lithiophorite is also observed in the form of thin films in the mass of concretions, composed of cryptomelane and ramsdellite. Unlike the Mokta deposit, the cap layers do not have any significant commercial value among the Tambao ores. This is due to the fact that the layers of manganese-bearing protore, distinguished by relatively large thicknesses and lateral extension, are developed among the substrate rocks of this region. As a result of their deep supergene alteration, the intensive enrichment in manganese, with the formation of more high grade ores of the residual nature, proceeded. At the same time, more sharp dissection of the relief did not favor the cap development, because its composing accumulations were disintegrated into blocks: fragments were transported and dispersed. Furthermore, these accumulations were weakly disclosed and exposed at the elevated emerged areas, where the accumulation of components of the oxyhydroxide manganese cap usually took place. The covering ferralite soils, not containing any significant concentrations of manganese, were destroyed by the deep chemical weathering. Their spotty relicts are enclosed in the considerably decomposed concretion ores (Grandin, 1976; Perseil and Grandin, 1978).

2.2.3.3.4. Nsuta. The cap ores of this deposit are characterized by a relatively moderate development in comparison with the extreme manifestations in Mokta and Tambao. But in some areas the cap is preserved rather well. The decrease of its lateral distribution is determined by a number of factors: (a) mechanical erosion; and (b) destruction as the result of deep chemical reworking of the underlying altered rocks due to action of draining waters. Under these conditions the cap remains within the ferralite soil, occurring on the rocks, covering the ancient base of slopes.

The main mineral of the ores is ramsdellite. The most distributed association is represented (Perseil and Grandin, 1978):

$$\text{pyrolusite} = \text{cryptomelane} \left\langle \begin{array}{l} \nearrow \text{ramsdellite} \\ \searrow \text{nsutite} \end{array} \right.$$

Lithiophorite is found only in some places.

2.2.3.4. Features of the Protore Alteration (Formation of the Primary Residual Ores)

The mineral parageneses of the ores considered, their history, and the sequence of mineral alteration with all other factors being equal, are controlled considerably by the nature, and composition of preliminary enriched, in manganese, parent rocks of the substrate (protores). The manganese components of the latter are represented essentially by garnet or carbonate (rhodochrosite).

Two types of transformations are distinguished in the alteration of Mn-carbonate. Nsuta:

$$\text{Mn-carbonate} \rightarrow \text{manganite} \rightarrow \text{pyrolusite} \rightarrow \text{cryptomelane}$$
$$\rightarrow \text{nsutite} \rightarrow \text{ramsdellite.}$$

Tambao:

$$\text{Mn-carbonate} \rightarrow \text{birnessite} (\rightarrow \text{pyrolusite}) = \begin{cases} \nearrow \text{cryptomelane} \\ \searrow \text{ramsdellite.} \end{cases}$$

The second transformation succession (Tambao) is developed with the protore alteration. The protore differs both in the relative significant pureness of carbonate and the considerably high contents of manganese. The alteration of Mn-carbonate directly into 7 Å-MnO_2 (birnessite), passing the manganite stage (γ-Mn OOH), may indicate the relatively hard oxidation regime of the environment. Under these conditions well-crystallized ramsdellite is sometimes formed after birnessite.

In the alteration of Mn-garnet, as a rule, the following sequence of mineral transformations is developed: Mn-garnet $\rightarrow$ lithiophorite $\rightarrow$ cryptomelane. The geochemical feature of the early (autochthonous) generation of the lithiophorite, formed after hydrolithically decomposed spessartine, and relatively late generation (filling in the voids and hollows in the leached ores) are shown in Table 12 (Perseil and Grandin, 1985). In the products of the alteration of the silicate protore material two varieties of cryptomelane are recognized. (1) Relatively early, developing as the patches of MnO_2 in the quartz streaks and grains, often in the form of botryoidal, concretion growths. These accumulations are characterized by relatively small contents of admixtures, for example, Fe_3O_3, SiO_2, Al_2O_3 (see samples 1 to 3, Table 13, Perseil and Grandin, 1985). (2) In cryptomelane varieties, developed after garnet, relict islands of spessartine are often encountered. In the relatively complete development of the process, the polyhedral aggregates of α-MnO_2 in the ground mass of quartzite are formed. The genetical specific feature of this variety is displayed by the presence of increased amounts of Fe_2O_3, SiO_2, and Al_2O_3 (see Table 13, samples 4 to 6), that may bear witness to the incorporation of these residual, relatively immobile components during substitution of deeply decomposed spessartines.

Under these conditions pyrolusite very rarely occurs. As it was mentioned above (Perseil and Grandin, 1978, 1985), the direct alteration of garnet into cryptomelane was accompanied in some cases by the subsequent development of nsutite and ramsdellite in this mineral (Mokta).

With all the empirical reliability of the considered mineral transformation rocks, they reflect only to some degree the evolution of the main parameters, controlling the formation processes of the laterite profile of the weathering crust. The spatial heterogeneity of primary rocks/protores, and the diversity and instability of the settings during the development history of the weathering crust profile, allow to mark only the general trends in the mineralization scheme of this type deposits.

2.2.3.5. *Formation History of the Major Manganese Ore Minerals*

Manganite history. In the most typical occurrences the manganite is formed in the initial stage of the carbonate protore alteration in the direct vicinity from the front of intensive oxidation. However at this alteration stage together with this mineral, formed due to the direct substitution of the carbonate shale matter, there are well-crystallized manganites,

Table 12. Chemical composition of lithiophorites of different nature, the Borumba and Mokta deposits, Côte d'Ivoire (after Perseil and Grandin, 1985).

Components	1	2	3	4	5	6
MnO_2	54.04	57.63	58.23	62.10	66.49	60.27
Fe_2O_3	0.14	0.09	0.29	–	–	–
SiO_2	0.54	0.18	0.10	0.18	0.20	0.30
Al_2O_3	23.85	23.87	24.12	19.10	19.40	20.40
Na_2O	0.03	–	–	0.10	0.05	0.10
K_2O	0.08	–	–	0.83	0.60	0.50
CaO	0.06	–	–	0.02	–	–
MgO	–	–	0.01	0.02	–	–
TiO_2	–	–	0.01	0.02	–	–
Cr_2O_3	–	–	–	–	–	–
NiO	–	0.11	–	–	0.30	0.03
CuO	–	–	0.09	2.98	2.80	3.10
CoO	0.17	–	–	–	0.05	–
ZnO	0.05	–	0.03	0.59	0.80	1.02
PbO	–	–	–	–	–	–

Samples 1, 2, 3 (sample EE10), the Borumba deposit. Neogenic accumulation of lithiophorite, formed on account of hydrolysis of garnet (spessartine).

Samples 4, 5, 6 (sample SS26), the Mokta deposit. Neogenic accumulations of lithiophorite in hollows, formed as a result of garnet leaching.

Table 13. Chemical composition of cryptomelanes, formed as a result of substitution of quartz (samples 1, 2, and 3) and garnet (samples 4, 5, and 6), the Borumba deposit, Côte d'Ivoire (after Perseil and Grandin, 1985).

Components	1	2	3	4	5	6
MnO_2	89.62	90.03	92.03	90.08	90.05	93.67
Fe_2O_3	0.15	0.10	0.09	0.33	0.36	0.14
CaO	0.59	0.60	0.47	0.38	0.50	0.29
SiO_2	0.05	0.05	0.09	2.39	0.21	0.41
TiO_2	–	–	–	0.08	–	0.08
MgO	0.02	0.01	0.08	0.03	0.03	0.14
Cr_2O_3	–	–	–	–	–	–
NiO	0.15	0.18	0.18	–	–	–
K_2O	2.49	2.50	1.66	1.48	3.06	0.46
Na_2O	0.06	0.04	0.06	0.09	0.31	0.07
Al_2O_3	0.63	0.60	0.74	2.92	1.74	1.08

that fill in voids of the cavernous ore. The latter is characterized by the contraction structure, determined by that the Mn oxyhydroxides occupy less volume than the corresponding carbonates, after which they were formed. The manganite, filling in the relatively late (secondary) streaks, appeared in the final formation stages of some oxyhydroxide ores, is of considerable interest. It is pointed out (Perseil and Grandin, 1978), that in the same thin section are observed sometimes several generations of manganite. To explain these observations it is assumed, that alongside with the events of the local autochthonous oxidation (*in situ*) with the formation of oxyhydroxide ore of the transitional composition and consequent its alteration into the ore of cryptomelane-nsutite composition, some part of manganese, leached from the carbonate matter, may migrate to considerable distances. In the result of such migration a new relatively late manganite generation may appear, and during the repetition of the appropriate conditions (for example, during new depression cycle of the ground water level) the formation of the whole sequence of oxyhydroxide minerals may take place (pyrolusite, cryptomelane, nsutite, and so on).

Pyrolusite history. In the zone of development of deeply oxidized ores, formed essentially in the carbonate protore, pyrolusite occupies the intermediate position in the general sequence of mineralization. Its predecessor is usually manganite, and pyrolusite is altered in its turn into cryptomelane, after which nsutite or ramsdellite are developed. In some cases pyrolusite is substituted by 10 Å-manganate (todorokite). In this case it is important to emphasize that in the ores of such type pyrolusite may be observed both as the product of Mn-carbonate alteration of the definite stage and as a mineral, filling in relatively late streaks.

In cap (cuirass) ores, occurring in the mature stage of development, as the relatively late accumulations, pyrolusite is present in rather limited amounts. It was mentioned above that the major components of these ores are cryptomelane and ramsdellite. In the formation dynamics of the ferralite profile of the weathering crust, pyrolusite may occupy a definite place. It is essential for understanding this event that the protore alteration begins at the lowest part of the profile, where pH is characterized by somewhat higher values. Later, as the relief surface depresses the solutions in these horizons, they become gradually more acid. At this relatively late stage the cap formation takes place in the soil horizon. Its manifestation in the relief during the Quaternary period is often followed by destruction and removal of the manganese-free material as the part of ferralite soils covered the cap. Ground and intersitial solutions of these soils are characterized by higher acidity, for example in the cap of the Mokta deposit pH of the soil solutions is 4–5 (Perseil and Grandin, 1978).

Considering pyrolusite in the light of the experimental studies (Brickes, 1965; Giovanoli *et al.*, 1975) as a manganese dioxide (β-MnO_2), characterized by relatively high stability in the oxidation medium at 25°C, it is necessary to point out (Giovanoli *et al.*, 1975) the relatively high stability of the γ-MnO_2 (nsutite) at more acidic values of pH. At the same time (Bricker, 1965) cryptomelane is characterized by higher stability in media rich in potassium. The Birrimian sericite schists, containing manganese ore layers and complexes may serve as the potassium source for such mineral-forming solutions.

Thus, pyrolusite occupies a rather definite place in the formation process of deeply oxidized ores, genetically associated with direct protore alterations and especially its carbonate varieties. Similar carbonate Mn rocks decrease and neutralize considerably the acidity of solutions, filtering from higher horizons. The surface ores, particularly the manganese cap, the major components of which nsutite (γ-MnO$_2$) and cryptomelane (α-MnO$_2$), are formed in these horizons.

A sharply different position is occupied by local streaks of large crystalline pyrolusite in ores of the exposing cap. Obviously the above neutralization mechanism of acidic solutions does not explain all the particular features of this mineral formation; but for the interpretation of this fact the locally limited circulation of corresponding Mn-containing solutions as derivatives of acidic surface waters may be assumed. The substitution of similar pyrolusite by cryptomelane, after which nsutite in turn is developed, is of interest for these cases.

Nsutite history. Except for Nsuta, the discussed mineral as the characteristic component of high grade (battery) ores (Mn > 52%) in relatively less amounts, is registered at the Tambao and Mokta deposits too.

The most considerable amounts of nsutite are observed in ores, formed after the carbonate protore with the moderately soft oxidation regime of its change. For example, at the Nsuta deposit the following sequence of alterations is determined (Perseil and Grandin, 1978): Mn-carbonate $\rightarrow$ manganite $\rightarrow$ pyrolusite $\rightarrow$ cryptomelane $\rightarrow$ nsutite. The nsutite, as a rule, is formed after cryptomelane; but the alteration of nsutite into pyrolusite has been never observed. Usually ramsdellite is developed after nsutite, which is typical for the cap.

The given data suggest a clearly displayed trend of alteration of chemical composition of the draining solutions, reflecting the formation dynamics of the weathering crust profile.

Lithiophorite history. Lithiophorite is the typical product of deep supergene alteration of Mn-silicates, in particular the spessartine quartzites. The above two genetic varieties (generations) of lithiophorites are characterized by distinct features of chemical composition (see Table 12 after Perseil and Grandin, 1985). The first variety: the lithiophorite formed due to hydrolithic decomposition of spessartine, is characterized by relatively high amounts of Fe$_2$O$_3$ and Al$_2$O$_3$ with extremely low contents of alkali and heavy metals. The second variety of the lithiophorite usually fills in voids and pockets; it is formed as the result of deep leaching of Mn-garnet and rather considerable transportation of mineral-forming components. It contains higher amounts of K$_2$O and heavy metals, in particular copper and zinc compared to the first variety. This data may be indicative of the dual geochemical history of the components, transported during the decomposition of Mn-silicates and other components of the protore. The early generation (autochthonous in considerable degree) is associated with the formation of lithiophorite in the matrix (substrate) of hydrolithically decomposed garnet, and the components of the primary material (Fe$_2$O$_3$, Al$_2$O$_3$), characterized by relatively slight mobility under these conditions, are distinctly displayed. The late generation of lithiophorite is enriched considerably in components, transported in the solution possibly with the notable Eh gradient, as the differences in Fe(III) concentrations may indirectly testify.

Chapter 3

MANGANESE ORES IN KARST

3.1. Introduction

3.1.1. MANGANESE ORE KARST IN THE GENERAL SYSTEM OF GROUND WATER ACTIVITY

The manganese deposits in paleokarst are characterized by certain genetic dualistic features. On the one hand, they may be referred to accumulations, associated with events of leaching, redeposition of ore components by surface meteoric waters, and it may correspond, according to the general classification scheme of Patterson (1971), to typical supergene associations. On the other hand, in a wide diversity of the ore accumulations, associated with the activity of subsurface waters of the Earth's crust (sedimentation, infiltration, regenerated, including metamorphogenous), the karst Mn ores are often characterized by relatively deep epigenetic nature that is caused by high dissolution and migration properties of the solutions under these settings.

It was established (Baskov, 1976, 1983; Baskov and Pavlov, 1985) that sedimentation ground waters with mineralization in the carbonate and terrigenous sequences of about 20–40 g/kg are distributed in the artesian basins of the zone of epigenesis (catagenesis, to the depth of 5–7 km). The genetic dualistic features of the manganese ores are determined by the considerable differences of hydrogeological conditions within the artesian basins for zones of supergenesis, diagenesis and epigenesis (catagenesis).

The deposits of karst manganese ores are usually closely associated with artesian basins, composed by slightly dislocated, weakly- or unmetamorphosed at all carbonate layers, containing accumulations of ground waters (often mineralized, chloride composition). Similar basins are developed within cratons, ancient and young platforms, plates, sometimes miogeosyncline regions, and intermountain troughs.

In the process of the geologic development of artesian basins and their general subsidence the accumulation of sediments, their compaction, accompanied by separation and migration to the surface of ground waters, took place. The largest masses of waters were squeezed during the stages of diagenesis-epigenesis. The processes of interaction in the water–rock system at the epigenesis (metamorphism)–supergenesis stages are essential for the understanding of the formation processes of manganese ore accumulations in paleokarst. Artesian basins are the best studied to depths of 4–6 km, where ground waters are characterized by temperatures of 100–200°C, that makes them hydrothermal solutions in many aspects. The hydrogeological settings change considerably below these depths: the chemically bonded waters of many hydroxyl-containing minerals are converted to the dissolved state, and the decarbonatization events are accompanied by the formation of carbonate-bearing high temperature solutions in limestone-dolomite sequences. The processes of the karst manganese mineralization are activated in those artesian basins where magmatism manifestations appear. Under these conditions the subsurface waters are usually enriched in fluids, supplied from endogenic sources (SO_2, H_2, CO_2, HCl, HF, etc.) that increases their reaction activity at the interaction with sedimentary, mainly, carbonate rocks. The large-scale karst manganese and iron ore deposits in the Proterozoic carbonate sequences of Postmasburg and Sishen, South Africa, considered below, (Beukes, 1983, 1986; Boardman, 1940, 1964; De Villiers, 1956, 1960, 1983; Grobbelaar and Beukes, 1986; Strauss, 1964; Van Schalkwyk and Beukes, 1986) suggest that these processes are rather efficient.

Intensive interaction of rocks with ground (predominantly stratal) waters, leading to the decomposition of organic matter, and the dissolution of relatively unstable carbonate and silicate minerals (biotites, pyroxenes, amphiboles, Ca, Na-plagioclases, etc.) takes place in the zone of epigenesis (catagenesis).

The most considerable deep alterations of rocks proceed in the zones of tectonic deformations and faults. In these settings manganese, together with other heavy metals, may be intensively leached by relatively hot chloride solutions and transported to the areas with higher oxidation values of Eh, where ore deposition goes on.

Under near surface conditions (to the level of local, regional erosional base) in association with the general development of karst, the supergene processes of manganese ore accumulation may be active. The geochemical essence of this type of mineralization is similar to the events, occurring in weathering crusts, developed in carbonate rocks (protores) initially enriched in manganese. As a rule, the considerable transportation of manganese and the accumulation of the ores in karst hollows give ground for separate discussion of these processes.

One of the most essential conditions of the karst development is the relatively active water movement in carbonate rocks (Bernard, 1972): (a) in fractures and (b) in pores, or in their combinations. The features of such solution movement determine the morphological varieties which formed the karst: (a) caves, caverns, channels, expanded fractures; (b) through the formation of cavernous-porous rocks – formation of carbonate dust of dolomite, dolomite-limestone composition – to residual siliceous-aluminous products, often containing accumulations of manganese and iron oxyhydroxides. The processes may take place in different hydrodynamic zones (Safronova, 1985): (1) modern aeration; (2) ancient aeration; (3) along the filtration pathways of water-bearing horizons: (3.1.) zone of the seasonal variation of karst waters; (3.2.) zone of the complete saturation in the sphere of draining interaction of the local hydrographical system; (3.3.) zone of deep circulation, where the water flow proceeds in the direction of distant discharge places beyond the direct influence of the local hydrographical system; (4) redeposition of the dissolved components and filling in the karst hollows.

In aeration zones the presence of fresh infiltration waters, whose aggressiveness is increased owing to biochemical phenomena and temperature variations, causing pore expansion, favors the processes of intensive leaching. A special place belongs to the acidic waters, formed during oxidation of pyrite-bearing sediments.

In ancient zones of aeration the considered processes are controlled by features of the geological structure of the region.

On filtration pathways of water-bearing horizons (that is in the zone of complete saturation) the aggressiveness of waters of upper horizons depends on the diffusion of CO_2 from the subsurface zone of aeration and the oxidation processes of the organic matter, that take place below the level of the subsurface waters.

In the zone of seasonal variation of the karst water level, the periodic deionization and temperature increase of the subsurface waters, saturated with CO_2 and O_2, favor the leaching processes.

In the zone of deep circulation of the subsurface waters a considerable role belongs to the rock leaching owing to the slow circulation of stratal waters during geologically long time.

3.1.2. FORMULATION OF THE PROBLEM

It is well known that certain types of manganese ore accumulations are associated with karst sequences. The ore matter in them usually fills in hollows and fractures, it is included into the composition of the karst deposits, forms independent stratified and vein deposits, composes the cement of the dissolution and caving breccias. The formation of similar deposits is often geologically long and of a periodically repeated or pulsation character. Multiple breccia formations, polychronous accumulations of the manganese ore matter may be the result of such events. The chemistry of the country rock has a special function: at relatively wide development of karst formation processes, the deposits are formed, as a rule, in those cases, when Mn contents in the parent, leaching carbonate sequences exceed the clark concentrations.

However, relatively slight attention has been focused on the general problems of the karst deposits in the literature on manganese ore formation.

The objective of this section is to elucidate the major aspects of the most representative karst type manganese deposits, to show the geochemical features of the ore formation processes.

3.2. Model Test Site: Zone of Oxidation and Accumulation of the Karst Ores, the Úrkút Manganese Deposit, Hungary (Mineralogy, Geochemistry, and Genesis)

3.2.1. INTRODUCTION

The current concepts, based on extensive research, recognize the formation of manganese accumulations in the region of the Úrkút deposit as a multistage process: (1) the Early Jurassic formation of hydrothermal-sedimentary layered carbonate-clay manganese ores;

and (2) the Late Jurassic-Cenozoic accumulation of manganese oxyhydroxide ores associated with supergene and karst processes (Cseh-Németh and Grassely, 1966; Cseh-Németh *et al.*, 1971, 1980; Grasselly and Cseh-Németh, 1961; Kaeding *et al.*, 1983; Szabo and Grasselly, 1980; Szabo *et al.*, 1981; Varentsov *et al.*, 1988). It is important to point out that the ores of the zones of oxidation and karst-formation at this deposit have th highest industrial value. This multistage process with apparent superposition of superg manifestation is globally common.

Examples of such deposits are the rich accumulations of manganese oxyhydroxide ores in Western Africa (Nsuta, Moanda, etc.) and South Africa (Postmasburg, etc.), which occurred in rather poor carbonate-silicate Proterozoic Mn-bearing rocks, and in Australia, in Groote Eylandt, where the rich oxyhydroxide manganese ores of the crust of weathering were formed in Cretaceous manganese-bearing formations. Less spectacular, but on a sufficiently large scale is the formation of fairly rich ores of the oxidation zone manifested in the Oligocene deposits in the south European section of the former U.S.S.R. (Nikopol, Chiatura), and in other regions.

In contrast to the indicated deposits, however, the oxidized karst ores in the Úrkút region experienced gradational development of oxidation processes with relatively restricted or minimal transport of the matter. These features characterize the relations between the initial hydrothermal-sedimentary ores and the products of their oxidation. It is important to note that for the part of the deposits of this type there is limited data mostly describing the behavior of the major components, the information about the geochemical behavior of heavy metals, trace elements and rare earth elements in the course of accumulation of oxidized ores is extremely fragmentary.

The purpose of this section is to reveal the basic characteristics of formation of relatively rich ores of the oxidation and karst-formation zone using the example of the Csárdahegy area of the Úrkút deposit by detailed mineralogical and geochemical research using data on the major components, heavy metals, trace and rare earth elements.

The investigation is based on field research data and on the analysis of samples of ores and of enclosing rocks collected in accordance with the Manganese project of the International Geological Correlation Program (IGCP). Great help in this study was provided by the late professor Gy. Grasselly (University of Szeged, Hungary) and Dr. Ing. Z. Szabo (the Úrkút Mine, Hungary).

The samples available were studied in thin sections under a microscope. The ore components, individual interbeds, patches, etc. were sampled under the binocular microscope. These materials were selected as samples for X-ray diffractometric and other methods of study. The X-ray analysis was carried out using the DRON-2 and DRON-3 diffractometers with radiation = CuK_α and CoK_α under 35 kV and 20 mA. The scanning rate was 2°/min and for detailed studies 1°/min. The infrared spectra of the analysed samples were obtained in 400–4000 cm^{-1} band.

The chemical composition of samples was determined at the Geological Institute, the Russian Academy of Sciences, and at several analytical laboratories in Hungary. The methods and laboratories where determinations of the chemical components was performed are shown in the notes for the tables of analysis.

The analyses, marked by indexes 'V' (Geological Institute) and 'Gr' (Szeged University, Hungary), were obtained from different samples. This is apparently the reason for the slight differences in the results of determinations.

3.2.2. GEOLOGICAL SETTING

The aspects of geological structure, stratigraphy and lithology of the Early Jurassic rocks and ores of the Úrkút deposit (Figure 42) are discussed chiefly by the Hungarian geologists (Cseh-Németh and Grasselly, 1966; Cseh-Németh *et al.*, 1971, 1980; Grasselly and Cseh-Németh, 1961; Kaeding *et al.*, 1983; Szabo and Grasselly, 1980; Szabo *et al.*, 1981). We have shown in our studies (Varentsov *et al.*, 1988) that the Liassic hydrothermal-sedimentary ores of this deposit are represented by three principal components in various proportions: (1) Fe-mica of the polytype modification IM (celadonite) containing up to 20% of Fe-smectite with interlayers of nontronite type; (2) Fe-smectite (nontronite), homogeneous as a rule, occasionally with a few Fe-mica packets; (3) Mn-carbonate very close to rhodochrosite, sometimes with a admixture of the $FeCO_3$ molecule.

Individual intervals of the section contain subordinate quantities of oxyhydroxides of Fe (goethite) and Mn (manganite). The composition of these deposits is similar to that of the nontronite-celadonite metalliferous sediments of the hydrothermal regions in the World Ocean (the Galapagos Rift Zone, South Pacific, the Red Sea deeps, etc.); these sediments were formed in moderately hot to low-temperature hydrothermal solutions. The formation of rhodochrosite with $FeCO_3$ admixture, sometimes with manganite, is associated with the diagenetic events in the reduction environments of local depressions of the Liassic basin in the western part of the Tethys.

The manganese oxyhydroxide ores of the deposit studied are mostly the products of oxidation of carbonate varieties (see Figure 42).

After the completion of the Jurassic–Neocomian sedimentary cycle, the Neo-Kimmerian orogenic movements became the essential feature in the geological evolution of the region. As a result, many areas of the deposit were lifted as blocks; this event was followed by the denudation of the Upper Jurassic sediments and often of the supra-ore rocks as well. The manganese ore accumulations were therefore exposed and subjected to supergene processes, in particular, rather intensive local denudation, karst-formation and redeposition. Consequently, the younger sediments overlaid the eroded surfaces of the Lower Jurassic deposits. These relatively young sediments are composed of clayey marls and limestones of Aptian to Albian age, and sometimes of nummulitic Eocene limestones. The Epleny deposit, situated 25–30 km from the Úrkút deposit, contains Miocene pebbles lying on the surface of the Lower Jurassic rocks (Cseh-Németh *et al.*, 1980). It is of interest that the erosion and denudation processes caused the appearance of a variety of karst forms where the residual products of dissolution of carbonate rocks accumulated (Figure 43).

In the larger part of the Úrkút deposit, the processes of supergene oxidation of manganese ores occurred, either with a minimal transportation of matter, or practically *in situ*. The oxidized ores are characterized by a finely banded layered texture which is similar to that of the initial carbonate-clayey varieties and which occasionally contains nodular oxyhydroxide accumulations alternating with deeply altered clays. The thickness of the

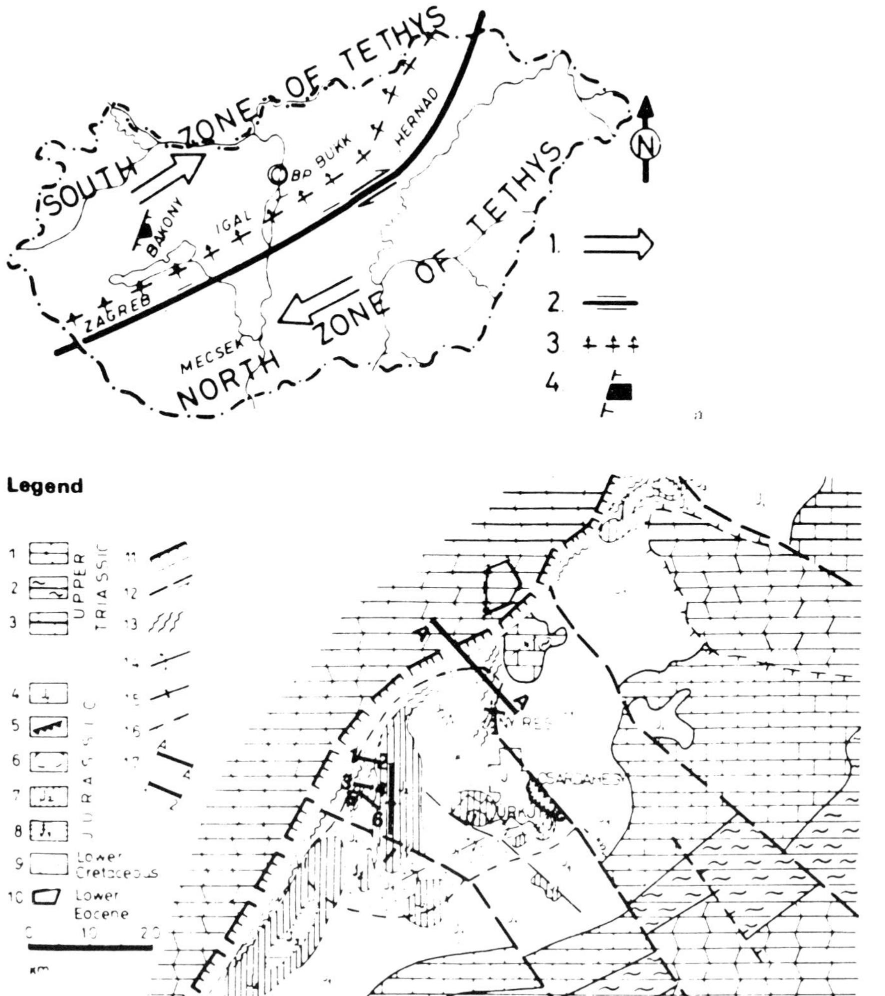

Fig. 42. (a) The Úrkút manganese deposit and the major structural lineaments of Hungary (after Szabo and Grasselly, 1981). 1. Direction of displacement; 2. Transform fault; 3. Eugeosyncline; 4. The Úrkút manganese deposit. (b) Geological sketch map of the region of the Úrkút manganese deposit, Hungary (after Szabo *et al.*, 1981). Legend. The Upper Triassic: 1. the Norian dolomite; 2. Neritic limestone, marl, Raetian dolomite; 3. Raetian 'Dachstein type' limestone; the Jurassic: 4. Liassic limestone sequence; 5. Karsted Liassic limestone; 6. Upper Liassic manganese deposit (tentative pre-denudation area is shown); 7. Dogger limestone; 8. Malm deposits; 9. The area covered by the reworked Jurassic deposits and leached cherty gravelites (Lower Cretaceous–Lower Eocene); 10. Bauxite ore manifestations (Lower Cretaceous–Lower Eocene); 11. The lines of large structural dislocations; 12. Horizontal displacements; 13. Zones of intensive crumpling; 14. Anticlines; 15. Synclines; 16. Faults; 17. Direction of geological profiles.

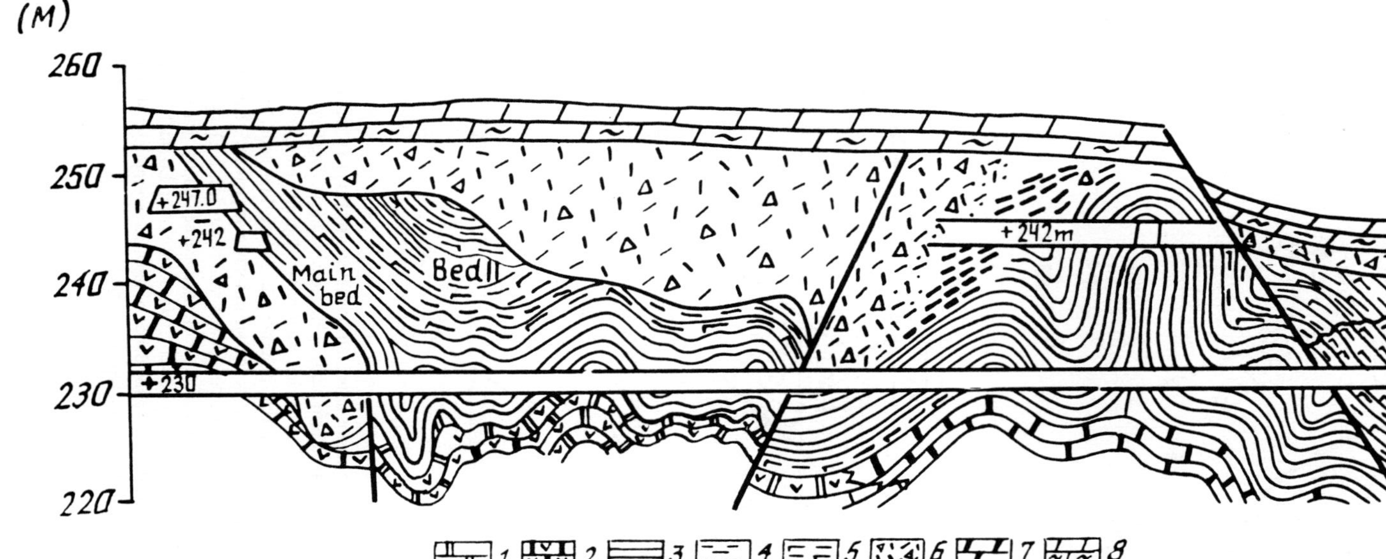

Fig. 43.   Section showing relationships between the carbonate-clayey manganese ores, oxidized varieties, karst accumulations and partially redeposited manganese oxyhydroxide ores, the Úrkút deposit, Hungary (after Cseh-Németh *et al.*, 1980). 1. Rocks of limestone complex, Middle Liassic; 2. Residual cherty limestone; 3. Finely banded manganese oxyhydroxide ore; 4. Loose, low density beds of finely banded manganese oxyhydroxide ore; 5. Radiolarian clay and clayey marl; 6. Redeposited ores containing siliceous inclusions; 7. Greenish-grey limestones and calcic marl, Dogger; 8. Requienian limestones of Cretaceous age.

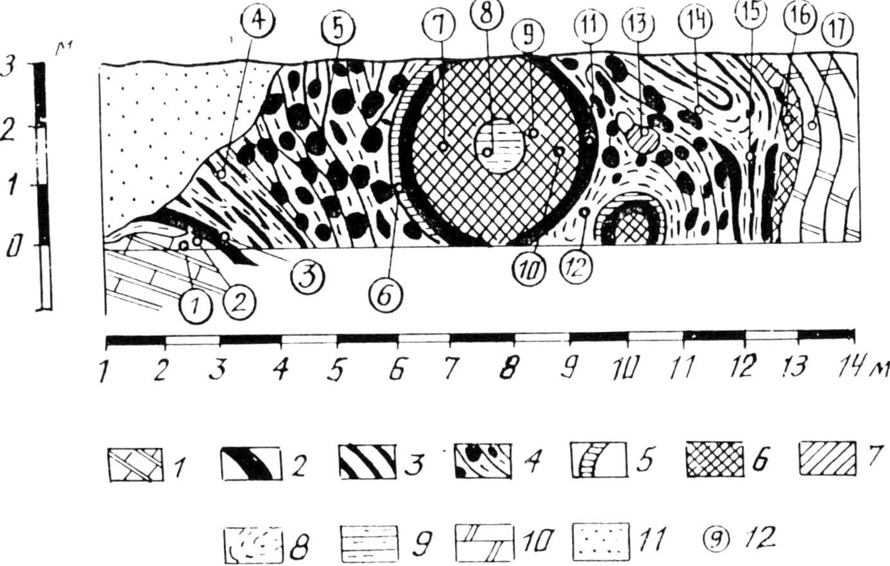

Fig. 44. Structural scheme of the Csárdahegy area (exploring opening) built up by highly oxidized manganese ores at the Úrkút deposit, Hungary. 1. Limestone of the 'Hierlatz type', Lower Liassic; 2. Manganese oxyhydroxide ore; 3. Banded manganese oxyhydroxide ore with clayey interbeds; 4. Nodular ferruginous manganese oxyhydroxide ore with clayey inclusions; 5. Limonitic crust; 6. Siliceous ferruginous manganese oxyhydroxide ore; 7. Cherty nodules; 8. Clay in manganese bed; 9. Fe and Mn clay from the inner part of a large nodule, ore boulder; 10. Weathered Upper Liassic limestone; 11. Ore-free redeposited material filling the cavity; 12. Sites of samples collection (no. 1 to 17). Description of samples see in 'Notes to Tables 14 and 15'.

oxidized zone is 3 to 5 m; it distinctly shows relict bedding, the thickness of some of the manganese oxide ore interbeds varying from 1–2 to 100 mm.

It should be emphasized, however, that the Úrkút deposit has not only the autochthonous oxidized ores, but the redeposited varieties as well.

3.2.3. Mineralogy of the Oxidized Ores

Among the autochthonous accumulations of the zone of oxidation in the Úrkút deposit the manganese oxyhydroxide ores of the Csárdahegy area occupy a particular position between the Lower Liassic limestones of the 'Hierlatz-type' and the Lower Eocene layers (Cseh-Németh *et al.*, 1980). On the whole, this deposit can be described as a fairly large karst cavity which was formed as a result of intensively folded tectonic structure. The lower beds of the deposit are built up by predominantly autochthonous oxidized ores, whereas the central and upper sequences contain locally displaced large blocks, boulder debris and smaller fragments enclosed in the clayey mass with different degrees of alteration and the color varying from green to yellow and brown. Large blocks of ore material have inclusions of silicate and limestone fragments.

Figure 44 shows the rather simple relationships of manganese oxyhydroxide ores whose formation is associated with low matter transport. In this case the altered ore sequence about 10 to 12 m thick is bedded between the Lower Liassic limestone of the 'Hierlatz-type' and the Upper Liassic weathered limestone. We may state that the lower (see Figure 44, samples 1 to 5) and upper (samples 12 to 17) parts of the deposit, on the whole, correspond to the general normal sequence of bedding of the Liassic ore-bearing deposits. However, the presence of the allochthonous boulder block (3.3 to 4.0 m) with a relatively symmetrical structure (samples 6 to 11) in the central part of this deposit allows us to characterize this accumulation as a karst cavity, the central part of which is filled with redeposited ore material.

Allowing for the indicated restrictions, we can suggest the following succession in the section of the deposit (from base upwards). Tables 14 and 15 present the chemical composition of these ores and rocks; the mineral composition was analysed by the X-ray diffractometry and infra-red spectroscopy.

1. Brachiopod limestone of the 'Hierlatz-type', Lower Liassic, insoluble residue is kaolinite with hydromica and smectite admixtures.

2. Red-brown ferruginous clay, overlying the limestone of the 'Hierlatz-type', composed of a mixture of hydromica and goethite with a notable admixture of kaolinite, hematite and quartz.

3. Manganese oxyhydroxide ore composed of a mixture of 10 Å-manganates (the todorokite-buserite-asbolan etc. group), pyrolusite, with cryptomelane admixture and traces of smectite.

4. Clayey manganese oxyhydroxide ore, low ferruginous, composed of a mixture of 10 Å-manganate and cryptomelane.

5. Manganese oxyhydroxide ore, crystalline grained. Predominant are pyrolusite, hydrogoethite with appreciable admixture of 10 Å-manganates, cryptomelane, probably with traces of the metastable phase $[x\ \mathrm{Fe(OH)_2} \cdot y\ \mathrm{FeOCl} \cdot z\ \mathrm{H_2O}]$. Pyrolusite dominates in some parts with patches composed of black sooty matter (sample 5/1/).

6. Predominantly limonitic, highly ferruginous encrustation composing the marginal part of the ore boulder block.

7. Ferromanganese oxyhydroxide ore composes the inner part of the ore boulder block. Hydrogoethite and quartz dominate; there is an admixture of 10 Å-manganates and pyrolusite and traces of the relict mixed-layered hydromica-smectite phase.

8. Highly ferruginized ore from the central part of the ore boulder block. The major components are: hydrogoethite and 10 Å-manganates; there is an admixture of cryptomelane and highly amorphous clayey matter (strongly degraded hydromica).

9. Ferromanganese oxyhydroxide ore, siliceous. The major components are: quartz and hydrogoethite with admixture of 10 Å-manganates, cryptomelane, and traces of relict smectite.

10. Manganese oxyhydroxide ore, ferruginous, siliceous, composed mostly of pyrolusite with subordinate amounts of hydrogoethite. Individual parts and patches of cryptocrystalline black matter are composed of hydrogoethite with an admixture of pyrolusite and traces of cryptomelane. Brown patches mostly contain hydrogoethite and quartz with traces of cryptomelane.

11. Ferromanganese oxide ore. Hydrogoethite dominates, 10 Å-manganates are present

in subordinate quantities with admixture of cryptomelane and traces of quartz. The parts, composed of dull black matter, contain X-ray amorphous hydroxides crystallized in 10 Å-manganates with cryptomelane admixture.

12. Clay, in some parts cherty, is composed mostly of deeply degraded hydromica and quartz with local appreciable admixture of hydrogoethite.
13. Ferruginous manganese ore with silicate inclusions. Pyrolusite with hydrogoethite and quartz admixtures dominate.
14. Clay, with green inclusions of manganese carbonate, is a mixture of quartz and deeply degraded hydromica.
15. Manganese oxyhydroxide ore, fallow, earthy, is composed of a fine mixture of pyrolusite and hydrogoethite.
16. Oxyhydroxide ore, cherty, ferruginous, is composed of pyrolusite with goethite and quartz admixtures. Some parts of mostly brown hydroxides (sample 16/1/) are composed of hydrogoethite, 10 Å-manganates with quartz admixture.
17. Weathered Upper Liassic limestone.

The determination of the mineral composition of the altered oxidized ores and enclosing rocks, and also the data of study under a microscope of their structural relationships, bring out several features:

1. On the whole, this deposit is not characterized by a developed weathering crust with zones of the carbonate protore transformation (Bouladon *et al.*, 1965; Grandin and Perseil, 1977, 1983; Leclerc and Weber, 1980; Perseil and Grandin, 1978, 1985; Weber *et al.*, 1979).

2. The general intensity of the supergene alteration is relatively low in rocks that are predominantly carbonate, siliceous, and clayey and in rhodochrosite – Fe-mica and Fe-smectite ores: the clayey components are altered to the stage of deeply degraded hydromica-smectite partially transformed into kaolinite with quartz and goethite admixtures in places of intensive circulation of karst waters (e.g., sample 2, Figure 44).

3. Throughout the entire section the minerals of Mn and Fe oxyhydroxides co-exist practically without any visible separation. The relationships between them are controlled mostly by the composition of the initial material and the intensity of the local circulation of karst waters.

There is a vague difference, however, between the lower and the upper parts. If the lower part (samples 3 to 11; Figure 44) contains an appreciable amount of minerals belonging to 10 Å-manganates (the todorokite-buserite-asbolan etc. group, and possibly, lithiophorite) and cryptomelane, then in the upper layers of the deposit (samples 13 to 16) the dominant manganese mineral is pyrolusite. The Fe oxyhydroxide compounds are represented chiefly by hydrogoethite, and only on the contact with the Lower Liassic limestone of the 'Hierlatz-type' by goethite.

As already mentioned, the X-ray diffraction methods can, as a rule, determine only the general presence of the 10 Å-manganate phase. In order to obtain a more reliable determination of minerals belonging to the 10 Å-manganates (at least 7 minerals with clearly different structural parameters), the methods of electron microdiffraction should be applied in combination with the microprobe determination of the composition (Chukhrov *et al.*, 1980a–1980c, 1982, 1989).

Table 14.   Chemical composition of manganese oxyhydroxide ores of the Úrkút deposit, Csárdahegy area, Hungary (wt. %, air-dry weight).

no./ comp.	1V 1	1/20Gr 2	2V 3	2/21Gr 4	3V 5	3/22Gr 6	4V 7	4/23Gr 8	5V 9	5/24Gr 10	6V 11	6/25Gr 12	7V 13	7/26Gr 14	8V 15	8/27Gr 16	9V 17	9/28Gr 18
SiO_2	1.31	0.80	27.06	22.48	5.36	8.99	4.60	17.48	2.16	9.50	7.92	11.09	36.52	30.58	4.77	14.25	40.96	40.30
TiO_2	0.04	0.10	0.82	0.60	0.15	0.30	0.12	0.60	0.02	0.35	0.13	0.35	0.00	0.35	0.14	0.40	0.00	0.40
Al_2O_3	1.34	1.40	12.36	11.33	3.16	2.95	3.04	5.06	1.28	1.37	3.90	3.95	1.24	2.43	2.95	3.53	1.08	1.37
Fe_2O_3 (total)	0.31	0.92	37.30	41.40	36.68	35.75	18.64	9.94	36.98	37.57	73.60	64.43	31.64	32.19	49.89	45.84	31.44	31.43
MnO (total)	0.13	0.25	0.58	1.63	32.66	31.88	50.41	45.50	38.20	30.49	1.28	1.44	15.62	19.11	21.30	14.85	16.33	14.85
MgO	0.61	0.51	2.22	1.74	1.78	0.83	1.04	1.07	1.34	0.72	0.54	0.56	0.54	0.95	1.17	0.74	0.52	0.45
CaO	53.70	52.46	3.10	2.68	0.98	1.96	0.72	1.85	0.94	1.85	0.56	1.85	2.00	3.99	0.89	1.85	0.32	1.85
P_2O_5	0.00	0.15	2.00	0.20	1.02	0.15	0.36	0.05	0.12	0.16	0.00	0.80	0.28	0.20	0.43	0.10	0.00	0.15
Na_2O	0.03	0.42	0.12	1.42	0.24	1.63	1.92	1.53	0.34	1.89	0.18	1.58	0.28	1.58	0.22	2.16	0.10	1.16
K_2O	0.13	0.35	1.94	1.43	0.92	1.25	–	1.46	0.60	1.63	0.52	1.33	0.52	0.36	0.78	2.54	0.44	0.50
L.O.I.	–	42.17	–	9.39	–	13.25	–	13.68	–	13.45	–	10.90	–	8.45	–	12.53	–	7.60
BaO	0.01	0.12	0.00	0.00	0.40	0.00	0.28	0.47	0.32	0.53	0.02	0.00	0.50	0.87	0.80	0.70	0.60	0.43
SrO	0.01	–	0.01	–	0.08	–	0.28	–	0.10	–	0.08	–	0.08	–	0.10	–	0.12	–
Cr_2O_3	0.03	–	0.00	–	0.00	–	0.00	–	0.00	–	0.00	–	0.00	–	0.00	–	0.00	–
NiO	0.01	–	0.00	–	0.01	–	0.00	–	0.00	–	0.00	–	0.00	–	0.00	–	0.00	–
ZnO	–	–	–	–	–	–	–	–	–	–	–	–	–	–	–	–	–	–
Fe_2O_3	0.31	0.78	37.30	41.34	36.68	33.70	18.64	6.28	36.98	35.82	73.60	64.29	31.64	31.03	49.89	44.38	31.44	30.70
FeO	no	0.13	no	0.06	n.d.	1.85	no	3.30	no	1.58	no	0.13	no	1.05	no	1.32	no	0.66
H_2O^+	1.82	0.12	7.99	8.20	8.84	5.80	6.53	3.21	8.57	5.23	10.74	10.53	6.43	4.16	10.04	9.03	5.76	4.67
H_2O^-	0.11	0.19	5.02	3.46	1.93	1.06	1.75	2.50	1.06	1.05	1.37	1.02	0.70	0.74	1.57	1.55	0.57	0.49
CO_2	42.20	42.06	0.40	1.16	no	0.00	no	0.00	0.60	0.00	–	0.00	no	0.25	no	0.00	no	0.00
SO_3 (total)	–	0.29	–	0.31	–	0.21	–	0.22	–	0.18	–	0.25	–	0.00	–	0.19	–	0.27
S	–	0.12	–	0.12	–	0.08	–	0.09	–	0.07	–	0.10	–	0.00	–	0.08	–	0.11
SO_3	–	0.00	–	0.00	–	0.00	–	0.00	–	0.00	–	0.00	–	0.00	–	0.00	–	0.00
$C_{org.}$	no	–	no	–	no	–	no	–	no	–	no	–	no	–	no	–	no	–
Total	101.79	100.02	100.92	98.16	100.77	98.53	100.11	99.11	92.63	98.90	100.04	99.95	96.35	100.26	98.98	100.03	101.12	100.10

Table 14. (Continued)

no./ comp.	1V 1	1/20Gr 2	2V 3	2/21Gr 4	3V 5	3/22Gr 6	4V 7	4/23Gr 8	5V 9	5/24Gr 10	6V 11	6/25Gr 12	7V 13	7/26Gr 14	8V 15	8/27Gr 16	9V 17	9/28Gr 18
Ca*)	37.97	20.00	3.419	2.94	2.609	0.340	1.138	0.860	0.689	0.430	0.297	0.226	0.191	0.20	0.538	0.300	0.128	0.133
Ti*)	<0.03	–	0.405	–	0.216	–	0.175	–	0.102	–	0.071	–	0.113	–	0.301	–	0.120	–
V*)	–	–	0.0219	–	<0.02	–	<0.02	–	<0.02	–	0.0210	–	<0.02	–	<0.02	–	<0.02	–
Cr*)	–	–	<0.01	–	<0.01	–	<0.01	–	<0.01	–	<0.01	–	<0.01	–	<0.01	–	<0.01	–
Mn*)	0.0619	–	0.3998	–	15.3500	–	23.3000	–	15.2700	–	0.6663	–	5.9100	–	7.8180	–	5.9240	–
Fe*)	0.691	–	24.8900	–	17.2900	–	5.5060	–	17.2000	–	39.2300	–	13.0800	–	25.9000	–	15.4300	–
Ni*)	<0.005	0.00073	0.0187	0.016	0.0225	0.0140	0.0067	0.0078	0.0270	0.0110	0.0241	0.0058	0.0137	0.0070	0.0109	0.0130	0.0145	0.0059
Co*)	–	0.00015	–	0.0124	–	0.0360	–	0.0360	–	0.0200	–	0.0029	–	0.0051	–	0.0174	–	0.0048
Cu*)	<0.005	0.00057	0.0056	0.0109	<0.005	0.0012	<0.005	0.0035	< 0.005	0.0042	<0.005	<0.00001	<0.005	<0.00001	<0.005	0.0013	<0.005	<0.00001
Zn*)	0.0048	0.0027	0.0124	0.0230	0.0022	0.0174	0.0013	0.0086	0.0022	0.0125	<0.003	0.0120	0.0064	0.0098	0.0058	0.0235	0.0211	0.0090
Rb*)	0.0005	–	0.0107	–	0.0012	–	0.0007	–	0.0005	–	0.0076	–	<0.0005	–	0.0016	–	0.0002	–
Sr*)	0.0191	0.0054	0.0140	0.0430	0.1080	0.1140	0.3491	0.9500	0.1042	0.2500	0.0164	0.0130	0.1574	0.2000	0.1013	0.1300	0.1162	0.1800
Y*)	<0.0005	–	<0.0005	–	<0.0005	–	<0.0005	–	<0.0003	–	<0.0005	–	<0.0005	–	<0.0005	–	<0.0005	–
Zr*)	0.0004	–	0.0132	–	<0.0005	–	<0.0005	–	<0.0005	–	<0.0005	–	<0.0005	–	<0.0005	–	<0.0005	–
Nb*)	<0.0005	–	0.0010	–	<0.0005	–	<0.0005	–	<0.0005	–	<0.0005	–	<0.0005	–	<0.0005	–	<0.0005	–
Pb*)	<0.003	<0.0001	<0.003	0.074	0.003	0.0058	<0.003	–	<0.003	–	<0.003	–	<0.003	–	<0.003	–	<0.003	–
Ga*)	–	–	0.0032	–	<0.003	–	<0.003	–	<0.003	–	<0.003	–	<0.003	–	<0.003	–	<0.003	–
K*)	<0.005	–	0.9230	–	2.0910	–	3.4090	–	1.2820	–	1.2210	–	0.8480	–	1.5500	–	0.9219	–
Ba*)	0.0027	0.0820	0.0107	0.0830	0.1477	0.1630	0.0991	0.1760	0.0910	0.3000	0.0042	0.0160	0.2481	0.5000	0.2666	0.4900	0.2234	0.3760
La*)	<0.01	–	0.0068	–	0.0025	–	<0.01	–	<0.01	–	<0.01	–	<0.01	–	<0.01	–	<0.01	–
Ce*)	<0.01	–	0.0155	–	0.0064	–	0.0055	–	<0.01	–	0.0031	–	<0.01	–	0.0036	–	<0.01	–
As*)	<0.003	–	0.0045	–	0.0082	–	0.0034	–	0.0090	–	0.0103	–	0.0112	–	0.0172	–	0.0225	–
MnO*)	–	0.0	–	0.0	3.58	4.60	4.25	6.55	–	0.27	–	0.14	–	3.38	3.88	1.64	3.58	4.35
MnO$_2$*)	–	0.31	–	1.99	35.64	33.53	56.58	47.61	–	36.88	–	1.23	–	18.39	21.35	15.76	15.63	12.81

Table 15.   Chemical composition of manganese oxide ores of the Úrkút deposit, Csárdahegy area, Hungary (wt. %, air-dry weight).

no./ comp.	10V 19	10/29Gr 20	11V 21	11/30Gr 22	12V 23	12/31Gr 24	13V 25	13/32Gr 26	14V 27	14/33Gr 28	15V 29	15/34Gr 30	16V 31	16/35Gr 32	17V 33	17/36Gr 34	18V 35	18/37Gr 36
SiO_2	3.44	16.86	5.48	8.30	34.20	13.11	58.02	67.93	53.53	48.67	5.54	10.30	44.18	51.08	11.36	21.04	30.74	23.43
TiO_2	0.04	0.75	0.12	0.50	0.61	0.45	0.01	0.40	0.79	0.60	0.14	0.45	0.04	0.65	0.10	0.62	0.30	0.60
Al_2O_3	1.12	0.85	4.44	6.12	13.64	6.85	1.06	1.37	15.03	21.92	3.50	0.32	1.80	2.32	2.30	2.36	4.78	6.53
Fe_2O_3 (total)	78.76	61.22	43.00	49.08	6.63	3.91	9.04	11.09	7.74	6.87	39.20	26.94	23.08	21.81	0.85	0.65	24.38	9.94
MnO (total)	3.24	6.35	25.91	18.25	1.24	0.38	24.14	11.22	0.00	0.50	29.82	45.17	19.80	13.15	0.21	0.50	10.90	28.47
MgO	0.00	0.38	0.20	0.60	2.30	1.40	0.00	0.29	2.94	1.86	0.98	0.49	0.04	0.33	0.46	0.41	5.48	2.86
CaO	0.56	1.85	1.48	1.85	15.96	37.57	0.00	1.85	0.07	1.85	1.68	1.85	0.26	1.85	46.53	38.16	2.22	2.18
P_2O_5	0.00	0.15	1.04	0.23	6.15	23.12	0.00	0.24	0.89	0.20	0.96	0.20	0.00	0.18	0.06	0.23	0.29	0.10
Na_2O	0.20	1.16	0.32	1.32	0.16	1.74	0.11	1.11	0.15	0.47	0.26	1.32	0.16	1.11	0.08	1.00	0.12	1.21
K_2O	0.56	0.60	0.72	0.74	2.37	1.47	0.56	0.68	3.18	1.55	0.74	0.90	0.34	0.67	0.33	0.48	2.82	1.95
L.O.I.	–	9.16	–	12.31	13.05	6.14	–	3.95	–	7.16	–	12.44	–	5.66	–	33.06	–	20.18
BaO	0.00	0.0	0.12	0.0	0.02	0.0	0.02	0.0	0.02	0.0	0.26	0.0	0.04	0.0	0.04	0.0	0.00	0.0
SrO	0.16	–	0.28	–	0.07	–	0.01	–	0.11	–	0.18	–	0.08	–	0.04	–	0.01	–
Cr_2O_3	0.00	–	0.00	–	0.00	–	0.00	–	0.00	–	0.00	–	0.00	–	0.03	–	0.00	–
NiO	0.00	–	0.00	–	0.01	–	0.00	–	0.00	–	0.00	–	0.02	–	0.00	–	0.00	–
ZnO	–	–	–	–	–	–	–	–	–	–	–	–	–	–	–	–	–	–
Fe_2O_3	78.76	60.94	43.00	48.43	6.63	2.71	9.04	10.88	7.12	6.08	39.20	25.78	23.08	21.38	0.85	0.51	24.01	8.56
FeO	no	0.26	no	0.59	no	0.33	no	0.19	0.56	0.72	no	1.05	no	0.39	no	0.13	0.33	1.25
H_2O^+	10.38	7.31	9.35	8.60	6.22	4.71	2.14	0.69	6.39	7.06	9.09	1.49	4.24	2.74	1.69	0.82	6.43	3.14
H_2O^-	0.40	0.28	1.64	1.69	6.79	3.12	0.27	0.12	6.98	6.55	1.79	1.40	0.78	0.26	0.83	0.16	4.13	2.27
CO_2	0.90	–	no	–	no	–	no	–	no	–	no	–	no	–	36.55	–	7.70	–
SO_3 (total)	–	0.30	–	0.24	–	0.50	–	0.24	–	0.33	–	0.19	–	0.28	–	0.31	–	0.26
S	–	0.12	–	0.10	–	0.20	–	0.10	–	0.13	–	0.08	–	0.11	–	0.12	–	0.11
SO_3	–	0.00	–	0.00	–	0.00	–	0.00	–	0.00	–	0.0	–	0.0	–	0.0	–	0.0
$C_{org.}$	no	–	no	–	no	–	no	–	0.52	–	no	–	no	–	0.22	–	no	–
Total	99.76	99.21	99.62	100.26	–	98.66	100.69	99.70	98.28	98.26	100.20	99.58	94.86	98.51	101.68	98.66	100.26	99.78

Table 15. (Continued)

no./ comp.	10V 19	10/29Gr 20	11V 21	11/30Gr 22	12V 23	12/31Gr 24	13V 25	13/32Gr 26	14V 27	14/33Gr 28	15V 29	15/34Gr 30	16V 31	16/35Gr 32	17V 33	17/36Gr 34	18V 35	18/37Gr 36
Ca*)	0.0342	0.0520	1.2730	0.4700	10.86	20.000	<0.05	0.0970	0.9962	0.6800	1.6900	0.2900	0.0855	0.1000	30.700	30.000	2.1390	1.7400
Ti*)	<0.03	–	0.0929	–	0.3022	–	<0.03	–	0.3650	–	0.1723	–	0.0004	–	0.0779	–	0.1343	–
V*)	<0.02	–	<0.02	–	–	–	<0.02	–	<0.02	–	<0.02	–	<0.02	–	–	–	<0.02	–
Cr*)	<0.01	–	<0.01	–	–	–	<0.01	–	<0.01	–	<0.01	–	<0.01	–	–	–	<0.01	–
Mn*)	1.3150	–	11.580	–	0.3893	–	8.8100	–	0.0643	–	13.8200	–	7.8690	–	0.6731	–	5.8760	–
Fe*)	39.760	–	21.390	–	4.9340	–	0.9037	–	5.3920	–	17.9300	–	8.8650	–	0.5560	–	15.2800	–
Ni*)	0.0281	0.0034	0.0236	0.0066	0.0063	0.0048	0.0046	0.0059	0.0084	0.0061	0.0235	0.0230	0.0121	0.0150	<0.005	0.0100	0.0112	0.0043
Co*)	–	0.0020	–	0.0080	–	0.0128	–	0.0028	–	0.0027	–	0.0072	–	0.0037	–	0.00066	–	0.0140
Cu*)	<0.005	<0.00001	<0005	0.00035	0.0094	0.0032	0.0053	0.0051	0.0146	0.0140	<0.005	0.0023	<0.005	0.0025	<0.005	0.00088	<0.005	0.0020
Zn*)	<0.003	0.0087	0.0039	0.0140	0.0066	0.0062	0.0007	0.0052	0.0078	0.0076	0.0045	0.0130	0.0023	0.0100	0.0033	0.0042	<0.003	0.0051
Rb*)	0.0075	–	0.0017	–	0.0109	–	<0.0005	–	0.0255	–	0.0016	–	0.0007	–	0.0013	–	0.0079	–
Sr*)	0.0011	0.0140	0.1949	0.2270	0.1365	0.0980	0.0252	0.0160	0.4057	0.2060	0.1439	0.1660	0.0189	0.0108	0.0258	0.0092	0.0066	0.0076
Y*)	<0.0005	–	<0.0005	–	<0.0005	–	<0.0005	–	<0.0005	–	<0.0005	–	<0.0005	–	<0.0005	–	<0.0005	–
Zr*)	<0.0005	–	<0.0005	–	0.0188	–	<0.0005	–	0.0324	–	0.0004	–	<0.0005	–	0.0012	–	0.0045	–
Nb*)	<0.0005	–	<0.0005	–	<0.0005	–	<0.0005	–	<0.0027	–	<0.0005	–	<0.0005	–	<0.0005	–	<0.0005	–
Pb*)	<0.003	0.0045	<0.003	0.0050	<0.003	0.0014	<0.003	0.0019	0.0049	0.0018	<0.003	0.0056	<0.003	0.0041	0.0001	0.00063	<0.003	0.0032
Ga*)	<0.003	–	<0.003	–	–	–	<0.003	–	0.0038	–	<0.003	–	<0.003	–	–	–	<0.003	–
K*)	<0.005	–	1.681	–	2.7080	–	<0.05	–	3.0080	–	1.8110	–	0.9906	–	0.7762	–	3.8980	–
Ba*)	0.0007	0.0220	0.0337	0.1650	0.0155	0.0720	0.0168	0.0165	0.0205	0.0500	0.0753	0.0750	0.0153	0.0140	0.0256	0.0840	0.0053	0.0186
La*)	<0.01	–	0.0029	–	0.0095	–	<0.01	–	<0.0105	–	0.0036	–	<0.01	–	<0.01	–	<0.01	–
Ce*)	<0.01	–	0.0105	–	0.0516	–	0.0022	–	0.0305	–	0.0074	–	0.0035	–	<0.01	–	0.0093	–
As*)	0.0112	–	0.0114	–	0.0026	–	0.0019	–	0.0066	–	0.0115	–	0.0027	–	<0.003	–	0.0031	–
MnO*)	–	0.0	1.66	4.91	–	0.22	0.58	0.0	–	0.0	2.94	5.28	–	2.79	–	0.02	–	28.23
MnO$_2$*)	–	7.75	29.77	16.28	–	0.19	28.87	13.75	–	0.60	32.94	48.67	–	12.65	–	0.61	–	0.38

The samples marked by 'V' index were analysed at the chemical-analytical laboratory of the Geological Institute of the Russian Academy of Sciences by a combined technique of the classical wet chemical analysis (H_2O^+, H_2O^-, CO_2, Mn and Fe forms, Cl, SO_3, S, and others) and by using the plasma spectroanalyser J.Y.-48 ('Joben Yvon', France) with light weights (0.1 g). The Mn forms (MnO and MnO_2) were determined by active oxygen with iodometric method. The heavy metals and trace elements were determined by X-ray fluorescent spectrometry (V*).

As in previous studies (Varentsov *et al.*, 1983), the analytical determinations were controlled by multiple measurements of reference international standards.

The samples marked by 'Gr' index were analysed at the laboratories in Hungary using the wet chemical analysis (major components) and inductively coupled plasma spectroscopy (ICP). The heavy metals and trace elements were determined by atomic absorption (A.A), plasma spectroscopy, and instrumental neutron activation analysis (INA).

The analyses shown by 'V' and 'Gr' indexes were carried out from different samples. The position of the samples and the site of their recovery are shown in Figure 44. The mineral composition was determined by X-ray diffractometry.

1. Sample 1V. Brachiopod limestone of the 'Hierlatz-type', Csárdahegy area. Calcite is dominant; the insoluble residue is composed of kaolinite with hydromica and smectite admixture.

2. Sample 1/20 Gr. See no. 1.

3. Sample 2V. Red-brown ferruginous clay bedded over the 'Hierlatz-type' limestone and composed of a mixture of hydromica and goethite with notable admixture of kaolinite, hematite, and quartz.

4. Sample 2/21 Gr. See no. 2.

5. Sample 3V. Manganese oxyhydroxide ore composed of a mixture of 10 Å-manganates (the todorokite-buserite-asbolan etc. group), pyrolusite, and hydrogoethite with an admixture of cryptomelane and traces of smectite.

6. Sample 3/22 Gr. See no. 5.

7. Sample 4V. Clayey manganese oxyhydroxide ore, slightly ferruginous, contains a mixture of 10 Å-manganates and cryptomelane.

8. Sample 4/23 Gr. See no. 7.

9. Sample 5V. Manganese oxyhydroxide ore, crystalline grained. Pyrolusite and hydrogoethite dominate with an appreciable admixture of 10 Å-manganate and cryptomelane and occasional traces of [x Fe(OH)$_2$ · y FeOCl · z H$_2$O] phase. In parts composed of black sooty matter (see sample 5(1) pyrolusite dominates).

10. Sample 5/24 Gr. See no. 9.

11. Sample 6V. Predominantly limonitic highly ferruginous crust, the marginal part of the ore boulder.

12. Sample 6/25 Gr. See no. 11.

13. Sample 7V. Manganese oxyhydroxide ore of the inner part of the ore boulder. Hydrogoethite and quartz dominate with a slight admixture of 10 Å-manganates, pyrolusite and traces of mixed-layered hydromica-smectite phase.

14. Sample 7/26 Gr. See no. 13.

Notes to Tables 14 and 15 (continued)

15. Sample 8 V. Highly ferruginous clay of the central part of the ore boulder. Major components are hydrogoethite and 10 Å-manganate with an admixture of cryptomelane and deeply amorphous clayey matter (highly decomposed hydromica).

16. Sample 8/27 Gr. See no. 15.

17. Sample 9V. Manganese oxyhydroxide ore, siliceous. Major components are quartz, hydrogoethite with an admixture of 10 Å-manganates, cryptomelane, and traces of smectite.

18. Sample 9/28 Gr. See no. 17. 19. Sample 10V. Manganese oxyhydroxide ore, ferruginous, siliceous, predominantly composed of pyrolusite with subordinate quantities of hydrogoethite. Individual parts of cryptocrystalline black matter contain hydrogoethite with an admixture of pyrolusite and traces of cryptomelane. In brown patches dominate hydrogoethite and quartz with traces of cryptomelane. 20. Sample 10/29 Gr. See no. 19. 21. Sample 11V. Manganese oxyhydroxide ore with the predominant hydrogoethite and subordinate amounts of 10 Å-manganates and an admixture of cryptomelane and traces of quartz. The parts of dull black matter contain the X-ray amorphous oxyhydroxides crystallized into 10 Å-manganates with an admixture of cryptomelane. 22. Sample 11/30 Gr. See no. 21.

23. Sample 12V. Clay, in parts silicified, mainly contains degraded hydromica and quartz with occasional admixture of hydrogoethite.

24. Sample 12/31 Gr. See no. 23.

25. Sample 13V. Ferruginous manganese ore with silicate inclusions, predominant pyrolusite with an admixture of hydrogoethite and quartz. The light-colored silicate inclusions are mostly composed of quartz with hydromica admixture.

26. Sample 13/32 Gr. See no. 25.

27. Sample 14V. Clay, with green inclusions of manganese carbonate, contains a mixture of quartz and highly degraded hydromica.

28. Sample 14/33 Gr. See no. 27.

29. Sample 15V. Manganese oxyhydroxide ore, brownish, earthy, contains a fine mixture of pyrolusite and hydrogoethite. Parts and patches, composed of finely crystalline shiny material, contain hydrogoethite, 10 Å-manganites, cryptomelane with an admixture of quartz.

30. Sample 15/34 G. See no. 29.

31. Sample 16V. Manganese oxyhydroxide ore, siliceous, ferruginous, contains pyrolusite with an admixture of hydrogoethite and quartz. Parts composed mostly of brown oxyhydroxides (sample 16) contain hydrogoethite, 10 Å-manganates with an admixture of quartz.

32. Sample 16/35 Gr. See no. 31.

33. Sample 17V. Weathered Upper Liassic limestone.

34. Sample 17/36 Gr. See no. 33.

35. Sample 18V. Finely banded clayey-carbonate ferruginous-manganese ore composed of fine alternating ferruginous celadonite-like clay and rhodochrosite, the Úrkút deposit, N-part, W-field, Schaft III, the principal ore bed, Hungary.

36. Sample 18/37 Gr. See no. 35.

4. Separate observations under a microscope of the structural relations of Mn minerals and their identification by X-ray diffractometry can be, to a certain extent, generalized by a scheme of stages:

$MnCO_3 \rightarrow$ manganite $\rightarrow$ pyrolusite (I) $\rightarrow$

$\rightarrow$ cryptomelane $\rightarrow$ 10 Å-manganates (mainly todorokite) $\rightarrow$ pyrolusite (II).

Moreover, manganite, being the unstable initial oxidation product of Mn carbonates, is mainly found in the autochthonous bedding of the Úrkút deposit, whereas in Csárdahegy this mineral is not in evidence (Cseh-Németh *et al.*, 1980; Szabo and Grasselly, 1980).

3.2.4. GEOCHEMISTRY

3.2.4.1. The Major Components

The basic information about the geochemical features of behavior of major components in the zone of formation of the oxyhydroxide manganese and ferromanganese ores of Csárdahegy is given in Tables 14 and 15 and Figures 45 and 46. The analysis of this data allows us to draw the following conclusions.

1. Extensive research has established that in the supergene zone the Ti and Al components are highly geochemically inert. These features are definitely revealed in the distribution of relative accumulation (removal) coefficients of these elements in the zone of oxidation of manganese ores and karst formation:

δTi and δAl. The coefficients were calculated from

$$\delta\text{El} = \frac{\text{El} - \text{El}_{18}}{\text{El}_{18}},$$

where El is the content of the element in a given sample; El_{18} is the content of the element in sample 18 composed of rhodochrosite-celadinite ore which is assumed as the primary unaltered substratum (Tables 14 and 15, Figure 45).

It should be emphasized that Ti and Al experience residual accumulation of one main type, in almost the same way, in rather intensively altered clayey, hematitic, silicitic (quartz) rocks, where the clayey matter is represented by deeply degraded hydromica with occasional appreciable admixtures of kaolinite (sample 2).

In manganese and ferromanganese ores of the oxidation zone the weight proportion of Al and Ti greatly decreases due to the massive supply and accumulation of the Fe and Mn oxyhydroxide phases. Moreover, the content of Ti in the authigenic accumulations proper of Fe, Mn oxyhydroxides sometimes does not exceed the determined minimum ($< 0.01\%$ Ti, samples 7V and 9V; Table 14).

2. On the whole, Fe and Mn are rather intensively accumulated in the oxidation zone. However, if Fe accumulation is fairly restricted in the lower and upper autochthonous parts of the deposit, the concentrations of this metal within the allochthonous boulder block are the highest. Contrarily, the largest Mn accumulations, associated with the supply of 250–360% of the primary amount of the metal, are observed in the lower part of the deposit (samples 4V and 5V; Figure 45). It is particularly remarkable that in these very

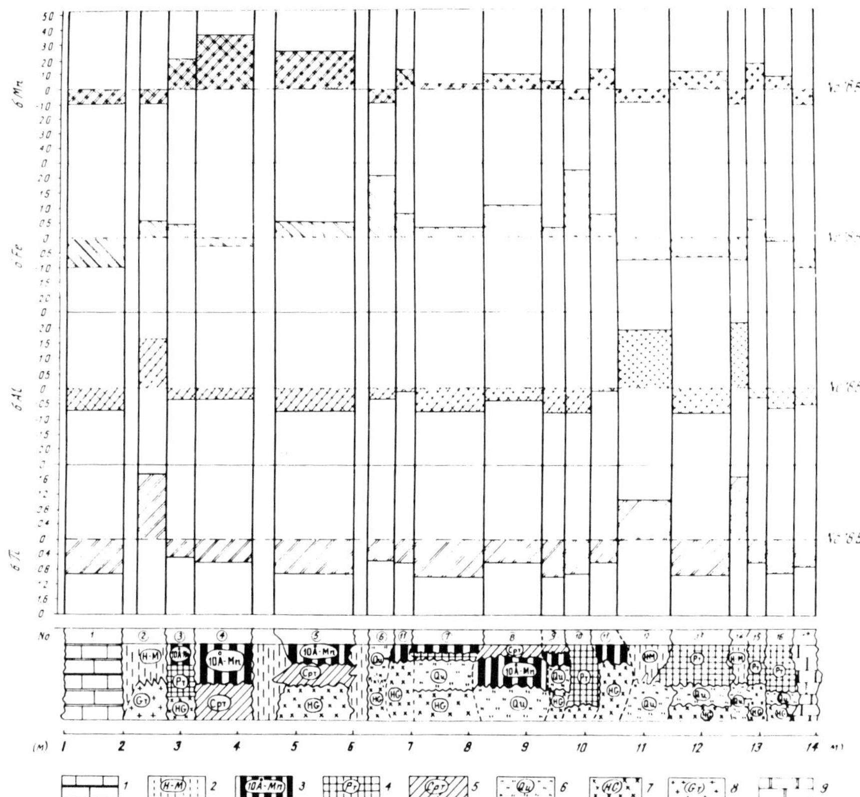

Fig. 45. Distribution of relative accumulation (removal) coefficients: δTi, δAl, δFe, δMn, in ores and rocks of the oxidized zone of the Úrkút deposit, Csárdahegy area. The coefficients were calculated according to: $\delta El = \dfrac{El - El_{18/37}}{El_{18/37}}$, where El is the content of the element in a given sample; $El_{18/37}$ is the content of the element in sample 18/37 (rhodochrosite-seladonite ore) assumed as unaltered primary material. The mineral content is given in 'Notes to Tables 14 and 15'. Lithology is shown in Figure 44. Legend: 1. Limestone of the 'Hierlatz type', Lower Liassic; 2. Degraded hydromica; 3. 10 Å-manganates; 4. Pyrolusite; 5. Cryptomelane; 6. Quartz; 7. Hydrogoethite; 8. Goethite; 9. Upper Liassic weathered limestone.

parts the products of relatively late and mature stages of supergene alteration dominate cryptomelane and 10 Å-manganates, whereas in the upper parts of the deposit pyrolusite is the principal Mn mineral.

These features of Mn and Fe behavior are confirmed by the distribution of Fe/Ti, Mn/Ti, and Mn/Fe ratios (Figure 46). As noted above, however, when interpreting Fe/Ti and Mn/Ti, the extremely low Ti contents should be taken into account as being caused by intensive Mn, Fe, and SiO_2 input to the upper parts of the deposit.

3. In the distribution of P/Ti ratio there is a marked tendency to phosphorus accumulation in the oxidation zone, in particular in its upper parts. The anomalously high

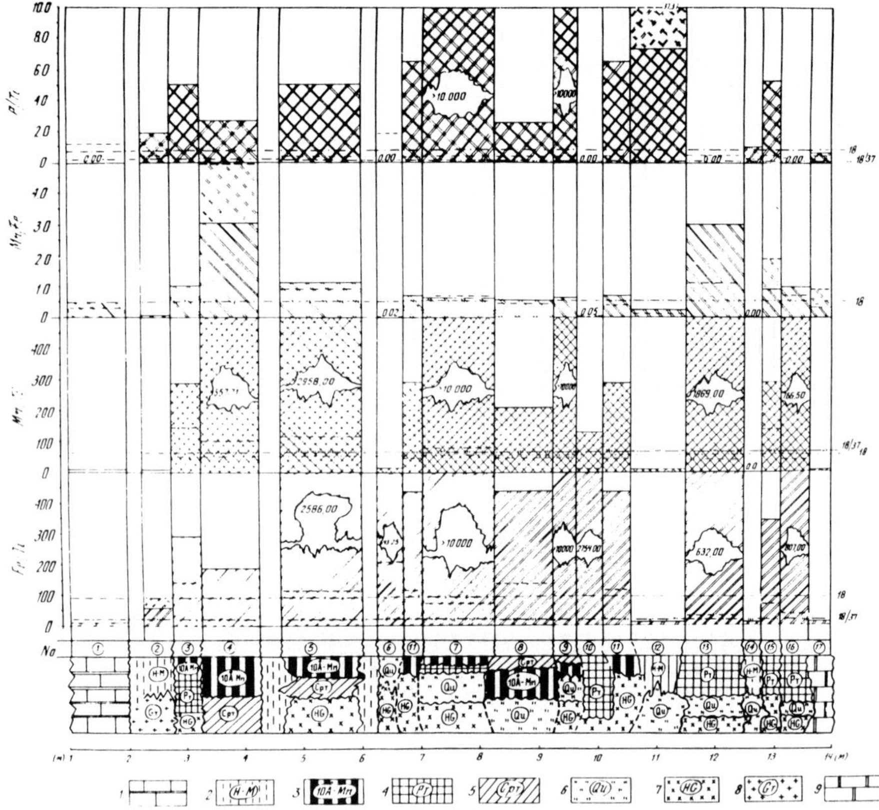

Fig. 46.   Distribution of Fe/Ti, Mn/Ti, Mn/Fe, P/Ti ratios in ores and rocks of the oxidized zone of the Úrkút deposit, Csárdahegy area.   The mineral composition is given in 'Notes to Tables 14 and 15'.   Lithology is shown in Figure 44.   Legend: 1. Limestone of the 'Hierlatz type', Lower Liassic; 2. Degarded hydromica; 3. 10 Å-Manganates; 4. Pyrolusite; 5. Cryptomelane; 6. Quartz; 7. Hydrogoethite; 8. Goethite; 9. Upper Liassic weathered limestone. Dashed line shows ratio values calculated from the analyses marked by 'Gr' index.

phosphorus contents (P > 10%) are of special importance, the phosphorus being represented by X-ray amorphous compounds in clayey rock (sample 12), chiefly composed of deeply degraded hydromica and quartz (Figure 46).

3.2.4.2. Heavy Metals

The distribution of relative accumulation (removal) coefficients by the example of Co, Ni, Cu, and Ba reveals notable differences reflecting the geochemical features of their behavior in the karst type of weathering.

Co is characterized by the most-distinctly defined distribution. The maximal accumula-

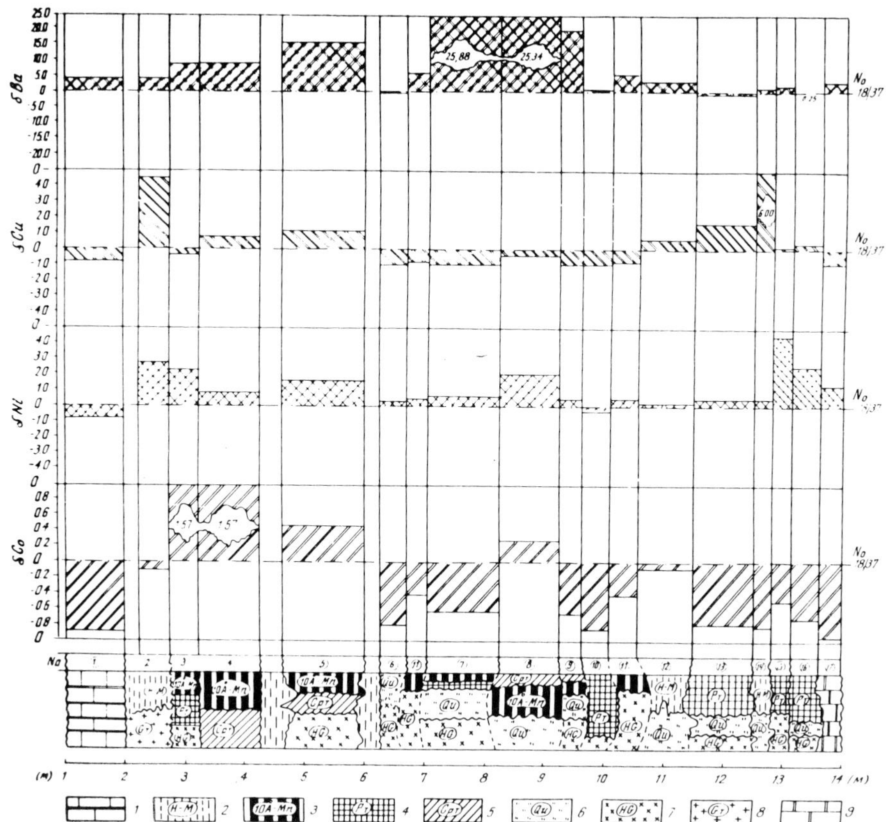

Fig. 47. Distribution of relative accumulation (removal) coefficients: δCo, δNi, δCu, δBa in ores and rocks of the oxidized zone of the Úrkút deposit, Csárdahegy area (see Figure 44). The mineral composition is given in 'Notes to Tables 14 and 15'. Lithology is shown in Figure 44. Legend: 1. Limestone of the 'Hierplatz type', Lower Liassic; 2. Degraded hydromica; 3. 10 Å-manganates; 4. Pyrolusite; 5. Cryptomelane; 6. Quartz; 7. Hydrogoethite; 8. Goethite; 9. Upper Liassic weathered limestone.

tion of metal (up to 157%) is observed in the lower part of the deposit (samples 3 to 5 and 8; Figure 47), in ores containing appreciable amounts of 10 Å-manganates and cryptomelane. Contrarily, in the upper parts, with dominating pyrolusite, hydrogoethite, and quartz, the removal of the metal is clearly displayed. We may presume that the distribution of Co, as the element sufficiently sensitive to changes in Eh and pH regimes, describes with greater clarity the zonality in the development of the oxidation and supergene transformation of the initial rhodochrosite-clayey ores. In this case, with a certain amount of justification, we may suggest the possibility of formation of asbolane varieties in the composition of minerals of the 10 Å-manganates (Chukhrov *et al.*, 1980a, 1980c; 1982; 1984; 1989).

Ni, in contrast to Co, is accumulated throughout the entire deposit, the richest beds of which are found in the upper parts (samples 15 and 16; Figure 47) in ores predominantly composed of pyrolusite and hydrogoethite.

Cu shows no distinct tendencies in its distribution, but the maximal accumulations (up to 400–600%) are found in the rocks composed of the deeply degraded hydromica, often with the admixture of goethite, quartz, and occasionally kaolinite (samples 2 and 14; Figure 47).

Ba is accumulated throughout the entire zone of oxidation. The ores of autochthonous bedding are characterized by higher concentrations of this element in the lower part of the deposit, which gradually grow toward the center: samples 3/22–4/23–5/24. Accordingly, δBa has values of 7.76–8.46–15.13 in the lower part, whereas in the upper part of the deposit (samples 12/31 to 17/35) δBa value does not exceed 3.52 (Tables 14 and 15; Figure 47). The maximal Ba accumulations, however, are observed in the central part of the allochthonous boulder block (samples 7 and 8).

It is important to note the clearly expressed mineralogical control over the distribution of the element, i.e., the largest Ba accumulations are in the ores with appreciable amounts of 10 Å-manganates and cryptomelane, whereas in pyrolusite varieties Ba concentrations are lower. Therefore, the most likely Ba carriers are tectonomanganites, whose tunnel structure is highly favorable crystallochemically for fixation of such large cations as Ba^{2+} (ion radius 1.38 Å) and K^+ (1.33 Å).

3.2.4.3. Rare Earth Elements (REE)

In spite of the extensive application of rare earth elements (REE) in the solution of various geological problems, the information on REE geochemistry in the crusts of weathering and particularly in the zone of oxidation of Mn ores is scarce and contradictory (Balashov, 1976; Basu *et al.*, 1982; Boiko *et al.*, 1986; Burkov and Podporina, 1967; Cullers *et al.*, 1979; Duddy, 1981; Geochemistry of elements-hydrolysates, 1980; Haskin and Haskin, 1966; Haskin *et al.*, 1966; Krainov, 1973; Nesbitt, 1979; Piper, 1974; Podporina and Burkov, 1977; Ronov *et al.*, 1967; Scherbina, 1959; Taylor and McLennan, 1981, 1985).

On the one hand, it was established with sufficient reliability that REE are geochemically highly inactive, almost immobile in supergene settings, and that they are generally accumulated in the profile of the crust of weathering even during profound lateritic decomposition of the parent rocks. It should be also emphasized that REE, as elements-hydrolysates, are fairly sensitive to pH regime, and individual REE change their modes of occurrence in accordance with the change of Eh values, and in the amounts of organic or inorganic complexing agents (Scherbina, 1959; Yatsimirsky *et al.*, 1966). These features are displayed with different intensity in individual lanthanides and in certain settings may result in their fractionation.

In this study it is interesting to document to what extent REE reflect the composition of the source rocks and the scope of their fractionation and of their relationships with the general evolution of oxidized karst weathering, which is accompanied by the formation of authigenic Mn and Fe oxyhydroxide compounds.

REE: Relationships in the Primary Ores and the Products of Oxidation, General Features of Distribution

The analysis of REE, normalized on North American shale and chondrite meteoritic matter (Baynton, 1984; Haekin *et al.*, 1966) shows that the general REE distribution corresponds to their composition in the parent rhodochrosite-clay ores. Moreover, REE distribution and abundance are very similar in North American and European clay shales, in Paleozoic clayey rocks of the Russian platform, and in the Post-Archean Average Australian Shale. The differences are, as a rule, not much above analytical errors (Balashov, 1976; Haskin *et al.*, 1966; Nance and Taylor, 1976; Ronov *et al.*, 1967; Taylor and McLennan, 1981).

Sample 18/37 (Table 17) of fine-layered rhodochrosite-clayey ore was chosen to characterize the initial parent rock. In a previous paper (Varentsov *et al.*, 1988), it was indicated that REE content in this sample is typical of the hydrothermal-diagenetic ores of the Úrkút deposit (Grasselly and Panto, 1987) and that REE distribution in these ores is of a dualistic nature, i.e., the behavior of light lanthanides resembles that of hydrothermal metalliferous sediments, whereas the heavy REE content reflects the seawater influence (De Baar *et al.*, 1985a, 1985b).

It is precisely these hydrothermal-diagenetic features of parent manganese ores that are fully exhibited in the distribution of REE of the weathering products. The ores and rocks of the oxidation zone are clearly subdivided into accumulations enriched or depleted in lanthanides, as compared to the source rocks which occupy the middle position in the plots of shale and chondrite normalized REE values (Figures 48 and 49). The extremely high accumulation of lanthanides in sample 12/13 is of particular interest, the sample having a fairly high content of X-ray amorphous supergenic phosphates (P > 10%; Table 17; Figure 48). These features are yet more apparent in the distribution of REE directly normalized on source ore (Figure 50).

The major statements are as follows:

1. There is an obvious relation between the extremely high accumulation of lanthanides and the formation of supergenic phosphates which, in general, characterizes REE geochemistry in the lithogenic zone (Arrhenius and Bonnatti, 1965; Blokh and Kochenov, 1964). In this particular case, however, it is interesting to note the increase in enrichment from the light (La, Ce, Nd) through intermediate (Sm, Eu, Tb) to heavy REE (Tm, Yb, Lu). In other words, the rather high mobility of intermediate REE, as a rule extremely inert in the crusts of weathering, can be regarded as a discriminating feature of the supergenic phosphate formation.

2. The products of supergenic alteration, characterized by the general REE enrichment (samples 14/33; 2/21; 15/34; 4/23; 11/30), are not enriched in Sm and Eu as compared to light and heavy lanthanides. This feature is particularly apparent in the autochthonous altered rocks with dominating contents of deeply degraded hydromica, quartz, goethite, occasionally of pyrolusite, and hydrogoethite. In the most typical cases, not only Sm and Eu are accumulated, but Ti and Al as well (samples 14/33, 2/31; Figures 45 and 50).

3. The oxidized Mn and Fe ores are the products of karst weathering rather enriched in REE. Moreover, the general decrease in lanthanides enrichment is observed in

Table 16.   Content of trace and rare earth elements in rocks and manganese ores of the Úrkút deposit, Csárdarhegy, Hungary, ppm, air-dry weight, instrumental neutron activation analysis.

Component	1/20	2/21	3/22	4/23	5/24	6/25	7/26	8/27	9/28	10/29
Sc	3.2 ± 0.1	9.2 ± 0.1	2.5 ± 0.1	4.8 ± 0.1	1.5 ± 0.1	2.1 ± 0.1	0.4 ± 0.1	3.0 ± 0.1	0.2 ± 0.2	0.6 ± 0.1
Cr	9 ± 1	70 ± 2	< 5	24 ± 3	< 5	24 ± 3	7 ± 1	11 ± 2	< 5	9 ± 2
Co	1.3 ± 0.1	133 ± 3	372 ± 4	374 ± 6	211 ± 4	29 ± 2	50 ± 2	170 ± 3	45.5 ± 1.5	21.2 ± 0.2
Zn	< 30	223 ± 6	154 ± 6	< 30	104 ± 4	42 ± 4	71 ± 5	174 ± 4	31 ± 4	< 30
As	< 10	1270 ± 10	198 ± 6	84 ± 4	149 ± 5	67 ± 3	88 ± 2	147 ± 4	71 ± 4	33 ± 3
Rb	< 50	< 50	< 50	< 50	< 50	< 50	< 50	< 50	< 50	< 50
Sb	0.8 ± 0.1	32.6 ± 0.4	6.8 ± 0.3	1.2 ± 0.2	2.0 ± 0.2	< 0.5	0.9 ± 0.1	4.5 ± 0.2	0.7 ± 0.1	< 0.5
Cs	0.5 ± 0.1	5.6 ± 0.3	< 0.4	2.4 ± 0.3	< 0.4	< 0.4	< 0.4	1.0 ± 0.1	< 0.4	< 0.4
La		90.0 ± 1.0	15.1 ± 0.2	35.5 ± 0.2	8.5 ± 0.3	7.6 ± 0.1	3.9 ± 0.1	28.5 ± 0.2	2.3 ± 0.1	5.3 ± 0.2
Ce		210 ± 7	55 ± 1.7	139 ± 3	28.1 ± 0.9	18.3 ± 0.7	9.2 ± 0.6	69.7 ± 1.8	6.4 ± 0.2	15.6 ± 0.3
Nd		62 ± 7	10 ± 2	26 ± 4	6 ± 1	5 ± 1	3 ± 1	18 ± 3	1.5 ± 0.3	3.5 ± 0.2
Sm		36.0 ± 0.2	5.3 ± 0.1	15.3 ± 0.1	2.9 ± 0.1	3.0 ± 0.1	0.8 ± 0.3	6.2 ± 0.1	2.2 ± 0.1	2.5 ± 0.1
Eu	0.28 ± 0.02	4.0 ± 0.08	0.52 ± 0.05	1.6 ± 0.05	0.30 ± 0.03	0.25 ± 0.03	0.17 ± 0.02	0.55 ± 0.04	0.33 ± 0.02	0.31 ± 0.02
Tb		2.4 ± 0.2	1.4 ± 0.2	2.1 ± 0.3	1.1 ± 0.1	1.1 ± 0.1	1.1 ± 0.1	3.2 ± 0.2	0.6 ± 0.1	0.7 ± 0.1
Tm		1.1 ± 0.1	0.6 ± 0.1	0.9 ± 0.2	0.5 ± 0.1	0.6 ± 0.1	0.6 ± 0.1	1.5 ± 0.1	0.3 ± 0.1	0.3 ± 0.1
Yb		5.8 ± 0.2	1.9 ± 0.2	2.2 ± 0.1	0.7 ± 0.1	1.3 ± 0.1	1.7 ± 0.1	3.5 ± 0.1	1.0 ± 0.1	0.9 ± 0.1
Lu		0.87 ± 0.05	0.34 ± 0.02	0.32 ± 0.02	0.16 ± 0.02	0.32 ± 0.02	0.27 ± 0.02	0.59 ± 0.03	0.15 ± 0.01	0.11 ± 0.01
Hf	0.5 ± 0.1	3.6 ± 0.2	< 0.5	1.3 ± 0.2	< 0.5	< 0.5	<0.5	0.7 ± 0.1	< 0.5	< 0.5
Ta	0.1 ± 0.03	1.1 ± 0.2	< 0.1	< 0.1	< 0.1	< 0.1	< 0.1	< 0.1	< 0.1	< 0.1
Th	1.1 ± 0.1	10.8 ± 0.3	1.3 ± 0.2	5.0 ± 0.2	< 0.5	1.3 ± 0.2	< 0.5	1.9 ± 0.1	< 0.5	< 0.5
U		4.6 ± 0.3	5.4 ± 0.3	4.6 ± 0.3	3.3 ± 0.3	3.8 ± 0.2	3.1 ± 0.2	5.7 ± 0.2	2.9 ± 0.2	2 ± 0.4
REE (La-Lu)		412.7	90.16	222.92	48.26	37.47	20.74	131.54	14.78	29.22
Ce/Ce**		1.39	2.20	2.27	1.94	1.46	1.34	1.49	1.70	1.78
(La-Sm)/(Cd-Yb)	–	39.17	21.90	38.56	19.78	11.30	4.97	14.93	6.53	15.38
La/Yb	–	15.52	7.95	16.14	12.14	5.85	2.29	8.14	2.30	5.89
LaN/YbN	–	1.35	0.70	1.40	1.05	0.50	0.20	0.71	0.21	0.14
Ce/La	–	2.33	3.64	3.92	3.31	2.41	2.36	2.45	2.78	2.94
Nd/La	–	0.69	0.66	0.73	0.71	0.66	0.77	0.63	0.65	0.66
Sm/La	–	0.40	0.35	0.43	0.34	0.39	0.21	0.22	0.96	0.47
Ce/Yb	–	36.21	28.95	63.18	40.14	14.08	5.41	19.91	6.40	17.33
En/Sm	–	0.11	0.10	0.10	0.10	0.08	0.21	0.09	0.15	0.12
δLa		1.719	-0.544	0.073	-0.743	-0.770	-0.882	-0.139	-0.931	-0.840
δCe		1.278	-0.403	0.508	-0.695	-0.802	-0.900	-0.244	-0.931	-0.831
δEu		3.040	-0.475	0.616	-0.697	-0.747	-0.828	-0.444	-0.667	-0.687
δYb		1.231	-0.269	-0.154	-0.731	-0.500	-0.346	0.346	-0.615	-0.654

Table 17. Content of trace and rare earth elements in rocks and manganese ores of the Úrkút deposit, Csárdarhegy, Hungary, ppm air-dry weight, instrumental neutron activation analysis.

Component	11/30	12/31	13/32	14/33	15/34	16/35	17/36	18/37
Sc	4.1 ± 0.1	13.2 ± 0.1	0.8 ± 0.1	16.9 ± 0.1	4.2 ± 0.1	1.7 ± 0.1	2.0 ± 0.1	4.2 ± 0.1
Cr	25 ± 3	69 ± 3	8 ± 1	92 ± 3	26 ± 2	8 ± 1	< 5	19 ± 2
Co	84.7 ± 1.2	128 ± 1	26.3 ± 0.2	24.1 ± 0.3	75.1 ± 0.9	36.4 ± 0.3	6.5 ± 0.1	151 ± 3
Zn	117 ± 5	177 ± 8	< 30	< 30	110 ± 3	68 ± 3	< 30	< 30
As	104 ± 6	15 ± 2	43 ± 4	84 ± 4	76 ± 5	54 ± 4	12 ± 1	16.1 ± 0.9
Rb	< 50	< 50	< 50	136 ± 8	< 50	< 50	< 50	95 ± 12
Zr		2350 ± 230						
Sb	< 0.5	< 0.5	2.2 ± 0.1	< 0.5	2.2 ± 0.1	1.5 ± 0.1	< 0.5	1.0 ± 0.2
Cs	0.9 ± 0.2	2.0 ± 0.3	< 0.4	8.8 ± 0.2	1.1 ± 0.1	< 0.4	0.6 ± 0.1	1.5 ± 0.2
La	32.8 ± 0.2	428 ± 12	2.3 ± 0.1	131 ± 3	66.8 ± 0.9	4.9 ± 0.1	14.1 ± 0.2	33.1 ± 0.4
Ce	122 ± 3	1104 ± 18	6.2 ± 0.2	299 ± 9	167 ± 3	13.4 ± 0.2	32.3 ± 0.4	92.2 ± 2.1
Nd	20 ± 4	273 ± 32	65 ± 6	45 ± 5	3 ± 1	9 ± 2	21 ± 3	
Sm	17.7 ± 0.1	180 ± 2	4.8 ± 0.1	51.1 ± 0.5	33.2 ± 0.2	5.4 ± 0.1	4.1 ± 0.1	9.4 ± 0.1
Eu	2.26 ± 0.05	22.2 ± 0.2	0.60 ± 0.11	5.09 ± 0.09	4.22 ± 0.06	0.73 ± 0.03	0.46 ± 0.02	0.99 ± 0.05
Tb	4.3 ± 0.2	124 ± 2	0.2 ± 0.05	3.9 ± 0.3	2.3 ± 0.2	1.1 ± 0.1	1.0 ± 0.1	2.3 ± 0.3
Tm	2.1 ± 0.2	60 ± 3	0.1 ± 0.03	2.0 ± 0.2	1.1 ± 0.1	0.6 ± 0.1	0.5 ± 0.1	1.1 ± 0.1
Yb	6.4 ± 0.1	159 ± 3	0.4 ± 0.1	3.4 ± 0.2	3.8 ± 0.1	1.1 ± 0.1	0.8 ± 0.1	2.6 ± 0.1
Lu	0.99 ± 0.02	2.9 ± 0.1	0.13 ± 0.01	0.81 ± 0.03	0.58 ± 0.03	0.16 ± 0.02	0.12 ± 0.09	0.39 ± 0.02
Hf	1.1 ± 0.2	1.6 ± 0.2	< 0.5	4.4 ± 0.2	1.1 ± 0.1	< 0.5	0.5 ± 0.1	1.5 ± 0.2
Ta	< 0.1	< 0.1	< 0.1	1.8 ± 0.1	< 0.1	< 0.1	< 0.1	< 0.1
Th	3.4 ± 0.2	4.0 ± 1.0	< 0.5	15.7 ± 0.4	2.8 ± 0.2	0.4 ± 0.1	1.1 ± 0.1	3.1 ± 0.2
U	5.6 ± 0.4	181 ± 7	2.0 ± 0.3	15.3 ± 1.8	6.9 ± 0.3	2.8 ± 0.2	< 2	< 2
REE (La-Lu)	208.55	2379.3	14.73	561.30	324.00	29.96	62.38	163.08
Ce/Ce**	2.32	1.58	–	1.54	1.50	1.70	1.40	1.71
(La-Lu)/(Cd-Yb)	15.04	5.79	19.00	58.72	43.33	9.54	25.87	25.95
La/Yb	5.13	2.69	5.75	38.53	17.58	4.45	17.63	12.73
LaN/YbN	0.45	0.23	0.50	3.36	1.53	0.41	1.50	1.42
Ce/La	3.72	2.58	2.30	2.28	2.50	2.73	2.29	2.79
Nd/La	0.61	0.64	-	0.50	0.67	0.61	0.64	0.63
Sm/La	0.54	0.42	2.09	0.39	0.50	1.10	0.29	0.28
Ce/Yb	19.06	6.94	15.50	87.94	43.95	12.18	40.38	35.46
En/Sm	0.13	0.12	0.13	0.10	0.13	0.14	0.11	0.11
δLa	-0.009	11.931	-0.931	2.958	1.018	-0.852	-0.574	
δCe	0.323	10.974	-0.933	2.243	0.811	-0.855	-0.650	
δEu	1.283	21.424	-0.394	4.141	3.263	-0.263	-0.535	
δYb	1.462	60.154	-0.846	0.308	0.462	-0.577	-0.692	

The determinations were carried out by the method of instrumental neutron activation analysis at the research centers of Hungary using international standards. The samples are described in 'Notes to Tables 14 and 15'. The cerium anomaly was calculated for ratio Ce/Ce** -[3Ce/Ce$_{shale}$]/[2La/La$_{shale}$ + Na/Na$_{shale}$], where Ce$_{shale}$ etc. is the content of this rare earth element in North American shale (Haskin $et\ al.$, 1966). The relative accumulation (removal) was calculated according to: $\delta E = \dfrac{E - E_{18/37}}{E_{18/37}}$, where E is the content of the element in the sample; $E_{18/37}$ is the content of the element in sample 18/37 (rhodochrosite-celadonite ore) assumed as the primary unaltered material.

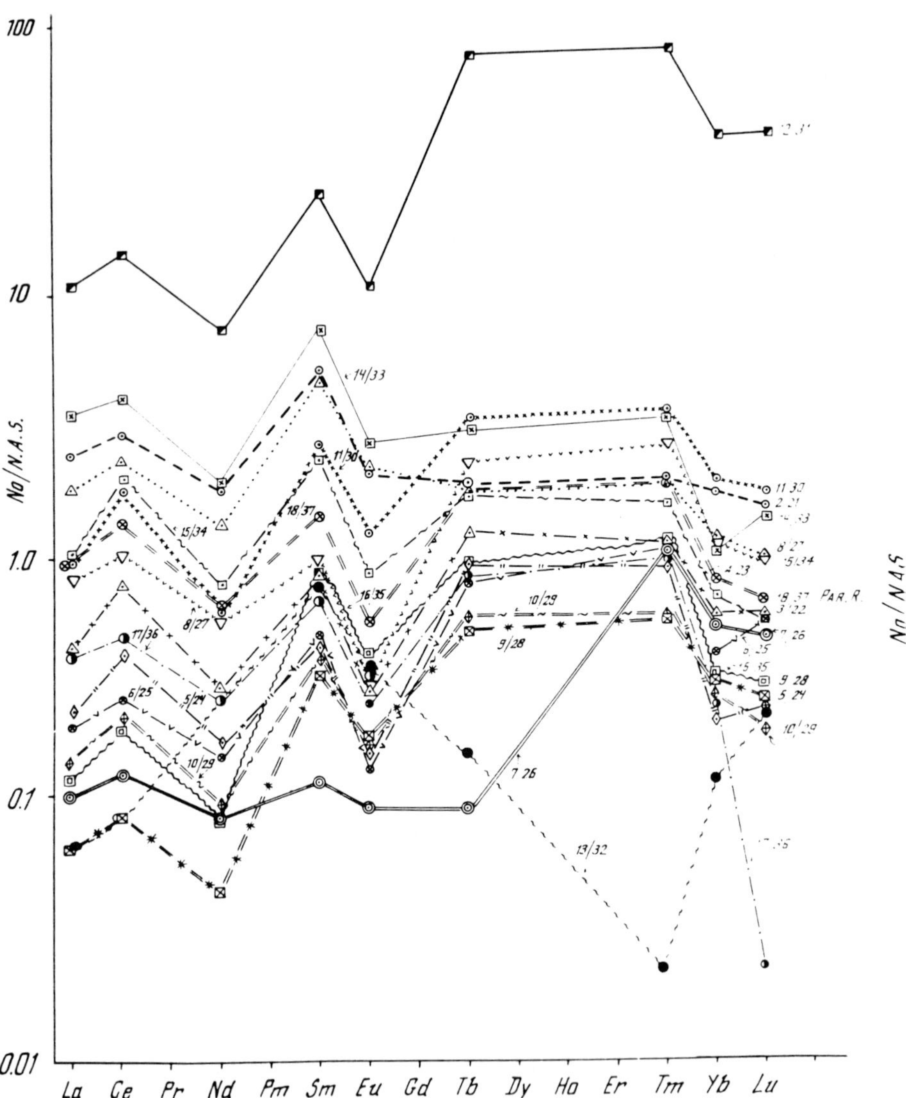

Fig. 48.   Distribution of shale-normalized rare earth elements (Haskin *et al.*, 1966) in ores and rocks of the oxidized zone of the Úrkút deposit, Csárdahegy area. The numbers of curves correspond to the analyses given in Tables 16 and 17. Lithology and mineral composition see in 'Notes to Tables 14 and 15'.

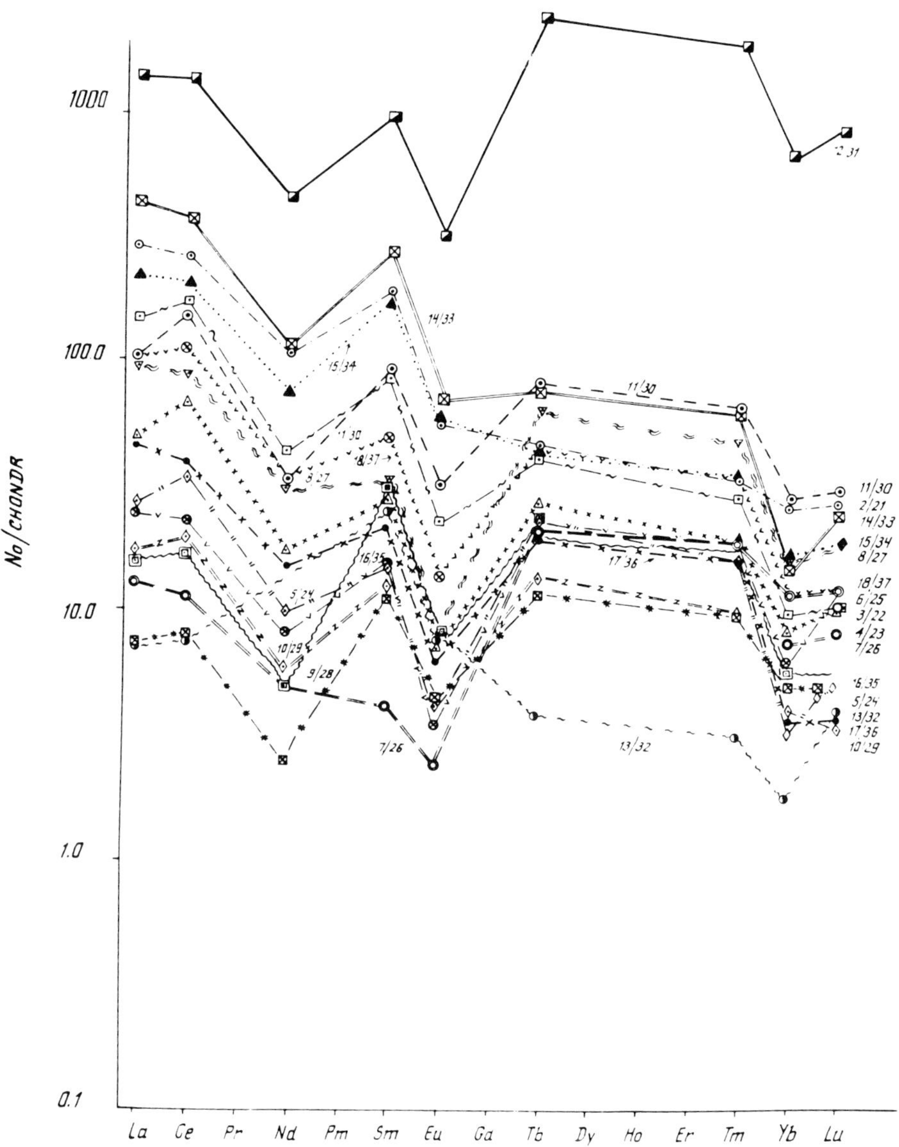

Fig. 49. Distribution of rare earth elements, normalized on chondritic meteoritic material (Baynton, 1984), in ores and rocks of the oxidized zone of the Úrkút deposit, Csárdahegy area. The numbers of curves correspond to the analyses given in Tables 16 and 17. Lithology and mineral composition see in 'Notes to Tables 14 and 15'.

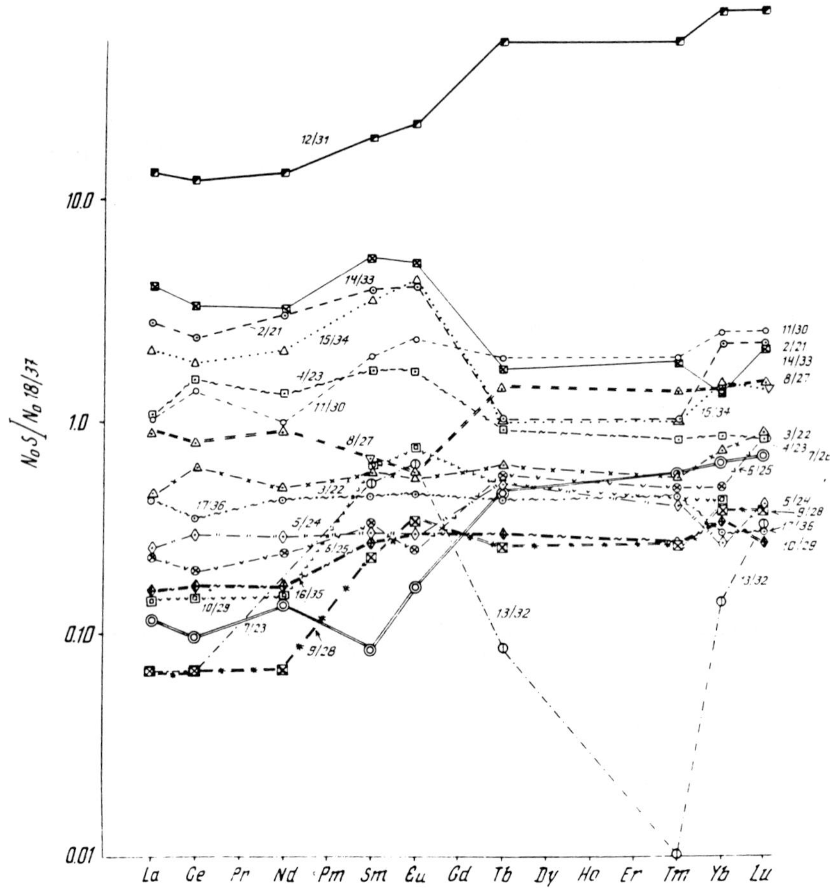

Fig. 50.   Distribution of REE normalised on unaltered primary material (rhodochrosite-celadonite ore, sample 18/37) in ores and rocks of the oxidized zone of the Úrkút deposit, Csárdahegy area. The numbers of curves correspond to the analyses given in Tables 16 and 17. Lithology and mineral composition see in Tables 14 and 15.

the transition from autochthonous ores to redeposited varieties. For example, it is evidently displayed by correlation of relatively low amounts of REE in ores in the central part of the boulder block (samples 7/23, 9/28, and 10/29) with the fairly rich autochthonous varieties (samples 16/35, Figure 50). Therefore, the process of authigenic formation of supergenic Mn and Fe oxyhydroxide phases is not associated with REE accumulation. In the process of general depletion, however, a tendency is observed in the distribution of lanthanides which was noted earlier in the products enriched in phosphates, i.e., a slight increase from light through intermediate to heavy REE fractions.

4. The features in REE behavior in the Mn oxidation zone of carbonate-clayey ores (items 1–3) discussed earlier are distinctly exhibited on the profile across the Csárdahegy area (Figure 51). The Σ(La–Lu) distribution implies that high accumulations of these elements are mostly associated with autochthonous altered rocks and are not typical of the authigenic Mn and Fe oxyhydroxide phases. The example of Σ(La-Sm)/(Gd-Yb), La/Yb, Ce/Yb ratios distribution shown that the autochthonous residual deposits of the oxidation zone generally accumulate not only ΣREE, but light fractions of lanthanides as well. However, the Mn and Fe oxyhydroxide ores, which are associated with an appreciable input of these metals (Figures 52 and 53), accumulate chiefly heavy REE.

REE: INDIVIDUAL FEATURES OF BEHAVIOR IN THE ZONE OF THE KARST MN ORE FORMATION

In the group of light REE, the measure of separation of La and Ce is geochemically rather informative (Figure 52). The analysis of Ce/La ratio distribution implies that the highest values of the ratio, as compared to source deposits, mostly characterize later oxidation products in the lower autochthonous part of the deposit, which are chiefly composed of the minerals of the 10 Å-manganates group and cryptomelane. In the ores of the upper part of the deposit and in the redeposited varieties, the Ce/La values are less than in the rhodochrosite-clayey sediments.

A similar distribution pattern is observed in the cerium anomaly (Ce/Ce∗∗) having higher positive values in the ores indicated above (Figure 52). This data can be interpreted as the evidence of Ce accumulation in a setting of rather high oxidation, in all probability, in the form of Ce with a high capacity for hydrolysis (Krainov, 1973; Schrbina, 1959).

The distribution of relative accumulation (removal) coefficients of δLa and δCe, on the whole, is in agreement with these conclusions (Figure 53). These values, however, distinctly reveal the extraordinary role of supergenic phosphates (sample 12) as concentrators of REE in the process of formation of their residual accumulations in the zone of karst supergenesis.

In the Ce/La–La/Yb plot (Figure 54), the dots indicating the karst oxidation products form a cluster around sample 18/37, the parent hydrothermal-diagenetic rhodochrosite-clayey ore, and notably differ from the average composition of the shales and the sedimentary rocks (Haskin *et al.*, 1966). Such relationships infer a high heritage of REE in the materials of supergenic alterations of source sediments which is likewise repeatedly observed in the crusts of more extensive weathering (Boyko *et al.*, 1986; Burkov and Podporina, 1967; Duddy, 1981; 'Geochemistry of elements-hydrolysates', 1980; Nesbitt, 1979; Podporina and Burkov, 1977; Ronov *et al.*, 1967).

Intermediate REE, as implied by Eu/Sm distribution, do not show appreciable deviation from the values typical of the parent rhodochrosite-clayey ores (Figure 52).

The Sm/La distribution, as a measure of separation of light and intermediate REE, reveals fairly high values of this ratio in the autochthonous oxyhydroxide ores of the upper part of the deposit (samples 13 to 16; Figure 52), which are mostly composed of pyrolusite, hydrogoethite, and quartz. Such relations can be attributed to residual passive accumulation of Sm with essentially higher migration and removal of La against the

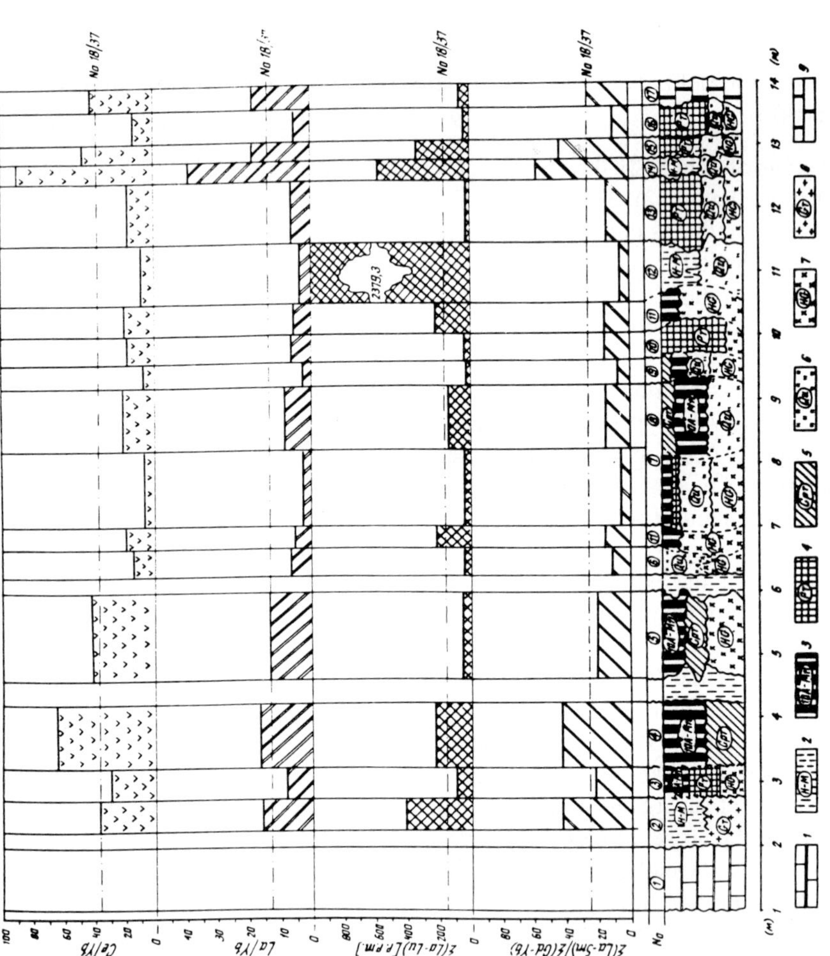

Fig. 51. Distribution of the ratio of the total sum of light rare earth elements (LREE) and the total sum of heavy REE (HREE), $[\Sigma(La\text{-}Sm)/\Sigma(Sd\text{-}Yb)]$, of the total sum of REE, $[\Sigma(La\text{-}Lu)]$, and La/Yb, Ce/Yb ratios in ores and rocks of the oxidized zone of the Úrkút deposit, Csárdahegy area. See lithology, mineral composition and legend to Figure 45; 'Notes to Tables 14 and 15'.

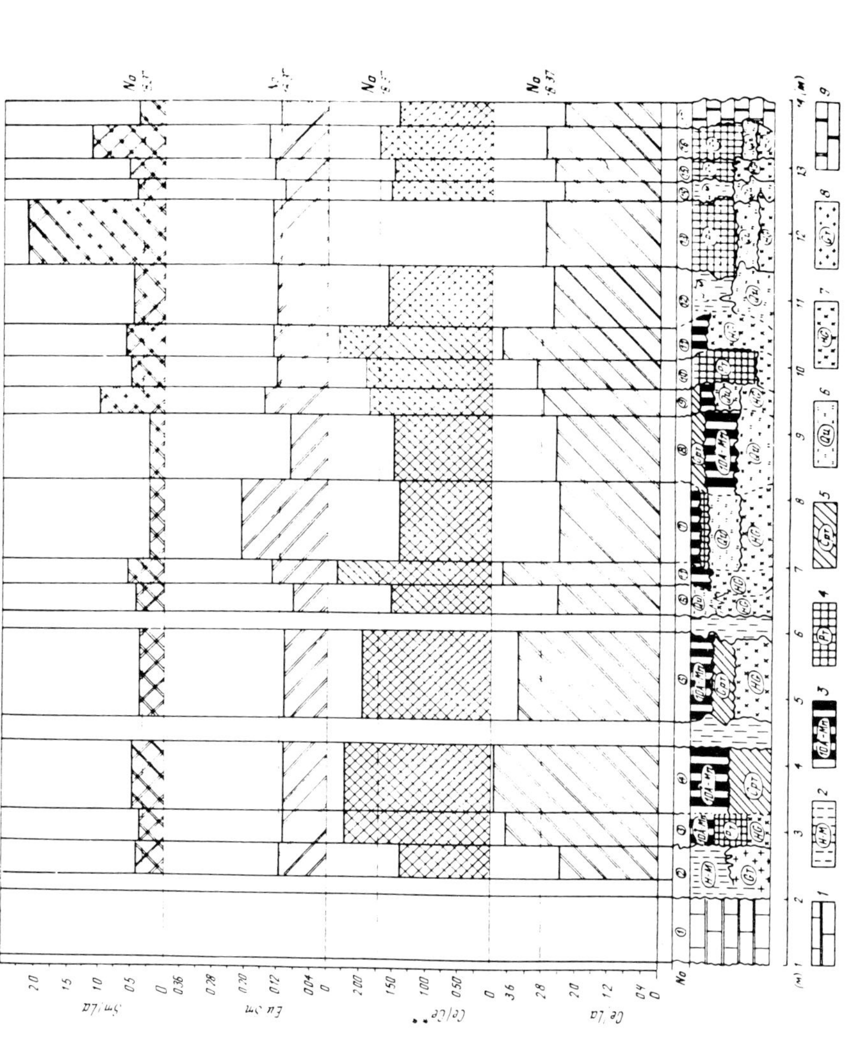

Fig. 52. Distribution of Ce/La, Ce/Ce**, Eu/Sm, Sm/La ratios in ores and rocks of the Úrkút deposit, Csárdahegy area. Lithology, mineral composition and legend see to Figure 45; 'Notes to Tables 14 and 15'. The cerium anomaly was calculated by: Ce/Ce** =

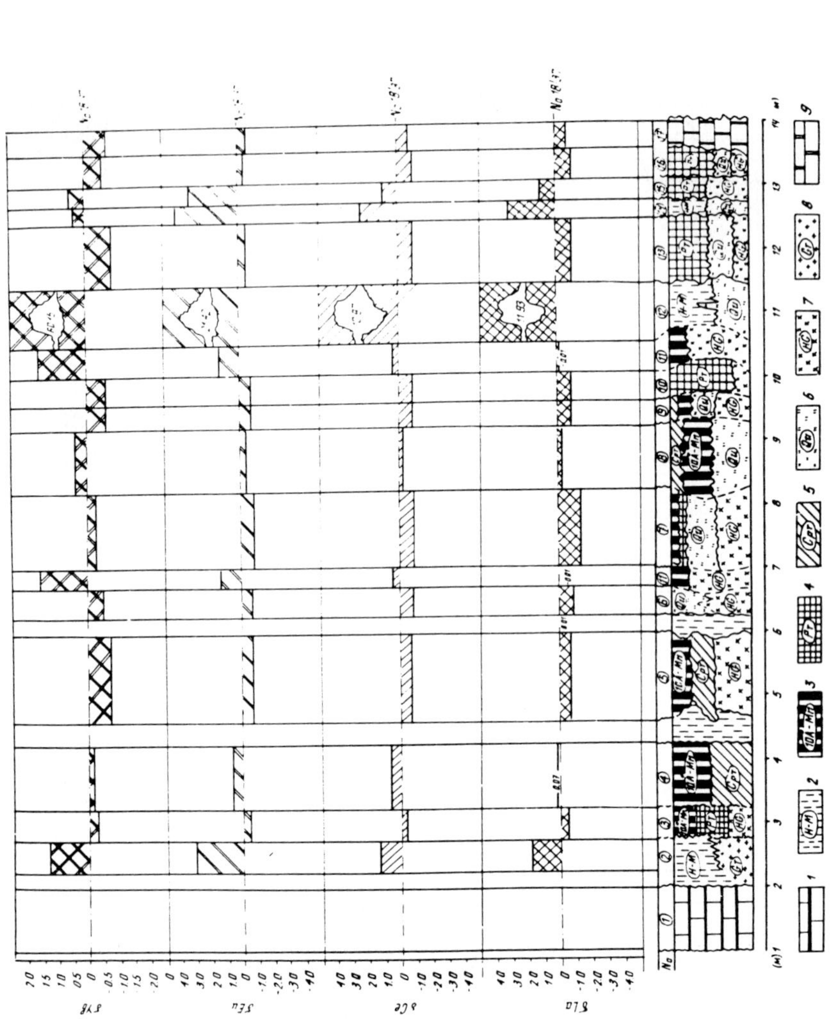

Fig. 53. Distribution of relative accumulation (removal) coefficients δLa, δCe, δEu, δYb in ores and rocks of the oxidized zone of the Úrkút deposit, Csárdahegy area (see Figure 45). Mineral composition is shown in 'Notes to Tables 14 and 15'. Lithology to Figure 44. Legend to Figure 45.

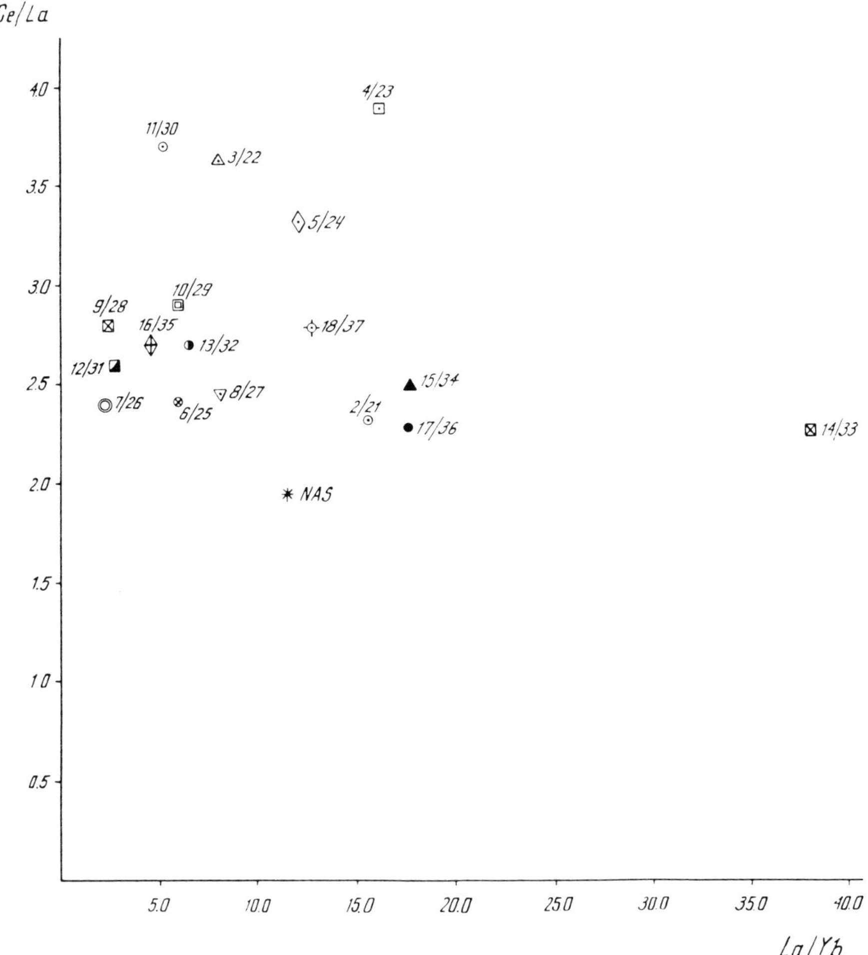

Fig. 54. The Ce/La and La/Yb ratios in ores and rocks of the oxidized zone of the Úrkút deposit, Csárdahegy area. The numbers of the dots correspond to the analyses shown in Tables 16 and 17. Lithology, mineral composition is given in 'Notes to Tables 14 and 15'.

background of the general accumulation of ΣREE (Figure 51). In particular, the common features in geochemical behavior of light and intermediate REE, by the example of La–Sm and Nd–Sm, and also the absence of Eu–Sm and Tb–Sm separation in the intermediate REE group, are exhibited in the direct relation of their concentrations in the products of oxide supergenetic alteration (Figures 55 to 58).

This statement is also confirmed by the distribution of coefficients of relative Eu concentration (removal) (Figure 53), i.e., their higher values are associated with autochthonous

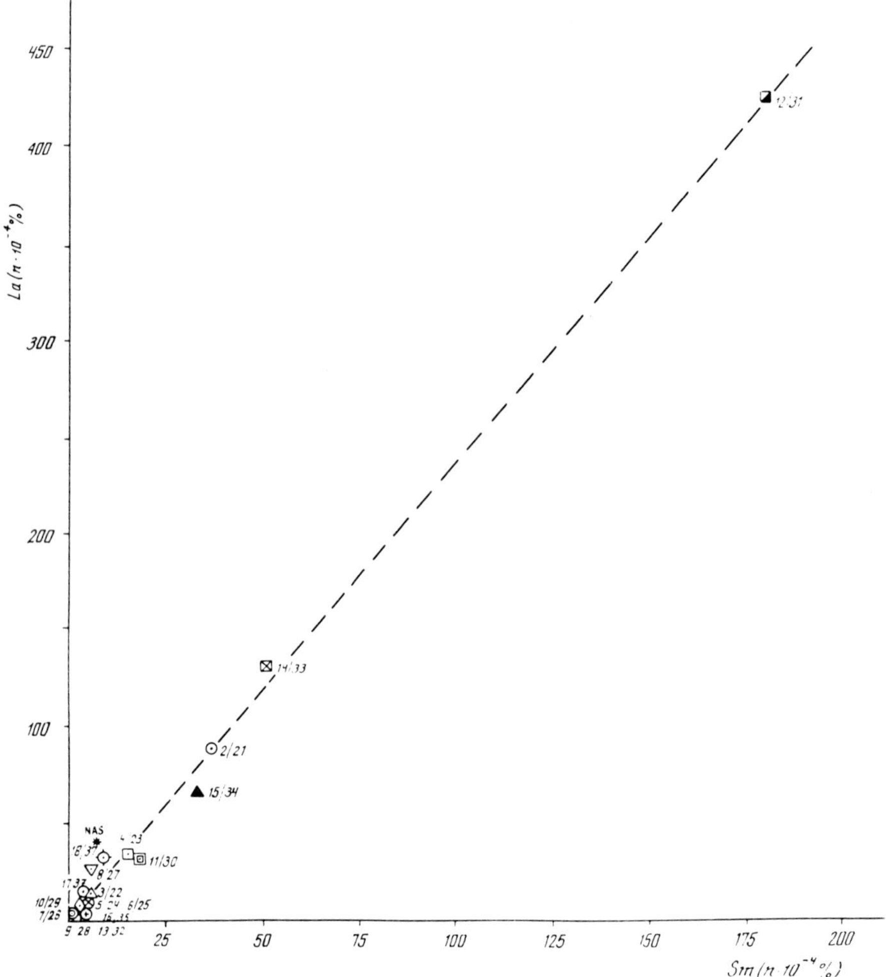

Fig. 55.   The ratios of La and Sm concentrations in ores and rocks of the oxidized zone of the Úrkút deposit, Csárdahegy area. See legend to Figure 54.

residual products of supergenic alteration and oxidation, whereas the maximal values are typical of supergenic phosphates.

Yb, as a typical heavy REE, is characterized by a certain dualistic behavior (Figure 53). On the one hand, like the rest of the lanthanides only at a less pronounced rate, Yb is accumulated in autochthonous residual products of supergenic alteration. On the other hand, this element, having greater migration capacity in this setting, is accumulated both in redeposited Mn and Fe oxyhydroxide products (samples 11 and 8) and particularly in authigenic phosphates (sample 12).

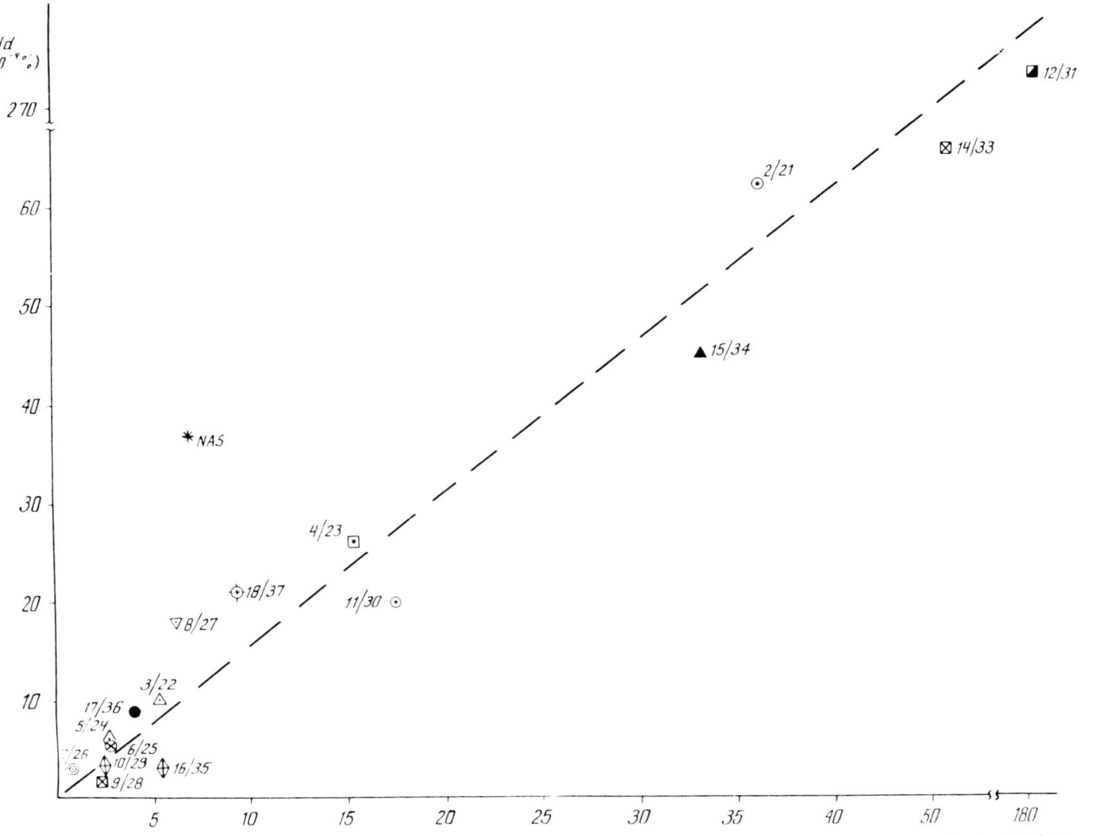

Fig. 56.   The ratios of Nd and Sm concentrations in ores and rocks of the oxidized zone of the Úrkút deposit, Csárdahegy area. See legend to Figure 54.

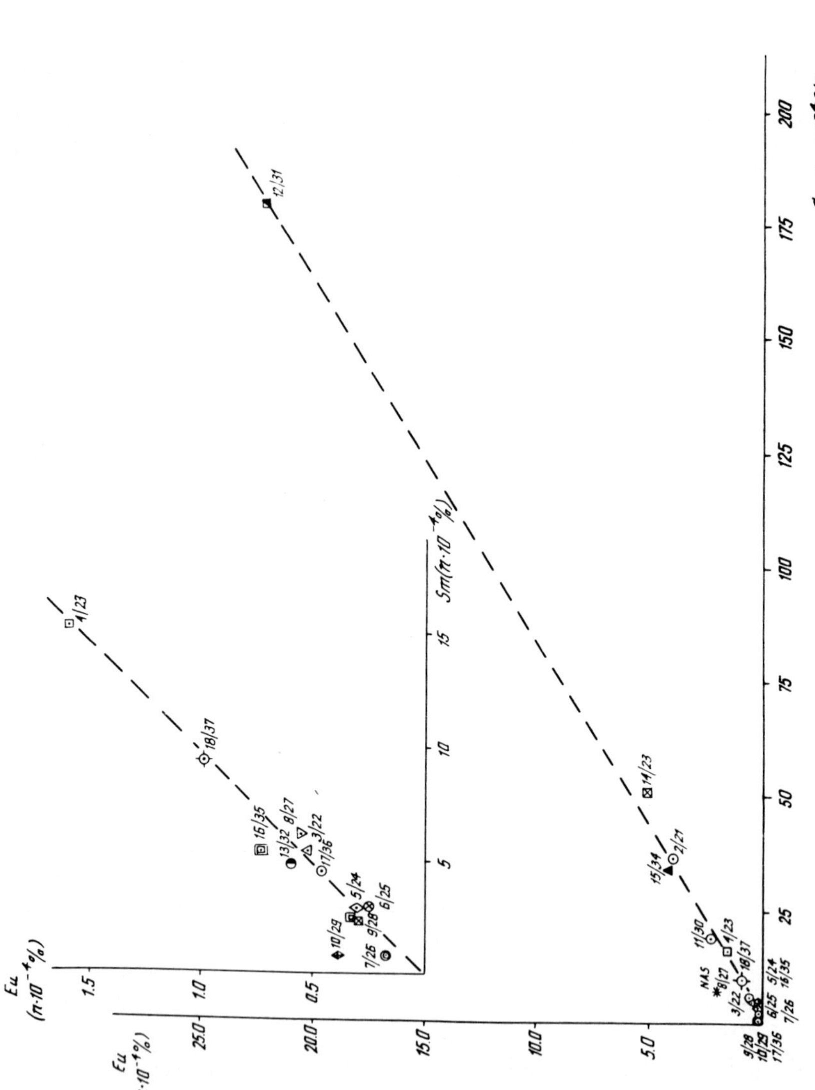

Fig. 57.   The relationships of Eu and Sm concentrations in ores and rocks of the oxidized zone of the Úrkút deposit, Csárdahegy area. See legend to Figure 54.

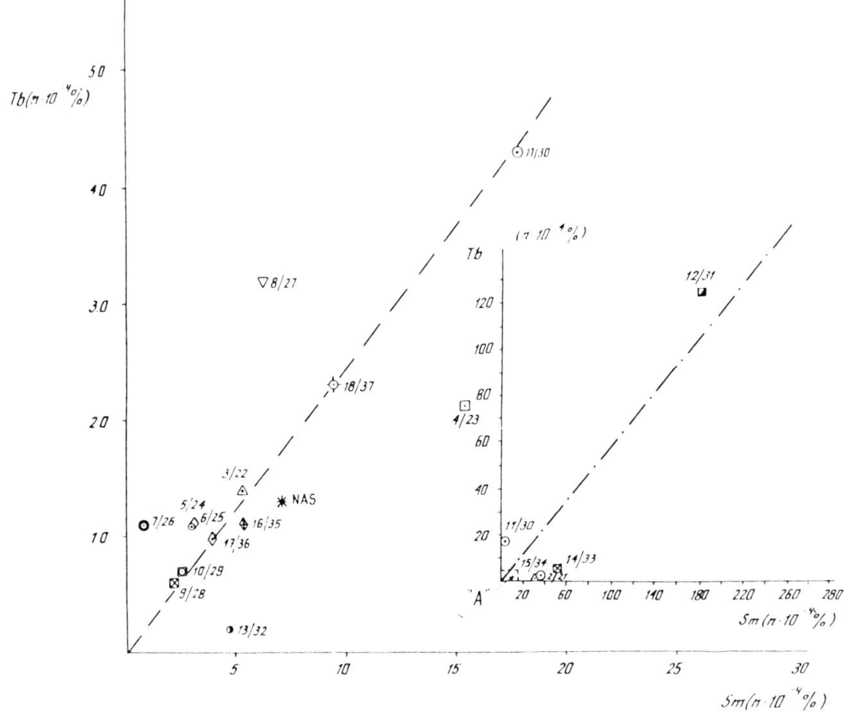

Fig. 58. The relationships of Tb and Sm concentrations in ores and rocks of the oxidized zone of the Úrkút deposit, Csárdahegy area. See legend to Figure 54.

3.2.5. ON GENESIS

The nature of manganese ores in paleokarst is characterized by a certain dualism. On the one hand, they have affinity with formations associated with leaching of ore components redeposited by surface meteoric waters, and correspond to typically supergene formations (Patterson, 1971). On the other hand, in the multitude of ore accumulations, associated with the activity of crustal ground waters, the karst Mn ores are characterized by the fairly deep epigenetic nature caused by the high solvent and migration capacities of solutions in these environments.

Such is the case that this dualistic nature of Mn ore accumulations is definitely exhibited in the Csárdahegy area of the Úrkút deposit.

The manganese ore deposit is a large karst cavity in the intensively folded tectonic structure. The studied section contains a supergenically altered ore sequence, 10–12 m thick, bedded between the Lower Liassic of the 'Hierplatz-type' and Upper Liassic weathered limestone layers. The lower and upper parts of the deposit correspond to the general normal sequence of the Liassic Mn ore sediments. The central part of the deposit has an allochthonous boulder block (3.5 to 4.0 m) and fragments of the redeposited ore material.

On the whole, the deposit has no well-developed profile of weathering. The general alteration of carbonate, siliceous, clayey, and hydrothermal-diagenetic rhodochrosite-clayey protores is not high. The clayey rocks are converted into deeply degraded hydromica-smectites, partly transformed into kaolinite, with quartz and goethite admixtures, in places with intensive karst water circulation.

The lower part of the deposit contains a considerable amount of minerals belonging to 10 Å-manganates group (todorokite-buserite-asbolane etc., probably, lithiophorite, etc.) and cryptomelane; in the upper part, the major Mn mineral is pyrolusite. The Fe oxyhydroxides are represented by hydrogoethite, and occasionally by goethite.

The following general succession of stages in Mn mineral formation is suggested:

$$MnCO_3 \quad \rightarrow \quad manganite \rightarrow pyrolusite(I) \rightarrow$$
$$\rightarrow \quad cryptomelane \rightarrow 10\,\text{Å-manganates} \rightarrow pyrolusite(II).$$

The accumulation of Fe was restricted in the lower and upper autochthonous parts, but the allochthonous boulder block contains the highest concentrations of this metal. Contrarily, the largest Mn accumulations, with enrichment up to 250–360% of the initial quantities of metal, are in the lower part of the deposit.

The ores of the lower part of the deposit are characterized by the maximal accumulation of Co and Ba, represented chiefly by 10 Å-manganates and cryptomelane. This means that, besides K, the most likely concentrators of these elements are tectonomanganates, whose tunnel structure, from the crystallochemical point of view, is favorable for the fixation of such large cations. Moreover, the formation of asbolan varieties can also take place. The maximal Ni concentrations are recorded in the upper parts, chiefly in the pyrolusite ores.

The REE distribution in the products of weathering is largely inherited from the parent hydrothermal-diagenetic Mn carbonate ores. The Σ(La–Lu) distribution indicates that high accumulations of these elements are mostly associated with the altered autochthonous rocks and are not typical of the authigenic Mn and Fe oxyhydroxide phases. The behavior of Σ(Le–Sm)/Σ(Gd–Yb), La/Yb, Ce/Yb ratios testifies that light lanthanides are accumulated at the same time with the general ΣREE concentration in the autochthonous sediments of the oxidation zone, whereas chiefly heavy REE are accumulated in the Mn and Fe oxyhydroxide ores with an appreciable supply of these metals. It is noted with interest that the highest values of Ce/La ratio, as compared to the parent rocks, are chiefly typical of later oxidation products in the lower part of the deposit, composed mostly of minerals of the 10 Å-manganates and cryptomelane. In these ores, the cerium anomaly (Ce/Ce∗∗) has a similar distribution and higher positive values. The obtained data may provide information about Ce accumulation in the setting of higher oxidation in Ce^{4+} form with an expressed preference for hydrolysis. In the process of such concentration the leading role belongs to chemisorption accumulation accompanied by the interphase autocatalytic oxidation: $Ce(II) \rightarrow Ce(IV)$. The similar reaction mechanism processes also control accumulation of some transition 3d metals (Mn, Fe, Co, and others) in various geochemical environments, from crusts of weathering to pelagic regions of the World Ocean.

The described data positively indicates that a greater part of the studied Mn ores can be ascribed to the supergene products of karst formation. The phenomena of karst formation are based on the fairly well-known interactions between the surface ground water and the carbonate rocks resulting in their dissolution (Thrailkill, 1968; White, 1984; Wood, 1985):

$$C_{org} + O_2 \rightarrow CO_2$$

$$CO_2 + H_2O \rightarrow H_2CO_3$$

$$H_2CO_3 + Ca, (Mn)\,CO_3 \rightarrow Ca^{2+}, (Mn^{2+}) + 2H_2CO_3.$$

The processes of karst manganese ore formation are active in those artesian basins, where attenuating exhalations take place and thermal waters develop into a system. In this setting, the ground waters are, as a rule, enriched with fluids supplied from endogenic sources (SO_2, H_2, CO_2, HCl, HF, etc.); this enrichment greatly increases their reactive capacity in interaction with the sediments, which are chiefly carbonate rocks. The thermal springs in the Bakony Mountains are an indirect indication of the role of these factors in karst formation. Moreover, the presence of SO_4^{2-} ion, the product of pyrite oxidation, in stratal waters essentially affects the dissolution process of carbonate rocks (Szabo and Grasselly, 1980).

The next period of geochemical evolution of Mn(II) can be described as a sequence of oxidation reactions in which the sorption-autocatalytic processes play an important part (Stumm and Morgan, 1970; Varentsov, 1989; Varentsov *et al.*, 1979).

It should be emphasized that in natural environments, where often the solutions with low to moderate concentrations of Mn and other transition metals are common, the reactions of oxidation are the most probable and their sorption and autocatalytic character is revealed in their pulsating, cyclic repetition. Moreover, at the same time with ion $Mn_{(hydr.)}$, the ions of heavy and alkalic metals, to a less degree of alkalic earths and REE, can participate in reaction with similar functions. This statement is particularly valid with regard to the formation of Co(III)–Ni(II)-bearing oxyhydroxide minerals of manganese-asbolanes in the crust of weathering (Chukhov *et al.*, 1980a, 1980b, 1982; Vitovskaya *et al.*, 1984).

The study of mineralogy and geochemistry of the major components, heavy metals and REE definitely attests that the Csárdahegy manganese ore accumulation, as a part of the Úrkút deposit, was formed after the termination of the Jurassic–Neocomian sedimentary cycle. At this stage of the geological evolution of the region, the essential role belonged to block movements which were accompanied by extensive supergene alterations and karst processes.

3.3. The Giant Manganese Ore Accumulations in Paleokarst: Deposits of the Postmasburg Region, Griqualand West, Cape Province, South Africa

3.3.1. GEOLOGICAL SETTING

Manganese deposits of the Postmasburg region are related to the largest ones in the world: reserves of about 3 billion tons recalculated as Mn for the ore, which is of extremely high quality (> 44% of Mn) (Boardman, 1940, 1964; De Villiers, 1960; Ellison, 1983; Roy, 1980).

The deposit area occupies a gently hilly area of about 400 square miles (1024 km^2) with the absolute altitude of about 4400 feet (1450 m) in the central part of the subcontinent (Figure 59). The main data on the deposit geology is given in a number of works by the South African geologists (Beukes, 1973, 1983, 1984a–c, 1986; Boardman, 1961, 1964; Button 1976a; Button and Tyler, 1981; De Villiers, 1956, 1960, 1967, 1970, 1983; Grobbelaar and Beukes, 1986; Sönge, 1977; Van Schalkwyk and Beukes, 1986).

The major role in the geological structure of this region belongs to the relatively slightly metamorphosed sediments of the Lower Proterozoic Transvaalian Supergroup (2500–2100 Ma) with an initial maximum thickness of about 12 km, developed on the Kaapvaal craton in the structural basin Griqualand West (Beukes, 1983, 1986). The stratigraphical scheme, based on relatively recent works and adopted by the Geological Survey of the South Africa, is used in our work (Beukes, 1983, 1984c, 1986, 1993; Grobbelaar and Beukes, 1986; Sönge, 1977; Van Schalkwyk and Beukes, 1986). For the deposit area the works by Beukes (1983, 1986) were used as a basis for the scheme. These works considerably defined as a whole the regions of Griqualand West and Transvaal stratigraphic units, suggested by the South African Committee for Stratigraphy (1980).

Wolkenberg and Buffalo Springs Groups (protobasin stage) are represented by basalt and acid volcanites, quartz clastites with subordinate carbonate and cherty rocks.

The Wryburg and Black Reef Formations (the beginning of the major transgression) are composed of shallow-sea quartzites and marine argillites of limited thickness.

The Proterophytic (Cloud, 1976) Transvaal Supergroup includes the Ghaap Group, a large chemical sedimentary unit, that is unconformably overlain by a mixed volcanic-sedimentary rock unit, the Postmasburg Group (Figures 59 and 60).

The Ghaap Group is subdivided (from the bottom upward) into the Schmidtsdrif Subgroup (interbedded siliciclastics and carbonate rocks), the Campbellrand Subgroup (limestone and dolomite), the Asbesheuwels Subgroup (iron formations), and the Koegas Subgroup (interbedded siliciclastites and iron-formation).

The Postmasburg Group consists of (from the bottom upward) the Makganyene Diamictite (unsorted terrigenous rock), the Ongeluk Formation (andesitic lava), the Hotazel Formation (interbedded iron-formation and sedimentary manganese) and the Mooidraai Formation (limestones and dolomites).

The Transvaal Sequence is overlain with an angular unconformity by red Paleophytic beds (Cloud, 1976), the Olifantshoek Group, composed by quartzites and shales, which builds the Korannaberg fold belt along the western margin of the Kaapvaal Craton.

According to Beukes (1983, 1986) the deposition of the Transvaal Sequence at the Griqualand West region was controlled by three main tectonosedimentary elements (Figures

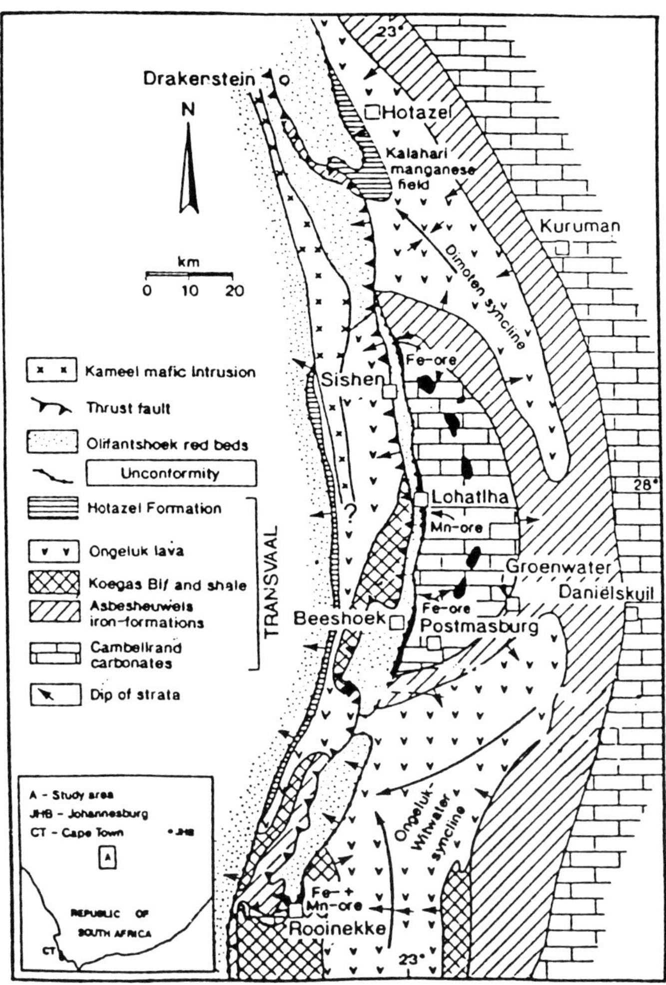

Fig. 59. Regional geological map of part of the northern Cape Province, showing distribution of manganese deposits, unconformity between Transvaal and Olifantshoek Sequences and general structure of the area (Beukes, 1993).

59 and 60): (1) basin type: the sediments were accumalated in the environment of the shallow water platform basin on the Kaapvaal Craton; (2) structural tectonic localization: sedimentation took place at the platform edge, located parallel to the Griquatown fault zone in the south, which then swung northwards parallel to the Korannaberg fold belt; (3) paleogeographic position: a basin formed along the western and southwestern margin of the Kaapvaal Craton.

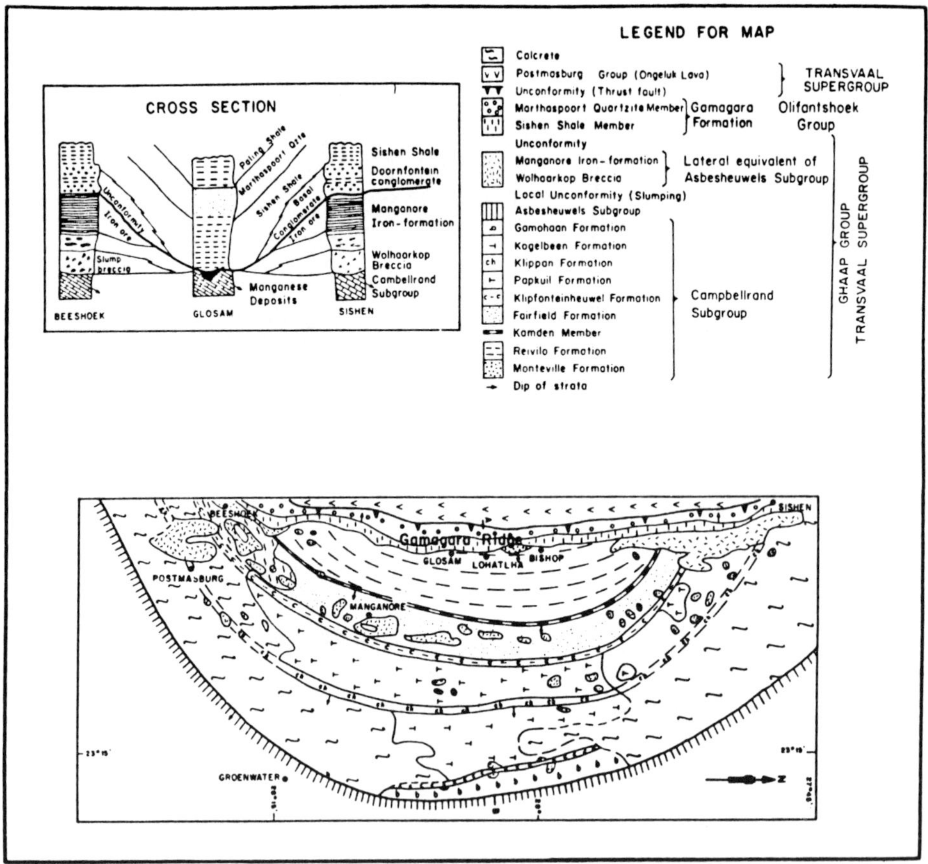

Fig. 60. Simplified geological map of the Maremane dome with a corresponding north-south section along the Gamagara ridge (Grobbelaar and Beukes, 1986).

The manganese deposits considered are associated with sediments of the Early Proterozoic Transvaal Supergroup in the northern Cape Province. The deposits of the sedimentary-diagenetic nature are developed in the Heynskop Formation of the Koegas Subgroup; exogenic sediments are associated with the unconformity between the Transvaal strata and the overlying late Early Proterozoic red beds of the Gamagara Formation of the Olifantshoek Group. The deposits of the hydrothermal-sedimentary nature are closely associated with the Hotazel Formation of the Postmasburg Group (see Figure 59).

Manganese accumulations in the Koegas Subgroup are represented by two types: (1) manganiferous marls associated with the stromatolite-dolomite bioherms and mudstones in the lower part of the Heynskop Formation; (2) two beds of oxyhydroxide Mn ores about 1 m thick, composed essentially of romanechite (psilomelane), and to a lesser degree pyrolusite, jakobsite, hausmannite, and hematite, intercalating with the Rooinekke iron-formation member at the top of the Heynskop Formation. It is supposed that Mn and Fe ores, altered by supergene oxidation, were initially represented by jakobsite and hematite. The limited mining of the ores of these deposits is performed in the neighborhood of Rooinekke. The content of Mn in the ores are 35 wt. % and of Fe 25–27 wt. %; the reserves do not exceed several thousand tons. The manganese marls contain 3–10 wt. % MnO and 5–15 wt. % FeO (Beukes, 1993). The manganese carrier is the diagenetic ankerite with the negative value of $\delta^{13}C$ equal to 11‰, that indicates the incorporation of organic carbon during the mineral formation (Beukes *et al.*, 1993). The associated limestones have an average value of $\delta^{13}C$ equalling 5‰, pointing to the fresh water environment of their formation; the value of $\delta^{18}O_{pos}$ for limestones is 14‰, for ankeritic marls 13%. Beukes (1993) assumes that these manganese-bearing marls were accumulated as the primary carbonate sediments in stromatolite bioherms ('reefs'). Similar marls were formed due to mixing of carbonate fragments and particles of reefs with terrigenous sediments, that rimmed these structures. The mudstone was deposited in distal areas from the source land, where the carbonate material was not supplied. The sedimentation of the banded iron-formation was an even more distal facies than the mudstones and they were accumulated at the extreme end of the sedimentary sequence, deposited after the main transgression.

The manganese ores, associated with the unconformity between the Transvaal Sequence and overlying Olifantshoek red beds, are developed at the Postmasburg area on the Maremane dome. This dome (Figure 60) is composed of the Campbellrand carbonates. Oxyhydroxide manganese ores were accumulated essentially as supergene oxyhydroxide sediments on the karst surface prior to deposition of the Olifantshoek red beds. Lens-like pods of braunitic ore were formed in association with chert breccia, that was developed in areas (Figure 61), where the pre-Olifantshoek erosion surface transected cherty dolomite units. Bixbyite ores, as a rule, are developed in areas where the paleo-erosion surface transected chert-free manganese-rich dolomite units. For example, near Lohatlha (see Figure 60) similar beds occur in the base of the Campbellrand sequence in the dome core. In some areas the Precambrian erosion surface was exposed for the influence of supergene factors during later geological epochs (till Holocene). In such environments due to the destruction of the Precambrian manganese ore beds a great number of clastic ores were formed, that were accumulated in the modern cavities of the dolomite dissolution (Figure 61). To the present the majority of similar ores have been mined out and the remaining reserves are 15 mln t with the content of 30–52% Mn and 9–14% Fe.

3.3.2. GENESIS

3.3.2.1. The General Features

Among geologists of the 1950–1960s and the following years, who studied the iron and manganese ores of the Postmasburg area, there were some contradictions regarding the genesis, that reflect the progress in the achievements of the last years. These works (Beukes, 1973, 1983, 1984a, 1984b, 1986; John de Villiers, 1956, 1960; P.R. de Villiers, 1967, 1970; J.E. de Villiers, 1983; Grobbelaar and Beukes, 1986; Van Schalkwyk and Beukes, 1986) made a great contribution into the study of the geology and mineralogy

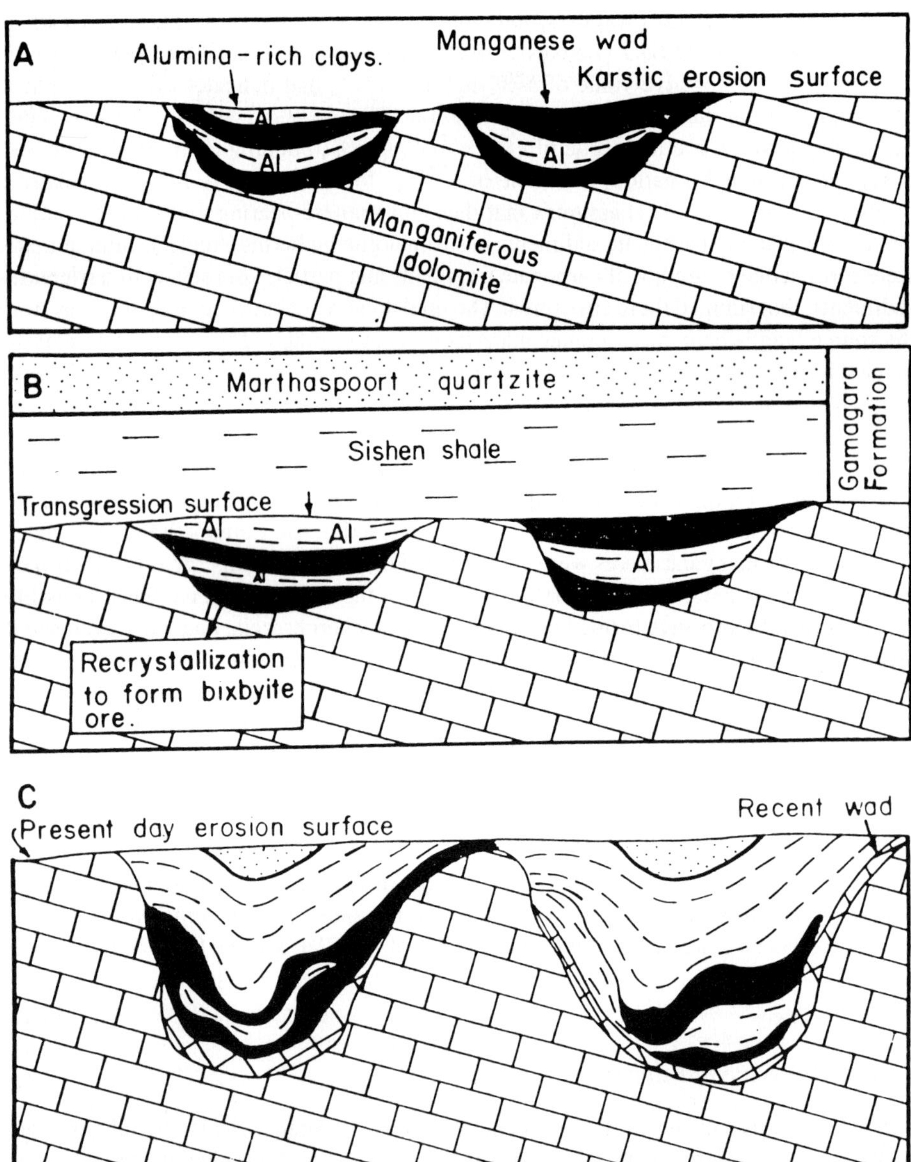

Fig. 61. Schematic illustration of the genetic evolution of the Glosam and Bishop manganese deposits (Grobbelaar and Beukes, 1986). A. Protomanganese ore formation under oxidizing atmospheric conditions during the periods of pre-Gamagara erosion. B. Formation of crystalline bixbyite ore after burial by Gamagara sediments. C. Cenozoic reactivation of sinholes with slumping and secondary enrichment of the ore by romanechite (psilomelane) precipitation.

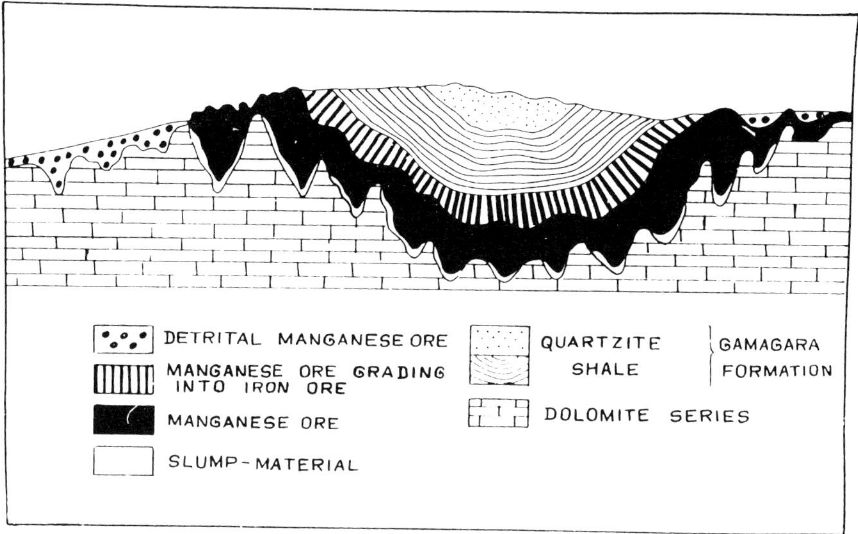

Fig. 62.   Schematic section across a typical deposit of manganese ore in the Western Belt, Maremane dome area, Postmasburg region, the Northern Cape Province (De Villiers, 1960).

of the deposit. If the factographical part of these works deserves the most attention, then the general conclusion (J.E. de Villiers, 1983) on the ore origin from so called hypogenic solutions needs, at least, precise consideration in the light of known geological information.

At the considered area of the iron ore deposits at Sishen (Van Schalkwyk and Beukes, 1986) and Beeshoek (Grobbelaar and Beukes, 1986), the manganese deposits at Lohatlha (Boardman, 1964; De Villiers, 1960; Grobbelaar and Beukes, 1986) and the iron and manganese deposits at Rooinekke (Beukes, 1986; Visser, 1954) are essentially supergene accumulations. As it was mentioned above they were associated with the period of weathering and intensive erosion that preceded sedimentation of the Gamagara and Mapedi Formations of the Olifantshoek Group. In the Maremane dome (see Figures 59 and 60) the paleokarst erosion surface was formed and regionally developed on the carbonate rocks of the Campbellrand Subgroup. The relatively hard rocks of the Kuruman iron-formation, which are resistant to weathering, slid down along this surface and were accumulated in paleokarst cavities prior to deposition of the Gamagara Formation. A siliceous slump breccia, called the Wolhaarkop Breccia, occurs at the base of the slumped iron-formation; the latter is distinguished as the manganese iron-formation (Beukes, 1983a, 1986), in order to distinguish it from the undisturbed Kuruman iron-formation (Figures 60 and 62).

In a number of works (Beukes, 1973, 1983, 1984a–c, 1986, 1993; De Villiers, 1956, 1960, 1983) it is shown that at the Griqualand West area the manganese and iron ore deposits form two belts (western and eastern), which extend for 130 km in a submeridional direction from Postmasburg in the south to Black Rock in the north. A small ore manifestation is known at Rooinekke which is 70 km to southwest from Postmasburg (see Figures 59 and 60).

In discussion of the problems of geology and formation of manganese and iron ore deposits it is necessary to emphasize the special role of the dolomite rocks of the Campbellrand Subgroup, bearing such important functions as: (1) distinct regional reference layer, over which the younger layers usually were shifted and thrust over during the sheet-slumping dislocations; (2) as the source of manganese and iron, that were leached from dolomites, they migrated and were accumulated in karst cavities of these and younger rocks; (3) dissolution and leaching of the considered carbonate rocks led, in numerous cases, to the development of slumpings and associated overthrust structures, that favored the preservation of the commercially important manganese and iron deposits.

3.3.2.2. The Main Manganese Deposits: Bishop, Lohatlha, Glosam, and Kapstewel, Postmasburg Region, Griqualand West

Manganese mines Bishop, Lohatlha, and Glosam are located between Sishen and Postmasburg in the center of the Maremane dome (the center of the western belt of manganese deposits), the Griqualand West region, Northern Cape Province (Figures 59 and 60). These deposits are rather similar in geological structure, mineral and chemical composition of the ores (Beukes *et al.*, 1993, De Villiers, 1960; Grobbelaar and Beukes, 1986).

It was mentioned above that manganese deposits of the Maremane dome of the Postmasburg region are directly associated with the unconformity between rocks of the Campbellrand Subgroup of the Ghaap Group. In those cases, when rocks of the Gamagara Formation overlay the manganese-bearing dolomites of the Reivilo Formation of the Campbellrand Subgroup in the central part of the dome (Figure 60), the supergene manganese ores of essentially bixbyite composition are developed at the base of the Sishen Shale Member of the Gamagara Formation (Beukes, 1983; Grobbelaar and Beukes, 1986). This relationship is distinctly distinguished at the Glosam, Bishop, and Lohatlha manganese mines (Figures 60 and 61).

As it was noted earlier, the main role in the formation of the manganese deposits belongs to formation processes of karst cavities and depressions and associated processes of leaching, extraction, transportation, and accumulation of manganese and associating metals during the erosion break that preceded the Gamagara Formation deposition.

Considerable, to gigantic, deposits of hematite ores were formed when the large fragments and blocks of the formation of deeply hematitized ferruginous quartzites of the Manganore Iron Formation was accumulated. This formation is the product of the exogene alteration of the Asbesheuwels Subgroup (i.e., Kuruman and Griquatown Iron-Formations), composed mainly of oxidized iron minerals, slumped down and loaded into karst depressions and graben-like basins. The process of slumping and intensive dissolution and leaching of the fragments of ferruginous quartzite rocks of the Kuruman and Griquatown Iron Formations was accompanied by the formation of the Wolhaarkop siliceous chert breccia, that occurred usually at the base of the Manganore Iron-Formation.

In the central part of the Maremane dome the Sishen Shale, as an integral member of the Gamagara Formation, occurs directly on the eroded dolomite of the Reivilo Formation. In this case the paired association of Wolhaarkop Breccia/Manganore Iron Formation is absent, that is typical for the iron ore deposits. Usually paleokarst basins, depressions and

Table 18. Chemical analysis of the Bishop, Lohatlha, and Glosam manganese ores (wt, %) (De Villiers, 1960; Grobbelaar and Beukes, 1986).

no./no.	Mn	Fe	SiO$_2$	P	Mn/Fe	Notes
1	52.2	8.6	5.5	0.066	6.1	Glosam
2	49.1	10.5	3.7	0.053	4.7	Bishop
3	46.5	12.6	4.4	0.046	3.7	Glosam
4	44.8	13.7	3.7	0.044	3.3	Glosam
5	39.8	17.5	2.6	0.038	2.3	Bishop
6	35.1	20.9	2.1	0.034	1.7	Glosam
7	30–32	22–24	4–6	max. 0.05	–	avg., Lohatlha

cones with the developed pre-Gamagara erosion surface are filled in with high alumina clays and low crystalline, structurally poorly organized manganese oxyhydroxides. In the subsequent geological history of the region these Mn oxyhydroxides were intensively recrystallized and altered into the accumulations of the bixbyite composition that, as a rule, occurred at the base of the Sishen Shale (Beukes, 1983; Beukes *et al.*, 1993; Grobbelaar and Beukes, 1986).

The Bishop, Glosam, and Lohathla manganese deposits (Figure 60) differ in irregularity forms of the karst depressions, developed in the dolomite substrate of the Campbellrand Subgroup formations. It is supposed (Grobbelaar and Beukes, 1986) that ore formation and slumping processes are probably associated with the different intervals of supergene alterations, including processes of supergene leaching, transportation, and accumulation of ore forming components.

The clearly crystalline bixbyite ores occur in the form of lens-like irregular bodies on the surface of large cavities after dolomite dissolution at the base of the Sishen Shale. The ores are distinguished by massive habit and slightly developed stratification. They were formed owing to intensive recrystallization of manganese, almost amorphous oxyhydroxides and the alteration processes of the clayey material of shales that filled in the pre-Gamagara karst erosion cavities (see Figure 61A).

During the later geological periods, after the accumulation of the Olifantshoek Group sediments and until the Mesozoic Karoo Supergroup sedimentation, a crystallization of manganese oxyhydroxides proceeded (see Figure 61B) (Grobbelaar and Beukes, 1986).

In the Cenozoic cycle of the regional geological history there was a wide reactivation of karst structures (see Figure 61C), which resulted in slumping, brecciation, and crushing with dissolution and redeposition of the earlier recrystallized ore, that led to the accumulation of romanechite (psilomelane) crusts and pyrolusite concretions (De Villiers, 1960; Grobbelaar and Beukes, 1986). The main part of the manganese ores is distinguished by a rather high quality (see Table 18). The reserves of the manganese ore of the deposits of the Western manganese belt are evaluated to 15 mln t with Mn contents of 28 to 50 wt. % (Grobbelaar and Beukes, 1986).

The Kapstewel manganese and iron ore deposits are located in the Klipfontein Hills region, 30 km to the northeast from the Postmasburg (see Figures 59 and 60) near the Fe–Mn deposits known as Manganore. They are typical representatives of deposits of the Eastern Belt of the Postmasburg Manganese Field (Beukes *et al.*, 1993; De Villiers, 1960). These deposits, in difference to deposits of the Western Belt at the common model of formation, are associated with the erosion and karst formation of Fairfield-dolomites (1–3% of MnO) of the Campbellrand Subgroup. Processes of Mn ore accumulations were manifested in the formation of Klipfontein Hills and essentially enriched with manganese Wolhaarkop chert breccia at the contact of eroded dolomites and overlying Manganore Iron Formation. The above features of the geological history of the region led to a considerable part of the manganese ores being represented by detrital type. Some parts of the Wolhaarkop chert breccia and the associated manganese ores were eroded and redeposited in the form of conglomerates. The thickness of the conglomerate beds is more than 5 m.

The iron ores belong to the Manganore Formation. This formation, as it was mentioned above, is a deeply supergenically altered equivalent of the Kuruman Iron Formation, the fragments of which were transported owing to slumping into large karst cavities. Due to these supergene processes, the accumulations similar to the iron ores of the Beeshoek deposit (Figure 63) were formed. As a whole, the mineral composition of the ores of the manganese deposits reflects relatively definitely the conditions of their formation.

Braunite usually crystalline, and fine-grained is the major component of ores of the Eastern Belt. This mineral is registered in the southern and northern parts of the Western Belt. It should be noted that braunite in difference to other manganese minerals of the deposit is characterized by relatively high contents of boron (B_2O_3 0.5–1.1%). However, braunite ores (De Villiers, 1960) of a number of other deposits of the world (Kabetsa de Ferro, East Minas, Brazil; Longban, Sweden; Sitapar, India; Thuringen) are also distinguished by similar boron concentrations.

According to the data by J. Gutzmer (Beukes *et al.*, 1993) the deep epigenetic alterations are distinctly manifested in manganese ores. A considerable part of the ores of the Eastern Belt is composed by fine- to midcrystalline granoblastic patches and grains of braunite and hematite, that enclose and substitute large (about 1 cm) grains (clasts) of quartz.

Bixbyite makes up the major part of the ores of the central and southern areas of the Western Belt. Lower supergene alteration of this mineral in comparison with braunite should be noted. In ores of the Eastern Belt, bixbyite is rare and always forms intergrowths with braunite (Beukes *et al.*, 1993).

Partridgeite differs from bixbyite ($Fe_2O_3 > 30\%$) by having a lower concentration of iron ($Fe_2O_3 < 10\%$). Partridgeite is little different from braunite at small contents of iron. The mineral is distributed throughout the deposit; its largest concentration are found at the southern part of the Western Belt. In the crystalline ore, partridgeite is associated with braunite, and in the shale it is partially substituted by the ore mineral, it is present together with braunite, hematite, and often bixbyite.

Hausmannite: Ores composed of this mineral differ by coarse granularity with the size of separate grains to several mm. They occur in the form of limited, isolated accumulations.

Jakobsite forms ores composing small deposits. Often this mineral is encountered

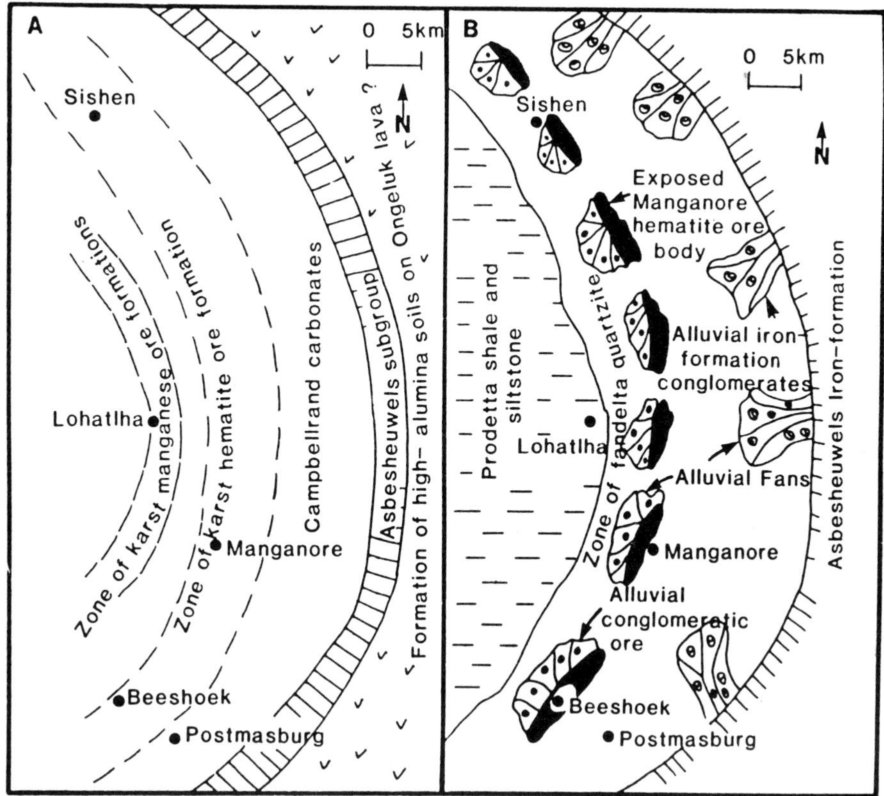

Fig. 63. A. Model illustrating the development and distribution of iron and manganese ore deposits of the Maremane dome during the period of weathering and erosion that preceded the deposition of the Gamagara Formation. B. Model illustrating the possible distribution of sedimentary environments during the early stages of deposition of the Gamagara Formation on the Maremane dome (Van Schalkwyk and Beukes, 1986).

in association with fine-grained hematite and romanechite (psilomelane), composing the oxyhydroxide matter substituted shale or breccia. But in the Postmasburg region the mutual occurrence of jakobsite with braunite and bixbyite was not found.

Pyrolusite and polianite. Fine-grained pyrolusite is developed in association with romanechite (psilomelane) everywhere in the deposit. Polianite crystals, their aggregates and streaks are found in separate, isolated ore accumulations.

The manganates with tunnel structures and the minerals of the cryptomelane, romanechite (psilomelane) group are widely distributed at the majority of the deposit and they are fairly definite products of surface weathering. The supergene alteration of granoblastic manganese ores was manifested in the formation of variable amounts of fine-crystalline accumulations of cryptomelane, manjiroite, pyrolusite, ramsdellite, and lithiophorite, replacing braunite and bixbyite. The largest accumulations of ores of such composition in association with pyrolusite are known in the very northern areas of the Western Belt.

Lithiophorite occurs in associations with braunite, bixbyite, and sometimes with hematite in the form of fine-grained small flakes or coarse-grained ore accumulations in some parts of the deposit.

Manganite is present in the form of an insignificant admixture in polianite-pyrolusite-partridgeite or pyrolusite ores.

Hematite is the major admixture of the ores of the Western Belt. The formation of the hematite composition is essentially associated with the substitution products of ferruginous shales or breccias. The gradational and gradual transitions between the unaltered rock enriched in iron and the pure manganese ore can often be traced (De Villiers, 1960). At the Eastern Belt, hematite is found in subordinate quantities.

Barite is widely distributed at the deposit. Especially great accumulations of this mineral are developed within the Eastern Belt. Spotty barite aggregates with idiomorphous inclusions of braunite and streaks, in which romanechite (psilomelane) is present, are characteristic. In the manganese ore deposits of the Eastern Belt the barite is usually associated with braunite ore, depleted by fragments and breccia clasts.

Diaspore and ephesite (Na-margarite) are remarkable components of bixbyite ores of the central part of the Western Belt (De Villiers, 1960). Usually diaspore fills in interstices between crystals and bixbyite patches, or it forms idioblastic acicular crystals. Ephesite occurrence conditions are the same as those for diaspore, but this mineral fills in streaks as well. The presence of these minerals in Mn and Mn-Fe ores of the Postmasburg region may indicate the depth of chemical leaching, the decomposition of alumosilicate material of dolomites and other primary rocks, and the intensity of the processes of their epigenetic alteration.

Amesite is observed fairly infrequently and usually in association with diaspore and ephesite. Akmite, albite, and apatite may be present in admixture quantities.

3.3.2.3. *Processes of the Manganese Ore Formation*

The concepts on the genesis of the ores of the Postmasburg deposits were proposed from the very beginning of its study (see review, De Villiers, 1960); but two sharply different points of view were actively discussed later:

1. on the hypogenic hydrothermal origin of the ore-bearing solutions (De Villiers, 1983);
2. on exogenic karst genesis of ores (Beukes, 1983, 1986; Beukes *et al.*, 1993 De Villiers, 1956, 1960; Grobbelaar and Beukes, 1986).

The data given above on the regional geology, ore deposit structure, and geological settings, and mineral composition of ores show that each of these concepts is based on a definite set of data. The attempt for generalized uncontradictory interpretation of available geologic information is made below.

As it was noted, the structure of manganese ore deposits and their tight association with the dolomite rocks of the Campbellrand Subgroup suggest that the latter is the main source of manganese. At the Griqualand West region the Mn content in the considered dolomites is from 0.32 to 1.62% (De Villiers, 1960; Klein and Beukes, 1981). It is necessary to emphasize that siliceous breccia, in many cases, includes ore bodies and it is represented by the insoluble residue of the tectonically crushed, brecciated and deeply leached manganese-bearing cherty dolomite and associated rocks, including ferruginous quartzites. Owing to tectonic crushing these rocks became highly porous and permeable. It favored the circulation of ground water in them. The general increasing of the reaction surface intensified the dissolution of these carbonate rocks, accompanied by leaching and deposition of manganese. The Mn deposits were formed both on the eroded surface of dolomites and in the residual siliceous breccia. The data on the deposit geology indicates that the composition of ores (Mn, Mn–Fe, and Fe) was sufficiently definitely determined by the lithology of primary rocks (Mn-dolomites, Si-dolomites, Fe cherty quartzites). Thus, the tremendous volumes of dolomite rock, altered into porous permeable accumulations, were subjected to dissolution with the removal of manganese that was accumulated with the formation of ore deposits in the corresponding environment (with relatively oxidizing Eh).

Major iron ore deposits of this region – Sishen and Beeshoek – were formed as a result of deep hematitization during slumping of the Manganore Iron-formation material under the effect of supergene solutions. In some places the slumped units of the iron formation were perfectly preserved between the cherty dolomite of the Klippan and Klipfontein-heuwel Formations of the Campbellrand Subgroup. In those areas of the Maremane dome core, where the unconformity surface between the Gamagara Formation and the Transvaal Supergroup transects the manganese-bearing dolomite of the Reivilo Forma-tion in paleosinkholes and large depressions, the manganese oxyhydroxide accumulation took place. These Mn oxyhydroxides were structurally low organized and almost amor-phous. Later they were deeply recrystallized and altered (Figures 61 and 62). Thus, the major manganese deposits of the region (Lohatlha, Bishop, and Glosam) were formed.

Transgression, that followed this significant break, resulted in the deposition of the sediments of the Gamagara Formation. Screes, that were supplied from the surrounding ridges and uplifts, were accumulated in the form of hematite pebbles, and alluvial pebble fans, occurring at the base of the Gamagara Formation. Such hematite conglomerates constitute an essential part of the iron ore reserves of the regarded region (Beukes, 1986, 1993; Button, 1976). During the following geological epochs to the present time there was a repeated activation of paleosinkholes, where slumping products were accumulated and the supergene enrichment of the earlier ore sediments and the Gamagara Formation rocks took place.

It is significant, that the Gamagara Formation is a correlation equivalent of the Mapedi Formation of the Olifantshoek Group (Beukes, 1983, 1986, 1993), that allows one to consider the iron and manganese ore deposits of the Postmasburg-Sishen region to be practically synchronous accumulations (see Figures 60, Section, and 63). The formation of these deposits was controlled by factors close in their nature, during the transitional interval, from the Proterophitic to Palaephytic Periods, which according to Cloud (1976) were the times of evolutional changes of the atmosphere from reducing to relatively oxidizing conditions. Such transition took place approximately in the interval from 2200 to 2100 Ga (Beukes, 1983, 1986, 1993), because the absolute age of the Ongeluk/Hekpoort lavas of the Transvaal Sequence is 2224 + 21 Ga, and the Hartley lava of the Olifantshoek Group is 2070 + 90 Ga.

Consequently the favorable combination of such factors as the composition of parent component-bearing rocks, structural position of the region, features of the geological development, and the extremely favorable time interval in the geochemical evolution of atmosphere for leaching, extraction, transport, and accumulation of Mn, Fe, and associating elements determined the development of such a tremendous and extended ore forming process in the Earth's history. In later works (Holland and Beukes, 1990) the dating of the transitional interval from the reduction to essentially oxidation atmosphere was loosely defined. The beginning of development of relatively deep oxidation of rocks of the Kuruman Iron Formation is evaluated to be about from 2.2 to 1.9 Ga. It should be noted, that in the unaltered form, the Kuruman Iron Formation is composed of siderite, ankerite, magnetite, greenalite, and quartz. Such relatively deep oxidation (hematitization) of these rocks may be explained by the presence of relatively increased O_2 content with a partial pressure of ≥ 0.03 atm in the atmosphere during the accumulation of red sediments of the Gamagara Formation. Such partial pressure of O_2, as considered by Holland and Beukes (1990), noticeably exceeds the value, calculated for paleosoils developed on basalt (age ≥ 2.2 Ga), but it corresponds to the relatively high oxidation of the soil of the Fun Flon. According to the data of these authors the partial O_2 value increased dramatically from 1% of the present atmospheric level (PAL) to $\geq 15\%$ PAL between 2.2 and 1.9 Ga (Holland and Beukes, 1990).

At the same time it is necessary to emphasize that the association of main ore minerals, in the composition of which braunite, bixbyite, and to a lesser degree hausmannite and jakobsite predominate, has a dualistic nature. It may be regarded as either relatively high temperature or as reflecting the relative reducing conditions, under which the Mn(IV) compounds could not be formed. If as in the iron ores of the Beeshoek and Sishen deposits (see Figures 60 and 63) the main mineral is Fe(III) oxide–hematite, then minerals of Mn(II) and Mn(III), but not the primary (Proterozoic) minerals of Mn(IV), dominate in the composition of manganese ores of the Precambrian age of the Postmasburg region. Such minerals as pyrolusite, romanechite, cryptomelane, and others are relatively late supergene products, formed in the place of older and less oxidized varieties. In the environment, when the content of O_2 in the atmosphere during the accumulation of the Gamagara Formation was about $\geq 15\%$ PAL (Holland and Beukes, 1990), oxidation of Mn(II) in karst solutions took place only to Mn(III), that corresponds well with experimental data (Hem and Lind, 1983; Murray *et al.*, 1985), thorough discussion of which is given in the following chapters.

Manganese compounds in comparison to iron are relatively sensitive to the Eh regime variation, determined mainly by the concentration of dissolved O_2 in component-bearing solutions. It is illustrated in the transitional zones from oxygen poor environments to water masses with normal aerated regimes of O_2 in oceans and other modern basins (Landing and Bruland, 1980, 1987). In paleobasins the illustration of such control may be the distribution of Mn forms in ores of the Famennian manganese ore basin of the Central Kazakhstan (Varentsov *et al.*, 1993). If in relatively deep water and stagnated environments, for example at the area of the Karazhal and Ushkatyn deposits the compounds accumulated were essentially of Mn(II) and Mn(III) braunite and hausmannite ores, then in the well-aerated shallow water environments were chiefly formed minerals of Mn(IV):

romanechite, cryptomelane, coronadite, and pyrolusite ores with similar hydrothermal sources of ore-forming components. At the same time it is obvious that for adequate genetic interpretation of the data on mineral composition it is necessary to take into account that ore enclosing rocks are altered to stages of deep epigenesis-early metamorphism.

Similar sets of minerals in Mn deposits in India indicate high stages of metamorphism that corresponds well with the mineralogy of enclosing rocks (facies of mica schists-granulites, and gneisses). However, in the regarded case, the enclosing rocks at the Postmasburg region are extremely weakly metamorphosed. Relatively high concentrations of lithium in bixbyite and lithiophorite, boron in braunite and other components may be considered as indirect indications of endogenic influence on the chemistry of ore-forming solutions that leached manganese from the dolomite layers. The mineral composition of ores that were not altered supergenically, allows one to consider that manganese in the solutions was transported in the form of Mn(II) and the oxidation of this metal to Mn(III) took place with relatively low oxidation values of Eh (relatively low concentrations of dissolved O_2). It is important to note too that ore deposition proceeded both as the result of substitution of the enclosing rocks and owing to the accumulation in different cavities, channels, or ore traps. The deposit structures indicate that the manganese mineralization was developed prior to and after the Carboniferous glaciations of Dwika-Karroo, including the Cretaceous or later erosion cycles.

As it was noted above, a rather considerable part of these ores may clearly be attributed to residual sedimentation, and infiltration deposits in karst cavities (Figures 61 to 63) of the carbonate Cambellrand Formation (Beukes, 1983, 1986; De Villiers, 1956, 1960; Grobbelaar and Beukes, 1986; Sönge, 1977; Van Schalkwyk and Beukes, 1986). The presence of such a mineral as diaspore in products of deep leaching may be indicative of the decomposition intensity of the primary alumosilicate components. The known geological data is not contradictory to the idea that the process of interaction of vadose solutions with the enclosing rocks and the formation of spatially separated accumulations of oxide minerals of manganese and iron could take place in a relatively wide geochronological span (about 2000 Ma), including after intrusions and the formation of the Bushweld intrusive complex. In the later cases the formation of ores, composed by some coarse-crystalline manganese and iron oxides, could take place from heated metal-bearing solutions, the bases of which were subsurface waters of the meteoric origin. The metamorphogenic alteration of ores to products, represented by braunite and bixbyite in association with specularite and muscovite-like mineral – ephesite – took place owing to the intrusion of 'bostonite' dykes and sills (Sönge, 1977).

Thus, the regarded type of karst sedimentation of the residual-infiltration manganese and iron ores is associated with the wide development of exogenic processes in the time predominantly of an erosion hiatus. During these intervals there was leaching and consequent concentration of Mn and Fe from dispersed amounts, available in underlying beds. In the dolomite formations, belonging to the Campbellrand Subgroup (Beukes, 1983, 1986, 1993; Klein and Beukes, 1989; Sönge, 1977) there is about 0.5% of Mn. In the overlying ferruginous-chert formation of banded jespillite ores of the Asbesheuwels Subgroup (Griquatown Formation) the concentration of Mn is 0.3%.

It is important to emphasize that in the optimum set of factors favoring the formation of considerable supergene deposits of manganese a rather essential role belongs to the primary parent rocks initially enriched with Mn as the source of ore matter (i.e., protore). The analysis of data, presented in the work by Klein and Beukes (1989) on average chemical composition of rocks of the Transvaal Supergroup and recalculated by us for the comparison with the average composition of sedimentary rocks of the Earth's crust (Tables 19 to 21), allows us to note a number of important geochemical features.

The carbonate rocks of the Campbellrand Subgroup (Gamohaan Formation), represented by carbonate, limestone (MC), and dolomite (MD) micrites are the sediments of relatively shallow water marginal basins. Relatively high average concentrations of Mn: 0.515–0.552 wt. %, are characteristic for them, that approximately in exceeds the contents of this element in carbonate and clay rocks of the Earth's crust by five times (see Table 19). The terrigenous clastic components of carbonate rocks of the Campbellrand sequence, judging from the ratio values of Ti/Al (< 0.022 to 0.051) and Zr/Al (0.003 to 0.004), differ slightly from the average composition of sedimentary rocks of the Earth's crust (Tavekian and Wedepohl, 1961). This conclusion is confirmed by the distribution of ratios of a number of heavy metals to Zr as the geochemically inert element in sedimentation and supergene ore formation environment (Cr/Zr, Co/Zr, Ni/Zr, Cu/Zr, Zn/Zr, see Table 20).

However, the relatively high values of the ratio of As/Zr (0.601 to 1.231) and Se/Zr (0.012) exceed in 1–2 orders of magnitude the similar characteristics in carbonate and clay rocks of the Earth's crust, and may be considered as an indication of significant role of hydrothermal exhalations in this sedimentary basin. In other words, the above increased concentrations of Mn are genetically associated with high concentrations of As and Se; this is characteristic for a number of intracraton manganese-bearing basins in geological history: Tertiary Mn deposits in the southwest of the U.S.A., Famennian manganese-bearing basin in Kazakhstan, etc. (Koski *et al.*, 1989; Varentsov *et al.*, 1993). The features of REE distribution in the regarded rocks (see Table 21) do not contradict these conclusions.

Rocks of the ferruginous quartzites (mainly the Kuruman Iron Formation) are represented by essentially oxide (rich in magnetite) and carbonate (rich in siderite) varieties. Extremely low concentrations of such clastogenic elements as Ti ($\leq$ 0.020 wt. %), Al (0.035–0.052 wt. %), and Zr (19.0–31.8 ppm) are characteristic for these sediments, that is in 1–2 orders of magnitude lower than in clays, and shales of the Earth's crust (see Table 19). In the distribution of REE the value of Σ(La–Lu) = 3.853–4.132 ppm is of special interest, and it is in almost two orders of magnitude lower than the similar parameter for clays, and shales of the Earth's crust (see Table 21). Values of the ratio of Eu/Sm = 0.386–0.401 are 1.5–2.0 times higher than in clay rocks of the lithosphere and they are close to the corresponding ratios for hydrothermal metal-bearing sediments of the World Ocean (Fleet, 1983; Klinkhammer *et al.*, 1983; Varentsov, 1993; Varentsov *et al.*, 1991).

Thus the regarded sediments of iron ore formations are chemogenic hydrothermal-sedimentary accumulations equivalent to the modern metalliferous sediments. Major parameters of the Early Proterozoic sedimentation are registered in the features of their occurrence, structure and mineral and chemical composition. As in numerous later basins

Table 19. Average compositions of the various lithologies of the Early Proterozoic Transvaal Supergroup, South Africa: oxide iron-formation, siderite iron formation (BS), limestone (MC), dolomite (MD), shale, ferruginous shale (SF), pyrite-rich shale (SP) and carbonaceous shale (SC) (Klein and Beukes, 1989). For comparison the average content of the elements in some major units of the earth's crusts is shown (Turekan and Wedepohl, 1961) (wt. %).

*	Oxide banded iron-formation	BS	MC	MD	Shale	SF	SP	SC	Clays	Carbonate rocks	Deep-sea clayey sediments
N**	9	16	8	4	10	2	2	6	–	–	–
Si	21.70	18.67	5.44	4.35	18.29	11.10	14.32	22.00	7.3	2.4	25.0
Ti	< 0.020	0.020	0.056	< 0.020	0.118	0.16	0.13	0.232	0.46	0.04	0.46
Al	0.035	0.052	1.088	0.915	5.050	6.020	3.788	5.147	8.0	4.2	8.4
Fe(III)	15.89	0.89	***						–	–	–
Fe(II)	13.64	18.54	2.14	2.43	11.51	19.13	22.77	5.21	–	–	–
Mn(II)	0.071	0.250	0.515	0.552	0.170	0.155	0.084	0.204	0.085	0.11	0.67
Mg	1.421	3.032	1.140	10.245	3.073	5.014	1.444	2.970	1.5	4.7	2.10
Ca	2.055	3.827	31.409	19.205	5.277	5.822	2.453	6.037	2.21	30.23	2.9
Na	0.015	0.013	0.024	0.020	0.132	0.022	0.075	0.188	0.96	0.04	4.00
K	0.022	0.017	0.510	0.510	2.369	0.033	1.291	3.507	2.66	0.27	0.25
P	0.048	0.027	0.037	0.024	0.033	0.032	0.031	0.034	0.07	0.04	0.15
$H_2O(^-)$	0.05	0.08	0.08	0.09	0.2	0.22	0.32	0.15	–	–	–
Vol****	7.42	23.89	35.10	39.85	15.67	22.16	7.8	16.22	–	–	–
S	0.012	0.048	0.571	0.406	4.236	0.415	16.15	1.245	0.24	0.12	0.13
$C_{org.}$	0.012	0.080	0.886	0.523	3.91	4.55	2.675	4.108	–	–	–
$Fe_{(tot.)}$	29.54	19.32	2.14	2.43	11.51	19.13	22.77	5.21	4.72	0.38	6.5
Mn/Fe	0.002	0.013	0.241	0.227	0.015	0.008	0.004	0.039	0.018	0.289	0.103
Ti/Al	< 0.571	0.385	0.051	< 0.022	0.023	0.027	0.034	0.045	0.057	0.009	0.055

*Facies code: B = banded iron-formation; BS = sideritic banded iron-formation; MC = micrite, calcareous; MD = micrite, dolomitic; S = shale; SC = shale carbonaceous; SF = shale, ferruginous; SP = shale pyritic.

**N = number of analyses used in average of most elements.

***Open space means value not determined.

****Vol. = Volatiles including $H_2O(^+)$ and CO_2 and corrected for oxidation of Fe(II).

Table 20. Average content of trace elements in various lithologies of the Proterozoic Transvaal Supergroup, South Africa: oxide iron-formation, siderite iron-formation (BS), limestone (MC), dolomite (MO), shale, ferruginous shale (SF), pyrite-rich shale (SP) and carbonaceous shale (SC) (Klein and Beukes, 1985). For comparison the average content of the trace elements in some major units of the earth's crust is shown (Turekian and Wedepohl, 1961) (ppm).

*	Oxide banded iron-formation	BS	MC	MD	Shale	SF	SP	SC	Clays	Carbonate rocks	Deep-sea clayey sediments
no.**	1	16	8	4	10	2	2	6	–	–	–
Sc	0.13	0.21	1.69	2.04	11.85	20.6	6.79	10.62	13	1.0	19
V	< 150	< 150	< 150	< 150	165	180	< 150	150	130	20	120
Cr	3.20	3.23	9.29	13.56	109.17	141.5	56.55	115.93	90	11	90
Co	0.37	0.59	4.36	3.99	31.95	29.05	34.05	32.22	19	0.1	74
Ni	30.0	29.3	49.4	36.5	142.6	158	140	138.3	68	20	225
Cu	< 15	< 15	26.8	20.0	79.0	15	141.5	66.8	45	4	250
Zn	25.0	25.5	57.8	< 20	182.4	127.5	135	216	95	20	165
As	1.57	11.7	52.8	16.4	133.5	44	325.5	99.3	13	1.0	13
Se	< 0.2	< 0.2	0.53	0.32	3.24	1.4	7	1.32	0.6	0.08	0.17
Br	1.85	2.83	2.80	18.22	1.10	0.8	–	1.15	4.0	6.2	70.0
Rb	23.2	18.9	37.0	25.2	83.1	19.5	75	107	140	3.0	110.0
Sr	7.9	9.3	51.9	40.7	18.4	9	11.5	23.8	300	610	180
Y	4.8	3.8	10.5	6.5	25.5	21.5	19	29	260	30	90
Zr	19.0	31.8	42.9	27.3	23.0	120	94	89.8	160	19	150
Nb	7.0	7.4	8.7	8.2	13.5	15	12	13.5	11	0.3	14
Sb	0.32	0.16	0.93	0.51	7.57	3.84	18.33	5.22	1.4	0.2	1.0
Cs	0.26	0.27	1.01	0.78	6.04	0.23	4.22	8.59	5.0	$0.1 \times n$	6.0
Ba	< 30	< 30	64.2	45.0	195.1	< 30	123	219.2	580	10.0	2300
Hf	0.040	0.040	0.874	0.416	3.002	3.69	2.605	2.905	2.8	0.3	4.1
Ta	0.030	0.025	0.207	0.092	0.828	1.09	0.575	0.825	0.8	$0.01 \times n$	$0.1 \times n$
Th	0.069	0.102	2.064	1.312	11.066	14.6	8.59	10.723	12.0	1.7	7.0
U	0.100	0.035	0.968	0.467	3.469	3.75	2.82	3.592	3.7	2.2	1.3
Cr/Zr	0.168	0.102	0.216	0.497	4.746	1.179	0.602	1.291	0.562	0.579	0.600
Co/Zr	0.019	0.018	0.102	0.146	1.389	0.242	0.362	0.359	0.119	0.005	0.493
Ni/Zr	1.579	0.921	1.151	1.337	6.330	1.317	1.489	1.540	0.425	1.053	1.500
Cu/Zr	< 0.789	< 0.472	0.623	0.733	3.434	0.125	1.505	0.744	0.281	0.210	1.667
Zn/Zr	1.315	0.802	1.347	< 0.733	7.930	1.062	1.436	2.405	0.594	1.053	1.100
As/Zr	0.082	0.368	1.231	0.601	5.804	0.367	4.673	1.106	0.081	0.053	0.087
Se/Zr	< 0.010	< 0.006	0.012	0.012	0.141	0.012	0.074	0.020	0.004	0.004	0.001
Br/Zr	0.097	0.089	0.065	0.667	0.048	0.007	–	0.013	0.025	0.326	0.467
Sb/Zr	0.017	0.005	0.022	0.019	0.329	0.032	0.195	0.058	0.009	0.010	0.007
Ba/Zr	< 1.580	< 0.943	1.496	1.648	8.478	< 0.250	1.308	2.441	3.625	0.526	15.333
Zr/Ti	> 0.095	0.159	0.077	> 0.136	0.019	0.075	0.072	0.039	0.035	0.047	0.033
Zr/Al	> 0.054	0.061	0.004	0.003	0.0004	0.002	0.002	0.002	0.002	0.0004	0.002

For notes: facies code, see Table 19.

Table 21. Average content of rare earth elements in various lithologies of the Early Proterozoic Transvaal Supergroup, South Africa: oxide iron-formation, siderite iron-formation (BS), limestone (MC), dolomite (MD), shale, ferruginous shale (SF), pyrite-rich shale (SP), and carboniferous shale (SC) (Klein and Beukes, 1985). For comparison is shown the average content of REE in the North American shale (Haskin, 1966), and in the sedimentary rocks (clays, shales) (Vinogradov, 1962).

*	Oxide banded iron-formation	BS	MC	MD	Shale	SF	SP	SC	North American shales	Sedimentary rocks (clays and shales)
no.**	9	16	8	4	10	2	2	6	–	–
La	1.326	1.421	9.864	6.140	28.560	33.85	26.7	27.417	39	40
Ce	1.852	2.059	18.209	11.220	53.540	61.5	49.35	52.283	76	50
Nd	–	–	7.757	5.350	21.030	25.75	17.65	20.583	37	23
Sm	0.217	0.228	1.428	1.056	4.150	4.9	3.32	4.177	7.0	6.5
Eu	0.087	0.088	0.269	0.262	0.697	0.911	0.532	0.680	2.0	1.0
Tb	0.048	0.042	0.190	0.135	0.492	0.563	0.303	0.533	1.30	0.90
Yb	0.261	0.242	0.648	0.313	2.053	2.7	1.045	2.173	3.4	3.0
Lu	0.062	0.052	0.102	0.050	0.332	0.458	0.169	0.344	0.60	0.70
Σ(La-Lu)	3.853	4.132	38.467	24.526	110.854	130.632	99.069	108.19	166.3	25.100
Ce/La	1.397	1.449	1.846	1.827	1.875	1.817	1.848	1.907	1.949	1.250
Ce/Ce*	–	–	1.006	0.963	0.755	1.305	1.909	1.052	1.000	0.739
La/Yb	5.080	5.872	15.222	19.617	13.911	12.537	25.550	12.617	11.471	13.333
La/Sm	6.111	6.232	6.908	5.814	6.882	6.908	8.042	6.563	5.571	6.154
Ce/Sm	8.535	9.031	12.751	10.625	12.901	12.551	14.864	12.517	10.857	7.692
Ce/Yb	7.096	8.508	28.100	35.846	26.079	22.778	47.225	24.060	22.353	16.670
Eu/Sm	0.401	0.386	0.188	0.248	0.168	0.186	0.160	0.163	0.286	0.154
(La-Sm)/(Eu-Lu)	–	–	30.817	31.271	30.017	27.202	47.350	28.005	22.781	21.339

Facies code see in Table 19. The value of crium anomaly was calculated by: Ce/Ce = [3Ce/Ce$_{shale}$]/[2La/La$_{shale}$ + Nd/Nd$_{shale}$], where Ce, La, and Nd are the contents of elements in the lithology; Ce$_{shale}$, La$_{shale}$, and Nd$_{shale}$ are their contents in the North American shale (Haskin *et al.* 1966).

the sporadic contribution of Mn and Fe to the sedimentation zone was determined by the dynamics of development and distribution of the major parameters of the supplying hydrothermal system: convection, temperature gradients, Eh, and pH. The supergenic alteration of rocks of the Kuruman Iron-Formation and the Griquatown Iron-Formation was the next most important factor in the formation of the tremendous iron ore (hematite) karst deposits Sishen and Beeshoek.

Cycles of essentially supergene leaching from dolomite rocks and deposition of manganese, and to a lesser degree, of iron were observed at the territory of the Transvaal basin repeatedly during 2000 Ma. Such a long history of the wide development of erosion breaks and the governing of supergene environments was reflected in that manganese oxyhydroxide minerals were distributed essentially to the depth of about 100 m in the hydrothermal-sedimentary accumulations of the Hotazel Formation, whereas the hausmannite predominated in ores, observed at depths, exceeding 400 m (De Villiers, 1983). In other words the wide development of supergene (or activity associated with the ground water) alterations of hydrothermal-sedimentary deposits of ores of the Hotazel Formation took place except manganese and iron mineralization associated with karst events (De Villiers, 1956, 1960, 1983).

3.4. Manganese Ore Deposits of North Africa: Imini, Morocco

3.4.1. GEOLOGICAL SETTING

The Imini deposit is located in the zone of Pre-African tectonical suture, dividing the region of the High Atlas in the north and the region of Anti-Atlas in the south in the form of a narrow band. This geosuture zone is characterized by the development of a number of Mn deposits of a supergene karst nature (Beaudoin *et al.*, 1976; Bouladon *et al.*, 1956; Pouit, 1964, 1976; Vincienne, 1956). Primarily these deposits of continental and sometimes marine genesis are represented by subhorizontal bodies, occurring unconformably on the Precambrian basement (western areas) or on the Paleozoic rocks (eastern areas).

A wide and contradicting spectrum of views has been suggested on the origin of oxyhydroxide Mn ores of this zone: from supergene, karst concepts (Beaudoin *et al.*, 1976; Bouladon *et al.*, 1956), sedimentary and hydrothermal-sedimentary (Jouravsky and Pouit, 1960; Pouit, 1964, 1968, 1976) to the recognition of the critical role of mixing of the ore solutions supplied from the zone of the supergenesis and ore-bearing solutions from the stagnant oxygen-free deep ocean basins (Thein, 1990). Such a situation makes the author adhere to the factographical basis in the interpretation of the available data.

The Imini ore deposits are situated at the southern margin of the geosuture band in the zone, characterized by several phases of secondary supergene mineralization, revealed in the lithofacies plan as the border of the transgression distribution in the northern province. The lower ore-bearing level is represented by the Cenomanian–Turonian dolomite sequence, occurring on the transitional sediments: Infracenomanian and Permo-Triassic, beds of clays and sandstones to conglomerates. Sometimes dolomites occur outside this transitional sequence on the basement, represented by the Ordovician crystalline schists

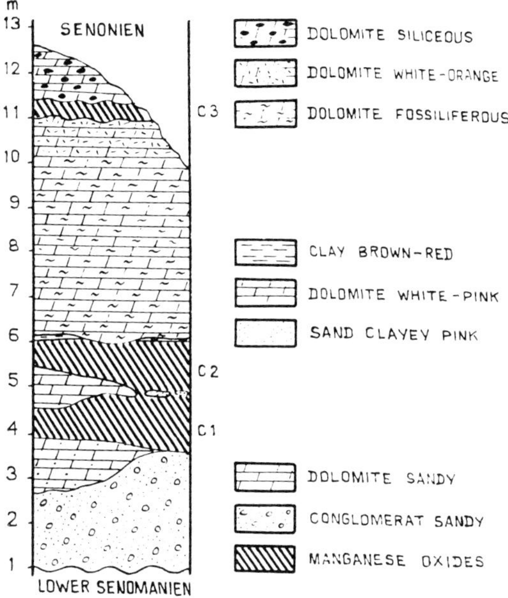

Fig. 64. Section of the Upper Cretaceous sediments, enclosing manganese ores, Imini deposit, Northern Africa (after Beaudoin *et al.*, 1976).

or on volcanogenic-sedimentary rocks (Precambrian III). The thickness of the considered Cenomanian–Turonian sediments increases gradually and steadily to the north from 10–12 m near the deposit to 20–25 m in 8 km to the north (Figure 64).

Ore deposit structure and occurrence features. The ore deposit is represented by three beds: the first and second beds are usually separated or form an almost single body in the section (see Figures 64 and 65), consisting of these two superimposed accumulations; they occur at the base of the dolomite sequence. The third bed is developed to the north of the area of characteristic manifestation of the first two beds.

Mineral composition. The ores are mainly composed of oxyhydroxides. The predominant mineral is pyrolusite, occurring in the form of clearly developed fibrous aggregates. The tectonomanganates of the tunnel structure are widely distributed minerals. Romanechite (psilomelane)–cryptomelane type minerals are usually represented by amorphous or cryptocrystalline accumulations: cryptomelane proper (with contents of K typical for it), hollandite (Ba), coronadite (Pb), and X-ray amorphous wads. Lithiphorite, hematite, goethite, and martite are present in subordinate amounts. The information on mineral and chemical composition of ores is generalized in Table 22.

In the work by Beaudoin *et al.* (1976) the supergene alteration of the Cenomanian–Turonian dolomite sediments is especially emphasized. It is fairly intensively developed at their base at the level of the first and second manganese ore beds. The authors point

Table 22. Mineralogical and chemical composition of ores and carbonate rocks of the manganese ore regions: Imini, Tasdremt, Cretaceous; and Tiouine, Precambrian (after Thein, 1990).

	Mn (%)	SiO$_2$ (%)	Pb (%)	Ba (%)	Ca (%)	Mg (%)	Fe (%)	Al (%)	K (%)	Cu (%)	Co (ppm)	Zn (ppm)	V (ppm)
Bou-Agguioun site, Imini; pyrolusite romanechite (psilomelane)	52.89	3.42	0.05	0.06	0.08	0.05	0.00	0.02	0.17	0.24	57	149	235
Bou-Agguioun site: pyrolusite, dolomite, romanechite (psilomelane)	31.33	0.85	0.03	0.08	8.40	4.65	0.05	0.02	0.03	0.13	26	116	129
Bou-Agguioun site: pyrolusite, romanechite (psilomelane), cryptomelane	46.75	3.75	0.37	0.25	0.14	0.04	0.10	0.22	2.37	0.19	162	402	149
Bou-Agguioun site: dolomite, (pyrolusite)	10.67	1.30	0.04	0.15	17.80	9.50	0.05	0.06	0.04	0.04	20	89	64
Ste. Barbe, Imini: pyrolusite	46.75	8.87	0.19	0.18	0.37	0.31	0.41	0.63	0.13	0.41	41	206	234
Ste. Barbe: pyrolusite	50.84	1.76	0.90	0.61	0.11	0.08	0.00	0.14	0.07	0.32	59	339	684
Ste. Barbe: dolomite (quartz, pyrolusite)	1.67	19.96	0.02	0.04	17.20	9.50	0.47	1.27	0.13	0.01	12	140	35
Section 28, bed 3: calcite (coronadite)	24.57	2.62	3.53	1.70	14.40	0.33	0.21	0.27	0.48	0.26	158	202	844
Section 28, bed 3: dolomite, calcite, romanechie, (psilomelane), quartz	2.74	5.25	0.51	0.42	22.80	6.10	1.68	1.05	0.15	0.04	58	138	269
Section 28, bed 3: dolomite, calcite, quartz, romanechite (psilomelane)	5.98	8.32	0.90	0.89	19.90	7.00	0.34	0.67	0.25	0.05	71	120	222

Table 22. (Continued)

	Mn (%)	SiO$_2$ (%)	Pb (%)	Ba (%)	Ca (%)	Mg (%)	Fe (%)	Al (%)	K (%)	Cu (%)	Co (ppm)	Zn (ppm)	V (ppm)
Section 28, bed 3: calcite, dolomite, coronadite, quartz	11.99	4.32	2.25	0.81	19.10	4.85	0.96	0.63	0.32	0.15	139	163	587
Section 36, bed 3: cryptomelane (dolomite)	42.37	2.13	0.01	0.95	3.60	0.31	0.36	0.79	4.41	0.09	188	1484	431
Section 19, upper bed: hollandite, dolomite	26.90	3.37	3.26	4.94	6.50	3.25	0.83	0.34	0.64	0.02	120	89	953
Section 44, upper bed: calcite (quartz, manganosite, romanechite (psilomelane)	8.35	9.28	0.01	2.07	24.00	0.79	1.84	1.67	0.34	0.04	46	1755	621
Section 19, upper bed: quartz, hollandite	12.53	42.58	1.88	2.07	0.35	0.66	3.21	5.02	1.15	0.02	103	103	546
Tasdremt, lower bed: dolomite, romanechite (psilomelane), quartz	12.55	10.12	1.35	2.32	11.80	7.35	0.31	0.58	0.41	0.02	68	78	165
Tiouine, Precambrian: cryptomelane, quartz (feldspar)	25.56	28.25	0.20	1.53	0.23	0.32	1.43	3.09	4.67	0.07	21	105	172
Tizi n'Bachkoum (vein, Precambrian): braunite, feldspar	24.23	29.51	0.01	0.64	0.85	0.45	3.55	5.02	3.76	0.08	22	64	169
Chemical ore, Imini (Bouladon and Jouravsky, 1952)	58.95	2.48	0.30	0.57	0.21	0.10	0.83	0.24	0.06	0.26	n.d.	400	not det.

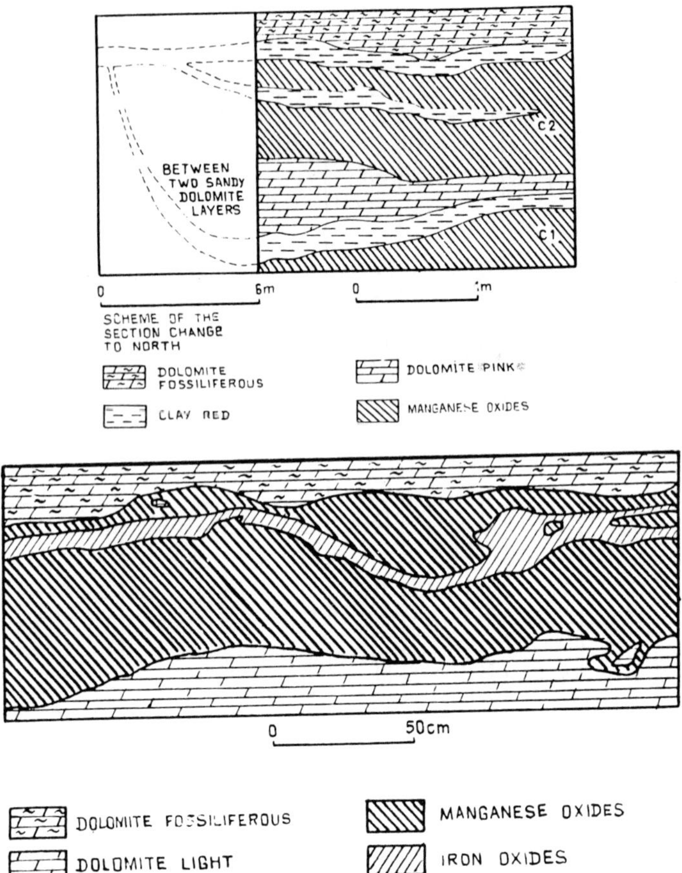

Fig. 65.   Scheme of manganese ore sequence structure, Imini deposit, Northern Africa (after Beaudoin *et al.*, 1976).

that the alteration has the karst nature. The development of karst structures controls the major features of manganese ore mineralization within the deposit. There are several types of alterations: (a) alterations with preservation of relatively unaltered rock structure; (b) alterations, associated with dissolution and filling in of cavities; and (c) alterations, associated with epigenetic (secondary) redistribution of manganese and iron. It should be noted however, that the concepts on the leading role of karst processes in the formation of manganese ores of the Imini deposit (Beaudoin *et al.*, 1976) contradict the views of some French geologists, adhering to traditional concepts on sedimentary genesis of these ores (Jouravsky and Pouit, 1960; Pouit, 1964, 1976). Pouit (1976) criticized the karst hypothesis. In the work by Thein (1990) on the basis of detailed geochemical investigations and the analysis of microfacies of the carbonate section of the Cenomanian–

Turonian region of the High Atlas, Morocco, there was an attempt to relate the origin of high grade manganese ores of the Imini–Tasdremt region to the processes that took place in the zone of mixing of oxygenized stratal waters of meteoric origin and reduced sea water near the shore. Ore forming metals were contributed by solutions, associated with continental weathering and owing to events of remobilization from oxygen-free deep water ocean basins. Recognizing the complexity of this problem and the possible role of sedimentation processes in the formation of relatively early generations of romanechite (psilomelane)–cryptomelane ores, it is necessary to display the essential data, on which the conclusions on karst genesis (Beaudoin *et al.*, 1976) is based.

3.4.2. ALTERATIONS WITH THE RETAINING OF THE RELATIVELY STABLE ROCK STRUCTURE

It is pointed out (Beaudoin *et al.*, 1976) that unaltered structure (see Figures 64 and 65) is usually observed in the dolomite interbed about 1 m thick, occurring between the first and second ore beds. Laterally this interbed is changed by the rock of the sandstone composition of irregular texture, composed of numerous thin intercalations several cm thick. The lateral transition between these two extreme varieties is represented by brecciated and altered dolomite. This breccia is a transitional rock in the continuous series between unaltered dolomite and the interbed, composed by sand material of a residual nature.

The formation of such deeply altered intermediate products may take place at relatively late stages after the accumulation of initial dolomites to the final products of the supergene epigenesis (in sequence):

1. Dolomite white, often somewhat sandy, weakly fractured, usually loose.
2. Dolomite white, fractured or microfractured with jointing of orange color.
3. Breccia, composed of light fragments or clasts, colored by iron oxyhydroxides into red color of different tints. As the development of this stage increases, the breccia cement becomes more ferruginous with a color transition from light orange to dark brown.
4. The formation of breccia similar to no. 3; however the manganese oxyhydroxides (mainly pyrolusite and X-ray amorphous wad-like phases) appear between the fragments.
5. The manganese oxyhydroxides at this stage form the breccia cement; contours of the composing fragments become relatively less distinct; thin sand intercalations of pink color of residual origin appear owing to the substitution of the dolomite breccia.
6. The breccia fragments are not distinguished at this stage, and the manganese oxyhydroxides are developed from the surface and fill in almost the whole volume of the primary rock; the pink sand intercalations from 2 to 15 cm thick substitute dolomite, they are the transitional material in the process of substitution of the dolomite by the manganese oxyhydroxides.

All these stages of the above alteration may be observed at a distance of several m.

3.4.3. DISSOLUTION, FORMATION OF CAVITIES, AND THEIR FILLING

The alteration of rocks of the dolomite series at a relatively late stage of development is represented by the formation of cavities and hollows and their consequent filling in by products of different types. Cavities and their combinations, formed owing to leaching, gave rise to irregular pockets that are often connected into general large sinkholes, developed essentially at the contact of the second ore bed and the fractured dolomite. The cavity or large pocket, formed in such a way, may be to 10 m long and several m high (Figure 65).

Owing to the wide development of such cavities and hollows the caving or subsiding of blocks and lumps of the fossiliferous dolomite proceeded and they were accumulated at the manganese bed level. They were subjected to dissolution and substitution by the components, transported in ground solutions. The forming cavities were filled to different degrees by the products, among which (Beaudoin *et al.*, 1976) are distinguished two main types: (1) manganese-free accumulations, represented by clays and dolomite breccias, enclosed in clayey matrix; (2) manganese accumulations, consisting of the aggregates of pyrolusite and breccias, composed of the fragments of dolomite and minerals of romanechite (psilomelane)–cryptomelane group with manganese oxyhydroxides cement.

3.4.4. GENESIS OF THE MANGANESE ORE ACCUMULATIONS

The data given above gives grounds to consider that the enclosing rocks of the Imini manganese deposit were subjected to the following alterations:

1. Dissolution of the carbonate material, leading to the formation of cavities and hollows, developed mainly in fractures and in some permeable horizons.

2. Due to closeness of formed cavities to the relief surface they, as a rule, were filled in by supergene clastic material.

3. The existence of oxidation environments of mineral-ore formation is confirmed both by the composition of formed products, represented by manganese and iron oxyhydroxides, and by their depositional sequence. The pyrolusite as a relatively late product indicates stronger oxidation environments than the preceding manganates of the tunnel structure group, romanechite (psilomelane)–cryptomelane. The features of pyrolusite occurrence indicate its secondary supergene epigenetic nature, recorded different manifestations of both pore circulation of stratal waters and proper karst alterations of the Cenomanian–Turonian dolomite rocks. It is essential that these alterations are associated mainly with the dissolution-deposition events due to flow of stratal waters of the surface origin. It is notable that pyrolusite fills in fractures or pocket-like cavities in dolomite; often it is a cement of karst breccia, and binds the fragments of romanechite-cryptomelane composition, or it forms diffusion aggregates in karst clays. Along with such obvious indications of epigenetic karstogenic origin of pyrolusite it is important to bear in mind that, according to Gaudefroy (1960) and Pouit (1976), all pyrolusite of the deposit is the product of manganite alteration, the relicts and structure features of which are preserved. The primary presence of manganite enables us to raise a question on the rather complicated history of epigenetic supergene alterations of manganese ore accumulations of this deposit.

The formation of minerals of the romanechite (psilomelane)–cryptomelane group took place prior to pyrolusite. Breccias, with the predomination of fragments of these minerals, are the mechanical products of karst formation. The concretions of this composition occur too. Their origin is associated with manganese-bearing karst waters. However, on the formation of fragments, composed of the initial tunnel manganates of the romanechite–cryptomelane group, contradicting concepts exist among investigations (Beaudoin *et al.*, 1976; Bouladon and Jouravsky, 1956; Vincienne, 1956; Pouit, 1964, 1976). On the basis of analysis of known data we may consider that as long as the fine banding of the romanechite–cryptomelane aggregates is conformable with syngenetic lamination of the primary dolomites, then these manganese accumulations are relatively early epigenetic stratimorphous products of stratal water activity. Later at intensive development of leaching and caving events, karst breccias of romanechite–cryptomelane were formed. Thus, the source of the regarded oxyhydroxide accumulations was the dispersed and relatively increased amounts of this metal in the carbonate form. At the same time, in the surroundings of the Bouarfa deposit, Eastern Morocco in the Liassic transgressive series, the unaltered beds of limestones and dolomites of chocolate color are characterized by Mn contents of from 2 to 3% (Jouravsky and Pouit, 1965). These sediments must also be the source of manganese that participated in the cycles of multiple redeposition of sediments of this region (Varentsov, 1964). The detailed discussion of physico-chemical features of formation of these minerals in supergene environments is given below.

The material testifies that the formation of the main part of the manganese ores proceeded in the environment of karst regime.

In order to imagine more completely the development features of the karst processes at this deposit it is important to emphasize that the Cenomanian–Turonian carbonate part of sediments occurs, usually on clayey sands and is overlain by a bed, represented by an alternation of argillites, marls, and evaporites of the Cenomanian age. The horizon of increased permeability occurs in the lower part of the Cenomanian–Turonian sediments. It is necessary to note an extremely essential role of widely developed fissures of the enclosing rocks, represented by the fossiliferous dolomites. They are associated with the development of deep leaching of considerable amounts of dispersed manganese and of iron to a lesser degree and their relatively limited transportation and accumulation.

Beaudoin *et al.* (1976) pointed out that available geological data indicates that karst formation began at the end of the Cenomanian and it continues to the present time. These authors state that they have no basis for the identification of some definite span of the geological history, corresponding to the most intensive karst ore formation.

Hence, the general concept on multistage formation of rich supergene Mn oxyhydroxide ores after initially depleted concentrations is supported by the examples of the formation of karst ores of Postmasburg, Imini, and other deposits. The examples of such accumulations, in the broad sense, may serve the deposits of the South Ukrainian manganese ore basin (Varentsov, 1962; Varentsov and Rakhmanov, 1974; Varentsov *et al.*, 1967), Groote Eylandt (Varentsov, 1982), and Moanda (Leclerc and Weber, 1980; Weber, 1968; Weber *et al.*, 1979) deposits.

3.5. The Other Mn Ore Accumulations in the Oxidation Zone and Paleokarst

3.5.1. ZONE OF OXIDATION AND PALEOKARST ACCUMULATION OF THE MANGANESE ORES IN THE DEPOSITS OF THE WESTERN UKRAINE

The Burshtynskoe deposit is relatively large and well studied among few manganese ore deposits and ore occurrences known in western Ukraine. It is related to the carbonate-gypsum manganese-bearing formation of the Ulu-Telyakskii subtype (Khmelevskii, 1968), included into the group of limestone-dolomite manganese-bearing formations of Imini-Tazdrem type (Varentsov, 1964). The deposit is situated in the jointing zone of the East European Platform and Predcarpathian marginal trough in the region of basins of left tributaries of the Dnestr river. The problems of geology, structure, composition of ores and the enclosing sediments of the Burshtynskoe deposit were considered in works of the Ukrainian geologists (Bobrovnik *et al.*, 1966, 1975; Yanchuk and Khmelevskii, 1976, 1981, 1982a, 1982b, 1985).

Manganese ore deposits are associated with the base of the Verbovetskii beds of the Kosovskaya Suite of the Upper Tortonian, Miocene. 'Primary' or relatively early sedimentary-diagenetic ores are composed of Mn carbonate minerals, constituting clay-marl rhythmic laminated accumulations. They are characterized by thin laminated structure, represented by alternation of the horizontal-linear laminated intercalations of light-grey and dark-grey varieties. Sometimes thin interbeds of vitroclastic plagioclase-liparite tuff are observed among clay-marl constituents. Major components (55–80%) of light-grey varieties are manganocalcites, rarely Ca-rhodochrosites. In dark interbeds the carbonates constitute 20–25%, the remaining part belonging to clay and to a lesser degree to silt constituents. Such carbonate material is represented essentially by montmorillonite and hydromica with an admixture of dispersed silica and finest pyrite spherulites (Bobrovnik and Khmelevskii, 1966, 1968; Khmelevskii and Gurzhii, 1975; Yanchuk and Khmelevskii, 1981). The average composition of carbonate ores of deposits (% wt., 10 analyses) is as follows: SiO_2 25.20; TiO_2 0.33; Al_2O_3 5.78; Fe_2O_3 3.58; CaO 18.92; MgO 3.69; MnO 11.43; K_2O 1.69; Na_2O 0.70; CO_2 22.92 (Yanchuk and Khmelevski, 1981).

The oxidized ores are formed in those parts of the deposit where Mn carbonate accumulations are relatively actively drained by ground and surface waters. At the initial stages of such supergene alterations along fractures and lamination surfaces, the spotty brown, black aggregates grow proportionately to the process development and constitute the brown-black earthy loose masses. At later stages of oxidation the volumetric weight of similar supergene products is 1.7–2.0 g/cm^3 with considerable density: 2.8–3.1 g/cm^3. Owing to alterations, CaO, MgO, Na_2O, and CO_2 are removed from Mn carbonates and the residual products of dissolution and oxidation of the following composition are formed (% wt., average from 10 analyses): SiO_2 22.70; TiO_2 0.36; Al_2O_3 6.02; Fe_2O_3 3.74; CaO 7.96; MgO 2.38; MnO 30.00; K_2O 1.79; Na_2O 0.23; CO_2 5.39 (Yanchuk and Khmelevskii, 1981).

On the basis of detailed observations of the wide scale of supergene alterations of Mn carbonates to relatively deep oxidized products, Yanchuk and Khmelevskii (1976, 1981) report the following formation sequence of Mn oxyhydroxide minerals. Their diagnosis is mainly based on the data of X-ray diffraction and DTA.

Manganocalcite (Ca-rhodochrosite) → rancieite (+buserite) →birnessite → vernadite.

The succession of the mineralization stages reflects the evolution of drainage intensity and the composition of surface and stratal waters, inherent to the area of the Burshtynskoe deposit. It may be interpreted as an indication that oxyhydroxide phases with the essentially decreasing structural ordering [7 $\overset{\circ}{A}$-MnO_2 (birnessite) → 2.4 $\overset{\circ}{A}$-MnO_2 (vernadite)] formed in the environments of relatively deep supergene alterations after relatively early structurally ordered oxidation products (rancieite + 10 $\overset{\circ}{A}$-manganate). It is worthy of notice that the most oxidized products are developed in layers, occurring near the surface, i.e., at the depression edges. Khmelevskii and Gurzhii (1975) point to the presence of distinct relation between manganese content and the ore oxidation degree, as well as to the inverse relation between the occurrence depth of the oxidized ore and its manganese-bearing grade.

It is shown by the examples of formation of oxidation zones of some Mn deposits (see other chapters of the work) that the formation sequence of Mn oxyhydroxide compounds may be essentially different depending on the composition of the substrate, hydrochemical features, circulation regime of surface and ground waters, including climatic features of alternation of humid and dry periods.

3.5.2. ZONE OF OXIDATION AND KARST ORE ACCUMULATION OF THE ENISEI RIDGE MN DEPOSITS (CENTRAL SIBERIA)

3.5.2.1. Geological Setting

The Enisei Ridge is a heterogenic structure, in the formation of which the Archean–Lower Proterozoic and Upper Proterozoic complexes take part. The main problems of geology in relation to manganese-bearing potential of this region are covered in a small number of works (Balitskii and Kornev, 1984; Divana, 1976; Golovko *et al.*, 1982; Gribov and Varand, 1984; Kavitskii *et al.*, 1980; Kornev, 1972; Mkrtych'yan *et al.*, 1979, 1980, 1982; Mstislavskii and Potkonen, 1990; Musatov *et al.*, 1980; Pasashnikova, 1984; Savan'yak *et al.*, 1984; Tsykin and Kostenenko, 1984; Ustalov, 1982; Varand and Andreev, 1984; Vladimirov and Gorovoi, 1972).

In the Precambrian sequences of the Enisei Ridge, five major stratigraphically isolated horizons of the regional development of manganese mineralization are recognized (Balitskii and Kornev, 1984; Golovko *et al.*, 1982; Kornev, 1972; Mkrtych'an *et al.*, 1979; Musatov *et al.*, 1980; Savan'yak *et al.*, 1984; Ustalov, 1982; Varand and Andrev, 1984):

1. The Middle Proterozoic, upper part Teiskaya Series (Penchenginskaya Suite): marble; quartzites, crystalline schists; the mineralization is represented by Mn carbonates.

2. The Riphean, the upper beds of the Sukhopitskaya Series (Sosnovskaya Suite): dolomites, limestones, volcanites of middle and acid composition. Manganese mineralization: manganocalcite, rhodochrosite.

3–4. The Riphean, the middle part of the Tungusikskaya Series (Potoskuiskaya and Tokminskaya Suites): carbonate-terrigenous sequence with diabase-albitophyre volcanites. Mn mineralization: rhodochrosite, manganocalcite, rhodonite, pyrolusite.

5.   The Vendian, the lower part of the Chapskaya Series (Pod'emskaya Suite): cherty-tuffogenic-carbonate-terrigenous sediments predominate.   Mn mineralization: rhodochrosite, manganocalcite.   The major manganese deposits and ore occurrences of the region are associated with the Prieniseiskaya eugeosynclinal zone. If the Penchenginskaya Suite of the Middle Proterozoic belongs to the ancient Preriphean platform cover and the Pod'emskaya Suite of the Vendian is attributed to the orogenic trough of the same age, then the manganese-bearing Riphean Suites (the Sosnovskaya, Potoskuiskaya, and Tokminskaya) are developed within the eugeosynclinal zone. These Riphean Mn-bearing sediments, the mineralization scales of which go beyond ore manifestations, are associated with spilite-keratophyre and spilite-diabase volcanism of basins of the eugeosynclinal zone. The largest accumulations of Mn ores are known in the Pod'emskaya Suite of the Vendian, developed in the northwestern part of the Enisei Ridge within the Vorogovski orogenic trough. The latter is a structure developed on the intensively folded and metamorphosed volcanogenic-sedimentary complex of the Baikalides. The Porozhinskoe deposit structure, observed at present, was formed at the end of the Vendian and it was overlaid by the platform cover in the beginning of the Cambrian (Figure 66).

3.5.2.2. MODE OF OCCURRENCE, LITHOLOGY, MINERALOGY, AND GENESIS OF PRIMARY (HYDROTHERMAL-METASOMATIC) ORES

The main manganese-bearing areas are located in the Porozhinskaya graben-syncline of the Mikheevskaya depression of the Vorogovski trough. The composition of manganese-bearing sequences (600–800 m), enclosed in molass-flysch sediments 5–6 m thick (Golovko *et al.*, 1982; Gribov *et al.*, 1984; Ustalov, 1982) (Figure 66) is distinguished by the following units:

The subore, predominantly dolomite unit (350–400 m), sometimes containing transecting streaks, and dendrites of Mn oxyhydroxides of the infiltration nature.

A manganese ore layer (10–120 m) occurs in synclines, conformably on the dolomite. Manganese-bearing rhyolite (liparite) tuffs compose the main part of the unit. They are represented by thin alternations of sand-silty pyroclastic material and interbeds, composed of slightly recrystallized siliceous matter, thinnest aggregates of Mn oxyhydroxides and fine-grained pyroclastics. In carbonate varieties the ore matter is represented by rhodochrosite and manganocalcite. The average content of Mn in tuffs is 8–10% and $C_{org.}$ 0.7%. In the upper part of the tuff beds the substitution of liparite and quartz volcanoclasts by rhodochrosite and manganocalcite is often observed (Golovko *et al.*, 1982). In tuffs there is common rhythmical, and cross bedding (Golovko and Nasedkina, 1982) that may be construed as an indication of their accumulation in shallow water basins with both calm and active hydrologic regimes.

The supraore unit (80–120 m) is composed of tuffosilicites, represented by slightly recrystallized cherty material and interbeds of fine-grained pyroclastics of the liparite composition.

The most concentrated sections of manganese ore beds are developed in synclines and on their flanks. In these cases, between the dolomites and Mn-bearing sediments, are

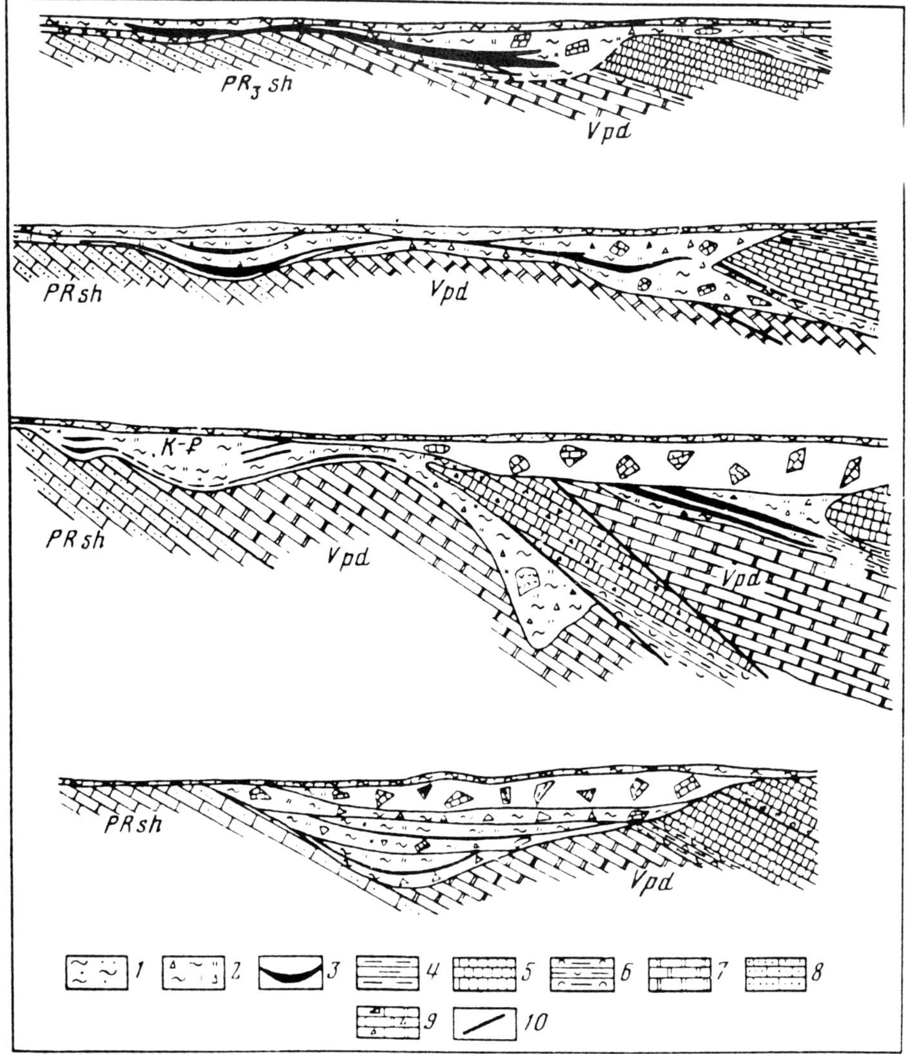

Fig. 66.   Typical sections of the Mokhovskoi area, Porozhinskii manganese ore region, the Enisei Ridge (after Pashashnikova, 1984).   1. silt-clay sediments of the Neogene-Quaternary age; 2. variegated Cretaceous–Paleogene clays with debris; 3. pyrolusite-romanechite (psilomelane) ores; 4–7. Vendian sediments: 4. silt, 5. tuffosilicites, 6. manganese-bearing silts; 7. dolomites; 8. carbonate sandstones of the Upper Proterozoic; 9. zone of brecciated rocks; 10. fault line.

gradual, but rather reduced in thickness, lithological transitions.   Within the anticlines between the dolomites and ore layer, according to M.M. Mstislavskii, there is a sharp contact and angular unconformity, or the manganese unit of the Upper Pod'emskaya Suite overly transgressively the sand-limestone rocks of the Sukhorechinskaya Suite (Vorogovskaya Series, lower part of the Vendian). It is also pointed out that at the contact of the dolomite unit and manganese ore tuffite sediments usually occur as brecciated dolomites with interbeds of the terrigenous material and traces of weathering (Varand and Andreev, 1984).

It is important to emphasize that manganese mineralization is closely associated with thick beds of tuffosilicites at areas of their most complete development. However, in marginal parts of the Vorogovskii trough at the general pinching out of the pyroclastic siliceous sediments the Mn accumulations and aggregates have disappeared.

In manganese ore units the superimposed relatively later events of infiltration, brecciation, and argillization are also developed. The most intensive secondary processes were manifested at the base of the Mn-ore unit near the contact of dolomites and cherty liparite tuffs. In these zones, to 6 m thick, the main alteration products are montmorillonite and rhodochrosite, and in subordinate amounts there are chlorite, sericite, hydromica of the muscovite type, kaolinite, goethite, hematite, siderite, and apatite (P_2O_5 to 4%). The development of similar mineralization is accompanied by deep decomposition and amorphization of the components of the enclosing pyroclastic rocks (Golovko *et al.*, 1982).

In manganese tuffites the thickness of ore interbeds, composed of Mn oxyhydroxides, varies from 0.1 to 3 m. In the lower part of the ore unit the manganese ores compose lens-like beds of 1–1.5 to 8 m thick with a Mn content of 17 to 27.4%. The spatial zoning is characteristic for the Porozhinskoe deposit. In its northwestern part the ore matter of manganese tuffs is represented essentially by manganite (to 80%), and in subordinate quantities pyrolusite, romanechite (psilomelane), vernadite, goethite, kaolinite, and quartz. The colloidal structure is highly typical for such early primary manganite ores. The breccia-like breaking textures that were formed during the substitution of tuffs by manganite matter, are of interest too. Several generations of manganites, indicating pulse-interrupted formation of similar metasomatic ores are distributed (Golovko and Nasedkina, 1982). Oxyhydroxide ores pinch out to the southeastern parts of the deposit and they are substituted by the crushing zones with ore mineralization at the base of the ore unit.

The differentiation of the primary oxyhydroxide ores and the oxidation products, associated with relatively late superimposed supergene processes, is of considerable difficulty for the Porozhinskoe deposit as well as for a number of other similar deposits. Tsykin and Kostenenko (1984) pointed to epigenetic alterations of subore dolomites: formation of dolomite powder, accumulations of clay and loam with fragments of enclosing rocks and impregnation of rocks with Mn oxyhydroxides. These observations, according to the authors' view, indicate the circulation of karst waters, and this is confirmed by the increased water permeability of manganite ore zones and in the spouting of a number of wells, that penetrated the fissured-stratal and karst waters. Carbonate ores predominate in the eastern part of the deposit. The ore matter within this zone is, in the main, represented by rhodochrosite and manganosiderite. The Mn carbonate substitute fragments of porous glass and cherty fragments and aggregates. The Mn contents in similar ores are to 5–10%. Siderite, manganocalcite, pyrite, apatite, and sometimes admixtures of vernadite and rancieite are observed in them in addition to rhodochrosites. Montmorillonite-rhodochrosite aggregates and lenses in the brecciated tuffs and tuffites are of special interest for understanding the deposit formation. The thick parts (5–6 m) of the altered rocks of similar type are observed (Golovko and Nasedkina, 1982) at the base of the ore unit at the contact with dolomite, as well as in the upper part of this section. Rhodochrosite mineralization is a relatively later process: rhodochrosite streaks transect the manganite ores. One may assume that relatively rare aggregates of manganosite and hausmannite (Potkonen, 1985) occupy intermediate positions in the row of the successive deposition of hydrothermal products of the Mn mineralization: Mn carbonate (I) – manganosite – hausmannite – manganite – Mn carbonate (II)–, reflecting the regime of Eh, pH, CO_2 of the ore formation environment. Similar relationships may be interpreted as the manifestations closely associated with the explosive rhyolite volcanism, but of later hydrothermal mineralization, displayed in several stages. In our works (Varentsov, 1964; Varentsov and Rakhmanov, 1974) it was shown that manganese ore mineralization, associated with rhyolite (liparite) volcanism, belongs to the group of volcanogenic-sedimentary formations of the porphyry series. The given formations, in difference to formations of the greenstone series, are characterized by relatively less dispersity of Mn among the composing accumulations and a correspondingly higher degree of concentration of this element. Formations of the porphyry series are widely developed in orogenic zones, regions of activation of ancient platforms, and marginal troughs, particularly in the regions of Siberia (Durnovskoe deposit, etc.). According to Dzotsenidze (1965, 1969), the relatively high manganese-bearing potential, associated with manifestations of rhyolite volcanism, is determined by the events of contrast differentiation of melts of such type and considerable enrichment in volatile components with Mn and other heavy metals. The similar petrologic feature of the liparite magmatism in the distinctly defined Mn mineralization is accompanied by hydrothermal activity. Thus, the Mn mineralization of the Vendian Vorogovskii trough, independent of its representation by oxyhydroxide or carbonate minerals, is a relatively late event of the hydrothermal nature. Its development is accompanied by essential alteration, dissolution, and substitution of the liparite pyroclastic and silicite material that took place after the completion of the explosive phase in relatively limited shallow water basins during diagenesis–epigenesis.

3.5.2.3. Zone of Oxidation and Karst Formation

The platform cover of the Vorogovskii trough, composed of Cambrian sediments during the subsequent geological history, was gradually destroyed. To the beginning of the Cretaceous–Paleogene time the Vendian structure of the Pod'emskaya Suite was exposed for the influence of supergene processes of crust and karst formation. The zones of grabens and structural scarps played the role of depression cavities, intensively drained by the meteoric waters. The features of the modern structure and hydrogeology of the deposit, as well as the composition of the substrate rocks, determined the development of supergene processes and the formation of Mn oxyhydroxide ores (Figure 66). According to data by Mstislavskii and Potkonen (1990) the formation of manganese ore zones of oxidation and accumulation of karst Mn ores did not proceed beyond the primary hydrothermal-metasomatic ore accumulations in the liparite tuff sequence. They were formed, as a rule, within the same structural facies zones, where the primary Mn mineralization was developed. Any distinct zoning is not developed in the kaolinite type weathering crusts reaching to 70–80 m in thickness. At the deposit, the thickness of the zone of oxidized Mn ores is about 8–10 m, and in some places the oxidation products occurred to the depth of 50 m and more (Figure 66).

It is established (Golovko *et al.*, 1982; Golovko and Nasedkina, 1982) that the oxidized products of pyrolusite composition were formed during the alteration of manganite ores. These ores are represented by large fragments and concretions, enclosed in a clayey loose mass, impregnated by manganese oxyhydroxides. The relict texture of the initial accumulations, for example, breccias often with residual fragments of manganite ores, is observed in them. The major minerals are represented by pyrolusite (to 80%), relict manganite (to 20%) with admixtures (to 10%) of vernadite, goethite, rancieite, and Al-containing varieties of MnO_2. In those cases, when the brecciated dolomites, impregnated with Mn oxyhydroxides and rhodochrosite, were subjected to supergene oxidation, rancieite-vernadite-romanechite (psilomelane) ores were developed. In the products of oxidation of this type are found the relicts of dolomite. They are composed of vernadite, romanechite (50–60%), and rancieite (40–45%) with admixtures of pyrolusite and goethite.

Therefore, the richest commercial ores are associated with the Cretaceous-Paleogene weathering crusts, developed in manganese-bearing sediments of the Vendian in the form of the autochthonous accumulations, oxidation products, and infiltration sediments in tectonically weakened zones and at the areas of karst development.

The average chemical composition of the oxidized Mn ores (25 samples), according to Mstislavskii and Potkonen (1990) is: (wt. %) Mn 10-39; Fe 0.4–7; P 0.1–0.4; (ppm), Ba 900; Sr 320; Ti 3282; Ni 176; Co 62; Cu 46; Pb 52; U 74; Sc 15; Y 37.

The deposits, associated with the karst sedimentary collector, i.e. exogenic structural form, produced as a result of the evolution of covered karst (Tsykin, 1985; Tsykin and Kostenenko, 1984), are of considerable interest among the supergene accumulations of Mn ores. The Mokhovyi area of the deposit (Figure 66) illustrates the development of ores of similar type. According to Pasashnikova (1984) this area is transected by a fault zone with a northwestern strike, accompanied by a gentle overthrust. The deep karst cavities and sinkholes are developed in the fault cavity in the carbonate rocks of the Pod'emskaya and Sukhorechenskaya Suites. The linear elongated karst depression (12 × 1.5 to 2.5 km), filled in by loose variegated clay, manganese-bearing sediments of the Cretaceous–Paleogene (Figure 66), occurs on steeply dipping carbonate sequences. The structure of the depression base is sharply inhomogeneous; there are local basins (from 400 × 200 to 1200 × 400 m) and dividing protrusions; absolute marks vary from -140 to -1 m. The areal weathering crusts are represented by variegated structural clays of hydromica-kaolinite composition up to 20 m thick. The linear weathering crusts are developed after manganese-bearing tuffites: different maturity stages of the crust to the structural clays of hydromica-kaolinite composition are observed along their contact with dolomites. It is noted that the crust formation process is not divided from the karst formation (Pasashnikova, 1984). The ore beds are 8 km long with the average width of 1 km (Varand and Andreev, 1984). The thickness of ore layers is from 0.5 to 3–4 m and their general thickness is to 30 m and more. The ores are represented by pyrolusite, romanechite (psilomelane) with admixtures of manganite and vernadite; contents: (%) of Mn are 10–43, average -18; P 0.2–0.8; Fe 2–10. Predicted reserves to the depth of 100 m are evaluated to be 30–40 mln t (Varand and Andreev, 1984). The Cretaceous–Paleogene sediments are distinguished by high contents of unsorted clastic material and accumulations of fine clays redeposited from the weathering crusts. Pasashnikova (1984) recognized subore beds, represented by coarse lumpy debris fragments of dolomites and silicites, bonded by grey clays and loams to 10 m thick. The beds of the ore sediments are composed of variegated clays and fragments of basement rocks, predominantly silicites. The thickness of these sediments is maximum in karst depressions: 30–40 m to more than 100 m. As a whole, karst sediments are characterized by a rhythmical nature. There are five rhythms (Tsykin and Kostenenko, 1984). The rhythm base is formed by subore coarse clastic sediments with high concentrations of fresh debris of tuffosilicites; the upper part is represented by clays, loams with weathered debris and enclosed manganese ore. The Cretaceous–Paleogene Mn ore deposits are covered by the Neogene–Quaternary silts and clays, at the base of which pebbles occur. The total thickness of ore sediments, controlled by the depression morphology, varies from 1–3 to 37 m.

According to Tsykin and Kostenenko (1984), formation of the karst sedimentary collector of the Mokhovoi area began in the Late Jurassic and it was developed at least to the Neogene with intervals of the washing and erosion of sediments and consequent eluvial weathering. Obviously, later in the Late Cretaceous the destructive processes were developed in the subsurface environment. Disintegration products of tuffaceous silicites, the rocks that are rather stable to weathering, were accumulated in karst cavities, covering the eluvium of tuffaceous siltstones, tuffites, and tuffaceous sandstones (Figure 67). It is emphasized that Mn ores in karst sediments have the postsedimentation nature, that is indicated by the substitution of silty-clay material and sometimes dolomites by Mn minerals. The lacustrine sediments of rhodochrosite, associated with subsurface, probably, thermal waters, are of special interest (Tsykin and Kostenenko, 1984, pp. 114–115). The redistribution of Mn took place in the Quaternary time: covering of sand-pebble sediments of the Pleistocene and modern flood plain alluvium by Mn oxyhydroxides.

Thus the most valuable commercial oxyhydroxide (pyrolusite–romanechite (psilomelane)) ores are the products of supergene alteration of the Vendian hydrothermal-matsomatic manganite, and rhodochrosite accumulations. They are accumulated in the residual weathering crusts of Cretaceous–Paleogene age, and sediments, filling karst cavities, and redeposited materials of the exogenic decomposition products. As in other deposits of a similar type, all the mentioned types of supergene manganese ore accumulations are distinguished by rather close relationships with the primary parent sequences, which are represented by relatively poor Mn ores.

3.6. Manganese Mineralization and Ores in the Regions of the Intensive Karst Development

Manganese mineralization, revealed in the accumulation of oxyhydroxide and carbonate phases, is usually widely distributed in the areas of karst development. The scales

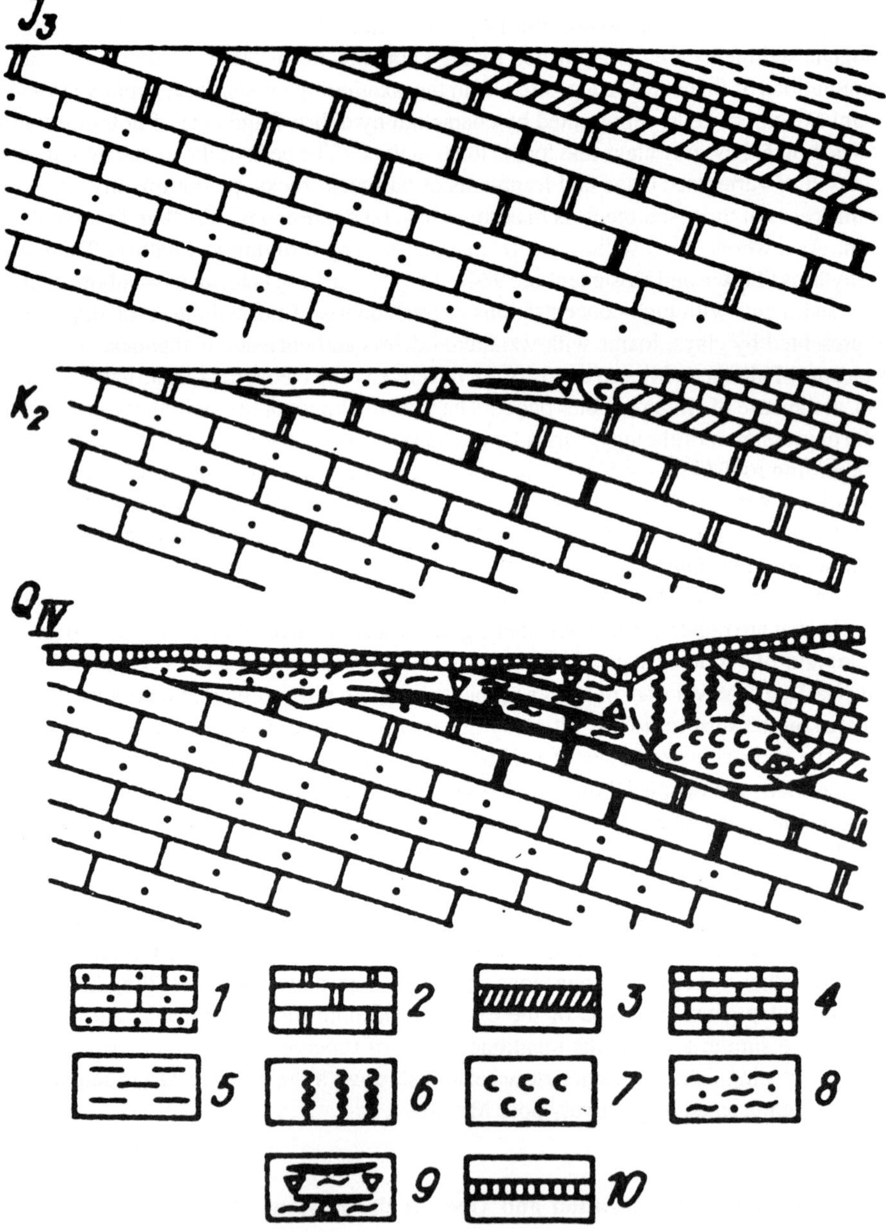

Fig. 67.   Scheme of formation of the Mn ore karst sedimentary collector of the Porozhinskoe deposit under the conditions of the monoclinal occurrence of the Vendian sediments (after Tsykin, 1985). 1. sand limestones; 2. dolomites; 3. tuffites, tuffaceous siltstones, tuffaceous sandstones; 4. tuffaceous silicites, 5. siltstones and sandstones; 6. slumps of tuffaceous silicites; 7. neoeluvium; 8. silt; 9. Mn ore lenses in loams and unsorted clastic accumulations; 10. Quaternary sediments.

of this process, as a rule, are controlled by factors responsible for the extraction of metal from drained substrate rocks, transportation intensity, corresponding setting of accumulation, and stability, and duration of the general functioning of these parameters during considerable geological time. In other words, the development of the accumulation process of Mn-bearing karstogenic accumulations, to a significant degree, is determined by the contents of Mn and the associated metals in the primary rocks, by the intensity of their draining, hydrochemical characteristics of circulating solutions, and favorable combination of mineral forming characteristics, and stability of the process development in the geological history of the region (Shvartsev, 1978, 1980, 1985; Sokolov, 1962).

3.6.1. MANGANESE MINERALIZATION IN MODERN KARST CAVES

The geochemical essence of the above processes is best determined in areas of modern karst development. Dublyanskii (1971) expounded the observation results on formation of the calcite sinter dams in the karst cavities, caves of the Crimea Mountains and associated events of accumulation of manganese oxyhydroxides of wad-romanechite (psilomelane) type ($\Sigma MnO + MnO_2 = 35\%$). Similar calcite dams are formed owing to chemical (i.e. chromatographic) deposition of calcite and Mn oxyhydroxides from flowing karst waters, flowing velocities of which are from 0.001–0.1 to 100.0 l/sec. The general mineralization of such solutions is from 190.0 to 508.0, average 275.0 mg/l, and their hydrogeochemical facies vary in narrow limits with predomination (66%) of HCO_3–Ca–SO_4 and somewhat subordinate values of HCO_3–Ca–Na; HCO_3–Ca–Cl; HCO_3–Ca–Mg types. It is interesting that the dams are characterized by a height of 0.05 to 5.0 m, though at greater velocities their height is significantly more: 0.2 to 7.0 m. It is important to point out that in the dams, growing from such high velocity flows, internal and especially external walls of the structure are usually covered by sinters of 0.2–0.3 mm, composed of admixtures of Mn oxyhydroxides (almost X-ray amorphous phases of wad-romanechite (psilomelane) type) and illite-calcite material. Contents of Ba, Ti, Mn, Co, and Sn in similar sediments are 2–4 times as high as in the enclosing rocks, and the general mineralization of waters, from which Mn oxyhydroxides are deposited, as a rule, is more than 300 mg/l (300 to 508 mg/l) with domination of HCO_3–Ca–Na composition. The characteristic feature of the structures is its rhythmical microlaminated texture. V.N. Dublyanskii (1971), using the thickness of the seasonal microlamination of the dams and its average height (1.9 m), evaluates the duration of the structure growth to be 9–10 thousand years. A rather significant feature of microlaminated accumulations is that the development history of karst cavities of the region is registered in the mineral and chemical composition of products. The enclosing rocks, in which karst caves of the Crimea mountains are formed, are essentially the Upper Jurassic organogenic, biochemogenic, clastic, and dolomitized limestones. In these rocks, according to the data of more than 380 analyses, the content of MnO varies in limited range: 0.01 to 0.04 wt. % (Dublyanskii and Polkanov, 1974). In addition to the above dam structures in karst caves, the accumulation of Mn oxyhydroxides are well displayed in the form of sinters, encrustations, and crusts, growing on the walls of karst cavities in the zone of periodical flooding by high waters. According to Dublyanskii and Polkanov (1974), in some samples the contents of $\Sigma MnO + MnO_2$ reaches 35% (average 20%). The composition of similar growths, determined by methods of X-ray diffraction and DTA, is represented by the mixture of slightly crystallized almost amorphous minerals of wad-romanechite (psilomelane), hydromica, and calcite type. The increased concentrations of Be, Ba, Ti, Co, Sn, and La relative to primary rocks are accumulated in these deposits too. Comparison of this data with the manganese growths of rhythmically laminated structures of cave dams points to the closeness of the mineral and chemical composition of these accumulations.

Similar growths are deposited from subsurface waters, the general mineralization of which is characterized by rather high variability: from 128 mg/l in areas of accumulations of snow and ice of karst cavities to 343 mg/l for waters of subsurface flows and 408 mg/l for waters of karst springs. Hydrocarbonate–calcium, rarer hydrocarbonate-potassium-calcium or hydrocarbonate-magnesium-calcium types predominate among them. The waters of karst water flows of the Crimea mountains recalculated to the oxide-anhydrite form contain MgO 6%, Na_2O 6%, SO_3 4.5%, CO_2 55.5%, CaO 27.5%.

Thus, the authigenic accumulations in karst cavities and caves reflect the well-defined geochemical features of circulating subsurface waters and, correspondingly, drained rocks. The oxyhydroxide growths in karst caves of Southern Siberia, in particular of the Krasnoyarsk region and Irkutsk Amphitheater may serve as examples, illustrating the justification of this concept. Rather clearly this process is distinguished in the Krest cave (Krasnoyarsk region), where Tsykin (1977) described the sinter patches and accumulations of romanechite (psilomelane), composing the central parts (0.5–2.5 cm) of stalactites with diameters of 12 cm. The brown rims (2–3 cm), composed of the mixture of calcite and limonite, grow around the black zone of Mn oxyhydroxides, and the external part of stalactites is represented by calcite. This data may indicate the essential separation of Mn and Fe, accumulated in oxyhydroxide form from bicarbonate solutions at different stages of the bedrock's drainage.

3.6.2. MANGANESE ORE OCCURRENCES IN THE KARST REGIONS

Manganese accumulation in the karst regions takes place with higher intensity, all other conditions being equal in those cases, when the rocks, containing increased amounts of this metal, are subjected to drainage. According to Vologodskii (1975), a rather large number (more than 50) of manganese ore deposits are associated with karst depressions of the Paleogene and Quaternary age, developed on the surface of the Archean marbles of the Priol'khon'e and the Ol'khon Island at the Baikal lake district in the region of the Irkutsk Amphitheater. The manganese accumulations occurred in the form of pockets and lenses of 0.4 to 4 m thick and from 5 to 20 m long. They are composed of loose sooty, sometimes hard or concretion ores, with Mn contents of from 2.7 to 24%. The brown iron ores (Fe, 2 to 17%) are associated with Mn accumulations. A number of deposits of infiltration-karst type are noted (Vologodskii, 1975) in carbonate sequences of the Karagasskaya Suite at the Prisayan'e region. Their formation is associated with the Mesozoic–Cenozoic weathering crusts, developed after the carbonate rocks, in which the initial contents of Mn are 0.05–0.3%. Manganese ores, essentially composed of cryptomelane and more rarely pyrolusite, form irregular lenses and pockets in the lower kaolinite zone of the weathering crust, which are formed mainly in the silty dolomites. Manganese contents in such ores vary from 14 to 40%. General reserves of Mn for the Nikolaevskoe deposit is evaluated to constitute several mln t (Vologodskii, 1975).

The Singaiskoe manganese ore occurrence is close in its nature; it is developed in the weathering products, filling in karst cavities in carbonate volcanogenic-sedimentary sequences of the Alambaiskaya Suite of the Lower Cambrian in the central part of the Salair Ridge (the divide of the Zeglovyi Sungai and Bol'shaya Rivers) (Portnyannikov and Ageenko, 1979). The formation of the karst cavities and associated weathering materials went on essentially in zones of tectonic crushing. Manganese, iron-manganese, and hematite-limonite types of ores of colloform, streak, and breccia texture are distinctly distinguished in the deposits of irregular lens-like form, substituted by the poor streaky iron varieties down the dip. The hydrothermally altered silicified and sulfidized carboniferous shales with the interbeds of chlorite schists and limestones are observed at the depth of 150–200 m under the weathering products. Manganese ores are composed of hollandite, cryptomelane, pyrolusite, lithiophorite, goethite, hydrohematite, and hematite. Chemical composition of Mn ores (wt. %) is: Mn 13.3; $FeO + Fe_2O_3$ 3.7; SiO_2 68.3; P_2O_5 0.054; Al_2O_3 2.7; Ca 0.6; Mg 0.07; L.O.I. 5.0; Ni 0.01 to 0.3, average 0.06; Co 0.03 to 1.0, average 0.16; Cu 0.01 to 0.3; Ba 0.05 to 5.0. We have every reason to believe that Mn ores in weathering crusts and karst of the central part of the Salair Ridge are the result of the Mesozoic–Cenozoic supergene alterations, that were widely developed not only in the south of the Siberia but as a global climatic event at least for the Northern Hemisphere.

The examples of wide development of Mn-bearing weathering crusts are known in Western Europe and at the Far East.

The accumulations of the Tertiary karst Mn ores in the area of the Dr. Geier Mine near Bingen city at the Rhine River, Germany (Bottke, 1969) are of interest in Western Europe, apart from the Mesozoic–Tertiary weathering crusts and karst accumulations, developed in depleted Mn ores in a number of Hungarian regions (see above). These ores form deposits in the karst cavities on the Middle Devonian dolomites. The intensity and scale of Mn ore accumulation development are controlled by a set of the karst relief, lithologic-facies features of karst sediments, and tectonic structures. The formation of karst cavities, depressions, and other forms is associated with distribution peculiarities of faults and fissured states as with distribution features of faults and jointing as well as bedding surfaces of dolomites within the isoclinal basin. Such structural, textural distortions, inhomogeneities in massive dolomites executed the role of drainage channels, that widened during the leaching of bedrocks at the period of uplifting under the conditions of warm humid climate, which dominated in the Eocene and Early Oligocene. The karst sediments, represented by clays, sands, breccia debris, and Mn, Fe ores, are most completely developed along the strike of faults of the synclinal structure. The ores are represented by Fe–Mn varieties with the ratio of Mn/Fe = 1/2. The localization of the richest ores near the contact with dolomites (0–4 m): Fe 30.17 to 32.78%, Mn 12.29 to 13.59%; SiO_2 11.62 to 14.88%, is of significant interest for the understanding of these sediments' genesis. Contents of ore components are decreased to the periphery of the karst deposit (22 to 38 m from dolomite), and ores become brecciated, clayey with fragments, boulders of crystalline schists and quartzites (Fe 6 to 12%; Mn 1.7 to 5.7%;

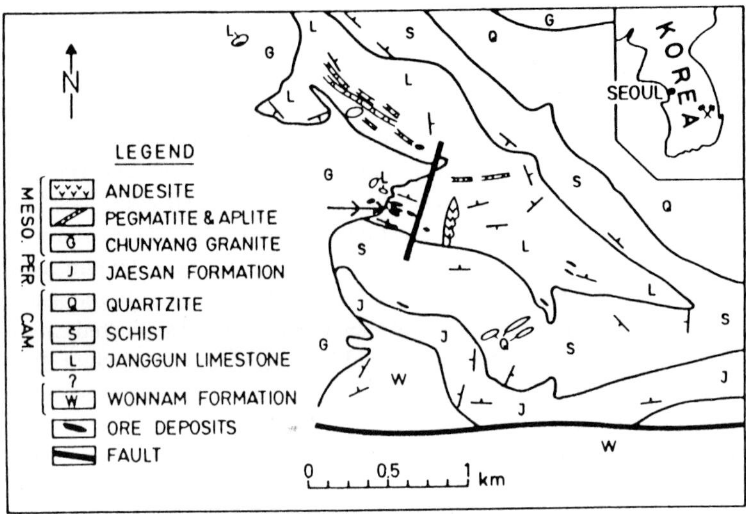

Fig. 68.   Geological map of the Janggun mining area, Korea. Arrow indicates the Nam ore deposit where the janggunite is found (Kim, 1977).

SiO_2 38 to 46%). It is important to note that the primary dolomites are distinguished by the increased contents of accumulated metals: Fe 0.5 to 2.5%; Mn 0.3 to 1.5%. Variable amounts of goethite (30 to 50%), X-ray amorphous Mn oxyhydroxides (20 to 30%) and muscovite (10 to 50%) are contained in ores of clay type, and hard concretion ores are composed of essentially manganite and pyrolusite (0 to 22%). The data (Bottke, 1969) may be interpreted as evidence for the Mn, Fe ore origin, associated with deep karst dissolution of the dolomite substrate and accumulation of oxyhydroxide Fe, Mn compounds from circulating solutions essentially near the contact with the parent rocks. In these environments there was rather limited transportation of Mn and Fe with slight separation of these metals. There are no reasons for accepting the concepts by Bottke (1969) on the leading role of mixing of colloidal solutions of Mn and Fe oxyhydroxides in the formation of the ores under the oxidation conditions of the karst relief.

The rather large Junggun mining area in South Korea (Choi and Kim, 1992; Kim, 1974, 1977, 1980, 1990, 1991, 1993) is of great interest among numerous supergene deposits and Mn ore occurrences of the Far East region. The primary Mn carbonate ores are closely associated with the Cambrian Janggun Suite, consisting of the limestone beds, crystalline schists, and quartzites, composing an anticlinal structure (Figure 68). Mn sedimentary-diagenetic mineralization is closely associated with the Janggun limestone sequence, represented by limestones (Mn 0.05 to 0.06%) by dolomites white (Mn 0.09 to 0.26%), grey (Mn 2.75 to 4.65%), dolomite limestones (Mn 0.06 to 0.43%), rhodochrosite rocks (Mn 25.38 to 38.56%), quartzites, jasperoids, and limestone-silicate rocks (Figure 69). Rhodochrosite rocks, consisting of rhodochrosite (70 to 97%) with admixtures of quartz

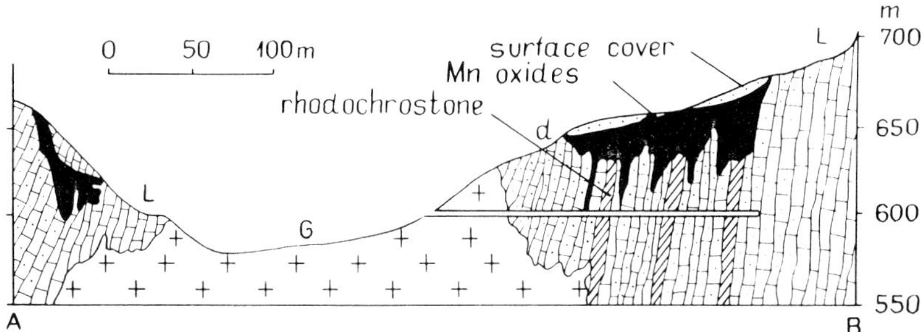

Fig. 69.   Geological section of the Janggun Mn deposit, Korea (Kim, 1980).

(2 to 5%), sericite (0 to 6%), and sulfides are most remarkable. Kutnahorite is often present among these components. Braunite, rhodochrosite, and calcite with a slight admixture of rhodonite and alletanite $[Mn_5[(OH)_2(SiO_4)]]$ are present in dolomites at the same stratigraphical level as the rhodochrosite rocks in the form of discontinuous lenses. This data (Kim, 1974, 1980) shows that irrespective of the complicating influence of the granite intrusion, in the Mn carbonate sequence, the sedimentary-diagenetic genesis of these Mn-bearing rocks is relatively obvious.

The deep oxidation zone and karst formation, developed in steeply dipping Mn carbonate beds, is characteristic for the deposit. Massive oxyhydroxide ores were formed mainly after beds of the carbonates rich in Mn. The discontinuous lens-like irregular concretion ore bodies were formed as a result of alteration of relatively depleted Mn carbonate ores. The contacts of the primary Mn carbonate ores and the supergene oxyhydroxide varieties are usually intricately winding and rather sharp with irregular borders. As a whole the supergene Mn ore accumulations fill in karst-like sinkholes, cavities of different form and a definite zoning is evident (downward):

1. Zone of earthy ores, composed of wad, X-ray amorphous Mn oxyhydroxides and similar minerals, fragments of dolomite, quartz, lumps of rocks. The thickness is from several cm to several m.

2. Zone of netted-branching streaky ores which is developed usually in carbonates depleted in Mn. The ores are represented by nsutite, 10 Å-manganate (Ca, Mg-buserite), and cryptomelane (Choi and Kim, 1992).

3. Concretion zone is formed, as a rule, in carbonates depleted in Mn or in Mn dolomites. Concretions consist of 10 Å-manganate (buserite), manganite, and nsutite. The zone thickness is from several dozens of cm to about 1 m.

4. Cementation zone is composed by hard and in some parts partially porous ores. Development of a mixture of secondary carbonates (calcite, rhodochrosite) with Mn oxyhydroxides directly over the Mn dolomites and rhodochrosite rocks is characteristic. Ores are composed of 10 Å-manganate, nsutite, manganite, and janggunite (janggunite is a Mn-oxyhydroxide mineral, found by Kim (1977, 1980)); according to the X-ray diffraction pattern it may be a mixture of 10 Å-manganate ((Ca,

Mg) buserite), cryptomelane, and pyrolusite. In the ore are also present calcite, and rhodochrosite (Choi and Kim, 1992).

5. Zone of compacted ores is formed directly over rhodochrosite bodies. Ores, retaining the textural features of primary rocks, are composed by manganite, birnessite, takanelite-rancieite, 10 Å-manganate (buserite), and rhodochrosite (Kim, 1990, 1991, 1993).

The following types of supergene karst Mn oxyhydroxide ores are recognized: (1) ores of substitution; (2) ores of filling and covering of the cavities; (3) ores, formed owing to leaching and consequent concretion growth. On the basis of structural relations observed under microscope, the following paragenetic sequence of Mn oxyhydroxide mineral formation in the supergene zone and karst formation of this deposit is proposed (Kim, 1980):

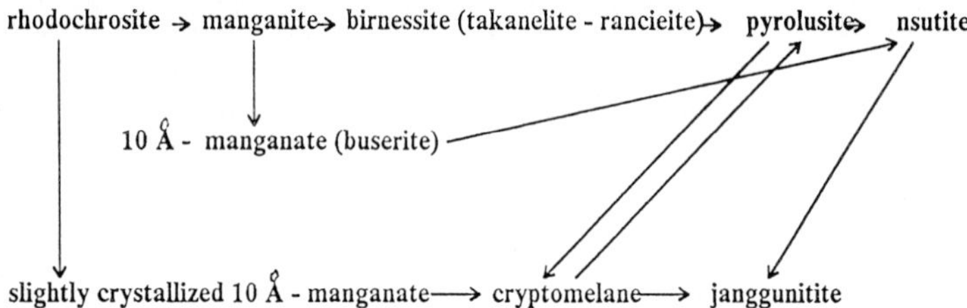

A comparison of this scheme of the supergene alteration of Mn carbonate with the mineralization sequence in the zone of oxidation, and karst of other deposits or areas, considered above, shows that their relatively similar character engages one's attention. Such similarity reflects the known closeness of processes, resulted in leaching, transportation, relatively narrow range of pH and Eh, mineralization, and macroionic composition of the solutions. We should mention also the chemisorption and autocatalytic oxidation and deposition, as well as the events of the consequent dissolution and redeposition, solid phase transformations, accompanied by alteration of the composition and their recrystallization.

3.7. On the Geochemical Model of the Karst Mn Ore Formation

In the process of karst formation, a consecutive general increase of decayed carbonate rocks proceeds. Against the background a number of components, and in first turn, the heavy metals, are bound into secondary new formed phases, and their amounts are increased too. During the leaching of manganese and iron in the supergene environment, an associated but spatially isolated formation of oxyhydroxide phases took place, with the accumulation of residual cherty, clay and aluminiferous components. Manganese ore karst processes differ from incongruent dissolution of alumosilicate rocks by essentially higher kinetic parameters. At the initial stages of dissolution the solutions, reflecting considerably the composition of the leached rocks, are formed: they are hydrocarbonate calcium solutions (main components: Na^+, Ca^{2+}, Cl^-, HCO_3^-, OH^-) for carbonate sequences. However, the long term of the formation of metamorphozed and more concentrated solutions is quite possibly rather highly aggressive to the country rocks ('Basis of Hydrogeology', 1982; Shvartsev, 1978, 1985).

The formation of karst manganese ore deposits takes place within the general system of the water–rock interaction which, as a rule, are irregular irreversible processes. Under these conditions the events of mass exchange occur. As a result of this, the non-equilibrium systems may regularly pass through the stages of partial and intermediate, or metastable equilibriums to the state of relatively stable chemical equilibrium (Krainov *et al.*, 1985). This concept may be illustrated as applied to the geochemistry of manganese in the karst formation conditions and exemplified by the successive deposition of the oxyhydroxide compounds of Mn(II) – Mn(III) – Mn(IV). This sequence is common at the gradational increase of concentrations of dissolved O_2 (general increase of Eh) proportionately to flowing of stratal waters to their discharge zones (Giovanoli, 1980; Hypoloto *et al.*, 1982a, 1982b, 1984; Murray *et al.*, 1985). The number of partial equilibriums with the formation of the transition products increases with the decrease of filtration velocities of subsurface waters and the increase of the duration of interphase interactions at the retaining and stability of the external environment. There are strong grounds to believe that, as a whole, the congruent dissolution in the water–rock system takes place for carbonate deposits. This system remains constantly in non-equilibrium at the dynamic filtration of waters relative to the primary rocks, but it passes through a consecutive number of equilibriums with corresponding reaction products. The dissolution of manganese and other heavy metals, with the change of the chemical composition of the solution, would be continued as long as the whole system (initial rocks – solution – products) does not pass to stable a thermodynamic state. The carbonate barrier is the major geochemical factor, hindering the accumulation of manganese and other heavy and transition metals in karst waters. That is, in oxygen-free environments with neutral–slightly alkaline pH, the maximum concentrations of the considered elements are controlled by their equilibrium values with carbonates and oxycarbonates in the waters of the given particular chemical composition. The degree of complexing of heavy metals is increased with the growth of sulfate and especially chloride concentration of the waters. It may lead to the growth of saturation deficit of these waters relative to the most probable solid phases.

The migration ability of manganese and iron, as the metals with variable valence in subsurface waters, is controlled by the redox regime of the environment (Eh), pH values, composition and concentration of macroions, availability of dissolved organic matter, composition, and contents of dissolved gases.

It was shown by direct investigations of natural objects, and computations, as well as by the experimental works (Shcheka and Zakutin, 1985; Stumm and Morgan, 1970) that these metals are present in the form of aquations of Mn(II) and Fe(II) in slightly mineralized waters with oxygen-free–sulfide-free regime in the field of neutral–slightly alkaline values of pH. In the carbonate systems with pH > 7.5 that are characteristic at least for karst zones of deep circulation (oxygen-free–sulfide-free environment), the concentration of hydrocarbonate forms of Mn(II) and Fe(II) is determined by dissolution values of rhodochrosite and siderite respectively. The redox potential values of the considered systems are controlled by relative amounts of potential-determining components of O_2/H_2O, S^0/S^{2-}, $Fe(OH)_3^0/Fe^{2+}$. Here the presence of dissolved Fe(II) is characteristic for oxygen-free–sulfide-free water environment in the interval of Eh: from +50 to +250 mV, whereas for Mn(II) at pH 7.5 this range is increased to +300, and to +400 mV. Thus, the variations in coordinates of pH and Eh at relative stability of other factors, in particular with the absence of the complexing of ions of these metals, determine the features of their migration and they may lead to spatial separated deposition of their ore accumulations.

In waters of oxygen-free–sulfide-free environments, there are relatively high concentrations of Fe (to 10 mg/l) and Mn. In many cases the ratio of their contents in the solution is close to the values of Mn/Fe in the enclosing rocks: about 0.02 (Shcheka and Zakutin, 1985). In subsurface waters of the source regions the content of O_2 was identified: from the first units to 10 mg/l at the values of Eh > +200 mV (the modal values of Eh: +250 to +350 mV). In these environments the supply of O_2 exceeds rates of its consumption for the oxidation reactions of the different components, including the organic matter (Shcheka and Zakutin, 1985). Fe(II) in the subsurface waters of these environments is, as a rule, absent owing to deposition in the form of $Fe(OH)_3$. The concentrations of Mn(II) in the solution reach to the first fractions of mg/l, and the formation of accumulations of oxyhydroxide phases of Mn(III) and Mn(IV) is possible at the local increase of Eh and with the leading role of chemisorption, autocatalytic processes. However, in the regions of subsurface water flows (filtration, circulation) the quantity of dissolved O_2 is considerably decreased to its complete disappearance. The oxygen-free–sulfide-free environment with Eh from 0 to +200 mV (the modal values of Eh: +50 to +150 mV) is created as a result of it. In some areas the oxygen-free–sulfide-free conditions of subsurface waters are changed by sulfide environments that are characterized by negative values of the redox potential (Eh: from 0 to -150 mV). But as a whole the sulfide waters are slightly typical for karst processes.

Three major types of zoning of the redox regime of subsurface waters (Shcheka and Zakutin, 1985) are common for many artesian basins: (1) The change of oxygen environment of the subsurface waters to an oxygen-free–sulfide-free one; (2) The change of oxygen environment to oxygen-free–sulfide-free and lower to sulfide one; (3) The change of oxygen environment to sulfide and lower to oxygen-free–sulfide-free one. With all possible varieties of geochemical regimes of the formation of karst manganese ores, the above examples of relatively large deposits indicate that the environments of free O_2 existence are most characteristic for them. The similar environments are considerably close to the supergene zone settings, in particular, weathering crusts in carbonate rocks.

Chapter 4

MODEL OF MANGANESE ORE FORMATION IN THE SUPERGENE ZONE

4.1. Formulation of the Problem

In geological science, in particular, in the geochemistry of sedimentary ore formation, relatively complete genetic models are created as a result of tight interaction of three methods:

1. Observation, and study of modern and ancient natural objects, accumulation, and understanding of empirical data, distinguishing of significant information, and factors that determine the investigated phenomena. In these investigations it is necessary to reduce the rather complicated natural system to some relatively elementary system.
2. Modeling, experimental simulation of separate parts or stages of processes with the inevitable simplification of the system.
3. The historical method is essential, because the history of geochemical events is manifested in the change of conditions, environments, and in the general trend of the basin development, i.e. in evolution of the proceeding processes.

It is significant that geochemical investigations of recent years convincingly indicate that in the zone of sedimentary-postsedimentary transformations the processes are subordinated to kinetic control and are in a rather restricted measure determined by the laws of equilibrium thermodynamics. It is necessary to add that in the zone of sedimentation a special role belongs to microorganisms and organic matter. More corroboration of Vernadskii's idea that in sedimentary environments the major geochemical processes including ore formation were to a rather significant degree controlled by different forms of life was obtained (Putilina and Varentsov, 1980).

Thus, for an adequate understanding of the geochemistry of low temperature supergene processes of ore formation it is necessary that non-equilibrium and irreversible features of these processes, controlled by kinetic and sorption factors, and relatively complete concepts on mechanism of these processes, including the nature of irreversible reactions, were reflected to a sufficiently complete degree in the specific and generalized models of these processes. It was noted above that in comparison with other ore-bearing weathering crusts, the geochemistry of the mineralization processes in manganese-bearing crusts is relatively scantily studied to the present time. Nevertheless for the derivation of the preliminary geochemical model of the weathering profile it is necessary to emphasize some items:

(a) Large laterite manganese deposits are associated with aerial weathering crusts and, to a considerably smaller degree, with linear crusts.

(b) Mn-ore crusts are usually characterized by zonal structure. Each zone is defined by its typomorphic mineral paragenesis, i.e. a definite mineral and chemical composition.

(c) The change of typomorphic mineral paragenesis together with features of substrata rocks reflects the chemical composition of filtrating ground solutions with definite regimes of pH and Eh and the value of their mineralization and contents of dissolved organic matter.

(d) In the weathering crust three main components-stages, often spatially weakly separated, may be distinguished: (1) source; (2) transportation medium; (3) ore formation environment. In the result of the activity of solutions, containing carbonic acid, humus and other acids and probably with participation of biochemical and bacteriological phenomena, the following phenomena take place: (1) dissolution, hydrolysis of primary minerals; (2) migration and autochthonous or with some transportation, accumulation of compounds of ore components.

(e) In processes of authigenic formation of oxyhydroxide mineral and ores of manganese and associated metals, the essential role belongs to sorption processes often accompanied by autocatalytic accumulation of transition metals from ultra-diluted solutions.

The manganese deposits in laterite crusts of West Africa, whose autochthonous and redeposited nature is sufficiently substantiated in most cases is accepted as a natural model (Bouladon *et al.*, 1965; Grandin and Perseil, 1977, 1983; Leclerc and Weber, 1980; Perseil and Grandin, 1977, 1983; Weber *et al.*, 1979).

4.2. Data of Natural Observations

4.2.1. ALTERATION OF THE SILICATE PROTORE

In some of the considered deposits of West Africa (Mokta, Ziemougoula, Côte d'Ivoire; Tambao, Burkina Faso; etc.), the regional manganese volcanogenic-sedimentary Precambrian rocks of the Birrimian Series is the Mn-protore. Irrespective of the variability of the series composition, the following generalized mineral composition may be accepted (Nahon *et al.*, 1983): tephroite (65%), spessartine (20%), manganocalcite (10%), Mn-bearing chlorite and sulfides (5%). Tephroite, manganocalcite, and Mn-chlorite are the most easily altered at the early stages of weathering to manganese oxyhydroxides, whereas the transformation of spessartine proceeds in the upper parts of the weathering profile, where manganese oxyhydroxides of the early generations are substituted by later ones.

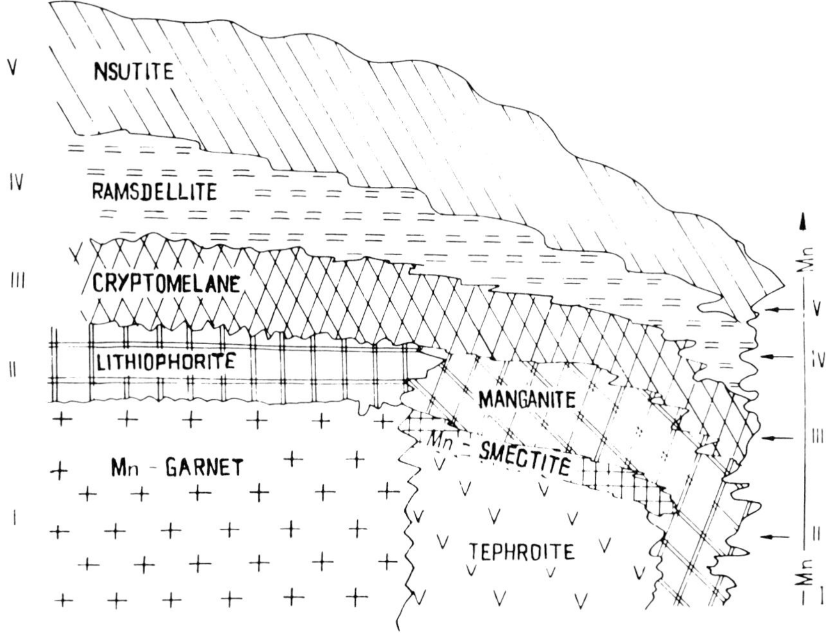

Fig. 70. Sequence of formation of Mn minerals in the weathering crusts developed after silicate parent rocks of the West Africa deposits (after Grandin and Perseil, 1977, 1983; Nahon, 1982, 1983; Perseil and Grandin, 1978, 1985).

4.2.1.1. Early Stages of Alterations (Transformation of Tephroite)

4.2.1.1.1. Stages of Mn(II)-smectite formation. The stages of Mn(II)-smectite formation are manifested at the most initial stages of tephroite weathering (Nahon *et al.*, 1982). The clayey matter is developed in the filling of the corrosion cavities and crystal fractures (Figure 70). Each of the patches of this matter, represented by Mn(II)-trioctahedral saponite, retains crystallographic (and crystalloptical) orienting of the initial tephroite grain, that may indicate a non-destructive (*in situ*) mechanism of transformation. According to data of microprobe analyses (Nahon *et al.*, 1982) the newly formed Mn(II)-smectite is somewhat depleted relatively to the initial tephroite in Mn and Mg, but is enriched in Al. The environment, in which the formation of Mn(II)-saponite proceeded, was not so open to the oxidation of Mn(II) to Mn(III) or Mn(IV).

4.2.1.1.2. Stage of manganite (γ-MnOOH) formation. The stage of formation of manganite (γ-MnOOH) is developed at relatively higher levels of the weathering profiles, where Mn(II)-smectite is destroyed or substituted by manganese oxyhydroxides formed mainly after unaltered tephroite (Nahon *et al.*, 1982, 1983). The manganite should be considered as the phase widely formed at early stages of transformation of slightly stable

Mn-silicate matter under the conditions of relatively soft gradational oxidation. The manganite is most often found filling in the microfractures transacting tephroite crystals and as mosaic idiomorphic crystals. It is interesting to note (Nahon *et al.*, 1983) that manganite is developed during Mg and Ca leaching and during incomplete removal of SiO_2, and to a lesser degree of Fe and Al, while the dispersed micronodules of SiO_2, and the highest concentrations of Si are observed near the borders of unweathered tephroite (by the data of microprobe scanning of the above authors).

During consideration of the manganite genesis as the initial product of the oxidation transformation, it is necessary to point out that its formation is not associated with any significant transportation of manganese released during complete dissolution of the tephroite. This feature essentially differentiates the formation of oxyhydroxide Mn minerals at the dissolution of manganese silicate from the formation of Mn oxyhydroxides after other minerals.

4.2.1.2. *Major States of Weathering (Alteration of Mn-Garnet)*

Alteration of Mn-garnet, essentially spessartine, in the laterite weathering crust was considered in previous chapters. On the basis of analysis of investigations (Grandin and Perseil, 1977, 1983; Perseil and Grandin, 1985) we may accept the following scheme of the alteration sequence of the manganese garnet (Figure 70) reflecting the directed change of the geochemical regime at different stages of the weathering crust development:

$$\text{Mn-garnet} \rightarrow \text{lithiphorite} \rightarrow \text{cryptomelane} \rightarrow \text{ramsdellite} \rightarrow \text{nsutite.}$$

4.2.2. ALTERATION OF CARBONATE PROTORE

In processes of interaction of ground solutions with the distinguished initial carbonate material, a number of alteration stages (Figure 71), are developed, controlled by properties of the composition, in particular, by presence of the organic matter and drainage intensity among general factors, determining the formation of the weathering crust profile (Boeglin *et al.*, 1980; Bouladon *et al.*, 1965; Grandin and Perseil, 1977, 1983; Kim, 1974, 1977, 1980, 1991; Leclerc and Weber, 1980; Perseil and Grandin, 1978, 1985; Weber *et al.*, 1979).

1. The earliest stage of alteration of the initial carbonate manganese-bearing rocks is distinctly manifested in a number of deposits of West Equatorial Africa (Moanda, Gabon; Ziemougoula, Côte d'Ivoire, etc.) and South America (deposits of the Lafayette region, Brazil). For example, in the carbonate metamorphosed protore of deposits of the Lafayette region (Boeglin *et al.*, 1980) four generations of rhodochrosites have been distinguished. Their composition is substituted and enriched by manganese exhibiting some regular trends from deeper and more ancient rocks to surface and relatively young beds. This stage is rather clearly represented in lower parts of the weathering profile of the Moanda deposit (Gabon). Initial rocks are represented by black carbonate clayey shales composed essentially of illite, complex Mn-carbonate, aggregates of pyrite, and

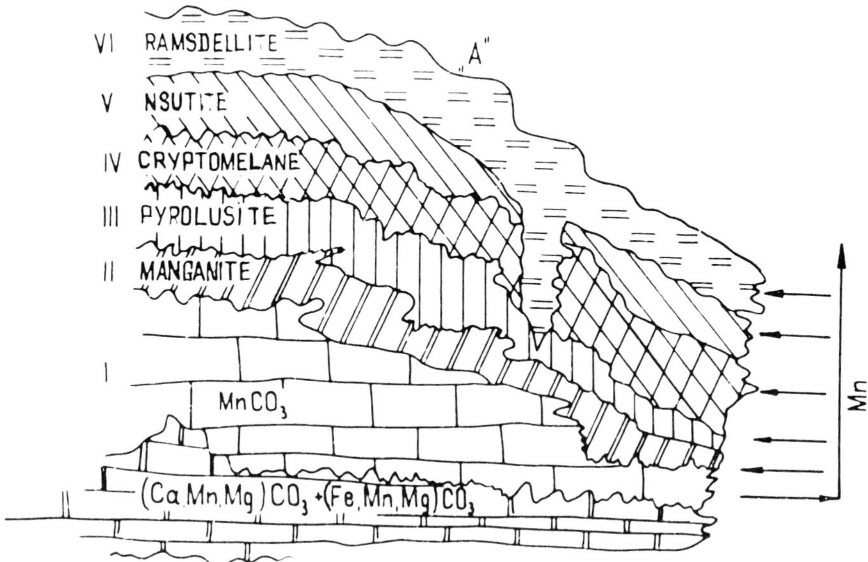

Fig. 71. Sequence of formation of Mn minerals in the weathering crusts developed after carbonate parent rocks of the West Africa deposits (after Boeglin *et al.*, 1980; Bouladon *et al.*, 1965; Grandin and Perseil, 1977, 1983; Kim, 1974, 1977, 1980, 1990, 1991; Leclerc and Weber, 1980; Nahon, 1982, 1983; Perseil and Grandin, 1978, 1985; Weber *et al.*, 1979).

admixtures of clastic quartz. It is important to emphasize the enrichment of these rocks by organic matter. The essence of alteration is that the initial multicomponent carbonate composed of $(Mn_{0.64}Ca_{0.22}Mg_{0.13}Fe_{0.01})CO_3$ and $(Mn_{0.45}Ca_{0.26}Mg_{0.29})CO_3$, near the weathering profile base, is transformed to well-crystallized almost pure rhodochrosite (nodules, intergrowths of crystals, and monocrystals) $(Mn_{0.991}Ca_{0.007}Fe_{0.001}Mg_{0.001})CO_3$ (Boeglin, 1981; quoted after Nahon *et al.*, 1983).

2. Slow, gradational oxidation, alteration of the initial material (Figure 71).

Rhodochrosite → (groutite), manganite →

→ pyrolusite → cryptomelane $\nearrow$ nsutite
$\searrow$ ramsdellite.

3. Fast, intensive oxidation and weathering (Figure 71).

Rhodochrosite → birnessite → pyrolusite → cryptomelane → ramsdellite.

4.2.3. FORMATION OF MANGANESE CAP (CUIRASS)

The formation of manganese ore cap is associated with the mature final stages of development of the laterite profile when Mn, leached from the bedrock and autochthonous

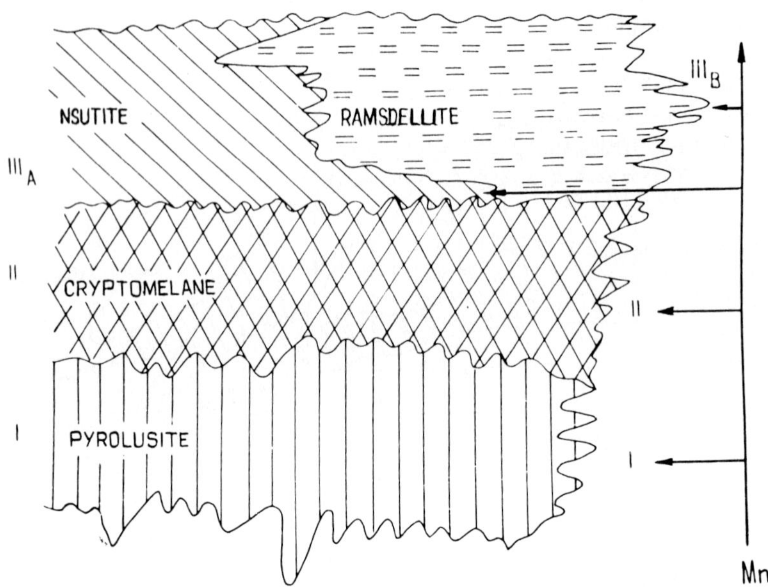

Fig. 72.   Sequence of formation of Mn minerals in the upper parts of the weathering crusts (caps) of the West Africa deposits (after Bouladon *et al.*, 1965; Grandin and Perseil, 1977, 1983; Perseil and Grandin, 1978, 1985).

rocks after considerable transportation, is accumulated with the formation of massive ores protruding in the relief. For deposits of West Equatorial Africa, and a number of environmentally similar regions, the definite sequence of mineralization has been established that may be accepted as sufficiently general (Grandin and Perseil, 1977, 1983; Perseil and Grandin, 1978, 1985):

$$\text{pyrolusite} \rightarrow \text{cryptomelane} \begin{cases} \nearrow \text{nsutite} \\ \searrow \text{ramsdellite.} \end{cases}$$

The comparison of this series with the sequence of the major and mature stages of mineralization of autochthonous, and lens-like laminated accumulations in the weathering crust of the silicate and carbonate protore indicates their similarities. This fact testifies to the definite similarity of the geochemical evolution of ore-forming solutions and conjugated character of the formation of upper horizons of non-redeposited crusts and cap accumulations.

4.3. Experimental Data

4.3.1. INTRODUCTION

The major parameters of the natural processes of the Mn and Mn–Fe ore formation in the *supergene* setting were considered above. Special attention was paid to chemical composition of ground solutions, subsurface and karst waters, as well as to solid Mn and Fe oxyhydroxide phases deposited from them.

Processes of removal of transition metals from such solutions and of their accumulation in the form of oxyhydroxides and other phases in the *supergene* zone are characterized by extremely low rates. It allows one to attribute them to unobserved phenomena. Studies of separate phases of Mn ore formation in weathering crusts (proper matter of ore accumulations, composition of ground solutions and subsurface waters, enclosing rocks, role of sources of metals and facies conditions) allow one to evaluate only rather approximately the parameters of processes and to discuss indirectly the chemistry of this phenomenon and the essence of its mechanism. Investigations, performed during the last two decades by the researchers of different fields of chemistry and geochemistry on the sorption extraction of transition metals from solutions of complex electrolytes by oxyhydroxides of iron, manganese, and other metals, significantly expanded the notions of these processes. The major results of the investigations were discussed in our previous works (Putilina and Varentsov, 1980; Varentsov and Pronina, 1973; Varentsov *et al.*, 1978, 1979a, 1979b, 1980, 1981, 1985).

4.3.2. SORPTION SYNTHESIS OF THE OXYHYDROXIDE PHASES OF MN, FE, NI, AND CO ON IRON OXYHYDROXIDES

Our previous works (Pronina and Varentsov, 1973; Pronina *et al.*, 1973; Varentsov, 1972, 1976; Varentsov and Pronina, 1973; Varentsov *et al.*, 1980) have shown that the formation of Fe–Mn ores in the Recent basins and in the weathering crusts involves chemisorption processes accompanied by the autocatalytic oxidation of transition metals sorbed from the component-bearing solution by the active surface of the sorbent. Several experiments have revealed the main stages of this complex interaction. They have also shown that the process may be proceeded in a wide range of concentrations, from ultramicroquantities to highly noticeable contents.

The purpose of the present study is to synthesize newly formed phases in amounts sufficient for identification by the existing techniques, using the process of chemisorption and autocatalytic oxidation of Fe, Mn, Ni and Co accumulations on the hydrous iron oxide (γ-FeOOH) and to examine the mechanisms of the process in more detail.

4.3.2.1. Main Results

Table 22a shows sorption dynamics data for the components studied. An analysis of this data indicates that the rate of sorption varied appreciably during each cycle, especially for Mn and less markedly for Ni and Fe. On the whole, there was a tendency for these elements

Table 22a.   Synthesis of Fe, Mn, Ni, and Co oxide compounds from seawater solution on hydrous iron oxide (γ-FeOOH) under static conditions* (Varentsov *et al.*, 1980).

No.	Initial concentration in solution, $\mu g/l$				Final concentration in solution, $\mu g/l$				Quantity sorbed from solution, $\mu g/l$				Total quantity of element sorbed from 5-litre solution, μg				Sorption rate, $\mu g/hr$				Total sorption rate, $\mu g/hr$	Sorption percent			
	Ni	Co	Fe	Mn	Ni	Co	Fe	Mn	Ni	Co	Fe	Mn	Ni	Co	Fe	Mn	Ni	Co	Fe	Mn		Ni	Co	Fe	Mn
1	2160	1920	1010	625	1040	812	80	35	1120	1108	930	590	5600	5540	4650	2950	77.75	76.95	64.55	40.95	260.20	56.69	57.68	92.05	94.44
2	1920	1840	1860	704	816	955	354	354	1104	855	1028	350	1120	9965	9790	4700	76.65	59.35	71.40	24.30	231.70	62.18	48.09	55.26	49.71
3	1540	1400	1620	740	790	1010	796	264	750	390	824	476	14870	11975	13910	7080	52.10	27.10	57.20	33.05	175.45	47.40	27.85	50.89	64.32
4	1260	1160	1460	820	983	935	674	600	277	225	786	220	16255	13100	17840	8180	19.25	17.70	54.60	15.25	106.80	22.02	19.40	53.83	26.83
5	790	1060	1060	618	600	860	87	572	190	200	973	46	17205	14100	22705	8410	13.20	13.90	67.55	3.20	97.85	24.05	18.80	91.89	7.44
6	1140	1100	1000	1180	555	630	111	555	585	470	889	625	20130	16450	27150	11535	40.60	32.65	61.75	43.40	178.40	46.50	42.70	88.90	52.96
7	1140	1190	1000	650	758	920	0	610	382	270	1000	40	22040	17800	32150	11735	26.50	18.75	69.45	2.75	109.45	33.35	22.66	100.0	6.15

*Solution volume 5 l, sorption time 72 h, sorbent weighing 0.300 g. Element content in the sorbent phase: (Me $= \frac{\Delta \text{Me}}{0.3+\text{Ni}+\text{Co}+\text{Fe}+\text{Mn}} \times 100\%$). The content of the metal sorbed in the solid phase: Ni = 19.46%, Co = 15.65%, Fe = 28.13%, Mn = 10.31%. Me is the quantity of the metal sorbed (Ni, Co, Fe, Mn) in mg.

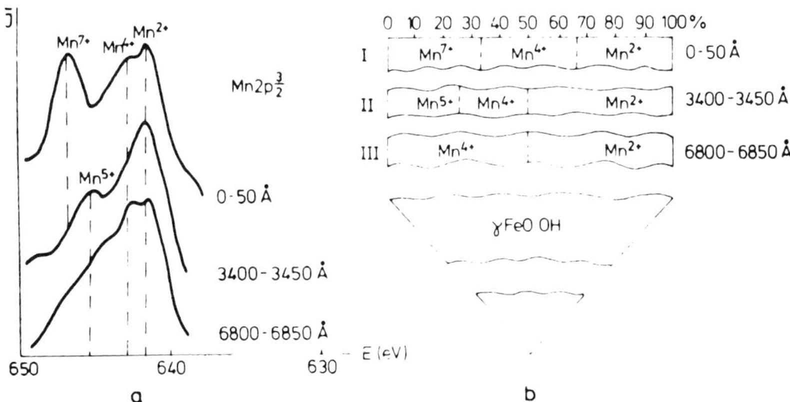

Fig. 73.   (a) Spectra $2p\frac{3}{2}$ of Mn electrons in the outside and inside parts of the newly formed layer; (b) The ratio (%) of Mn valence states in the outside and inside parts of the newly formed layer (Varentsov *et al.*, 1980).

to show a significant reduction in the rate of sorption. As a result of the seven changes of the original solutions the following relative quantities of the metals were accumulated on the sorbent: Ni = 19.46%, Co = 15.65%, Fe = 28.13%, Mn = 10.51%.

From the study of the initial phase of the sorbent by means of a scanning electronic microscope it can be stated that lepidocrocite occurs as microglobular aggregates with characteristic needle-shaped crystallites on the surface. High-resolution photographs show the general globular structure of these formations having characteristic fine needle-shaped crystallites developed on the surface. On the photograph of the final product the globular structure is seen to have been preserved only for relative coarse aggregates. At the higher resolution it is possible to distinguish the relative flat microlaminate stepped forms of the outside parts of the sorbent aggregates that are defined as epitaxial forms of successively growing new phases.

X-ray photoelectron spectroscopic methods make it possible to study the features of the valence states of atoms only in surface layers (with the depth of photoelectron ejection averaging 50 to 70 Å). so that the surface of the particles were etched by argon ions with an energy of 0.7 KeV at 6 mA (the argon in the camera having a pressure of 0.14 torr during sample preparation) to examine deeper sections. Two rounds of etching with a step of 3400 Å were performed on the sample under investigation. The structural and chemical characteristics of the sorbed layer were evaluated from the spectra $2p\frac{3}{2}$ of the electrons of Mn, Fe, Ni, and Co, which are illustrated in Figures 73a, 74a, 75a, and 76a. As shown, the spectral pattern for all the atoms studied in the outside portion of the layer is very distinct from that for the atoms in the inside portion of the layer.

Manganese. The spectrum $2p\frac{3}{2}$ of the electrons of Mn in the outside portion of the layer has a complex structure showing a distinct subdivision into two maximum groups: a low-energy group (with a bond energy range of 640 to 644 eV) and a high-energy group

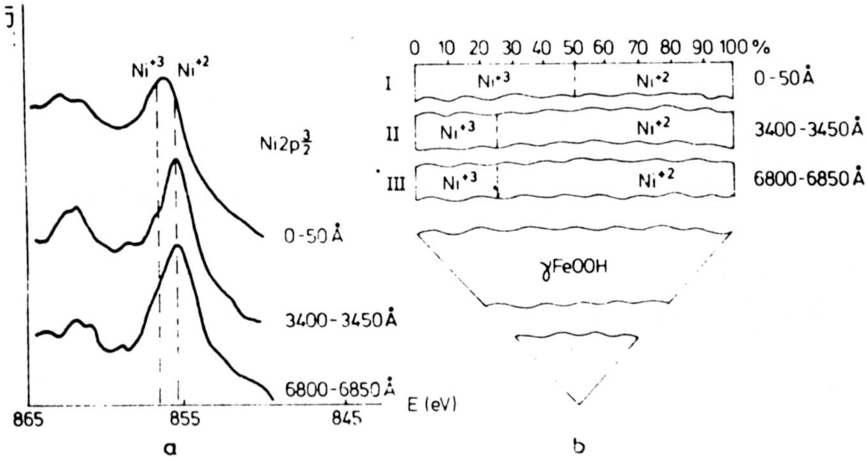

Fig. 74. (a) Spectra $2p\frac{3}{2}$ of Ni electrons in the outside and inside parts of the newly formed layer; (b) The ratio (%) of Ni valence states in the outside and inside parts of the newly formed layer (Varentsov *et al.*, 1980).

having a bond energy range of 645 to 647 eV (Figure 73a). The low-energy portion of the spectrum has a substantial breadth and is asymmetrical to allow one to assume the superposition of several manganese charge states. The same is indicated by the step configuration of the top part of the spectrum in this bond energy range. A comparison of the spectrum thus obtained with the spectra known for various manganese oxides indicates that the 641.6 eV peak refers to Mn^{2+} and the 642.7 eV peak to Mn^{4+}. The 646.6 eV peak, which shows intensity and high resolution, is associated with the manganese in the uppermost oxidation valent state (Mn^{7+}). However, the valence states Mn(VII) and Mn(V) should be regarded as transitional and metastable. It is significant that in the present case the ratio $Mn^{2+} : Mn^{4+} : Mn^{7+}$ approaches 1 : 1 : 1 (see Figures 73a, b). The peak responsible for Mn^{7+} disappears altogether at a depth of 3400 Å from the surface in the sorbed layer and the most oxidized form of manganese here is Mn^{5+} (with a bond energy of 645.2 eV), while there is simultaneously a considerable reduction in the proportion of Mn^{4+} (the ratio $Mn^{2+} : Mn^{4+} : Mn^{5+} = 2 : 1 : 1$). The inside portion of the layer (6800 Å) is represented only by lower manganese oxidation states, Mn^{2+} and Mn^{4+} having approximately equal quantities.

Nickel. The spectra $2p\frac{3}{2}$ of the nickel electrons in the outside part of the layer also has a substantial breadth (3 eV) (Figure 74a). In addition, satellite lines are found at intervals of 6 to 7 eV in the high-energy portion of the spectrum. These satellite lines can be distinctly subdivided into two narrow components: one with a bond energy of 862.5 eV and the other with a bond energy of 863.0 eV. The transition in the intermediary portion of the layer (3400 Å) causes the basic line $2p\frac{3}{2}$ of the nickel electrons to become much narrower and move toward the low-energy portion, up to 855.5 eV, compared with the bond energy of the basic peak in the outside portion of 856.3 eV. The 855.5 eV value is characteristic of divalent nickel (Frost *et al.*, 1972; Minachev *et al.*, 1975; Nemoshkalenko and Aleshin,

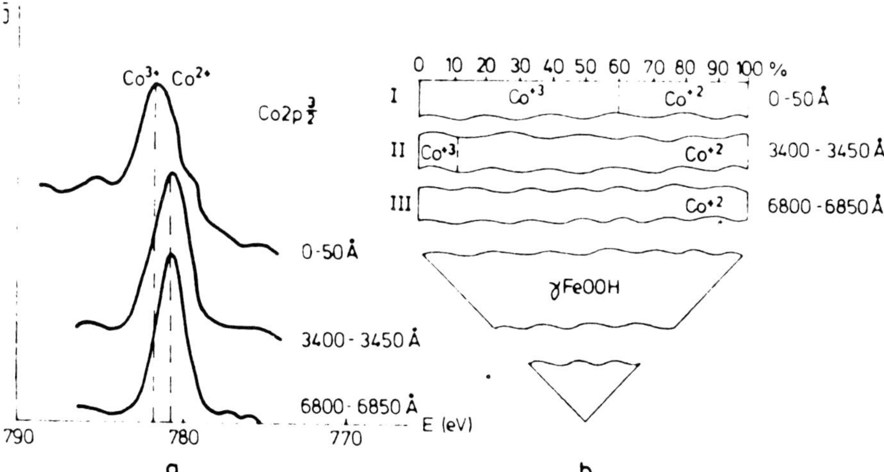

Fig. 75.   (a) Spectra $2p\frac{3}{2}$ of Co electrons in the outside and inside parts of the newly formed layer; (b) The ratio (%) of Co valence states in the outside and inside parts of the newly formed layer (Varentsov *et al.*, 1980).

1976; Rosencwaig *et al.*, 1971), the satellite lines having a bond energy of 861.5 eV. In other words, the significant breadth of the nickel spectrum in the outside portion of the layer and the double structure of the satellite lines give reason to suppose that this spectrum is a superposition of the two nickel states: Ni^{2+} and Ni^{3+}, quantitatively proportioned as 1 : 1 (see Figure 74). As stated above, the intermediary portion of the layer is largely represented by divalent nickel, but the amount of trivalent nickel, as evidenced from an influx in the bond energy region of 857.0 eV, is 1/4 of the total. It is interesting that a corresponding redistribution of intensity also occurs on the satellite lines in this case. The spectrum $2p\frac{3}{2}$ of the nickel electrons in the lower layer (6800 Å) does not differ from the spectrum of the intermediary layer.

Cobalt. The structure of the $2p\frac{3}{2}$ lines of the cobalt electrons is similar to that of the nickel lines (Figure 75a, see Figure 74a): the cobalt spectrum in the outside portion also has a substantial breadth and tends to shift into the high-energy region, compared with the inside portion of the sorbed layer. By analogy with nickel, this spectrum is a superposition of Co^{2+} and Co^{3+} in the ratio $Co^{3+} : Co^{2+} = 3 : 2$ (see Figure 75). It is of interest to note that the cobalt spectrum has a satellite structure only in the outside portion. The satellite in the 785.2 eV region is probably due to the discrete losses associated with the secondary transfer of the electrons 3d–4s to the ion Co^{3+}. The cobalt spectra at 3400 Å and 6800 Å are absolutely identical and have rather symmetrical lines. The peak of the cobalt basic line at 3400 Å and 6800 Å has a bond energy of 780.0 eV, which exactly corresponds to the presence of Co^{3+}. The remnant amounts of Co are noticeable only in the intermediary layer from a certain asymmetry in the high-energy portion of the basic line.

The presence of a satellite in the Co spectrum for the outside portion of the layer is most likely to be connected with the specific nearest surroundings of Co^{2+} and not of Co^{3+}, since Co^{3+} possesses no satellite structure. This is also supported by the fact that

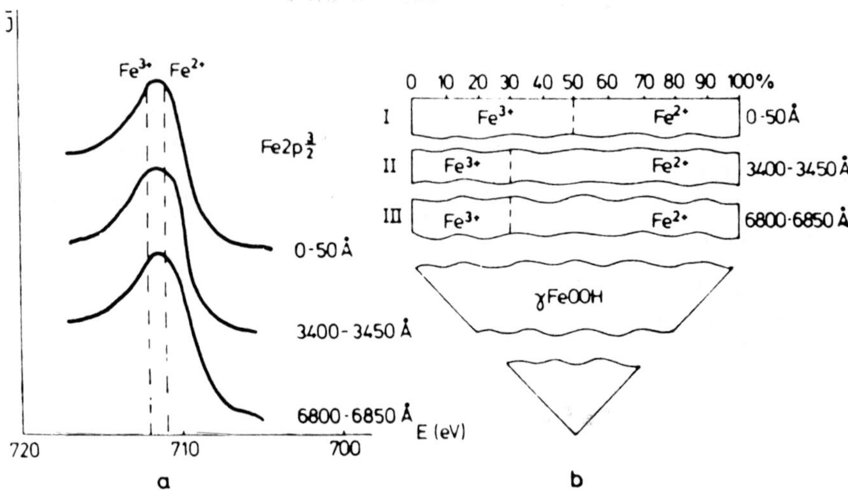

Fig. 76.   (a) Spectra $2p\frac{3}{2}$ of Fe electrons in the outside and inside parts of the newly formed layer; (b) The ratio (%) of Fe valence states in the outside and inside parts of the newly formed layer (Varentsov *et al.*, 1980).

the satellite distance is very small: Co^{3+} $2p\frac{3}{2}$: 3.3 eV. instead of the commonly found 4.5 eV.

As known from experimental studies, satellites are mainly characteristic for octahedral configurations with Co as the ligand, while the absence of satellites indicates the transfer of Co into tetrahedral surroundings, i.e. into the inner parts of the sorbed layer Co^{2+} are distributed largely over tetrahedral vacancies.

Iron. All the three levels of the studied layer of the spectrum $2p\frac{3}{2}$ of the iron electrons are close both in the configuration of the lines and in the magnitude of the binding energies (Figure 76a). Nevertheless, in the outside portion of the layer the reduction of the maximum of the $2p\frac{3}{2}$ lines of the iron shows somewhat higher magnitudes of binding energy (711.6 eV), while in the intermediary and lower part of the layer this energy falls to 711.2 eV. This indicates that the outside portion is characterized by a higher effective iron charge than the inside portion. Since the present iron spectrum has a relatively great breadth, then we can assume that this spectrum reflects the concurrent presence of Fe^{2+} and Fe^{3+}, with the trivalent form of iron prevailing in the outside portion of the layer. A comparison of the iron line in the intermediary portion and the shift of this line into the low-energy region points to a relative reduction of the trivalent form. The subsequent broadening of the iron lines on the low-energy side to a magnitude of 6800 Å is most likely to be attributed to the hydrate form of $Fe(OH)_2$ and this points to an increase in the relative quantities of Fe^{2+} (see Figure 76).

From a study of the sorbent phase and the product synthesized on its basis by using X-ray diffraction powder methods one can determine that a number of hydroxide phases have synthesized on the surface of lepidocrocite in the form of epitaxial growths, as evidenced from scanning microscope photographs. The most pronounced crystalline

Table 23. X-ray powder diffraction data for initial material and final product, experiments on chemisorption with autocatalytic oxidation interaction of hydrous iron oxide (γ-FeOOH) with dissolved forms of Fe, Mn, Ni, and Co (Varentsov *et al.*, 1980).

Initial hydrous iron oxide (γ-FeOOH)		Newly formed phase		γ-FeOOH lepidocrocite		NiO CoO solid solution		NiO		K_2MnO_4		Mn_2O_3		$Ni(OH)_2$	
ASTM		–		8–99		10–188		4–0835		12–264		10–69		14–117	
1	d(A)	1	d(A)	1	d(A)	1	d(A)	1	d(A)	1	d(A)	1	d(A)	1	d(A)
6	6.27	9	6.271	100	6.27	–	–	–	–	–	–	–	–	–	–
–	–	–	–	–	–	–	–	–	–	20	5.17	–	–	–	–
7	4.51	5	4.302	–	–	–	–	–	–	40	4.29	–	–	–	–
0.5	3.57	–	–	–	–	–	–	–	–	–	–	–	–	–	–
–	–	5	3.859	–	–	–	–	–	–	40	3.83	25	3.84	–	–
10	3.31	10	3.277	60	3.29	–	–	–	–	10	3.59	–	–	–	–
–	–	2	3.079	–	–	–	–	–	–	100	3.08	–	–	–	–
–	–	2	2.81	–	–	–	–	–	–	35	2.978	–	–	100	2.805
–	–	2	2.85	–	–	–	–	–	–	20	2.596	100	2.72	45	2.707
8	2.46	10	2.46	30	2.47	40	2.44	90	2.41	–	–	–	–	–	–
2	2.34	1	2.35	15	2.36	–	–	–	–	30	2.483	11	2.35	100	2.334
–	–	1	2.196	–	–	–	–	–	–	14	2.291	–	–	–	–
0.5	2.10	6	2.11	12	2.08	100	2.11	100	2.09	16	2.147	13	2.01	–	–
6	1.93	8	1.93	30	1.93	–	–	–	–	35	2.037	–	–	–	–
–	–	1	1.82	10	1.84	–	–	–	–	12	1.918	13	1.84	–	–
4	1.73	3	1.73	15	1.73	–	–	–	–	4	1.824	–	–	35	1.754
–	–	0.5	1.67	8	1.57	–	–	–	–	6	1.821	30	1.66	25	1.563
–	–	2	1.42	15	1.52	100	1.50	57	1.48	8	1.729	15	1.42	16	1.48
1	1.38	2	1.39	3	1.39	–	–	–	–	4	1.678	5	1.39	–	–
–	–	2	1.36	8	1.37	–	–	–	–	4	1.675	3	1.36	8	1.325
1	1.30	1	1.26	–	–	100	1.28	16	1.26	–	–	3	1.26	10	1.299
–	–	2	1.20	–	–	–	–	13	1.20	–	–	3	1.19	8	1.167

phase is that of the manganate of potassium (K_2MnO_4) which is distinctly identifiable from the series of reflexes: 4.29, 3.83, 3.08, and others (Table 23). In addition, there are the oxyhydroxide phases of some other metals whose low crystallization and weakly pronounced reflexes do not match their true composition in the newly formed layer. In particular, the iron compounds are represented by an X-ray amorphous phase ($FeOOH$, $Fe(OH)_2$). Among the compounds of Mn, Ni, and Co one can identify a solid solution of $CoO \cdot NiO$, NiO, $Ni(OH)_2$, Mn_2O_3, and trace amounts of Co_2O_3 may also be present (see Table 23). It should be stressed that X-ray photoelectronic spectroscopy data on the interlayer distribution of the valence forms of these metals should be taken into account when considering the distribution of these phases in the newly formed layer. On the whole, the results of X-ray photoelectronic spectroscopy agree well with the phases identified on X-ray diffraction patterns.

4.3.2.2. Discussion

Although the conditions for each cycle of sorption of Mn, Fe, Co, and Ni from the solution by the dispersed phase of the sorbent (lepidocrocite) were kept as identical as possible, the degree of sorption was found to vary from component to component. For Co and Ni, and to a lesser extent for Mn, there is a distinct reduction in the degree of sorption in the final stage (5–7) as compared with the beginning of the interaction (see Table 22a). For iron no such tendency is in evidence. In Table 22a it is shown how the rates of sorption (μg/hour) vary and illustrate the dynamics of absorption. An analysis of these data leads to the conclusion that the rates of sorption during the final stages (5–7) are four to five times lower than those during the initial stages. Since the joint sorption of the hydrous iron oxides and Mn, Fe, Ni, and Co present in the solution took place during the process of the interaction, it is necessary to consider changes in the rates of the overall sorption of these metals during the experiment (see Table 22a). It is obvious that reduction in the overall sorption rates for the metals under investigation is to be compared with the specifics of the distribution of their valence forms at different levels of the sorbed layer. This reduction in the sorption rates can be correlated with the distribution of the metals over three depth levels in the newly formed layer. Characteristically, the behavior of Mn, Ni, Co, and Fe follows a general pattern: high-oxide forms of these components predominate substantially in the outside portion of the layer. Manganese within the range of 0–50 Å is characterized by the presence of the highest oxidation from Mn^{7+} (33%) and the high degree of oxidation -Mn^{4+} (33%), only one third of the total amounts of this metal being present as Mn^{2+}. Also characteristic of the outside portion is a considerable content of the high-valence forms of Co, Ni, and Fe (see Figures 73 to 76). In the intermediate portion of the layer (3400–3450 Å) the relative amount the high-valence forms of Mn, Co, Ni, and Fe falls markedly. In the interval of 6800–6850 Å, the low-valence forms of Co, Fe, and Ni are significantly predominant and it is only for Mn that the ratio between Mn^{4+} and Mn^{2+} is found to be approximately equal. Thus, the extensive occurrence of the high-oxide forms of the metals in the outside portion of the newly formed layer may be attributed to a significant reduction in the rate of sorption. The composition of certain zones in the newly formed layer may be considered to be controlled by the relationship between the rates of the two processes: the initial ion-exchange sorption process and the subsequent autocatalytic oxidation process at the surface-solution interface. However, as the experiment results indicate, the ion sorption process is predominant at high accumulation rates, while the development of the interphase catalytic oxidation reaction is blocked. It is only during the final stages of the experiment (5–7) when the sorption rates drop sharply that the interphase oxidation reaction could be developed. This is also facilitated by the prolonged washing of the final product by deionized water from Cl-ions, increasing the overall length of the oxidation stage. It can be supposed that on the surface of the sorbent there occurs sorption of relatively high concentrations of dissolved oxygen, leading to the formation of the high-valence compounds of Mn, Ni, Co, and Fe.

The most probable cause controlling the decrease of the rate of sorption in the course of the experiments is the reducing of the total surface of the sorbent (see Table 22a). The relatively large specific surface of the initial sorbent (its increasing is favored by wide development of needle-shaped crystallites on globular particles) is essentially reduced at the final stages of the experiments. The SEM photomicrographs of the final products shows the development of microlaminated epitaxial overgrowths, relative enlargement and aggregation of the particles.

The data described herein indicates that when the kinetic parameters of the stages of the process are in a certain relationship the sorption of Mn, Fe, Co, and Ni by hydrous iron oxides is of a pronounced chemisorption nature with autocatalytic oxidation. The interaction is a complex, multi-stage one and a tentative model of it has been described in previous works (Pronina and Varentsov, 1973; Pronina *et al.*, 1973; Varentsov and Pronina, 1973). It is to be noted that under natural conditions the sorption of Mn, Fe, Ni, Co, and other metals proceeds at extremely low rates, as these rates are controlled by microquantities of these components in natural solutions. Where the quantities of dissolved oxygen are sufficient and the magnitude of Eh is high, the conditions may favor the development of autocatalytic oxidation, a leading process in the formation of Fe–Mn ores.

4.3.3. SORPTION SYNTHESIS OF THE OXYHYDROXIDE PHASES OF MN, FE, NI, CO, AND CU ON MANGANESE OXYHYDROXIDES

Goldschmidt (1937) was the first to consider sorption as being the process responsible for Me removal from seawater as well as the natural ground waters. Goldberg (1954) and Krauskopf (1956, 1957) contributed to this idea and Goldberg and Arrhenius (1958) suggested that the transition metal sorption from seawater on amorphous $FeOOH$-xH_2O might be autocatalytic.

Evidence for the accumulation of transition metals together with amorphous $FeOOH$-xH_2O in Mn oxyhydroxides has been put forward since (Grütter and Buser, 1952) and related to earlier evidence in laboratory experiments (Feitknecht and Marti, 1945).

More recent results have shown:

1. that Me sorption occurs in a wide range of concentrations of which the lower limit is frequently unreliable to detect due to complex formation phenomena;
2. that sorption specificity at low and high concentrations ($10^{-1} \ldots 10^{-2}$ μg/l and above

10^3 μg/l, respectively) is important;

3. that at low concentrations the ion exchange nature is less prominent in the Me accumulation by amorphous FeOOH-xH$_2$O and Mn oxyhydroxides than at higher concentrations;

4. that below 10^4 μg/l the Co and Zn accumulation does not follow the Langmuir equation (as opposed to K and Na which follow the equation in a wide concentration range);

5. that the Me accumulation depends on ionic strength and other matter present in the electrolyte (such as seawater, the most complicated electrolyte found as yet and other natural waters also);

6. that the anion present is of particular importance for the Me accumulation, and that also pH plays an important part as the solid-solute interface depends on it;

7. that Mn oxyhydroxides show little effect of the electrolyte at pH 6–8 for Me accumulation but that the competition of 10^{-7} M turns up at 2 to 10 (Murray, 1975a, 1975b);

8. that the interface sorption of Me as opposed to ion exchange varies little with the crystallinity of the substrate and has merely an influence on the sorption rate (Anderson *et al.*, 1973; Sipalo-Žuljević and Wolf, 1973);

9. that Mn oxides and oxyhydroxides depend in their surface charge considerably from the pH (Gabano *et al.*, 1965) but that in natural waters (pH 5 to 11) Mn oxides and oxyhydroxides are always carrying a negative surface charge;

10. that contact oxidation Mn^{2+} $\rightarrow$ Mn^{3+} might explain the higher sorption capacity for Mn^{2+} as compared to Ag$^+$, Ba^{2+}, Ca^{2+}, Mg^{2+}, Nd^{3+}, Sr^{2+} (Posselt *et al.*, 1968);

11. that hydrolysis increases Me accumulation drastically since at low pH only sorbent OH-groups exchange their proton against the Me ion, while in the alkaline region hydrolysis overtakes this surface reaction;

12. that evaluation of chemical and Coulomb adsorption (Murray and Brewer, 1977) leads to unsatisfactory results which, however, may also be due to the extremely poor characterization of the sorbent used by these authors;

13. that Stumm *et al.* (1976) can explain surface phenomena without having to resort to hydrolysis while other authors (James *et al.*, 1975; Novikov, 1972; Novikov and Gonchrova, 1972; Novikov and Knyazev, 1977) came to conflicting views.

All these models and ideas suffer from poor characterization of substrate and hence, poor evidence. It is, therefore, necessary to concentrate the efforts on using defined Mn oxyhydroxides such as synthetic Mn$_3$O$_4$, birnessite, and buserite (Giovanoli *et al.*, 1975). We are not aware of the use of similar substrates by other authors and it is no wonder that as a consequence they suffer from the inadequacy of existing sorption models.

Our own studies (Pronina and Varentsov, 1973; Pronina *et al.*, 1973; Varentsov and Pronina, 1973, 1976; Varentsov *et al.*, 1978) revealed once again a high sorption selectivity in seawater as well as in other mineralized natural waters, e.g. karst waters, and a sequence of several stages from surface exchange of OH groups to formation of new phases.

The purpose of the present study (Varentsov *et al.*, 1979a, 1979b) is to elucidate the principal factors in the sorption of Mn, Fe, Ni, and Co from seawater or similar mineralized water by Mn$_3$O$_4$.

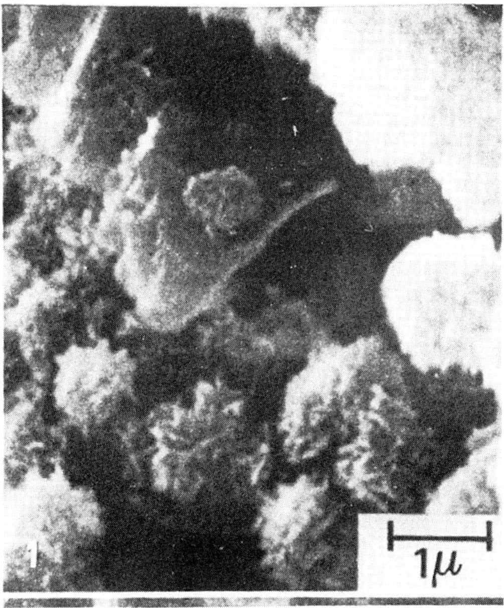

Fig. 77. Initial Mn_3O_4. Scanning electron microscope. The microglobular topography of the aggregates is visible (Varentsov *et al.*, 1979a).

4.3.3.1. Experimental

Substrate; metal solutions. Synthetic, pure Mn_3O_4 (Feitknecht and Marti, 1945) in bipyramids about 2000 Å size was used as substrate (Figure 77).

The synthetic seawater (Bruevich, 1946) was prepared from deionized water: 19‰ Cl^-, 35‰ S; pH 8, 10 : 8, 15. Me ions were complexed with a 10-fold molar excess of citric acid to prevent hydrolysis, Fe was added as analytical grade $NH_4Fe(SO_4)_2 \cdot 12H_2O$, and Co as analytical grade 'Ni free' $CoCl_2 \cdot 6H_2O$. Ni was used as analytical grade 'Co free' $NiSO_4 \cdot 7H_2O$.

Initial concentrations are shown in Table 24.

Experiment. 5-litre polythene bottles with the solution and 300 mg Mn_3O_4 were shaken for 72 h. Mn_3O_4 was then isolated by centrifuge. Controls were carried out to account for sorption by the bottle wall and for possible hydrolysis.

After each run the isolated Mn_3O_4 was exposed to a new portion of starting solution. Seven cycles proved necessary to get sufficient Me accumulation (Table 24).

The final precipitate was washed free of Cl^- with deionized water.

Analytical. Ni and Co contents were determined by polarography (Vinigradova *et al.*, 1968) and Fe by photometry of 1×10-phenanthroline (Marchenko, 1971). Mn was determined as MnO_4^- by photometry (Marchenko, 1971).

X-ray diffraction, electronmicroscopy, X-ray photoelectron spectroscopy, Mössbauer. Routine diffraction has been made with a 57.3 mm powder camera and V filtered Cr radiation (Sokolova). For better resolution a Guinier–de Wolff camera Mk.I with focusing quartz monochromator and $FeK_{\alpha1,2}$ radiation was also used.

The morphology was observed with a scanning electron microscope HITACHI-AKASHI MS M-2 and a JEOL transmission electron microscope (Serebrennikova). A HITACHI HU-12A transmission electron microscope also was used.

The valence of accumulated Me was studied by X-ray photoelectron spectroscopy (spectrometer VIEE-15) with MgK_{α} excitation of 1.2536 keV energy. The accuracy of +0.1 eV and resolution of 1.2 eV were attained *in vacuo* (3×10^{-7} torr). The carbon 1-s line with binding energy 285.0 eV was used for calibration. Samples were prepared to cover a finely corrugated Al cylinder.

Photoelectrons penetrated about 50 to 70 Å into the sample. For deeper zones the substances were ion etched (Ar ions, 0.7 keV, 6 mA, at 0.14 torr) in steps of 3400 to 6000 Å.

The Fe valence was determined by Mössbauer spectroscopy, using an electrodynamic spectrometer and Co^{57} as source, in a Pd lattice. A NaJ(Tl) crystal scintillation counter served for recording. Spectra were taken at room and liquid nitrogen temperature.

4.3.3.2. Results

Analytical.   Table 24 shows the following features:
(a)   The amount of sorbed Co decreases from 40.06 to 15.50% in the final stage of the experiment.
(b)   Ni shows a less pronounced trend.
(c)   Fe is sorbed gradually and reaches 100% after stage 5.
(d)   Mn shows a decrease from 21.14 to 14.40% up to stage 3, and from stage 4 to 7 even a desorption of Mn occurs. The released Mn in solution drastically exceeds, in the final stage, the initial Mn concentration, i.e. by 2535 μg.

After 7 changes of the initial solutions 17.67% Ni, 19.50% Co and 27.64% Fe, respectively, were accumulated on the sorbent; while from this data it may be formally concluded that there is no Mn accumulation on Mn_3O_4. A manganese amount equivalent to 16.14% of sorbed Co or 17.96% of the Ni was removed from the Mn_3O_4 substrate during the surface exchange reaction.

Morphology.   Figure 77 shows the globular aggregates of Mn_3O_4. Figure 79 presents the actual Mn_3O_4 crystals as bipyramids.

After the reaction, the globular aggregates have changed into microlaminated steps of epitaxial outgrowths of the new phase (Figure 78).

Photoelectron spectra.   The $2p^{\frac{3}{2}}$ photoelectron spectra (Figures 80 to 82) give the following evidence:

Table 24. Sorption of Fe, Mn, Ni, and Co under static conditions from seawater solution on Mn_3O_4.[a]

No. of runs	Initial concentration in solution (μg/l)				Final concentration in solution (μg/l)				Amount of element sorbed from solution			
	Ni	Co	Fe	Mn	Ni	Co	Fe	Mn	Ni	Co	Fe	Mn
1	1640	1580	1320	700	910	938	864	552	730	642	546	148
2	1500	1540	1000	600	995	927	492	396	505	613	508	104
3	1420	1600	1360	400	980	857	805	342	440	743	555	58
4	1280	1360	960	500	811	894	665	624	469	466	295	-124
5	1060	1320	1540	620	885	935	7	703	175	385	1533	-83
6	1140	1260	1000	630	847	934	575	790	293	326	425	-160
7	1000	1000	940	565	600	850	0	1015	400	150	940	-450

Total amount of element sorbed from 5 litres of solution				Element sorption rate (μg/h)			Total element sorption rate (μg/h)	% of sorption			
Ni	Co	Fe	Mn	Ni	Co	Fe		Ni	Co	Fe	Mn
3650	3210	2280	740	50.69	44.58	31.67	126.94	44.21	40.06	34.54	21.14
6175	6275	4820	1260	35.07	42.57	35.28	112.92	33.33	39.63	50.80	17.33
8375	9990	7595	1550	37.50	51.59	48.54	127.63	30.98	43.12	40.84	14.50
10720	12320	9070	930	32.57	32.36	20.49	85.42	36.61	34.18	30.07	-24.80
11595	14245	16735	515	12.15	26.74	106.46	145.35	16.50	28.22	99.00	-11.16
13060	15875	18860	-285	20.35	22.64	29.51	62.50	25.73	25.91	42.50	-25.39
15060	16625	23560	-2535	27.78	10.41	65.28	103.47	40.00	15.50	100.00	-79.65

[a] Volume of the solution = 5 l. Time of interaction = 72 h. Sorbent = 0.300 g Mn_3O_4 (weighed). Content of the sorbed metals in the solid phase (%): Ni = 17.67, Co = 19.50, Fe = 27.64. Me = $\frac{Me}{0.3+Ni+Co+Fe}$ 100%. Me = amount of the sorbed metal (Ni, Co, and Fe in g).

Fig. 78.   Product, obtained after the sorption experiment. Scanning electron microscope. The aggregates have grown and the spherical form has disappeared (Varentsov *et al.*, 1979a).

Mn. In the external part of the sorbed layer there is a clear cut separation of the low-energy peaks, i.e. binding energy 640 to 644 eV. The considerable breadth and the symmetry suggest that several valence states of Mn are present. This is supported by the top portion in this binding energy region.

From reference obtained with defined Mn oxides (Frost *et al.*, 1972; Minachev *et al.*, 1975; Nemoshkalenko and Aleshin, 1976; Rosencwaig *et al.*, 1971) it follows that the 641.5 eV peak represents Mn^{2+} and the 642.7 eV peak Mn^{4+}.

The relative amounts of Mn^{4+} and Mn^{2+} will then be 60 and 40% in the top layer (Figure 80). With depth this relation alters to 50–50% in 6000 Å and in 12.000 to 15.000 Å depth the Mn^{4+} peak disappears entirely in favor of a new peak at 641.4 eV corresponding to Mn^{3+}. At 12.000 to 12.050 Å the ratio $Mn^{3+} : Mn^{2+}$ is 70 : 30 and decreases below 15.000 to 15.050 Å to 40 : 60.

Ni. Figure 81 shows again a considerable peak breadth, i.e. 3 eV. In the high energy region, at a distance of 6 to 7 eV, satellites turn up which can be separated into two components of binding energies 862.5 and 863.0 eV, respectively. At a depth of 6000 Å the main line becomes narrower and is shifted to 855.5 eV (cf. main peak at the surface with 856.3 eV). 855.5 eV are characteristic for Ni^{2+} (Frost *et al.*, 1972; Minachev *et al.*, 1975;

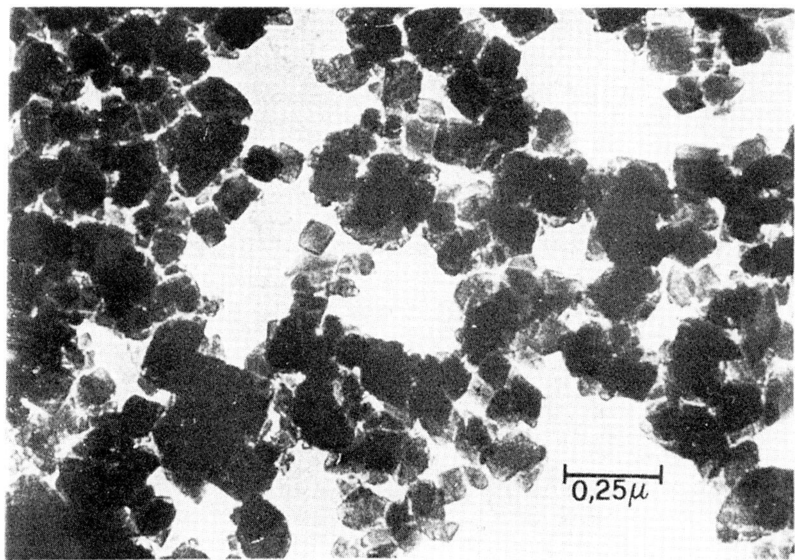

Fig. 79.   Product, obtained after the sorption experiment. The initial Mn_3O_4 bipyramids show indications of corrosion, and a new phase has grown on the crystals (Varentsov *et al.*, 1979a).

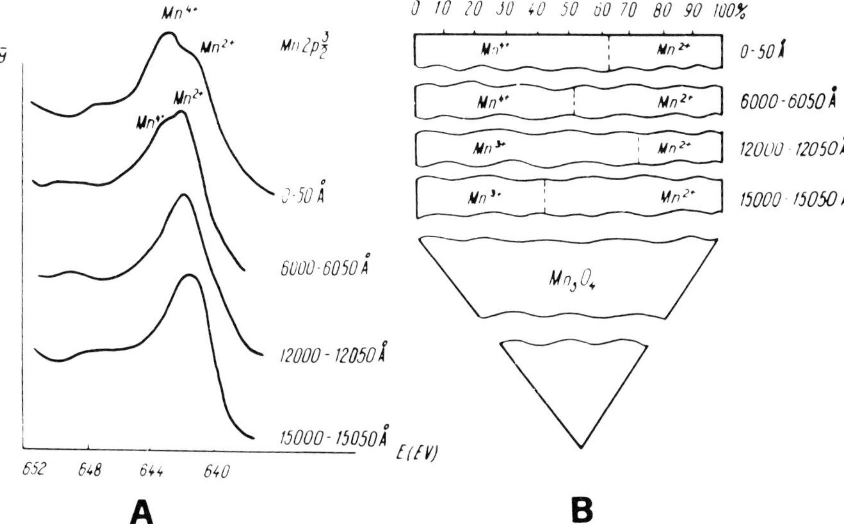

Fig. 80.   (A) Mn X-ray photoelectron $(2p\frac{3}{2})$ spectra for the external and internal parts of the newly-formed layer. (B) Relationships (%) between the Mn valence states in the external and internal parts of the newly-formed layer (Varentsov *et al.*, 1979a).

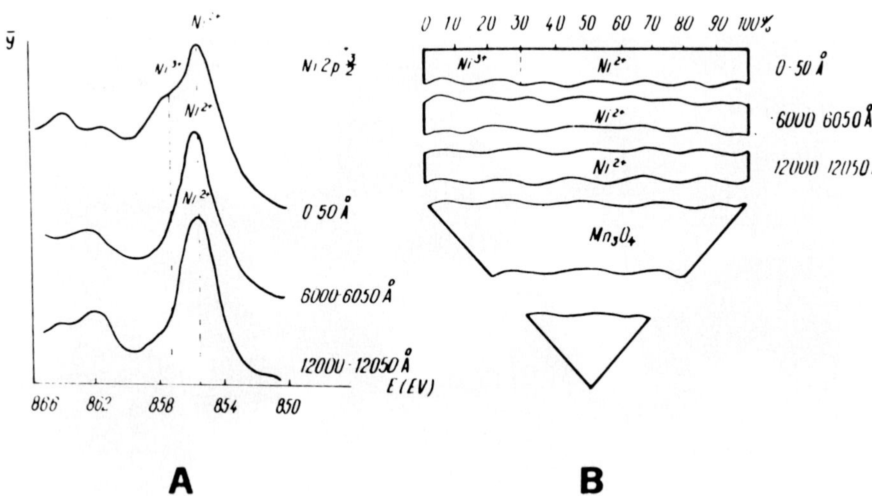

A   **B**

Fig. 81.   (A) Ni X-ray photoelectron ($2p\frac{3}{2}$) spectra for the external and internal parts of the newly-formed layer. (b) Relationships (%) between the Ni valence states in the external and internal parts of the newly-formed layer (Varentsov *et al.*, 1979a).

Nemoshkalenko and Aleshin, 1976; Rosencwaig *et al.*, 1971). The satellite indicates a binding energy of 861.5 eV.

In short, there are two valence states of Ni (+2 and +3) in the outer layer, with a ratio of 70 : 30. In the depth Ni^{2+} is prominent. In this case the corresponding intensity redistribution on the satellites can be observed as well.

In the depth of 12.000 Å the spectrum remains unaltered.

Co. Figure 82 shows very much the same spectrum as Figure 81, i.e. as in the case of Ni there is a superposition of Co^{2+} and Co^{3+} in a ratio 70 : 30.

There are satellites, however, in the spectrum of the outer part of the new formed layer. At 782.5 eV, most probably, lies an indication of discrete losses due to secondary 3d-4d electron transition in the Co^{3+} ion.

At 6000 and 12.000 Å depth the spectra are identical and show fairly symmetrical lines. The Co main line indicates a binding energy of circa 780.5 to 781.5 eV which corresponds to Co^{2+} (Frost *et al.*, 1972; Minachev *et al.*, 1975; Nemoshkalenko and Aleshin, 1976; Rosencwaig *et al.*, 1971).

The satellite occurring in the outer layer spectrum is due to special features of the nearest surrounding of Co^{2+}, not Co^{3+}. The latter produces no satellites (Rosencwaig *et al.*, 1971). The small distance of 3.3 eV instead of the usually observed 4.5 eV supports this. Satellites are typical for octahedral Co (Frost *et al.*, 1972; Minachev *et al.*, 1975; Nemoshkalenko and Aleshin, 1976). Their absence implies tetrahedral Co. In the deeper parts Co^{2+}, thus, is mainly occupying tetrahedral sites.

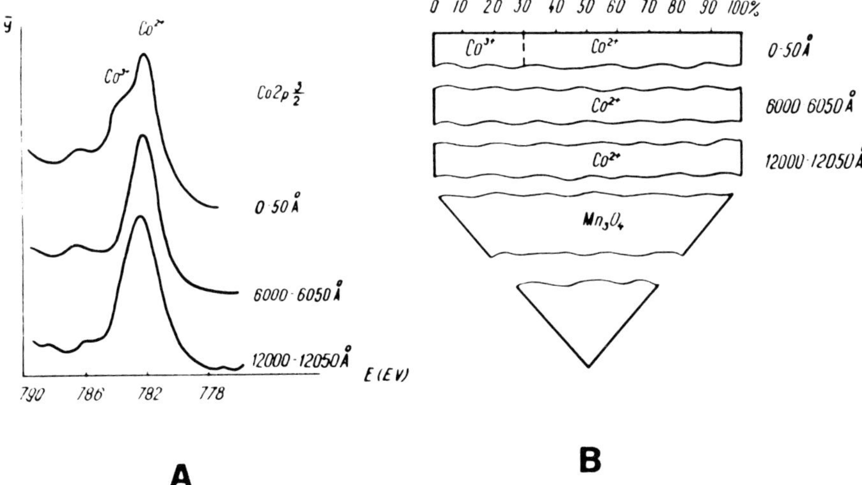

Fig. 82. (A) Co X-ray photoelectron $(2p\frac{3}{2})$ spectra for the external and internal parts of the newly-formed layer. (B) Relationships (%) between the Co valence states in the external and internal parts of the newly-formed layer (Varentsov *et al.*, 1979a).

Table 25. Mössbauer spectra parameters of $FeOOH_x H_2O$ amorphous on accumulated Mn_3O_4 (Varentsov *et al.*, 1979a).

Sample	T ($^\circ$K)	Isometric shift δE (mm/s)	Quadrupole splitting ΔE (mm/s)	Half-width of $r\frac{1}{2}$ lines (mm/s)	Effective magnetic field intensity (kOe)	ε (%)	$\frac{\varepsilon 300}{\varepsilon 80}$
a	300	0.18	0.68	0.61	–	1.00	
	80	0.17	0.66	0.48; 0.42	–	1.80	0.55
b	300	0.18	0.71	0.61	–	0.53	
	80	0.17	0.69	0.45; 0.60	–	1.50	0.34
c	300	0.18	0.66	0.61	–	0.20	
	80	0.17	0.69	0.54; 0.39	–	1.16	0.15
Synthetic	300	0.18	0.66	0.62	–	2.50	
$FeOOH$-xH_2O	80	0.17	0.66	0.54; 0.60	–	3.80	0.66
'Goethite' from	300	0.15	0.57	0.46	–	7.70	–
the Red Sea	80	0.15	0.69	0.45	454	–	–
(X-ray amorph.)							
Goethite synth.	300	0.20	0.55	0.44	–		
α-$FeOOH$ 200 Å	80	0.08	0.58	–	–	–	–

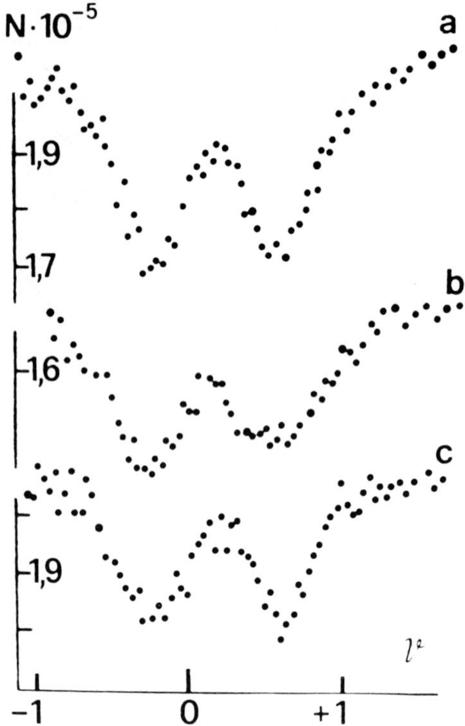

Fig. 83.   Mössbauer spectra of the product obtained as a result of the sorption of the dissolved Ni, Co, Fe, and Mn from the solution of seawater by Mn_3O_4. The spectra were taken at a temperature of liquid nitrogen. The samples a, b, and c are the products of parallel experiments. The parameters of these spectra correspond to amorphous $FeOOH\text{-}xH_2O$. The samples of the a-b-c series show an increase of disorder in amorphous $FeOOH\text{-}xH_2O$ (Varentsov *et al.*, 1979a).

Mössbauer spectra.   Spectra of three samples a, b, and c are shown in Figure 83 and summarized in Table 25.

Sample a. The quadrupole doublet with an isometric shift of 0.17 mm/s and quadrupole splitting ΔE = 0.66 mm/s is characteristic for Fe^{3+}, the line width of 0.48 mm/s is 1.5 times the instrumental line width. The magnitude of the effect is 1.8%.

Sample b. The quadrupole doublet has the same order of isometric shift but a somewhat higher quadrupole splitting (ΔE = 0.69 mm/s). The lines are slightly asymmetrical with 0.45 mm/s and 0.60 mm/s width, respectively, for the left and the right component. This is 1.40 and 1.87 times the instrumental line width, respectively. The magnitude of the effect is 1.5%.

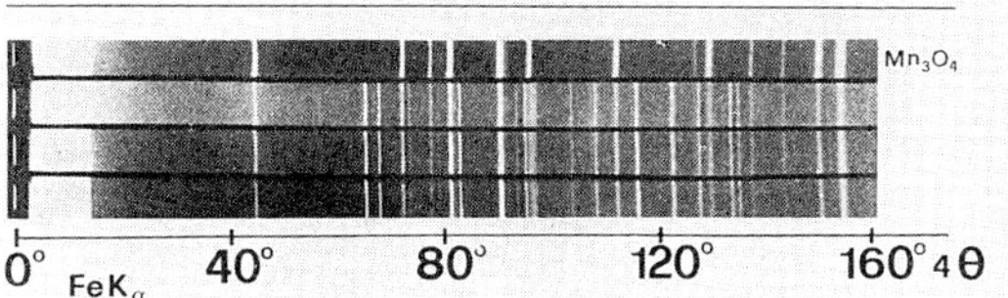

Fig. 84. X-ray powder pattern of initial Mn_3O_4 (top) and products of the sorption experiments. New lines appear on account of the intensity of the Mn_3O_4 reflections. Guinier–de Wolf camera; FeK_α radiation.

Sample c. The quadrupole doublet is similar to that of b, with reverse asymmetry and a magnitude of the effect 1.16%.

The constant isometric shift (1.17 mm/s) and the magnitude of quadrupole splitting (0.66 to 0.69 mm/s) indicate Fe^{3+} for all three samples and correspond most closely to amorphous $FeOOHxH_2O$.

The difference from a to c in the magnitude of the effect (1.80% to 1.16%) as well as the line width correspond to a decrease of crystallinity.

The parameters given above were obtained at liquid N_2 temperature and there is virtually no difference when passing from room temperature to liquid N_2 temperature. However, there is a prominent increase of the effect, characterizing the crystallinity. For comparison, finely divided α-FeOOH and amorphous $FeOOHxH_2O$ are included in Table 25 (temperature of liquid N_2) to underline the splitting effect. Our evidence indicates the presence of amorphous $FeOOHxH_2O$ in the sorbed layer.

X-ray diffraction. Table 26 and Figure 84 show the X-ray diffraction results. After the experiment there appears a new set of reflections (3.39; 3.26; 2.69; 1.97; 1.88; and 1.74 Å amongst others) in addition to the Mn_3O_4 pattern. From comparison of the d values only, phases like γ-MnOOH, α-CoOOH, Co_2O_3, Ni_2O_3, and $NiSO_42H_2O$ could be found in reference files (ASTM file and others). The direct X-raying of our samples with these substances side-by-side reveal that γ-MnOOH and α-CoOOH can be discarded and also the other compounds listed are most improbable under the conditions of our experiments. A purely numerical coincidence of some of the d values is not sufficient proof for these phases.

The exact identification of the additional reflections has therefore still to be solved.

4.3.3.3. Discussion

The main features of the sorption experiment have turned out as follows:
1. The amount of sorbed Co decreases from 40.06 to 15.50% during the experiment. The same holds for Ni but is less pronounced.
2. Fe shows a substantial increase of sorption, especially towards the end of the experiment.
3. Mn is accumulated less – as opposed to Fe – by each step, and in the end is even released into the solute phase. This can be compared to the morphological evidence:

Table 26. X-ray powder patterns of the starting material Mn_3O_4 and final product (Guinier–de Wolff camera Mark I, FeK_α radiation) (Varentsov *et al.*, 1979a).

Initial Mn_3O_4		End	Product	Reference Mn_3O_4	Hausmannite ASTM 16-154
d(Å)	I	d(Å)	I	d(Å)	I
4.91	20	–	–	4.94	30
–	–	4.37	0.5	–	–
–	–	3.93	0.5	–	–
3.56	0.5	3.56	0.5	–	–
–	–	3.43	0.5	–	–
–	–	3.39	10.0	–	–
–	–	3.33	0.5	–	–
–	–	3.30	0.5	–	–
–	–	3.26	8.0	–	–
3.06	0.6	3.02	0.5	3.09	50
2.87	1.0	2.82	0.5	2.89	30
2.76	10.0	2.72	0.5	2.77	90
–	–	2.69	7.0	–	–
2.48	9.0	2.45	0.5	2.49	100
–	–	2.40	1.0	–	–
2.31	2.0	2.37	3.0	2.36	40
–	–	2.35	0.5	–	–
–	–	2.33	1.0	–	–
–	–	2.32	1.0	–	–
–	–	2.18	1.0	–	–
–	–	2.11	0.5	–	–
2.04	5.0	2.10	2.0	2.04	40
–	–	1.97	6.0	–	–
–	–	1.88	5.0	–	–
–	–	1.81	4.0	1.82	20
1.782	5.0	–	–	1.795	50
–	–	1.72	0.5	–	–
–	–	1.74	4.0	–	–
1.696	5.0	1.72	2.0	1.706	30
1.634	2.0	1.64	0.5	1.642	20
1.579	7.0	–	–	1.575	80
1.541	10.0	1.55	0.5	–	–
1.44	6.0	1.49	0.5	1.468	10

Table 27. Relationship between the intensities of the ($2p\frac{3}{2}$) lines of Co, Ni, and Mn for the end product on the substrate Mn_3O_4 and on γ-FeOOH (Varentsov *et al.*, 1978, 1979a).

Depth from the surface of the accumulated material (A)	Co	Ni $\dfrac{I(2p\frac{3}{2})\ Mn_3O_4}{I(2p\frac{3}{2})\ \gamma\text{-FeOOH}}$	Mn
0–50	0.74	0.68	0.44
6000–6050 (6800–6850)	1.27	1.58	0.93

the initial spherical aggregates of Mn_3O_4 (Figure 77) show the sorbed material as outgrowths (Figure 78).

The active surface of the initial Mn_3O_4 diminishes therefore during the runs which explains the behavior for Ni and Co. Similar relations were observed for Ni, Co, and Mn on Fe oxyhydroxides (Varentsov *et al.*, 1978).

From spectroscopy we conclude that the ions accumulated on Mn_3O_4 prefer the higher of two possible valences. This differs from what we have observed with Fe oxyhydroxides (Varentsov *et al.*, 1978). The presence of much Mn^{2+} in the accumulated layer (Figure 80) is remarkable. The Ni and Co spectra resemble those produced on γ-FeOOH at least in the depth of 3400 to 3450 Å on Mn_3O_4.

It is noteworthy that the Co, Ni, and Mn accumulation on Mn_3O_4 decreases with time (Table 27) as opposed to similar processes on γ-FeOOH.

The intensity ratio of the ($2p\frac{3}{2}$) lines for accumulated material on Mn_3O_4 on the one hand and γ-FeOOH on the other indicates that the reaction does not proceed in a uniform manner; the starting phase shows high adsorption but then it slows down. These data are consistent with chemical analysis (Table 24) and a consequence of the vanishing active surface.

The decrease of Ni^{3+} and Co^{3+} with depth is more pronounced than in material accumulated on γ-FeOOH. There must, therefore, be a specific influence of the substrate. However, the accumulation rate seems to have an influence as well: the outer part with low rate shows the highest amount of Ni^{3+} and Co^{3+}.

The redox reactions during the diffusion of accumulated ions into the substrate will presumably account for the apparent inconsistencies of the Mn data (Table 24) and the Mn valence distribution: Mn^{2+} and Mn^{3+} leave the Mn_3O_4 lattice when Ni and Co are adsorbed and there occurs a kind of ion exchange at the lattice interface. This exchange, however, is not stoichiometric.

It should be underlined that the accumulated Fe forms exclusively amorphous $FeOOHxH_2O$ spectra.

4.3.3.4. Geochemical Interpretation

In supergene environments we must consider
— metal sources,
— transportation,
— ore formation.

Our present data deals with the third item listed. Their main results are that Mn, Fe, Ni, and Co accumulate in several distinct stages, involving autocatalytic oxidation, interaction of active surfaces with the solute phase, sorption, and ion exchange. We find a very selective sorption of transition metal which is explained partly by the particular properties of transition metal ions and partly by the substrate. A third factor is hydrolysis.

Later stages include postsedimentation phase transformations, particularly in the geological time scale.

The autocatalytic stage is distinct at pH values slightly beyond 7 (e.g. 8.20) and under aerobic conditions, i.e. in the presence of free O_2. What is actually deposited on the substrate will also depend on the solute phase, where the dissolved Me can be supplied.

The autocatalytic stage controls the entire process while the high sorption rate blocks interface oxidation.

The role of micro-organisms in recent basins and other supergene environments has recently been stressed and Ehrlich again makes a distinction between sorption and biocatalytic oxidation (Ehrlich, 1972, p. 66). The former takes place on an active surface while the second involves specific enzymes such that the oxidation rate depends on enzyme concentration.

Schweisfurth *et al.* (1978, p. 66) have found that with *Pseudomonas manganooxidans* Mn^{2+} oxidation occurs only in concentrations below 3×10^{-5} M and is favored by complexing agents which decrease the ion activity of Mn^{2+}.

Thus, the reactions of microbacterial oxidation of transition metals (biocatalytic oxidation) occur in complete agreement with the principal laws of chemical transformations (Bruce and Benkovich, 1970; Jencks, 1969; Thomas and Thomas, 1967). For the phenomena of microbacterial oxidation of transition metals, the mechanisms of chemical transformations is likely to be based on the reactions of autocatalytic oxidation considered above. Moreover, the autocatalytic oxidation is the leading process in the formation of oxide metalliferous sediments.

It is supposed that in natural environments the autocatalytic oxidation is the main process of the phenomena discussed (Morgan and Stumm, 1965; Varentsov, 1976) and our own data is consistent with this assumption.

4.3.3.5. Sorption Synthesis of the Oxyhydroxide Phases of Mn, Fe, Ni, Co, and Cu on 7 Å MnO_2 (Birnessite)

These experiments are the continuation of the studies on the synthesis of oxyhydroxide compounds of a number of transition metals (Mn, Fe, Ni, and Co) on hausmannite considered above. The conditions of these experiments were identical.

It is important to emphasize that the most distributed Mn compounds in ore accumulations of different environments of the supergene zones, as well as the modern basins, are essentially represented by 7 Å MnO_2 (birnessite), 2.4 Å MnO_2 (vernadite), and 10 Å MnO_2. Hausmannite (hydrohausmannite) is found among supergene Fe-Mn ores comparatively more rarely than the minerals mentioned above. In the environments of the distinct oxidation regime the hydrohausmannite is usually a metastable compound, transformed into MnO_2 modifications. Owing to such distribution features of the studied phases, the problem of how specifically the processes of the sorption synthesis of the studied transition metals on these two oxyhydroxide Mn minerals (Mn_3O_4 and 7 Å MnO_2) proceed inevitably appears.

We have every reason to assume that the considered sorption interaction is controlled mainly by the chemical properties of the component-bearing solutions: pH and Eh values, concentration, set of sorbed metals, state of their occurrence, composition and the component contents of the background electrolyte, kinetic parameters, that jointly and considerably determine the character and the charge value of the solid phase surface, i.e. the active functional groups:

$$- \text{Me-OH} \quad , \qquad =\text{Me}\overset{\textstyle /\text{OH}}{\underset{\textstyle \backslash\text{OH}}{}} \qquad \text{or} \qquad =\text{Me-O}$$

We can assume from this assumption that the essence of the mechanism of chemisorption process during sorption of the considered transition metals is common for the sorbents represented both by hydrohausmannite and birenessite. The differences in values of the specific surface, distribution features of the active centers, functional groups, and the surface chemistry may be manifested in kinetic characteristics of the process and the features of interactions of the sorbed ions with the components of these sorbent structures.

The experimental data obtained on the sorption of Ni, Co, Cu, and Fe from seawater and other mineralized solutions by biressite allow the following conclusion.

1. The sorption of Ni, Co, and Cu does not considerably vary during eight cycles of the experiment. The variation of the sorption ratio (in %) is relatively limited: Ni 15.38 to 50.00; Co 13.95 to 45.83; Cu 12.19 to 34.43.
2. The sorption of Fe during all eight cycles of the experiment proceeds almost entirely.
3. The sorption of the studied transition metals during the most part of the experiment was of a distinctly defined ion exchange nature. In the result of sorption of these cations the noticeable amounts of the hydrogen ions were released into solution, that was confirmed by noticeable reduction of pH of the final solution: ΔpH to 0.95 (third cycle). However, along with the exchange releasing of protons for the most stages during the second and fifth cycles of interaction was an insignificant alkalization of the final solution that indicates the ion exchange releasing of alkaline cations during the sorption of the given transition metals.

The study of the initial sorbent under transmission and scanning microscope shows that the used birnessite is characterized by the finest microglobular structure of grain aggregates. Such developed surface of the sorbent resulted in relatively high extraction of Ni, Co, Cu, and Fe during all eight cycles of the experiment. We may note that in the experiments described above with γ-FeOOH and Mn_3O_4 was a decrease of rates of sorption to the final cycles that are associated with considerable reduction of the specific surface of the sorbent, accompanied by the development of step-like microlaminated epitaxial growths of the newly formed layer on the sorbent. The comparison of the initial sorbent (7 Å MnO_2) and the final product under the transmission electron microscope indicates the appearance of distinct signs of enlargement, aggregation of the sorbent grains, and the development of the newly formed margins on their periphery, representing the sorbed layer.

The study of the final product with the Mössbauer spectroscopy shows that iron is present in the form of Fe(III). The obtained Mössbauer spectra do not practically differ from those for the iron compounds synthesized on Mn_3O_4. Thus, the compound of the FeOOH type was formed in the result of the Fe sorption by birnessite on its surface in the newly formed layer (Varentsov *et al.*, 1979a, 1988a).

The content of other transition metals in birnessite is relatively low, %: Ni 0.59; Co 0.59; and Cu 0.42. With such concentrations the known methods of the X-ray diffraction structural analysis and X-ray photoelectronic spectroscopy do not allow one to determine the phase composition of the newly formed compounds and the distribution of the valence states of atoms in the sorbed layer. In the separate experiment definite amounts of Mn (300 to 500 $\mu g/l$) were introduced into the system in addition to the metals, studied in the experiment (Ni, Co, Cu, and Fe) considered above. The sorption dynamics of these components in many features is close to the observed one for the other experiments of this series. At the powder X-ray diffraction patterns of the final product obtained in the experiments with 7 Å MnO_2 and with Guinier–de Wolff camera (FeK_α radiation and MoK_α, performed by R. Giovanoli, Bern University), the new lines corresponding to reflections of cryptomelane-hollandite (α-MnO_2) type are distinctly observed.

The experimental data on the synthesis of the oxyhydroxide compounds of some of the transition metals on 7 Å MnO_2 are well compared with the results of the described studies. They are satisfactorily described by the model of the process discussed above, consisting of two main stages of the interaction: (1) the sorption proper (ionic exchange and hydrolytic accumulation); (2) interphase autocatalytic oxidation of some transition metals, sorbed during the first stage.

4.3.4. THE EFFECT OF THE MAJOR IONS OF THE BACKGROUND ELECTROLYTE SOLUTION ON SORPTION OF THE HEAVY METALS

4.3.4.1. Formulation of the Problem

The processes of sorption extraction of heavy metals by manganese and iron oxyhydroxides from ground waters and the waters of recent basins play a most important role in the control of the main features of their geochemical behavior (Goldberg, 1954, 1961; Goldberg and Arrhenius, 1958; Murray and Brewer, 1977; Turekian, 1977; Vernadskii, 1954; etc.).

The results of the study of Fe–Mn ores, features of their occurrence, distribution, structure, mineral and chemical composition, and the data of their experimental research allow us to believe that the basic process in the formation of these ores is chemisorption with autocatalytic accumulation of heavy metals from natural waters or interstitial solutions.

A characteristic aspect of this process is an extremely high selectivity in the sorption of ultramicrogramme amounts of heavy metals from the solution of such a relatively strong electrolyte as seawater, or other mineralized natural waters.

The concentration of major cations (Na^+, Mg^{2+}) in standard seawater is 10^6 to 10^7 times higher than the concentration of heavy metals. In this connection there appears an unavoidable problem of the influence of the macrocomponents of such solutions (cations and anions) on the sorption of heavy metals by manganese and iron oxyhydroxides.

Numerous publications present rather conflicting data evaluating the influence of major ions of the background electrolyte solution on the process of sorption of heavy metals by manganese and iron oxyhydroxides.

The data currently available makes it possible to think that the phenomenon of sorption of heavy metals from electrolyte solutions is a rather complex multifactor process that is controlled by the following parameters:

1. the nature of the heavy metal that is sorbed, its form in the solution and its concentration;
2. pH of solution;
3. the nature of the sorbent, features of its surface chemistry, structure, electric and crystal-chemical characteristics, its relative amount in the solution;
4. the composition and concentration of major ions in the solution of the background electrolyte, molecular-kinetic properties of their hydration, features of the chemical reactions and complex formation with the surface of the sorbent and the sorbed metals;
5. the dynamics of the sorption process involved;
6. the temperature and pressure: when they change (for instance, in natural environments), these parameters may cause an essential effect on the sorption processes.

The aim of this study is to present experimental data on microamount sorption of heavy metals on the example of copper (II) by hydrated manganese dioxide from solutions of NaCl and the mineralized natural waters, like e.g. artificial seawater as a complex electrolyte, and to estimate the role of major ions in the process (Varentsov *et al.*, 1985).

4.3.4.2. Sorption of Cu(II) from Seawater and NaCl Solutions

Sorption of Cu(II) from seawater

The findings of the experiments on copper sorption by birnessite from seawater in the presence of complexing agent (citrate ion) are shown in Table 28 (Figure 85). The time of interaction between the solution and the sorbent varied from 10 min to 720 h (30 d). About 80% of the initial concentration of copper (about 100 μg/l) was extracted from the solution in 10 min. A maximum uptake of about 92% was reached in 24 h after which it was practically stable.

The dynamics of the process is shown by the change in the mean rates of Cu(II) sorption: from 480 μg/h during the first 10 min to 0.1 to 0.2 μg/h during the longest experiments (see Table 28). A certain decrease in pH of the final solution was observed in the process of Cu(II) uptake: ΔpH was 0.17 in the shortest experiment (10 min), while in the longest

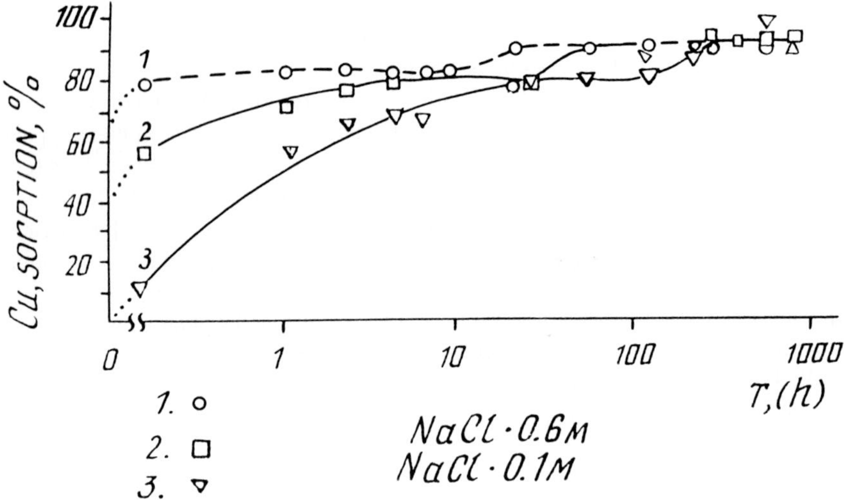

Fig. 85.   Dynamics of Cu sorption by manganese dioxide (synthetic birnessite) from electrolyte solutions: (1) artificial seawater; (2) 0.6 M NaCl solution; (3) 0.1 M NaCl solution (Varentsov *et al.*, 1985).

Table 28.   Cu uptake by birnessite from seawater in the presence of a complexing agent (citric acid, 5.2×10^{-6} M).

Exper. No.	Interaction time (h)	Initial solution pH_i	Initial solution Cu concent. ($\mu g/l$)	Final solution pH_f	Final solution Cu concent. ($\mu g/l$)	Cu extraction degree (%)	Mean rate of Cu uptake by birnessite ($\mu g/h$)	$\Delta pH = pH_i - pH_f$
1	1/6	8.14	96	7.97	19	80	480.0	0.17
2	1	8.12	94	7.93	18	81	81.0	0.19
3	2	8.12	110	7.92	18	84	40.5	0.20
4	4	8.12	100	7.91	19	81	20.2	0.21
5	6	8.10	90	7.88	16	82	13.7	0.22
6	8	8.12	100	7.86	18	82	10.2	0.26
7	18	8.04	94	7.80	10	89	4.8	0.24
8	24	8.10	100	7.90	8	92	3.7	0.20
9	48	8.13	90	7.90	10	89	1.8	0.23
10	96	8.10	92	7.85	10	89	0.9	0.25
11	192	8.13	92	7.88	10	89	0.5	0.25
12	240	8.13	90	7.84	9	90	0.4	0.29
13	480	8.10	100	7.85	8	92	0.2	0.25
14	720	8.11	110	7.81	10	91	0.1	0.30

Table 29. Cu(II) uptake by birnessite from 0.6 M NaCl solution in the presence of a complexing agent (citric acid, 5.2×10^{-6} M).

Exper. No.	Interaction time (h)	Initial solution		Final solution		Cu extrac-tion degree (%)	Mean rate of Cu uptake by birnessite (μg/h)	Δ pH = $pH_i - pH_f$
		pH_i	Cu concent. (μg/l)	pH_f	Cu concent. (μg/l)			
1	1/6	8.20	96	7.60	42	56	336	0.60
2	1	8.12	106	7.10	29	73	72	1.02
3	2	8.14	106	6.90	26	75	37.5	1.24
4	4	8.10	110	6.80	24	78	19.5	1.30
5	18	8.25	92	6.85	22	78	4.3	1.40
6	24	8.10	96	6.64	22	76	3.2	1.46
7	48	8.14	106	6.65	9	89	1.8	1.49
8	96	8.05	94	6.55	10	91	0.9	1.50
9	240	8.00	94	6.65	7	92	0.4	1.35
10	312	8.10	96	6.63	8	92	0.3	1.47
11	480	8.20	96	6.55	8	92	0.2	1.65
12	720	8.15	94	6.56	7	92	0.1	1.59

experiment (720 h) it reached 0.30 (see Table 28).

Of interest is a comparative evaluation of the participation of the major ions of seawater and Cu(II) in the ion-exchange displacement of protons from birnessite. While there are no differences in the change of pH with and without addition of Cu(II) in the early stages (to 96 h) of interaction between birnessite and seawater, the relative contribution of Cu(II) amounted to about 0.1 pH during the longer runs (more than 240 h).

It should also be noted that Mn discharge from birnessite in its interaction with major ions of seawater and in Cu(II) uptake differs quite negligibly, i.e., the displacement of structural Mn(III) into solution in the sorption of the microamounts of Cu(II) is so small that it is not pronounced against the background of a tangible (10 to 20 μg/l) ion-exchange displacement of Mn(II) by major cations of seawater from the sorbent involved.

Sorption of Cu(II) from NaCl solutions. The concentration of salts dissolved in seawater, and the ionic strength of seawater with due regard for the relative quantities of the ionic forms of its macrocomponents correspond approximately to a 0.6 M NaCl solution. It is accepted in chemical oceanography to simulate seawater properties with 0.5 to 0.6 M NaCl solutions (Horne, 1969). We used 0.6 and 0.1 M NaCl solutions in the experiments conducted to evaluate the influence of the major ions of seawater on Cu(II) sorption.

The 'pure water–birnessite' system, where Na^+ prevailed among the uptaken cations, showed increased peptization: the release into solution of fine stable colloid manganese dioxide occurring during experiments.

In the experiment with a 0.6 M NaCl solution, 56% of the initial amount of Cu(II) was

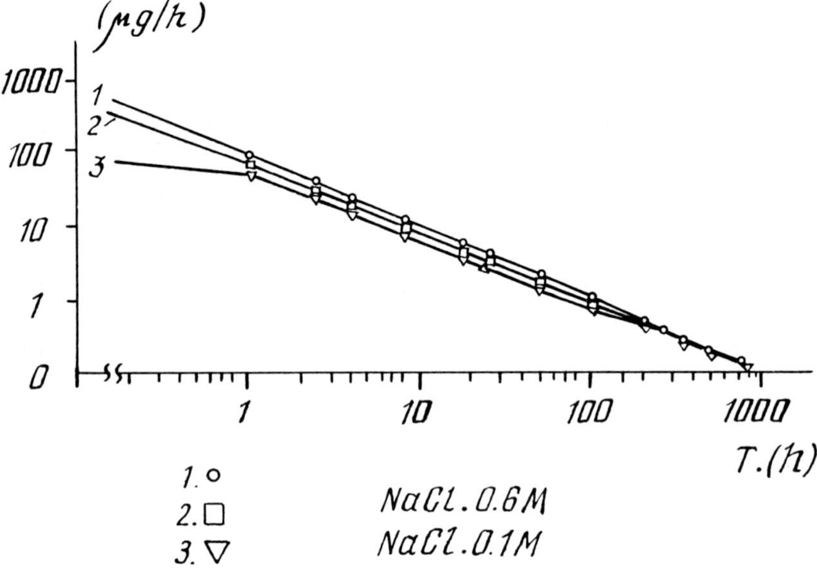

Fig. 86.   Dynamics of mean rates of Cu sorption (μg/h) by manganese dioxide (synthetic birnessite) from electrolyte solutions: (1) artificial seawater; (2) 0.6 M NaCl solution; (3) 0.1 M NaCl solution (Varentsov *et al.*, 1985).

extracted in 10 min, the initial concentration being 100 μg/l. A maximal sorption of 92% was reached in 240 h and it had not decreased by the end of the experiment after 720 h (Table 29, Figures 85 and 86). Just as in the experiments with seawater, the maximum value of the mean rate of sorption was registered within the first 10 min (336 μg/h); by the end of the experiment this value decreased to 0.1 to 0.2 μg/h.

Under similar conditions, the experiments with a 0.1 M NaCl solution showed that only 12% of the initial amount of Cu(II) was uptaken by birnessite in 10 min (Table 30, Figures 85 and 86). On the whole, Cu(II) sorption from this solution is characterized by appreciably smaller rates than from a 0.6 M NaCl solution. While 92% of Cu was uptaken from a 0.6 M NaCl solution in 240 h, the same amount of Cu was uptaken from a 0.1 M NaCl solution in 480–720 h of interaction.

It is striking in the experiments with NaCl that there is an essential, ion-exchange release of protons, compared with seawater solutions: ΔpH is about 1.0 in the early stages of the interaction (up to 1 h) and increases to 1.65 in 480 h (see Tables 28 to 30). Since NaCl solutions, unlike seawater, are not relatively stable buffer systems, the exchange release of H$^+$ ions in these experiments is observed very clearly. This noticeable decrease of pH in the earliest stages of interaction in the NaCl solutions caused a certain drop in the rate of Cu(II) sorption by birnessite at the later stages. Despite the values of pH change being comparatively close and the dynamics of ion-exchange release of protons in 0.6 and 0.1 M NaCl solutions being similar, there are considerable differences in the degree

Table 30. Cu(II) uptake by birnessite from 0.1 M NaCl in the presence of a complexing agent (citric acid, 5.2×10^{-6} M).

Exper. No.	Interaction time (h)	Initial solution		Final solution		Cu extraction degree (%)	Mean rate of Cu uptake by birnessite (μg/h)	Δ pH = $pH_i - pH_f$
		pH_i	Cu concent. (μg/l)	pH_f	Cu concent. (μg/l)			
1	1/6	8.05	92	7.05	81	12	72.0	1.00
2	1	8.10	88	7.00	40	54	54.0	1.10
3	2	8.25	88	7.05	33	62	31.0	1.20
4	4	8.10	94	6.85	31	67	16.5	1.25
5	6	8.04	92	6.90	31	66	11.0	1.14
6	18	8.16	106	7.10	20	81	4.3	1.06
7	24	8.14	90	7.00	19	79	3.25	1.14
8	48	8.12	100	6.80	21	79	1.6	1.32
9	96	8.05	90	6.55	20	78	0.8	1.50
10	192	8.14	106	6.85	17	84	0.4	1.29
11	240	8.20	90	6.70	8	91	0.4	1.50
12	480	8.25	90	6.60	6	93	0.2	1.65
13	720	8.20	90	6.55	7	92	0.1	1.65

of Cu(II) uptake and the mean rate of sorption during respective time intervals, and they reveal the role of the solution macroions with certainty.

Forms of Cu(II) in solutions. The forms of Cu(II) occurrence in solutions are important for the proper interpretation of the experimental data. Metal-ligand complexes in solutions can be characterized by their individual behavior in the sorption processes depending on their strength, charge, steric properties, relationship to the structure and physicochemical properties of the sorbent surface: they can be sorbed like a metal ion, a ligand, or a specific compound that are considerably transformed after their fixation.

We have mentioned above that the concentration of Cu(II) in our experiments amounted to about 100 μg/l (1.6×10^{-6} M), and citric acid was added as a complexing agent with 10 times the weight concentration, i.e., 1 mg/l (5.2×10^{-6} M). Approximations based on stability constants of Cu(II) complexes for such solutions as seawater show that at pH = 8 there can be a predominance of $Cu(OH)_2^0$ in this medium, there can be less Cu_{aqv}^{+2}, $CuOH^+$, and significantly less $Cu(C_6H_5O_7)^-$, $CuCl^+$, $(CuCO_3)^0$ (Balistrieri and Murray, 1982b; Reuter and Perdue, 1977; Ruppert, 1980).

The results of calculations for the forms of Cu(II) in 0.1 and 0.6 M NaCl solutions and in seawater are presented in Table 31.

The analysis shows that the main role in the studied solutions of NaCl and seawater belongs to hydroxy complexes, ionic forms, less so to citrate complexes of Cu(II). The content of the citric form in NaCl solution is 32%. However, it is important to bear in

Table 31. Forms of Cu(II) in 0.1 and 0.6 M NaCl solutions, artificial seawater at pH = 8, Cu(II): 100 $\mu g/l$, $H_3C_6H_5O_7$: 1 mg/l.

Cu(II) forms	Amount of Cu(II) forms in solution (%)			
	0.1 M NaCl	0.1 M NaCl	artificial seawater	0.1 M NaCl Cu(II) 40 $\mu g/l$
Cu^{2+}	8.75	8.62	13.00	8.80
$CuCl^+$	0.40	2.37	0.60	0.40
$CuCl_2^0$	1×10^{-2}	0.37	0.01	1×10^{-2}
$CuCl_3^-$	8.75×10^{-4}	0.20	1×10^{-3}	9×10^{-4}
$CuOH^+$	3.13	3.10	5.0	3.20
$Cu(OH)_2^0$	55.63	54.30	80.00	55.7
$CuCit^-$	31.80	31.20	0.10	32.0
$CuHCit^0$	2.6×10^{-3}	2.5×10^{-3}	6×10^{-3}	3×10^{-12}
$CuSO_4^0$	–	–	0.90	–
$CuHCO_3^+$	–	–	1×10^{-4}	–
$CuBr^+$	–	–	2×10^{-3}	–

mind that the relations between different forms of copper can change significantly in the dynamics of its sorption. During the initial stages (up to 8 h), predominantly $Cu(OH)_2^0$, $CuOH^+$, Cu^{2+}, $CuCl^-$ are actively uptaken (see Tables 28 to 31, Figures 85 to 87). Later, against the general background of sorption decrease, a certain activation of sorption can be seen in the sharp maxima of its rates at some stages (see Figure 87). This fact can be interpreted as an evidence of the development of copper hydroxide sorption layers and of the preserved domination in the solution of such forms as $Cu(OH)_2^0$, and to a lesser degree $CuOH^+$, Cu^{2+}, and $CuCl^+$. These forms can be actively sorbed by birnessite. Thus, when the amount of Cu was decreased down to 40 $\mu g/l$ in a 0.1 M NaCl solution, there was no essential change in the forms of this metal (Table 31).

pH of solution.   It has been established by now that the sorption of heavy metals by oxides is most active in a relatively narrow range of pH. A number of researchers believe that sorption of cations is very much similar to hydrolysis of metals in solution, since both sorption and hydrolysis intensify with an increase of pH, and both are accompanied by releasing of protons into the solution (Balistrieri and Murray, 1982b; James and Healy, 1972). However, in view of the known data this analogy seems to be of a somewhat formal nature, because it does not take into account the chemical properties of the sorbent surface and the reactions that may possibly to occur at the interface between the phases.

Very interesting for the comprehension of the processes of sorption of heavy metals by oxyhydroxides is the conclusion made by Balistrieri and Murray (1982b) on that the onset of sorption of heavy metals can be shifted towards lower pH with an increase of the relation of the solid phase to the solution, and towards greater pH with a decrease of

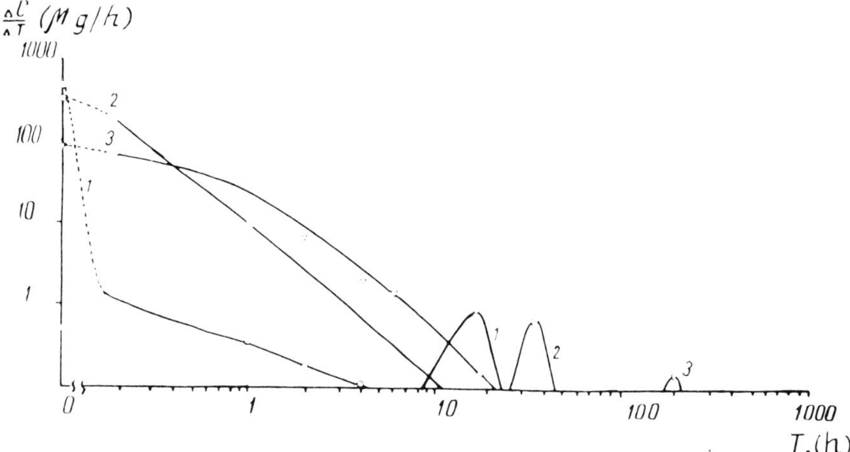

Fig. 87. Dynamics of mean rates of sorption of Cu (μg/h) by manganese dioxide (synthetic birnessite) from electrolyte solutions: (1) artificial seawater; (2) 0.6 M NaCl solution; (3) 0.1 M NaCl solution (Varentsov *et al.*, 1985).

the solid phase/solution relation. In other words, the relative concentration of protons for a certain amount of the solid phase (its surface) essentially influences the pH range of the onset of sorption of heavy metals because of the competition for active sorption sites with other components of the solution. This conclusion is true for static conditions of sorption. The data of our earlier studies on the dynamics of sorption, and the findings of the experiments described above, provide the grounds to believe that because of the great selectivity of uptake with respect to many heavy metals by oxyhydroxides in the systems with diverse solid phase/solution relations, the results of interaction, as a rule, are very similar in the long run.

Physicochemical properties of the surface of hydrated MnO$_2$. On the examples of δ-MnO$_2$ and 0.7 nm (7 Å) manganate-birnessite, a number of authors established in their research that pH of the point of zero charge (PZC) was less than 3.0 (within 1.40 to 2.25), the titratable charge of the surface was significantly higher than that of other oxides: 27.2 ($\pm$2.2) mol/kg for δ-MnO$_2$, 1.3 mol/kg for α-FeOOH, 10.9 mol/kg for Fe(OH)$_3$, and the value of the surface area could vary within 74 to 350 m^2/g depending on the techniques used to synthesize the phase and to determine the parameter (Balistrieri and Murray, 1982a, 1982b; Healy *et al.*, 1966; Morgan and Stumm, 1964; Murray, 1974; Loganathan and Burau, 1973).

The low value of pH(PZC) for MnO$_2$ modifications is the reason why a number of anions (Cl$^-$, SO$_4^{2-}$, etc.) cannot be sorbed on this phase in the pH range within 2 to 9 (Balistrieri and Murray, 1982a).

Unlike MnO$_2$, such typical representatives of iron hydroxides in various supergene environments as α-FeOOH (goethite) are characterized by pH(PZC) amounting to 7.5

(Balistieri and Murray, 1982a, 1982b). The findings of these researchers show that while MnO_2 sorption capacity in conditions of seawater for Na, Mg, Ca, and K is almost an order greater than goethite sorption capacity (recalculated for weights), the relative distribution of potential sorption sites (determined by the tritium exchange technique) for both hydroxides is approximately similar: 85 to 90% of surface sorption sites are occupied by protons and 10 to 15% of them predominantly by Mg (Balistrieri and Murray, 1982a).

For proper interpretation of the obtained data it is essential to bear in mind the main distinction between the surface properties of these hydroxides in seawater: anions are sorbed by the hydroxides of iron but not of manganese, which results from the different values of pH of the point of zero charge (pH(PZC)).

4.3.4.3. *The Role of the Major Ions of the Background Electrolyte Solution*

It has been shown in the sections above (see Tables 28 to 30, Figures 85 to 87) that birnessite can sorb essentially different amounts of Cu(II) from artificial seawater, 0.6 and 0.1 M NaCl solutions in the course of the same time periods.

Since Cu(II) sorption is accompanied by a release of protons into solution (see the drop in pH, Tables 28 to 30), which can have its effect on the intensity of copper sorption during later periods, it is expedient to compare the value of sorption in the course of the first 10 min of interaction, when the initial solutions are of very similar pH. These are the stages that are characterized by the maximal rates of sorption (see Tables 28 to 30, Figures 86 to 87).

During the first 10 min the degree of Cu sorption amounts to 12, 56, and 80% of the initial quantity, and the rate of sorption was 72, 336, and 480 μg/h for 0.1 and 0.6 M NaCl solutions and seawater respectively. In other words, the intensity of sorption in this sequence increases as 1, 4.7, and 6.7, taking 0.1 M NaCl solution as unity. The calculations that have been carried out allow us to believe that Cu(II) forms in the studied systems, in particular 0.1 and 0.6 M NaCl solutions, differ insignificantly: the predominant ones are hydroxy complexes, Cu^{2+}.

A relatively easy comparison of Cu(II) sorption from 0.1 and 0.6 M NaCl solutions show that Cu(II) sorption increases with the rise in the concentration of this electrolyte. Still more pronounced is the increase in the sorption of copper if we take seawater, ions of Mg^{2+} and SO_4^{2-} are also present.

The available data on the relations between sorption of Cu, Pb, and other heavy metals on hydroxides of manganese, iron, and other substrates with a change in the concentrations of the components of the background electrolyte are very contradictory (Balistrieri and Murray, 1982b; Beevers, 1966; Benjamin and Leckie, 1981; Chuiko and Gavrilyuk, 1969; Gavrilyuk, *et al.*, 1970; Idzikowski, 1971, 1972a, 1972b; Makatova and Imanalieva, 1977; Murray, 1975; Novikov, 1972; Novikov and Goncharova, 1972; Novikov and Knyazev, 1977; Novikov *et al.*, 1977; Swallow *et al.*, 1980; Tewari *et al.*, 1972; Volkhin *et al.*, 1979).

Without going into an analysis of each publication's conclusions we can say that they tend to fall into three groups:

1. sorption of microamounts of heavy metals gains essentially with the rise in the con-

centration of the major ions of the background electrolyte;

2. there are no indications of a competition between ions of heavy metals and the electrolyte components;
3. sorption of heavy metals is suppressed by the ions of electrolytes.

This complicated and contradictory data on the role of major ions of the background electrolyte can be adequately interpreted considering the following:

1. Structural features of the electrical double layer at the solution/sorbent (oxyhydroxide) phase interface.
2. Modern concepts on the structure of aqueous solutions of electrolytes, on hydration of ions, structural features of water near the solution/solid phase interface.
3. Features of sorption of Cu(II) and other heavy metals, their forms in solution, relations between various kinds of sorption: ion exchange, physical sorption, specific sorption, chemisorption, etc.

Evidently, these characteristics of a sorption system are not manifested separately while determining a certain reaction mechanism.

Structural features of the electrical double layer at the solution/oxyhydroxide interface. The model of electrical double layer (Bockris *et al.*, 1963; Grahame, 1947; Lyklema, 1968, 1971) is very efficient for the comprehension of the sorption processes. Unlike mercury, the electrical double layer on the surface of most oxides is produced owing to adsorption of potential-determining ions. Ions H^+ and OH^- are potential-determining for all oxides in a wide range of pH, with due respect for the individual features of the oxides. Typically, most oxides are able to adsorb protons at low pH, producing a positively charged surface; at higher pH the surface becomes negatively charged due to adsorption of hydroxyl groups.

Lyklema (1968, 1971) described the features of electrical double layer of oxides Al_2O_3, Fe_2O_3 (hematite, goethite), MnO_2, ZnO, TiO_2 (rutile), SnO_2, and SiO_2 (various modifications), and others; it is shown that unlike metallic Hg and Ag(I), a characteristic property of oxides is a steep rise of the surface charge with a gain in pH. For most of the diverse oxides, a general regularity was established: the greater the surface porosity (in a wide sense, including colloformity, gel-like quality), the greater the surface charge (δ_0) at given pH, where Δ pH = pH $-$ pH^0, pH^0 corresponding to pH of the point of zero charge. The absolute value of the surface charge is related to the density of OH-groups on the surface, and the maximal surface charge is determined by the general number of OH-groups per surface unit. This author emphasizes, as an essential point, the fact that the experimentally measured surface charge of oxides with a relatively high porosity of the surface can exceed the charge produced by complete dissociation of all hydroxyl groups on the surface without taking porosity into consideration.

These conclusions laid down the foundation of the concept of Lyklema (1971), the essence of which is in that the surface charge, as well as the countercharge, is not limited by the surface proper, and can be located in the solid phase, at a distance behind the surface, since the surface layer is 'porous' and permeable for ions.

This experimentally substantiated model of porous double layer explains the most important features of sorption on oxides:

1. the more permeable such a layer is for ions, the greater the charge density (per unit of surface) that can be reached;
2. sorption is not limited by the surface proper, and the values of experimentally measured charges were more than the value of charges corresponding to the number of OH-groups on the surface;
3. since counterions can also pass through the surface, the charge and the potential of the double layer from the side of solution remain low.

An extremely important problem in the use of this model is the depth to which counterions can penetrate and their affinity for the solid phase.

It has been established that porosity decreases exponentially with the distance from the surface; it is shown on the example of highly selective sorption of Li^+ (as compared with K^+ and Cs^+) on $\alpha\text{-}Fe_2O_3$ (hematite) that the ionic radius of lithium 0.068 nm (0.68 Å) is almost identical with the ionic radius of Fe 0.064 to 0.067 nm (0.64 to 0.67 Å). The author believes that Li^+ ions can occupy positions of Fe^{3+}. At the same time, the solid phase in this case appears to be permeable with respect to counterions but not water molecules. It means that other ions prove to be too large to occupy a position of Fe^{3+}, and the distribution of the components beyond the limits of the system under consideration is governed by the features of the Stern layer.

Ideas similar to the conceptual model of Lyklema (1971) were also developed by Perram *et al.* (1974). These authors showed that the model of electrical double layer of Gouy–Stern–Grahame, with distinct plane interface and the surface potential described by the Nernst's equation, cannot be applied to the oxide/solution system. Improving the concept of porous oxide surface (Lyklema, 1971) the authors provide the data on the boundary gel layer on the surface of oxide. The structure of the interface with a gel layer 2.0 to 4.0 nm thick was determined quantitatively for oxides with high values of surface potential: TiO_2, Fe_2O_3, SiO_2, Al_2O_3, etc. The presence of a relatively thin surface layer that is distorted, deformed due to penetration of water and various ions, is a characteristic feature of the oxide/solution interface.

Benjamin and Leckie (1981) cite very interesting data on a model of multiple heterogeneity of the adsorption sites on amorphous iron oxyhydroxide ($Fe_2O_3 \cdot H_2O_{am}$) in phenomena of a sorption of Cd, Cu, Zn, and Pb ions. It is emphasized that the surface of this sorbent possesses several types of active sites with different affinities for adsorbed ions. It is indicated in the paper that the estimates of the charge/potential relationship of the surface area can vary depending on the selected model of electrical double layer.

Though not directly, the authors write that they made use of the Stanford generalized adsorption model, i.e., the general model of electrical double layer with related active sites (Davis and Leckie, 1978; Davis *et al.*, 1978), that had been worked out on 'model' colloids, for instance, quartz, rutile, goethite, etc. In the opinion of the authors, these are non-porous, crystalline oxides, and their phase interfaces can be described in terms of plane, non-porous models.

The concept of 'porous electrical double layer' was not considered in the model of surface complexing (Stumm *et al.*, 1970). However, it is known that crystalline dispersed oxides with undisturbed non-porous surfaces occur very rarely in natural sedimentation basins. As it has been noted above, the most important role in other models elaborated for oxides (Lyklema, 1971; Perram *et al.*, 1974) belongs to the porous or gel layer. The basic (experimentally supported) postulate of these models is that the surface charge and a portion of countercharge are localized in the surface, external, porous part of the solid phase, since the surface layer (or the gel layer) is permeable for these ions. Consequently, there is a chance for the distribution of greater surface charge with a low electric potential and small charges of the diffuse layer.

4.3.4.4. *Structural Aspects of Electrolyte Aqueous Solutions, Ion Hydration, Salting out; Structural Properties of Water near the Solution/Solid Phase Interphase*

Structural aspects of electrolyte aqueous solution and ionhydration. The most essential manifestation of hydration is the interaction of ions with water molecules accompanied by a transformation of water structure into solution structure. Two kinds of hydration are distinguished: near and far hydration. The essence of far hydration is that the field of ions polarizes water molecules surrounding it. According to Mikhailov and Syrnikov (1960), the electrostatic ion field can bring disorder in the structure of water, accompanied by the relative number of molecules occupying cavities in the water structure increasing.

The investigations carried out by Vdovenko *et al.*, (1967) show that water structure is destroyed, or, more exactly, disordered to a certain extent in an electrolyte water solution under the influence of ions. Similar inferences on a certain destruction in the model of double-layer hydrate shell are noted by Frank and Wen (1957), Frank (1965) and others.

The problems of near hydration are most essentially covered by Samoilov (1957a, 1957b, 1972, 1980) in the works on the molecular-kinetic theory of hydration. Taking into account the properties of an ion (its charge, radius, structure of the electron shell) and the parameters of solution, one can estimate the mean time of the presence of water in the nearest vicinity of the ion: that is determined by the change of the energy of near hydration of the ion in solution.

Two types of near hydration are distinguished. In one of them (positive hydration) the exchange of the molecules of solution water nearest to ions occurs less often than the exchange of water molecules in water, i.e., there is an effect of relative binding by the ions of the nearest molecules of the solution water (e.g., Li^+, Na^+, Ca^{2+}, Mg^{2+}, Fe^-, etc.).

In the other type (negative hydration) the molecules of water nearest to the ions exchange more often than molecules of water in water (e.g., K^+, Cs^+, Cl^-, Br^-, SO_4^{2-}, etc.).[*]

---

[*]It should be noted that there are contradictions in the estimations of the sign of ion SO_4^{2-} near hydration. Goncharov *et al.* (1967) showed that for SO_4^{2-} in diluted aqueous solution at 21.5°C it is typical that ΔSO_4^{2-} = -0.15. Making use of a magnetochemical method, Ergin and Kostrova (1969) established that the boundary between positive and negative hydration at 25°C, atmospheric pressure, and infinite dilution corresponds to an ion with crystal–chemical radius $r_i = 1.1$ Å. This is in agreement with earlier data obtained by Samoilov (1967). Note that SO_4^{2-} ionic radius is 2.30 Å (Krestov, 1973). Ions with such ionic radii r_i should certainly be characterized by negative hydration. However, Erdey-Gruz (1976), while analyzing the viscosity of electrolyte solutions, came to a conclusion that SO_4^{2-} ion holds the place close to Ca^{2+} ion. A calculation of the change in SO_4^{2-} partial mole volume carried out by Stunzhas (1979) reveals negative SO_4^{2-} ion hydration. The problem of establishing the nature of the relations between SO_4^{2-} ions and water molecules is a complex one, and was discussed by Musinu *et al.* (1982).

It has been established that the effect of ions on the translation of the nearest water molecules in solution can be seen by a change in solution viscosity. Negative hydration is characterized predominantly by a disordering effect in the structure of water with concomitant drop in its viscosity. Positive hydration effect is the reverse.

Analysing the data of the studies on dielectric constants of electrolyte solutions, Yastremsky and Samoilov (1963) showed that the introduction of a diluted organic component (for instance, methylene alcohol) into the solution brings about stabilization of the aqueous solution structure. Such an effect is similar in its action to a decrease in temperature, and is due to lessening of hydration (i.e., dehydration) of ions in solutions.

Investigations into the changes of entropy characteristics of both near and far hydration at various temperatures allowed Krestov and Abrosimov (1967) to show a difference on their relation to water structural changes. It was made clear that the effect of temperature on far hydration is unimportant. This parameter considerably influences near hydration: the higher the temperature, the lower is negative hydration, and the ions become positively hydrated when a certain temperature is reached.

There is conflicting data on the role of pressure on near hydration. Nevolina *et al.*, (1969) provide some data showing that the rise of pressure leads to destruction of the water structure accompanied by an increase in near hydration of sodium and potassium ions. This conclusion supports earlier data of Samoilov (1959) on the existence of certain critical pressures at which negative hydration of ions in solution gives way to positive hydration and favors the general findings that the destruction of the structure of free solvent intensifies solvation (Samoilov, 1967).

However, Horne (1963) draws a conclusion that a relatively high pressure does not only result in structural destruction of water in solution, but also leads to a decrease in ion near hydration.

The problem of relationships between anions and cations is very prominent in the works of Samoilov and his colleagues. Areas of a relatively elevated density of water molecules appear around cations with positive hydration, while anions are distributed in the areas of lowered density of water molecules. These relationships are reversal for cations with negative hydration, i.e., an intensification of cation hydration results in a displacement of anions from the vicinity of cations. Moreover, cations with positive hydration suppress the translation of the nearest water molecules, while negatively hydrated cations, on the contrary, intensify such motion.

These findings are in agreement with conclusions of Frank (1972) on the subdivision of ions into structure-reinforcing ones (with positive hydration) and structure-destroying ones (hydrated negatively).

Consequently, the molecular mechanism of near hydration in solution permits us to qualitatively describe a continuous sequence from relatively weak hydration to strong hydration, resulting in the formation of stable aqueous complexes. Numerous experimental studies using thermodynamic, magnetochemical, and spectroscopic methods confirmed the validity of Samoilov's model (Ergin, 1983).

4.3.4.5. Problems of Salting Out

Of great interest for the comprehension of heavy metal sorption in the light of the theory of ion hydration are the phenomena of salting out the components of electrolytes by electrolytes. Samoilov and his colleagues point out in their works that salting out is the process of decreasing near hydration of the salted out ions due to the effect of the salting out agent (Ionov *et al.*, 1969; Samoilov, 1966a, 1966b; Samoilov and Tikhomirov, 1960; Varshal and Sennyavin, 1964).

Nosova and Samoilov (1964) showed that while dehydration of salted out ions promotes their salting out and hydration promotes their salting in, the end result depends on which of these effects prevails in the solution with due regard for the overall action of the cations and anions of the salting out agent. Samoilov (1966a, 1966b) writes in his works on the theory of salting out that the molecular mechanism of salting out, in particular the change in the value of the potential barrier ($\Delta E_{s \cdot out}$), depends on the nature and composition of the salting out agent, as well as on the changes in concentration, temperature, and pressure. On the whole, the established dependencies of the molecular mechanism of salting out are valid for solutions of low or moderate concentrations, since with the approach to complete solvation the molecules of the solvent come into the hydrate shells of ions as fully as possible.

The experimental findings of Nevolina *et al.* (1969) are of essential interest for our understanding of the processes at significant depths in recent basins, or deep-seated ground waters. The data reveal that with the rise of pressure there occurs an essentially greater lessening of the distance between oppositely charged ions than between ions of the same sign, since in the latter case the electrostatic repulsion forces come into play, which brings about a predominance of the phenomena of hydration over those of dehydration, with a prevailing influence of anions of the salting out agent.

As it is emphasized by Samoilov, the structural theory of salting out is of a qualitative nature, and it can be used for the interpretation of natural phenomena only on a clear enough phenomenological basis. This mainly concerns natural electrolyte solutions of low or moderate concentrations (e.g., waters of inland basins, seawater, ground and karst waters, etc.) where the role of salting out agents belongs to major ions (macroions).

4.3.4.6. Structural Properties of Water near the Solution/Solid Phase Interface

Comprehensive research carried out has established with certainty that the structure of the boundary layer of water near the solution/solid phase interface is essentially different from the structure of water in the bulk of solution (see reviews by Deryagin, 1983; Drost-Hansen, 1969; etc.). Summarizing ongoing studies, Drost-Hansen (1969) suggested a three-layer structural model of boundary water. This author shows that near the surface of the solid phase (especially when it has pronounced polar properties) there is a layer of water with ordered structure. The ordering decreases exponentially with the distance from the surface in connection with lessening of the influence of the surface field and greater effect of the thermal motion. Between the structurally ordered layer and the bulk water (with an essentially different structure) there is a layer of disordered structure that is only several molecular diameters of water molecules thick. It is pointed out that the boundary water is characterized by a greater density than the bulk water. The gain in density can be related to filling the structural cavities by monomer water molecules (Samoilov, 1957), as well as to the existence of a disordered structure with more compact packaging of water molecules.

Later research (Schufle *et al.*, 1976) on further elaboration of the boundary water model, on the basis of studying temperature dependencies of surface conductivity, showed that the thickness of the layers with modified structure can vary from 0.05 to 0.2 μm.

Of interest is the clarification of the influence of various components of the system on the peculiarities of changes in water structure: the role of the nature of the solid phase, composition and concentration of electrolytes and temperature (Drost-Hansen, 1977). In the context of the problems under consideration, the estimation of the effect of the structural properties of water in the surface layers and the electrical double layer of the solid phase acting together is of interest. In a critical analysis of a relatively small number of experimental works, Drost-Hansen (1977) indicates that the electrical double layer, according to preliminary estimations, does not affect the formation of structural properties of the boundary water layer. However, the influence of the substrate through the process of 'energy delocalization' reveals itself in the protection of the boundary, specially structured, water layer from the impact of the thermal motion and the extensive development of structures of the bulk water.

In recent years, the most convincing data on the structural properties of boundary water layers was obtained in the course of studies of the viscosity of thin films with the use of various techniques (Deryagin, 1983). It has been established that the influence of a hydrophilic surface on viscosity (i.e., its appreciable increase) can be felt at distances up to 10 nm. In cases of highly hydrophilic surfaces, e.g., of crystal quartz, their structuring influence can reach as far as 20 to 30 nm. The action of structural forces of repulsion in electrolyte solutions can be followed in interlayers no more than 5 nm thick. As it has been noted, a characteristic feature of specially structured thin boundary layers of liquids is their relatively elevated density.

Radiospectroscopic research (NMR and EPR) on the structure of the boundary water layers (Mank and Ovchrenko, 1983) reveals some comparative data on the condition of water molecules in the near vicinity of the surface: they are in their intermediate form, between the states in the bulk liquid and in the solid. Proceeding from the analysis of NMR and EPR spectra, the authors identified the specific role of the exchangeable (sorbed) ions (copper in particular) in the formation of the boundary layers: when the exchangeable ions are localized on a solid surface, water molecules are found around them in a certain order. In order to develop our understanding of sorption phenomena, particular attention should be drawn to the findings of the authors obtained as a result of studying chemical NMR shifts of water protons in various salt forms of ion exchangers and chloride solutions. The authors found out that ions with both positive and negative hydration interact in a similar way with water molecules both in aqueous solutions and in the adsorbed phase of ion exchangers. It is emphasized that there are distinct changes in water structure near the phase interface as compared with water structure in the bulk solution.

However, it might be well to point out that the studies of Lyaschennko (1968, 1972,

1973) have established that water structure undergoes, relatively, the least possible changes when an electrolyte solution is obtained with a concentration of up to 6 to 7 molecular per cent. Similar concepts are presented by Mischenko and Poltaratsky (1976). These studies allow us to believe that water layers with modified properties cannot extend far from the surface into the bulk solution.

Unlike the indirect estimations of the relationships between the layers of boundary water and the electrical double layer (Drost-Hansen, 1977) that were mentioned above, Mank *et al.* (1983) state the example of their study of dispersed silicates with certainty that the boundary water layers are within the electrical double layer and 'apparently very much depend on the condition of the latter' (p. 130). Referring to the studies of Churaev (1973), Mank *et al.* (1983) (taking a metal electrode as the example) write that adsorption of neutral molecules increases with the rise in the population of the charged monolayer in the dense portion of the electrical double layer.

The authors registered anisotropy in the mobility of water molecules by means of EPR and NMR techniques at the distance of up to 20 nm from the surface. Boundary layers of two types were distinguished within this range: (a) layers 1 to 3 water molecules thick; (b) relatively thick layers (from tens to the first hundreds Å) with comparatively poorly changed structural characteristics.

Similar conclusions on the structure of the boundary water layers on model phases of Na- and Li-montmorillonite are presented by Tarasevich (1983). This author points out in his critical analysis the diversity in their methods of study that the ratio of molecules with broken hydrogen links in the boundary water layer 10 nm thick is greater than in the water bulk. The properties of boundary water may be most adequately explained proceeding from the existence of a destructured transition layer between the bulk water and the ordered layers of adsorbed water (Efremov *et al.*, 1980) that, on the whole, supports the validity of the Drost-Hansen's (1969) three-layer model of vicinal water.

Tarasevich (1983) stresses that in the interpretation of sorption phenomena it is extremely important that the viscosity of the water of boundary layers near Na- and Li-montmorillonite (layers 3 and 5 nm thick) is approximately one order greater than the viscosity of the bulk water; this can be seen distinctly in the difference between diffusion rates. The author provides quantitative parameters of the three-layer model of boundary water that he develops for the example of the above-mentioned layered silicites:

1. The structure of a 1.0 nm layer of strongly bound (adsorption) water is governed by the distribution of the active sites on the basal and side faces of the mineral.
2. An intermediate 7 to 9 nm layer is formed under the influence of the field of the surface and the adsorbed water layers; comparatively greater amounts of water molecules with broken hydrogen links and less concentration of exchangeable cations are characteristic for this layer.
3. The formation of the third layer (15 to 20 nm) of osmotically bound water is caused by hydration of exchangeable cations that are within the diffuse portion of the electrical double layer.

4.3.4.7. Features of Uptake of Cu(II) and Other Heavy Metals: The Relationships between Various Kinds of Sorption

A general term 'sorption' is used in this work since there are many cases when it is rather difficult to determine the specific type of uptake that takes place: adsorption, ion exchange, or chemisorption. Moreover, considering the dynamics of sorption of heavy metals on the example of Ni, Co, Cu, etc. (McKenzie, 1980; Varentsov and Pronina, 1973; Varentsov *et al.*, 1979), one becomes convinced that one kind of uptake substitutes another with time. In particular, the ion exchange of the early stages becomes replaced by chemisorptive (or catalytic) accumulation of the uptaken metal at the later stages. The widely used term 'specific sorption' is related to such a type of uptake that cannot be described on the basis of the interaction between charges. This can be illustrated by an example from Grahame, *et al.*, (1952) showing that the specific sorption of halide ions (I, Br, Cl, and F) on the surface of mercury is caused by the covalent bond of the anions to mercury. However, this inference is disputed by the studies (Bockris *et al.*, 1963) that regard the degree and type of hydration as the main factors defining specific sorption.

It has been established that specific sorption occurs only within the Stern layer and the diffuse portion of the electrical double layer nearest to the surface (the external Helmholtz plane); however, non-specific sorption takes place both in the diffuse and in the fixed part of the double layer.

Specific sorption is to no less extent characteristic for cations. According to Bockris *et al.* (1963), at a given charge cations reveal a greater tendency to specific sorption than anions of the same radius. However, these conclusions were drawn on the grounds of the study of sorption phenomena for the mercury/solution phase interface that is essentially different from the system oxide/solution of Lyklema (1968, 1971).

Another example of specific sorption that is closer to the considered systems is the uptake of heavy metals on negatively charged surfaces of clayey particles. It is noted in one of the earlier publications dealing with this phenomenon (Elgabaly, 1950) that besides ion exchange reactions in zinc uptake, strong binding of zinc ions to montmorillonite occurs in such a manner that this sorbed metal does not participate in ion exchange under the action of ammonium acetate.

A striking example of specific sorption is the uptake of cobalt into montmorillonite (Hodgson, 1960) in the presence of a large excess of calcium chloride. The uptake reveals itself in two specific reactions, one of which is related to chemisorption, the other occurs in connection with the introduction of cobalt ions into the cavities of the crystal lattice that are formed following isomorphous substitution of $Si(IV)$ and by $Al(III)$. The cobalt fixed in this way does not exchange.

Benjamin and Leckie (1981) draw particular attention to the specific, very high selectivity of heavy metal sorption (on amorphous iron oxyhydroxide) as compared with that of alkaline and alkaline-earth metals (in particular Ca and Mg) that are present in solution in much greater concentrations. The authors explain this selectivity of uptake by the solid phase ($Fe_2O_3 \cdot H_2O_{am}$) possessing a wide variety of active sorption sites with different affinity for various metals.

Despite the reliability of the cited experimental data, the interpretation of the physical background of this phenomenon would be more convincing if spectroscopic studies of the sorbed layers were undertaken.

It has been established that ion exchange occurs only with counterions of the double layer, while coions play an insignificant role in exchange reactions if they are non-specifically sorbed.

Most of the publications indicate that at least a portion of, if not the entire, specific sorption can be regarded as chemisorption. However, in a number of cases the criteria for unambiguous detection of chemisorption phenomena are rather conditional. It is widely accepted that chemisorption is the process of producing chemical bonds between particles of a gas or a liquid and the surface of a solid or a cluster (Smith, 1980). It should be noted that experiments in chemisorption research are considerably advanced leaving the theory behind. This is due to the appearance of the techniques of photoemission spectroscopy, diffraction of slow electrons, and a number of spectroscopic methods to obtain reliable characteristics of chemisorbed layers. This is particularly characteristic for chemisorption on oxides of transition metals (Wolfram and Ellialtioglu, 1980).

These authors show that the concepts of 'active' sites on the surface are often used to explain chemisorption and catalysis on oxides. It is assumed that these sites interact with the reacting molecules producing either a 'surface complex' or the end product of chemisorption, or an intermediate product of catalysis. It is conditionally supposed that the active sites are either coordination-unsaturated surface ions of the transition metal or such defects as oxygen vacancies. It is very important to have a clear understanding of the nature of the active sites, but in most cases this is unknown. Nevertheless these authors write that it is generally accepted that an important part in the processes of chemisorption and catalysis is played by d-electron surface states and the states of various surface defects (for instance, surface oxygen vacancies). There are numerous coordination-unsaturated d-orbitals on the surface of perovskites. It follows from considerations of geometry and energy that at the points of localization of these so-called free links there are favorable conditions for chemisorption of molecules and the transfer of charge between the solid and the interacting molecules. The authors especially emphasize the role of the electron structure of the states of surface oxygen vacancies (SSOV). When bonds are formed between SSOV and molecules, there can occur transitions of electrons from filled SSOV to adsorbed molecules. And vice versa, unfilled SSOV can receive electrons from electrophobic substances. This means that both donors and acceptors of electrons can take part in reactions with a transfer of SSOV charge.

In the light of the presented data, the features of Pb(II) uptake on α-Al$_2$O$_3$ can be seen in that the prevailing form of Pb(II) fixed on the surface at pH $\geq$ 5 is a hydrolytic complex (a monohydroxy complex) SO$^-$PbOH$^+$, where SO$^-$ is the functional active group of the surface (Davis and Leckie, 1978). These authors point out that adsorption of monohydroxy complexes usually starts near those pH values at which hydrolysis of metal ions begins. The conclusions drawn from the example of uptake of Pb(II) on iron hydroxide (as well as on other hydroxides) hold true with respect to sorption of Cd(II), Cu(II), and Ag(I); these metals are sorbed mainly in the form of monohydroxy complexes. However, Davis and Leckie (1978) make an important remark: they think that a thorough investigation into the kinetics of surface reactions is needed in order to determine whether hydrolysis occurs first in the solution or if this reaction takes place on the surface following sorption of metal aqueous complexes.

This remark is singularly essential because the studies of the dynamics of sorption of heavy metals on hydroxides (McKenzie, 1980; Varentsov and Pronina, 1973; Varentsov *et al.*, 1979, 1980) allow us to distinguish the following three main stages:

1. the initial stage, when the uptaken metals are fixed to a considerable extent, in the exchange form;
2. the hydrolytic stage, when the hydrolytic reactions occurring on the surface result in the formation of strongly bonded hydroxy complexes of transition metals;
3. following the processes of dehydration, the formation of oxyhydroxide forms of the sorbed metals, as a sorption layer, on the surface of the sorbent.

It should be stressed that these processes in many cases proceed on low-crystalline, amorphous phases. This allowed Swallow *et al.*, (1980) to use the terms 'surface' and 'adsorption' somewhat vaguely, taking uptake of Cu and Pb by amorphous iron hydroxide as an example. They believe that ions of heavy metals can diffuse freely into the structure of the sorbent and the process is not restricted to external surface localizations. The loose, highly hydrated structure of hydrous ferric oxide can be penetrated easily by ions of heavy metals that become incorporated into the solid in the process of sorption when they are hydrolyzed. Phases that contain ferrites can form as the result of such interaction.

Identification of newly-formed sorption phases and determination of the appropriate characteristics of their crystal lattice provide reliable proofs of the chemisorption processes on the sorbent surface. Appelbaum and Hamann (1980) indicate that diffraction of slow electrons is a very efficient method for identification of chemisorbed phases on semiconductor surfaces. These authors point out that these surface phases are lattices coordinated with the structure of the substrate, i.e. an elementary cell of the adsorbent is a multiple of an elementary surface cell that corresponds to the ideal volume lattice of the substrate.

Identification of mixed-layer phases presented by disordered alternating interlayers of manganese dioxide and oxide of cobalt or other heavy metals in oceanic Fe–Mn nodules (Chukhrov *et al.*, 1983a, 1983b) can be considered as substantiation of the existence of the third stage of the discussed mechanism in the natural environment. This conclusion is indirectly supported by Davis and Leckie (1978), when they write that for many dissolved metals the chemical component of the binding energy with active sites of oxides is approximately the same as from one oxide to another. However, these authors note that with the present level of knowledge on the nature of the bonds it is impossible to correlate the differences in the stability of surface hydroxy complexes and free aqueous complexes (the former are considerably greater) with physical or chemical binding energies. They explain that the heavy metal ions are considerably easier to hydrolyze on the surface than in the bulk water due to a change in the entropy of this environment. In conclusion Davis and Leckie (1978) remark that chemical models can be more practical for several other areas of application, e.g. for modeling adsorption of metals dispersed in waters.

4.3.4.8. On the Effect of Major Ions of Electrolytes on Cu(II) Sorption by Manganese Oxyhydroxides

In the previous section we have considered the concepts that are essential for the clarification of the effect of major ions of electrolytes on sorption of heavy metals.

It has been pointed out that a characteristic feature of manganese hydroxides, as well as of a number of other hydroxides, is a relatively well-developed porosity of the surface layers of these particles. Taking this into account, Lyklema (1968 and 1971) showed experimentally that the surface charge of such substances essentially exceeds the charge that would be produced by surface active sites (predominantly OH-groups) only. It means that the surface charge, just as the countercharge, is not restricted to the surface proper, but can be located in the solid phase off this surface, since the surface layer is porous and permeable for ions. The model of an electrical double layer elaborated by the author on this basis allows us to explain the most important features of sorption on hydroxides:

1. the more this layer is permeable for ions, the greater is the charge density (per unit surface);
2. sorption is not limited to the surface only;
3. since counterions are capable of passing through the porous surface into the solid phase, both the charge and the potential of the double layer on the side of solution are relatively small.

Lyklema stresses that this porous double layer can be permeable for certain ions that possess high enough crystalchemical affinity for the solid phase, but not permeable for water molecules. It is essential that the distribution of the components (including major ions of electrolytes) that are beyond this system is governed by the features of the Stern layer.

The studies by Perram *et al.* (1974) supported the structural concept of such a boundary 'gel layer' on the surface of oxides whose thickness for oxides with high values of surface potential was estimated to be about 2 to 4 nm (Figure 88).

Hydration of ions with a transformation of water structure into solution structure is a characteristic feature of electrolyte solutions. Two types of hydration are distinguished: near and far hydration (Blokh, 1969; Frank, 1965; Frank and Wen, 1957; Mikhailov and Syrnikov, 1960; Samoilov, 1957a, 1957b, 1972; Stunzhas, 1979; Vdovenko *et al.*, 1967).

Near hydration is most significant for the understanding of the structure of solutions where ions that are positively hydrated, such as Li^+, Na^+, Ca^{2+}, Mg^{2+}, etc., and negatively hydrated, such as K^+, Cs^+, Cl^-, Br^-, SO_4^{2-}, stand out. It has been established that positively hydrated ions are structure-reinforcing ones, increasing the viscosity of solution. Negatively hydrated ions possess structure-destroying properties, they lower viscosity, increase the electric conductivity of solution, and intensify the diffusion of particles.

The presence of a dissolved organic component (e.g., methylene alcohol) in solution provides a stabilizing effect on the structure of aqueous solutions, which is similar to the effect of a temperature drop that results in a lesser hydration of ions (Yastremsky and Samoilov, 1963). This point is essential in the interpretation of the features of viscosity and structure of such an electrolyte as seawater, where dissolved organic substances are

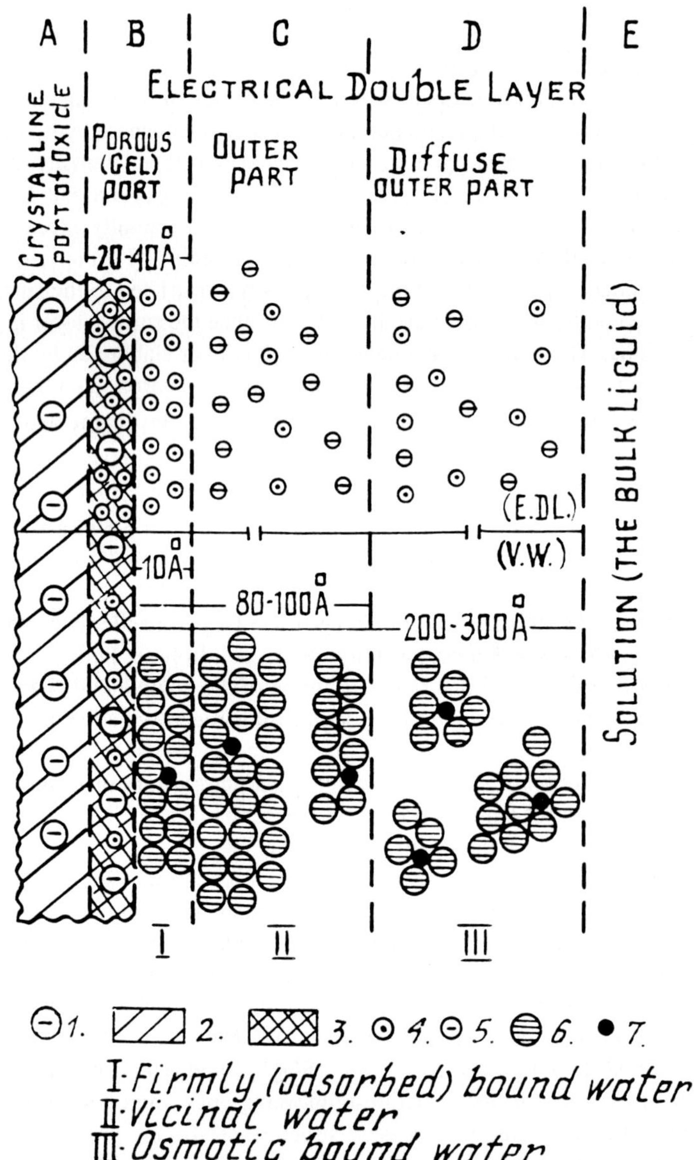

Fig. 88.   Diagrammatic representation: – (E.D.L.) – 'porous' electrical double layer (Lyklema, 1971) – 'gel' layer (Perram *et al.*, 1974) and – (V.W.) – layers of vicinal water around oxides (hydrate shell) (Deryagin, 1983; Drost-Hansen, 1969; Mank *et al.*, 1983; Tarasevich, 1983). I. Strongly (by adsorption) bound water; II. Structurally bound water of the boundary layer; III. Osmotically bound water. (1) Oxygen ions of the sorbent structure; (2) internal crystalline-portion of the sorbent; (3) external porous portion of the sorbent; (4) positively charged ions; (5) negatively charged ions; (6) water molecules; (7) ions, structure of the electrical double layer; 'A = crystalline portion of the sorbent; 'B' = porous portion (20 to 40 Å); 'C' = external portion; 'D' = diffusion portion; 'E' = bulk solution (Varentsov *et al.*, 1985).

invariably present (up to several mg C_{org}/l).

At the same time, electrolyte ions can exert a salting out effect with respect to the heavy metals of the system (Samoilov, 1966a, b), which can bring about appreciable dehydration of the sorbed components and a rise in their effective concentration.

The findings of the studies of the structure of the vicinal water layer near the solid phase/solution interface (Deryagin, 1983; Drost-Hansen, 1969, 1977), i.e., the structure that is essentially different from water structure in the bulk solution, are of outstanding importance for the comprehension of the sorption of heavy metals.

It has been established by means of NMR techniques (Mank *et al.*, 1983) on the example of chlorides that ions with both positive and negative hydration interact in a comparatively similar way with water molecules both in the bulk solution and in the adsorbed phase of ion exchangers in various salt forms of ion exchangers and electrolyte solutions. A distinct change in water structure is marked near the phase interface as compared with the structure of the bulk water.

Recent investigations (Efremov *et al.*, 1980; Tarasevich, 1983) corroborate the qualitative three-layer model of vicinal water developed earlier by Drost-Hansen (1969). Proceeding from the study of Li- and Na-montmorillonite, three main boundary layers are distinguished:
1. the structure of a 1.0 nm layer of strongly bonded (adsorption) water is governed by the distribution of the surface active sites of the solid phase;
2. the intermediate boundary layer (7 to 9 nm) is formed under the influence of the field of the surface and the adsorbed water layers;
3. the formation of the third layer (15 to 20 nm) of osmotically bonded water is caused by hydration of exchange cations in the diffuse portion of the electrical double layer.

Evidently, the three-layer model of vicinal water that is quantitatively substantiated on the basis of layered silicates can be applied to manganese hydroxides of layered structure, in particular to the 0.7 nm manganate-birnessite (Giovanoli, 1980) that was used in our experiments.

This three-layer model can also be valid for a large number of oxyhydroxides, e.g., for hydrophilic surfaces of crystalline quartz type (Deryagin, 1983) whose structuring influence on boundary water is estimated to be present within 20 to 30 nm (see Figure 88).

Consequently, the phenomena of sorption of Cu(II) ions that exist in electrolyte solutions (including seawater and other natural waters) mainly in the form of a hydroxy complex on 0.7 nm MnO_2 are, to a large extent, governed by the factors cited above. Owing to the developed porous structure of the electrical double layer, Cu(II) ions can be selectively hydrolytically sorbed both on the surface and in the porous portion of the solid phase (see Figure 88).

Major cations of the electrolyte (Na^+, Mg^{2+}, and Ca^{2+} in the case of seawater and other natural waters) can be fixed, as positively hydrated counterions, within the Stern layer and the diffuse portion of the double layer.

The presence of boundary water layers near the solid/solution interface that are characterized by elevated structural ordering, viscosity (by an order higher than that of the bulk water), and in many cases increased density (the second layer in particular) can impede a diffuse influx of Cu(II) ions.

However, the presence of negatively hydrated ions of electrolyte (Cl^- and SO_4^{2-}) reveals itself in their disordering effect on water structure. In other words, the presence of negatively hydrated ions in solution, primarily of such anions as Cl^- and SO_4^{2-}, results on the one hand, in a decrease of the viscosity of solution and promotes diffusion of heavy metals in its bulk, and, on the other hand, acts destructively on the boundary water layers and enhances the diffusive influx of these metals directly to the sorbent surface. It is appropriate to note here that Balistrieri and Murray (1982b) also noticed the promoting effect of SO_4^{2-} ions on Cu, Pb, Zn, and Co sorption by goethite from electrolyte solutions of seawater type.

At the same time, an appreciable role in the promotion of heavy metal sorption can belong to the salting-out action of the components of electrolytes in the bulk solution, in particular to the action of such positively hydrated ions as Na^+ and Mg^{2+} within a range of relatively moderate concentrations of, for instance, seawater.

The final stages of this interaction include sorption accumulation of Cu(II):

1. when on the surface (or in the porous portion of the solid), following hydrolytic reactions, strongly bonded insoluble hydroxy complexes of transition metals appear;
2. because of ageing, after this stage comes dehydration of oxyhydroxide forms of Cu(II) (or other heavy metals) in the sorbed layer.

It seems that such a mechanism allows for a consistent explanation of the phenomenon of a significant rise in the sorption of heavy metals on hydroxides, in connection with an increase of the electrolyte concentration within the range of their relatively low and moderate values characteristic for natural waters.

4.3.5. ON THE ROLE OF ORGANIC MATTER IN THE SORPTION PROCESS OF TRANSITION METALS

During the last decade, different aspects of geochemistry of organic matter in natural waters were studied rather thoroughly (Putilina and Varentsov, 1980). When studying the migration and concentration of a number of chemical elements, mostly heavy metals, Vernadskii repeatedly referred to the unique role of organic matter in the distribution of chemical elements. In fact, many elements, such as uranium, germanium, vanadium, molybdenum, iron, copper, manganese, and others, form complex compounds with organic matter in the waters of recent basins (Linnik and Nabivanets, 1977; Varshal *et al.*, 1979).

In the process of vital activity and posthumous decay, organic matter decomposes into more simple compounds. Horne (1969) gives a scheme of decomposition using the example of proteins.

The decomposition process can be accompanied by an intermediate synthesis of compound substances from simple ones. The most stable compound formed as the result of synthesis of products of decay is the carbonic-proteins complex which is a complicated polymeric compound composed of amino acids referred to as humus in literature; its solution is called 'humus water'. On average about 60–80% of dissolved and suspended organic matter consists of humus compounds (Reuter and Perdue, 1977).

Humus is divided into humus compounds, soluble in alkalies and insoluble in acids, and fulvo-compounds, soluble both in acids and alkalies. Chromatographic analysis of

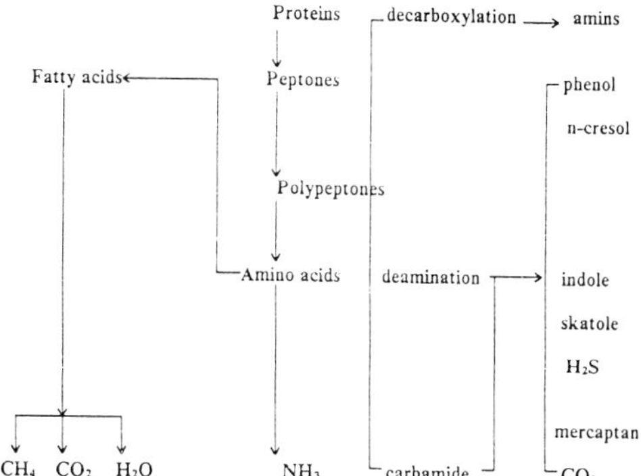

Decomposition of organic matter in the process of vital activity and posthumous decay using the example of proteins (Horne, 1969).

products of the hydrolysis of humus and fulvo-compounds revealed that up to ten kinds of amino acids and more take part in their compositions; these amino acids belong to basic, neutral, acid, and aromatic groups: lysine, histidine, glutamic acid, serine, proline, glycine, alanine, valine, isoleucine, and phenylalanine. Glycine, asparagine, and alanine comprise the greater part of amino acids (up to 44%).

Fulvo- and humus compounds in solution tend to associate and form polydispersion systems with the range of molecular mass from 300 to 6000 and more (Varshal *et al.*, 1979).

Humus compounds with high molecular mass can flocculate and accumulate on fairly shallow bottom. The nature of formation of compounds with high molecular mass is unknown as yet, and open to discussion. It is not ruled out that coextraction of compounds with low molecular mass occurs, or polymerization processes take place. Reuter and Perdue (1977) obtained data on the content of humus compounds in the rivers of the United States. The concentration of humus dissolved compounds is on average 3.4 ± 0.6 mg/l and that of suspended compounds is 3.8 ± 1.6 mg/l. However, higher and lower values are also known. Evidence is available which suggests that the content of humus compounds can be up to 100 mg/l (Reuter and Perdue, 1977).

Apart from humus and fulvo-compounds, natural water contains a great amount of other organic substances. Varshal *et al.* (1979) illustrate the most common organic compounds by the example of the water in the Moscow River (Table 32).

Table 32 can be supplemented by a number of organic compounds found in the surface waters of the States of California and Washington, U.S.A. (Hem, 1965). The identified constituents included acetic, propionic, butyric, valeric, catroic, lactic, malonic, maleic, succinic, and other carboxyl acids. The concentration of these substances varied from 4 to 592 μg/l; in organic substances containing nitrogen most of them belong to free amino acids: glycine, serine, ornithine, and alanine (Siegel, 1971).

The products of decay of organic substances may enter into reactions. Thorstenson

Table 32. Content of the major classes of dissolved organic matter (after Varshal *et al.*, 1979).

Class of natural organic substances	Concentration
Carboniferous and oxycarboniferous acids of aliphatic series: formic acid, proteins, oxalatic, citric, tartaric, etc.	0.01–10 mg/l
Humic acids	0.01–30 mg/l
Fulvic acids	1–100 mg/l
Alcohols	up to 2 mg/l
Aldehydes, ketones, polyfunctional compounds	0.05–2.2 mg/l
Phenols	0.001–0.060 mg/l
Polyphenols	up to 10 mg/l
Reducing sugars	0.1–0.2 mg/l
Polysaccharides	0.2–0.6 mg/l
Compound ethers, lipids. mono-, di-, and triglycerides of aliphatic acids (C_1–C_{18})	10–200 μg·equiv./l
Aliphatic amines	15–50 μg/l
Amino acids (20 varieties)	2–25 μg/l
Petrol products	0.05 mg/l
Polyaromatic carbohydrates	$(1.5–2) \times 10^{-4}$ mg/l
Pesticides	1×10^{-3}–1×10^{-1} g/l
Synthetic surface-active substances	0.5–2.0 mg/l
Phenols	1×10^{-3}–2.0 mg/l

(1970) studied the decomposition reactions of amino acids and carboxyl acids using the example of alanine and carbohydrate. He distinguished several types of reactions:

(a) oxygen reactions:

$$CH_2O + O_2 \rightarrow CO_2 + H_2O$$

$$4C_3H_7O_2N + 15O_2 \rightarrow 12CO_2 + 2N_2 + 14H_2O$$

(b) sulphate reactions:

$$2CH_2O + CO_4^{2-} \rightarrow CO_2 + HCO_3^- + HS^- + H_2O$$

$$2C_3H_7O_2N + 3SO_4^{2-} + H_2O \rightarrow CO_2 + 5HCO_3^- + 2NH_4^+ + 3HS^-$$

In the decomposition reactions under consideration, an essential part belongs to the equilibrium between SO_4^{2-}, HS^-, N_2, and NH_4^+:

$$3SO_4^{2-} + 8NH_4^+ \rightleftharpoons 4N_2 + 3HS^- + 5H^+ + 12H_2O$$

$$4H_2O + 2C_3H_7O_2N \rightleftharpoons 2NH_4^+ + 2HCO_3^- + CO_2 + 3CH_4.$$

4.3.5.1. The Interaction between Organic Matter and Metals

The interaction between organic matter dissolved in natural water and trace metals is a problem which has not yet been studied sufficiently.

According to the results, obtained by several researchers (Gardner, 1974; Tkach, 1974), simple amino acids and hydrocarboxyl acids are less active in the process of dissolution of metals than bisulphides and polysulphides. However, the principal role in dissolution of trace elements of metals belongs to such polymeric compounds as humic acids (Picard and Felbeck, 1976), because the stability of metal-humus complexes in natural waters is much higher than the stability of complexes of metal-ion-organic compounds (Reuter and Perdue, 1977). By the example of humus, Siegel (1971) has demonstrated its ability to bind copper and lead. Gardner (1974) showed that humus compounds bind Cd(II) to a greater extent than the major inorganic ligands CO_3^{2-}, SO_4^{2-}, Cl^-, OH^-.

Experiments on ion exchange testify that ions of Ca, Zn, and Mg are combined by humus matter in cation form, while iron and zirconium make anion and non-ionic complexes (Desai and Ganguly, 1970; Linnik and Nabivanets, 1977; Varshal *et al.*, 1979). Pillai *et al.* (1971) studied complexing between the metal ions and dissolved organic compounds; the obtained data give evidence that within 16 weeks 65% of amphoteric zinc, and also 70% of manganese, pass from the cation form to non-ionic form. In 4 weeks, 71% of Co passes to non-ionic form and remains at that state later. It was also noted that an extra amount of metal ions added to solutions, containing humus and other organic compounds, results in exceeding their solubility. For example, the study of iron solubility in the coastal waters of the State of Washington (Desai and Ganguly, 1970) revealed that the concentration of ionic iron is 4 μg/l, whereas the general concentration after oxidation by perchloric acid is 17 μg/l, implying that 75% of dissolved iron was in a bound form. Analysis of algal filtrate, containing a great amount of dissolved organic constituents (Desai and Ganguly, 1970), shows higher solubility of a number of elements: the amount of dissolved manganese is 5,779 μg/l, of zinc 380 μg/l, of iron 4.79 μg/l, of zirconium 0.78 μg/l. The study of some of the elements (Sc, Ti, V, Cr, Mn, Co, Ni, Cu, Mo, Au, Ag, etc.) in the ash of sea plants and in the coastal sediments of the Sea of Okhotsk testify that they are accumulated in amounts 10^2–10^6 times exceeding their average content in the waters of the World Ocean (Saenko *et al.*, 1977). The maximum accumulation coefficient for Mn is 1.8×10^5, for Ti 4.4×10^4, for Fe 2.4×10^4, for Cr 1.2×10^6 (Saenko *et al.*, 1975).

Picard and Felbeck (1976) studied the formation of complexes of iron with humic acid. It was observed that by adding humus to the solution the amount of iron in the precipitate decreases. Experiments, and the study of natural processes, brought the authors to the conclusion that humic acid facilitates the transportation of iron in seawater. This conclusion is in accordance with the results obtained by Rashid (1972a) and Koshy *et al.* (1969).

Duursma (1968) studied Co and Zn sorption in seawater. The sorption of these elements was reduced by adding leucine. The author attributes this effect to the formation of complexes between leucine and the studied elements, and emphasizes the necessity to take into account the concentration of elements and of organic matter. Complexing did not occur at low leucine concentrations. Only iron (Fe^{3+}) and chromium (Cr^{3+}), having high stability constants, easily form complex ions. Duursma (1968) believes that iron regulated the binding of all other metal ions at low concentrations of organic matter.

The work by Varshal *et al.* (1979) presents data on the nature, composition, and stability of complex compounds formed by metal ions, i.e. calcium(II), strontium(II), cerium(III), yttrium(III), iron(II) and (III), ruthenium(III) and (IV), copper(II) and some other elements, with fulvo-acids, polyphenols and other constituents of dissolved organic matter. Interaction between the metal ions and fulvo-acids is accompanied by the formation of a number of soluble cation, anion, and neutral complex compounds and of insoluble forms.

Three groups of metal ions have been distinguished according to their interaction with fulvo-acids. Cu(II), Sr(II), and Mg(II) form soluble complex compounds and are not hydrolysed at the pH of surface waters. Fe(III), Ru(III) and (IV) and other easily hydrolysed elements are ascribed to the second group. Solubility of these elements increases owing to complexing by fulvo-acids and, consequently, their migration ability. The formation of soluble complexes with associated forms of fulvo-acids is the principal feature of these elements. The third group comprises Ce(III), Y(III), and Yb(III), which form, depending on pH and metal-ligand, insoluble and soluble complex compounds. Solubility increases from cerium to ytterbium in the process of interaction with fulvo-acids of a definite molecular mass.

Linnik and Nabivanets (1977), in the course of their studies, determined different forms of manganese in natural waters. These authors show that the main amount of dissolved forms of manganese are represented by high molecular complex compounds. Between 8 and 15% of the amount of $Mn_{comb.}$ are cation complexes with molecular (ion) masses of 500 to 5000. In waters of the Dnieper River, manganese is combined in complexes of anion nature; their molecular (ion) mass is about 70 000, 15 000 to 5000, and 500 to 1500. In the waters of the Desna River, the complexes have molecular (ion) masses of 1500 to 5000 and 500 to 1500. This combination of metal ions, according to the data by Rashid (1972b, 1974) and Duursma (1968), belong to the reaction of chelation:

$$Me^{n+} + n\ H_2N\text{-}C\text{-}COOH =$$

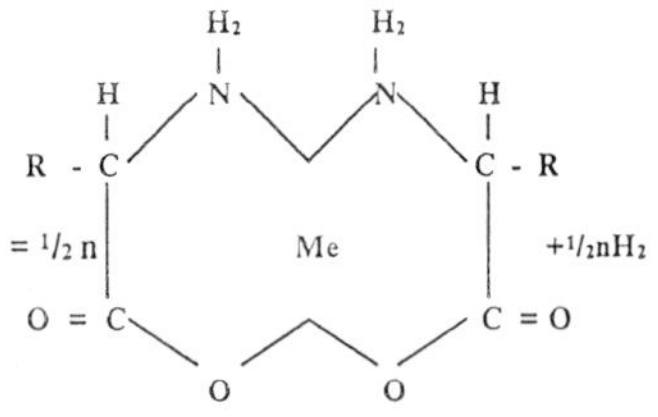

$$= \tfrac{1}{2} n\ \left[\begin{array}{c} \text{(chelate ring structure with two } N H_2 \text{ groups, R-C-H and C-R, O=C and C=O, O and O, Me in the centre)} \end{array} \right] + \tfrac{1}{2} n H_2$$

where Me^{n+} is a metal ion, R is the non-polar aliphatic group:

$CH_2NH_2\cdot COOH$	GLYCINE
$CH_3\cdot CH_2\cdot NH_2\cdot COOH$	ALANINE
$(CH_3)_2\cdot CH\cdot CHNH_2\cdot COOH$	VALINE
$(CH_3)_2CH\cdot CH_2\cdot CHNH_2\cdot COOH$	LEUCINE
$CH_3(C_2H_5)\cdot CH\cdot CHNH_2\cdot COOH$	ISOLEUCINE

or hydroxyl-containing aliphatic or aromatic group, aspartic, and glutamic acids, sulphur-containing group:

$$CH_2OH \cdot CHNH_2 \cdot COOH \quad \text{SERINE}$$

$$CH_3 \cdot CHOH \cdot CHNH_2 \cdot COOH \quad \text{THREONINE}$$

$$OH \cdot C_6H_4 \cdot CH_2 \cdot CHNH_2 \cdot COOH \quad \text{TYROSINE}$$

$$C_6H_5 \cdot CH_2 \cdot CHNH_3 \cdot COOH \quad \text{PHENYL ALANINE}$$

$$NH_2 \cdot CH_2 \cdot (CH_3)_3 \cdot CHNH_2 \cdot COOH \quad \text{LYSINE}$$

$$HS \cdot CH_2 \cdot CHNH_2 \cdot COOH \quad \text{CYSTEINE}$$

$$COOH \cdot CH_2 \cdot CHNH_2 \cdot COOH \quad \text{ASPARATIC ACID}$$

$$COOH \cdot CHNH_2 \cdot (CH_2)_2 \cdot COOH \quad \text{GLUTAMIC ACID}$$

$$C_4H_8N \cdot COOH \quad \text{PROLINE}$$

According to Wershaw (1976), lead forms two types of complexes. The first type is chelate complexes, where lead forms ionic compounds. Compounds, covalently combined with the atoms of carbon, can be regarded as the second type.

Besides the reaction of chelation, all the amino acids can have separate reactions in amino and carboxyl groups (Williams, 1975).

Many other groups, found in the side chains of organic molecules, can also interact with metal ions, for example, OH^- in serine and pyrosine, the second group COO^- in aspartic and glutamic acids: $-SH$ in cysteine assumes the role of polydentate ligands in relation to the metal ion. The size and form of side chains of amino acids are essential too, because they determine steric effects in chelation. In addition to these factors, the effect of the nature of elements is essential; e.g. in the presence of equal concentrations of Cu, Co, Ni, Mn, and Zn, adsorption of copper is 50% of the total adsorption capacity of organic substances. Copper forms relatively more stable complexes with humus compounds. The following order of adsorption of elements is observed: Cu > Ni > Co > Zn > Mn. Among trivalent elements, iron is characterized by higher adsorption by humic acids than aluminium (Rashid, 1974). Fulvo-acids form complexes in the molar relation 1 : 1 with ions of Cu^{2+}, Ni^{2+}, Co^{2+}, Pb^{2+}, Mn^{2+}, Zn^{2+}, Ca^{2+}, Mg^{2+} at pH = 3–5 and ionic strength $\mu = 0.1$. However, at the same ionic strength, but at pH = 1.7–2.3, fulvo-acids form a complex 1 : 1 with the ions of Fe^{3+} and Al^{3+}. If pH is more than 5, and ionic strength is decreased from 0.1 to 0, the relation of a fulvo-acid to the metal ion becomes less than 1; this fact indicates the formation of mixed polynuclear complexes. With pH = 3, stability constants decrease in the following order: Fe^{3+} > Al^{3+} > Cu^{2+} > Ni^{2+} > Co^{2+} > Pb^{2+} = Ca^{2+} > Zn^{2+} > Mn^{2+} (Schnitzer and Hansen, 1970).

The formation of complex compounds of metal ions with organic substances in natural water is confirmed by IR-spectrometric research (Rashid, 1972c). Comparison of infrared spectra of amino acids, and of their complexes with metals, implies the participation of atoms of oxygen and nitrogen in these reactions. For instance, the IR-spectra of the aspartic acid show a series of bands of –NH group of amino acid at 2,080, 1,640, and 1,520 cm^{-1} and of carboxyl group at 1,690, 1,610, and 1,420 cm^{-1}. In the IR-spectra of the complex of aspartic acid with copper, the bands disappear at 2,080 and 1,640 cm^{-1}, whereas the intensity of the band at 1,520 cm^{-1} increases considerably. The bands of the carboxyl group disappear at 1,690 and 1,420 cm^{-1} and simultaneously the band at 1,630 cm^{-1} develops. The new band, which appears at 1,390 cm^{-1}, is extra evidence for transformation of the carboxyl group into carboxylate. For natural amino acids the bands corresponding to NH-group are in the region of 3,000–3,300, 1,630, 1,500 cm^{-1}; for the carboxyl group they are 1,730 and 1,430 cm^{-1}. When a complex with metal is formed, then changes occur as described above.

A number of factors affect the formation of complexes. Molecular mass of organic matter and valency of metal are the most essential ones. It has been determined (Picard and Felbeck, 1976; Rashid, 1972a) that those fractions of organic matter with relatively low molecular mass bind metal 2 to 6 times more than fractions of higher molecular mass. Equal amounts of divalent metals are complexed by these fractions 3 to 4 times more than trivalent metals. Hem (1972) has defined six factors controlling the solubility of iron and manganese in water systems: (1) changes in pH of solution; (2) Eh of systems; (3) increase or decrease of activity of some of the anions with other different factors being constant; (4) increase in the activity of the complexing ligand results in higher solubility of elements; (5) change of temperature and pressure as compared to normal conditions; (6) effect of various sources of energy, for example, solar. Calculations show that at given Eh values, changes of pH by one tenth of unit, at pH $\simeq$ 7, change the solubility of iron approximately by a factor of two. Effect of pH on migration properties of elements is studied by the example of aluminium (Matveieva *et al.*, 1978). It has been established that oxalic acid reduces precipitation of Al to pH = 8; this is associated with the presence of soluble complex oxalates of aluminium in the acid pH region. These complexes disappear at pH > 8. Humic acids encourage quicker precipitation of aluminium in media with pH = 4–8, but produce more stable forms at pH > 8. Antimony and titanium in different pH regions form differently charged complexes with humic and fulvo-acids (Ostroumov, 1956; Udodov *et al.*, 1975). The degree of Cd(II) combination, by the same organic compounds, considerably increases with an increase of pH (Reuter and Perdue, 1977).

The effect of background ions on complexing reactions is essential. Ferguson and Bubela (1974) note that the presence of relatively small quantities of Na$^+$ and Mg^{2+} cations in the solution reduces the sorption of zinc by *Ulothrix* algae to zero, whereas the sorption of Pb(II) and Cu(II) increases considerably. Determination of the content of polyvalent metals in seawater, depending on the season of the year, has shown that iron and manganese are subjected to the greatest seasonal variations (Dobrosmylova *et al.*, 1979). Their maximal content has been observed in December and the minimum in August; this distribution is connected with the period of intensive production of plankton. According to data of Crerar *et al.* (1979), in summer 60 to 70% of iron is associated with organic matter, while in winter this value does not exceed 5 to 10%.

During the last decade, much attention was drawn to the concentration of organic matter and heavy metals in the surface microlayer of waters in basins. For example, the concentration of lead, iron, nickel, and copper increases by 0.5 to 50 times in the surface layer, within a range of 100 μm, in relation to the basic mass of water 20 cm below the surface (Duse *et al.*, 1972). On the surface of the air–water interface, the surface-active fractions of organic colloids accumulate forming films. Such films affect the surface properties of seawater, including surface potential, viscosity, and attenuation of capillary waves; evaporation is reduced. The films enhance warming of surface water (Horne, 1969).

Garrett (1968) believes that surface films are monomolecular, since only polar molecules can adhere directly to the surface of the air–water interface. Adsorption is carried out by the hydrophilic functional groups. Duse *et al.* (1972) are of opinion that the film can be several molecules thick.

The concentration of heavy metal occurs in films in all the cases. Williams (cited after Garrett, 1968), as the result of his studies of the process of formation of such a film, comes to the conclusion that this process involves lipids, proteins, carbohydrates, and phosphorous compounds. The lipids of high molecular mass are, however, the most important and long-lived on the surface of the water–air interface. For example, fatty acids, alcohols, and fatty ethers are highly polar and insoluble in water. They can have a strong tendency to selective adsorption, as compared with other less active substances; they can also stay longer on the sea surface. As the result of selective adsorption on the surface, water can be enriched by combined insoluble compounds of heavy metals.

4.3.5.2. *Role of Dissolved Organic Matter in the Process of Fe(II) Sorption by Birnessite*

The aim of this study was to obtain experimental data elucidating the sorption mechanism of Fe(II) compounds by 7 $\overset{\circ}{A}$ MnO_2 (birnessite) from seawater in the presence of different amounts of the organic component (citrate-ion). The latter is characterized by values of the formation constants of complex (K_{form}) compounds which are similar to the most part of compounds of the dissolved organic matter (DOM) of seawater. Special attention was paid to the distribution of the Fe(II) forms in seawater solution. The main results of these studies are presented in our previous works (Putilina and Varentsov, 1980, 1982, 1983, 1984).

The performed calculation and experimental investigations on studies of the sorption dynamics of Fe(II) from seawater by birnessite, in the presence of organic matter (sodium citrate), allowed us to determine a number of features of the sorption mechanism.

(1) The behavior of Fe(II) in seawater in the presence of different quantities of organic complex-forming component is determined by competitive reactions, essentially between Fe^{2+}, Ca^{2+}, Mg^{2+}, and citrate-, sulphate-, and chloride-ions. The main forms of Fe(II) in seawater with pH 8.0 to 8.2 are aqua-, sulphate-, and chloride-ion.

(2) In alkaline aerated media, Fe(II) is oxidized to Fe(III). The oxidation rate is increased due to the oxidizing function of 7 $\overset{\circ}{A}$ MnO_2.

(3) The adsorption rate of Fe(II) by birnessite is controlled by the concentration of citrate-ion. With low concentration of the organic component, the oxidized iron is transformed into a relatively insoluble form and is sorbed by birnessite. At high concentrations of sodium citrate, Fe(II) is bonded into complexes with the organic matter and occurs in a dissolved state. However, in the course of time the iron complexed by organic matter is

transformed into a more stable form of iron hydroxide and is sorbed by the active surface.

(4) The process of iron sorption is accompanied by a relative decrease of pH of the solution that may be explained by substitution of counterions at the sorbent surface by different ionic forms of iron.

Under the conditions of extremely low concentrations of DOM, there is a relatively quick formation of hydroxy forms of Fe(III). They may possess the tendency to coagulate and formate suspended hydroxide compounds of iron that, together with other suspended components of seawater, are deposited with sediments.

4.3.5.3. Role of Dissolved Organic Matter (Citrate-Ion) in the Process of Cu(II) Sorption by Birnessite

It was shown in the previous section that in the basis of the sorption processes of a number of transition metals (Mn, Fe, Ni, Co, Cu, Pb, etc.) in supergene environments there are processes of a chemisorption nature, in a number of cases, that are accompanied by autocatalytic oxidation. The important role of DOM in the process of sorption of the transition metals is emphasized. The aim of studies was to obtain experimental data on the role of organic matter in the sorption process of Cu(II) on 7 $\overset{\circ}{A}$ MnO_2, as well as in geologic interpretation of the results for natural systems.

The performed investigations allow one to define the main propositions to the model of sorption of Cu(II) on 7 $\overset{\circ}{A}$ MnO_2 in seawater in the presence of DOM. The process of Cu sorption by birnessite in the presence of different amounts of citrate-ion may by presented in a generalized form as follows. At a low concentration of citrate-ion, the copper is essentially in the form of aqua-ion. In seawater, Cu(II) is easily hydrolysed $(pK_{Cu(OH)_2} = 12.2)$. Hydroxy complex is sorbed by the surface of 7 $\overset{\circ}{A}$ MnO_2. With an increase in the concentration of sodium citrate, the quantity of copper aqua-ions in the initial solution is considerably decreased, leading consequently to a change of copper sorption. The lower the contents of aqua-ions of Cu(II) in the initial solution, the less is the quantity of copper sorbed during the first stage. With an increase in the concentration of citrate-ion, the amount of copper, associated with organic matter, is increased. The equilibrium of $CuC_{it}^- \rightleftharpoons Cu(OH)_2^0$ is attained, and the dominating form becomes hydroxy form of Cu(II) essentially sorbed by birnessite. The higher the concentration of $CuCit^-$ complex in the initial solution, the slower the equilibrium is shifted to the right, reflected in the moderation of the Cu(II) sorption process. Simultaneous with this process, some sorption of the citrate ion by MnO_2 surface may be admitted. Citrate-ion is a polyfunctional ligand comprising three carboxyl and one OH^- groups and it may be sorbed on the sorbent surface. Presence of Ca^{2+} and Mg^{2+} ions influences greatly the sorption of organic matter by hydroxide surfaces. The experimental studies show that during sorption from seawater the decrease of pH of the solution occurs due to exchange sorption of Na^+, Ca^{2+}, and Mg^{2+} ions. At the same time, the calculations have shown that practically all citrate-ion is consumed in the formation of complexes with Ca and Mg. It may be supposed that $CaCit^-$ and $MgCit^-$ complexes are concentrated at the surface where Ca^{2+} and Mg^{2+} ions fulfil the role of 'bridges' between the surface and the organic component. The location of citrate-ion in the external part at the surface of 7 $\overset{\circ}{A}$ MnO_2 probably increases its negative charge. At early stages of interaction, the sorption process of the organic complex hinders temporarily the sorption of copper from the solution. There is a decrease of the rate of copper sorption, and the amount of copper sorbed is approximately constant. The consequent stage of the sorption process is associated with copper sorption increase, in which a definite role belongs to uncompensated functional groups of the organic matter. They form complex compounds with Cu_{aqv}^{2+} and $CuOH^+$ and promote additional copper sorption. Here the formation of the composite surface complex is supposed: birnessite – Ca(Mg) – citrate – copper hydroxy form. The reality of such hydrolytic accumulation of copper is confirmed by the data of studies of solid phase composition, both with the example of Cu and other elements (Ni and Co).

Thus, in the presence of relatively high contents of dissolved organic matter with relatively moderate values of constants of complex formation, the rate of Cu sorption may be considerably decreased due to intermediate stages in which an important role belongs to surface reactions between a sorbent and organic matter.

In natural environments, during long periods, the definite amounts of sorbed organic matter may be oxidized under the influence of MnO_2 and other oxidants.

4.3.5.4. Experimental Data on Cu(II) Sorption by Birnessite at Different Concentrations of the Complexing Component (EDTA)

The diverse role of dissolved organic matter (DOM) was considered above. It is able to bind metal ions into complex compounds, participate in redox processes, and act as surface-active matter. Depending on some or other properties, the DOM may considerably influence the sorption processes of transition and heavy metals.

The aim of the experimental investigation was to study the influence of DOM concentration, forming relatively stable complex compounds ($pK > 10$), on Cu(II) sorption by MnO_2 (7 Å MnO_2). It is known that in natural water a considerable part of Cu(II) is bonded into relatively strong metallorganic complexes. At the same time, sorption interaction of such sea and pore waters with manganese hydroxides leads to the formation of ore concentrations of copper in iron-manganese concretions.

The obtained results indicate that the trend of Cu sorption process is determined by what properties of DOM, to a greater degree, are manifested in the given environments. In the considered experiment, etylenediaminetetracetate (EDTA) may display properties of both surface-active matter and a rather strong complexing agent. It is supposed that having sorbed on the surface of 7 Å MnO_2, the EDTA may increase the number of its active centres favoring additional sorption of metal ions, Cu(II) in the given case.

But the main mechanism of Cu(II) sorption on birnessite from seawater and similarly in mineralizing natural waters, is characterized by hydrolytic reactions. Cu hydroxy forms are sorbed directly on the surface of 7 Å MnO_2. There is also sorption of definite amounts of Cu on modified surfaces of this sorbent with the possible consequent formation of a mixed complex of $\{Cu(OH)^+HR^{3-}\}\cdot(Ca, Mg)MnO_2$ type, in which ions of alkali-earth metals play the role of bridges between the surface and the organic component.

Copper sorption by birnessite from seawater, and similar solutions comprising dissolved organic matter, is a multistage process, the main place in which belongs to: (1) reactions of hydrolysis and sorption of metal hydroxy forms; (2) sorption of dissolved organic matter by oxide surface; (3) additional sorption of metal ions by modified surface with the formation of a mixed surface complex. In natural environments, during postsedimentation transformations due to redox processes, there may be decomposition of the organic matter and formation of mixed oxyhydroxide phases.

4.3.6. GEOCHEMISTRY OF FORMATION PROCESSES OF MN OXYHYDROXIDE MINERALS

4.3.6.1. *Early Stages: Initial Oxidation of Mn(II) to Mn(III)*

Numerous studies have established that in supergene environments, in particular in the weathering crust, the leaching processes and transformations of Mn-bearing minerals usually proceed in the range from low to neutral values of pH and substantially in normal partial pressures of O_2 relative to that of the atmosphere. Under similar conditions Mn(II), supplied to ground waters, as a rule, is transformed to relatively stable forms: Mn(III) and Mn(IV) (Giovanoli, 1976, 1980; Ruetschi and Giovanoli, 1982; Stumm and Giovanoli, 1976). Nominally, these transformations proceeding in acid medium may be presented by equations of reactions:

$$Mn^{2+} \rightleftharpoons Mn^{3+} + e^- \qquad \text{(soft oxidation)} \tag{1}$$

$$Mn^{3+} \rightleftharpoons Mn^{4+} + e^- \tag{2}$$

$$Mn^{2+} + H_2O \rightleftharpoons MnO_2 + 4H^+ + 2e^- \qquad \text{(rigid oxidation)} \tag{3}$$

$$Mn^{3+} + 2H_2O \rightleftharpoons MnO_2 + 4H^+ + e^- \tag{4}$$

$$Mn^{4+} + 2H_2O \rightleftharpoons MnO_2 + 4H^+. \tag{5}$$

In surface environments with absence of catalytic phenomena, and in acid–neutral ranges of pH, Mn(III) is a rather stable form that represents the initial stage of Mn(II) oxidation. Mn^{3+} formation may take place as the result of a reaction with the participation of Mn^{2+} and MnO_2, i.e. direct reaction (1) or reverse reaction (4), or owing to the combination of two reactions: reverse (4) and direct (3) ones. Finally the reaction may proceed:

$$MnO_2 + Mn^{2+} + 4H^+ \rightleftharpoons 2Mn^{3+} + 2H_2O. \tag{6}$$

But it is necessary to note that in natural media the redox transformations of manganese are described by these equations only in extremely general schematic form.

A more justified understanding of the behavior of this element in the supergene zone may be obtained by the data from the study of natural phenomena and experimental modeling.

Mn(II) oxidation in neutral–slightly alkaline media. At early stages of development of the weathering crust profile, the neutral and slightly alkaline solutions are formed during solving the initial manganese-bearing minerals, due to neutralizing reactions and the transition of ions of alkaline metals and alkaline-earth metals into solution. The behavior of Mn(II) in ground waters of similar types may be considered by the way of the example of the experimental data presented below.

The investigations of oxidation dynamics of Mn(II) in artificial and natural lake water (Giovanoli, 1980; Stumm and Giovanoli, 1976) have shown that in the absence of Fe, suspended organic matter, and other components, slow oxidation of Mn(II) into Mn(III) proceeds. The authors have established that ion $[Mn(H_2O)_6]^{2+}$ present in the solution, introduced in the form of $[Mn(NO_3)_2]$, the initial concentration of which is 10^{-5} to 10 M^{-6}, is oxidized and precipitated after occurring in the medium with pH 8.06 to 8.99 within approximately 350 days. The observed situation is similar to results that proceed during $Mn(OH)_2$ oxidation in humid environments (about 70 to 85% of relative humidity). Under these conditions, rather rapid oxidation of Mn(II) takes place with a number of consecutive transformations: $Mn^{2+} \rightarrow Mn_3O_4 \rightarrow \beta\text{-MnOOH} \rightarrow \gamma\text{-MnOO}$. Thus, the authors illustrated that the precipitation, resulted due to long relatively soft oxidation, is represented by acicular crystals of manganite ($\gamma\text{-MnOOH}$) as the phase which is stable in these environments. It is interesting that in microphotographs made under electronic microscope (Giovanoli, 1980) we could easily see laminated crystals of 7 $\overset{\circ}{A}$ MnO_2 (birnessite) and 10 $\overset{\circ}{A}$ MnO_2 (buserite), the identification of which was confirmed by electron microdiffraction.

Later studies have defined that under conditions at 0°C, Mn(II) may be oxidized to feitknechtite ($\beta\text{-MnOOH}$), whereas at 250°C Mn_3O_4 (hausmannite) is mainly formed (Hem and Lind, 1983). In the systems with sulphate-ions, these authors have observed direct formation of $\gamma\text{-MnOOH}$ (manganite). Moreover, Hem and Lind (1983) do not consider that the initial products of oxidation of Mn(II) in aerated water systems are metastable solid phases in which the oxidation state of manganese does not exceed Mn(III). They think that in the result of disproportionation reactions the transformation of Mn(III) to Mn(IV) takes place by the scheme:

$$2MnOOH + 2H^+ \rightarrow MnO_2 + 2Mn^{2+} + 2H_2O, \tag{7}$$

that is essentially similar to the reverse reaction (6) considered above.

The problem of the existence of long duration lowest forms of manganese oxides in media with neutral or slightly alkaline values of pH has needed thorough experimental testing. Such investigation has been conducted with modern methods of study of fine dispersed phases, including X-ray photoelectron spectroscopy (Murray *et al.*, 1985). It has been established that from aerated solution of ammonium chloride with pH 9.0, and containing Mn^{2+} (1.6×10^{-2} M), of Mn_3O_4 (hausmannite) initially precipitated, with admixtures of $\beta\text{-MnOOH}$ (feitknechtite) and specific MnOOH (Marti modification). After approximately eight to nine months they are quantitatively transformed to $\gamma\text{-MnOOH}$ (manganite). It is of interest that the initial stage of Mn^{2+} oxidation in the solution proceeds rather quickly: the precipitate, deposited in 30 min, was characterized by ratio 0/Mn ≈ 1.367 that, in combination with the X-ray diffraction data, pointed to the presence of hausmannite, and to small amounts of $\beta\text{-MnOOH}$, and traces of $\gamma\text{-MnOOH}$. Further transformations proceeded according to the mechanism of dissolution-precipitation. The main transformation is associated with the transition of $Mn_3O_4 \rightarrow \gamma\text{-MnOOH}$ described by reactions (Giovanoli *et al.*, 1976b):

$$Mn_3O_4 + 2H^+ \rightarrow MnOOH + Mn^{2+} \tag{8}$$

with consequent oxidation of Mn^{2+} into hausmannite

$$3Mn^{2+} + 1/2\,O_{2(aq)} + H_2O \rightarrow Mn_3O_4 + 6H^+. \tag{9}$$

It is necessary also to recognize the possibility of the disproportionation reaction of Mn_3O_4 and the formation of MnO_2. However, as the experiment and calculations show, under these conditions the activity of dissolved manganese, that may admit the disproportionation occurrence, is lower than the activity, with which the protonation reaction (8) in this system proceeds (Hem and Lind, 1983). Thus, the presented data (Hem and Lind, 1983; Murray *et al.*, 1985) indicate that during this experiment $MnOOH$ is most probably formed from Mn_3O_4. Concentrations of dissolved manganese are too high to promote the formation of γ-MnO_2. We have every reason to believe that disappearance of Mn_3O_4 (the reverse course of reaction (9) may take place with the decrease of dissolved Mn^{2+}.

Presented results unambiguously demonstrate the possibility of γ-$MnOOH$ (manganite) existence as a relatively early product of Mn(II) oxidation in neutral to slightly alkaline solutions, with normal (atmospheric) values of the partial pressure of O_2 and relatively high concentrations of the initial Mn^{2+}. Nevertheless, the results also indicate that with an insignificant change of these parameters, for example, with some decrease of the system pH and the concentration of Mn^{2+}, a more stable compound (γ-MnO_2) may be formed instead of γ-$MnOOH$.

4.3.6.2. *Major Stages: Oxidation of Mn(II) to Mn(III) and Mn(IV)*

Lithiophorite formation.   The characteristic feature of a series of alteration sequence of Mn silicates differentiating it from products formed on carbonate or oxide (braunite-hausmannite) protore, is the formation of lithiophorite. It is developed mainly after residual aluminiferous matter, from incongruent dissolution, that play the role of matrix or active surface. Numerous experimental and field investigations have established that under conditions of acid values of pH (3 to 5) and the oxidizing regime of ground waters of the initial stages of weathering, Al is characterized, as a rule, by extremely limited mobility in comparison with manganese. It may be accepted that the spessartine formula is $Mn_3Al_2(SiO_4)_3$ with the parameter of the cubic cell to be $a_0 = 11.61$ Å, and the lithiphorite formula is $[Mn_5^{4+}Mn^{2+}O_{12}][Al_4Li_2(OH)_{12}]$ with $a = 5.06$ Å; $b = 8.70$ Å; $c = 9.61$ Å, and $\beta = 100°7'$. Simple calculations show that in spessartine the ion of Al and the ion of Mn occupy similar volumes and their relation is approximately 1. Similar relations of volumes of Mn and Al ions are characteristic for lithiophorite too. Thus, we may safely assume that autochthonous transformation of spessartine to lithiophorite may proceed on a residual aluminiferous matrix without any considerable addition of Mn. In this case, Al forms octahedral layers in lithiophorite structure $[(Al_2, Li)(OH)_6]$ alternating with octahedral layers of $[MnO_6]$ (Burns and Burns, 1979; Giovanoli, 1985; Giovanoli *et al.*, 1973). However, along with autochthonous pseudomorphs after spessartine in these parts of the crust profile there are also transecting streaks composed of lithiphorite. Such data may be interpreted as an indication of a significantly inhomogeneous and spotty pattern of pH regime, when in relatively more acid solutions the transportation of mineral-forming components takes place.

In works by Giovanoli (Giovanoli, 1985; Giovanoli *et al.*, 1973) it is emphasized that in experimental conditions lithiphorite synthesis is possible only when 7 Å MnO_2 phyllo-manganate of the analytical composition of $Mn_7O_{13}5H_2O$, as well as γ-$Al(OH)_3$-gibbsite and $LiOH \cdot H_2O$, was used as the initial manganese oxyhydroxide. The reaction proceeds during 2 to 3 days at 300°C and 90 to 130 atm. In other words, the application of manganese oxyhydroxide of layered structure as an initial component is of critical significance. The attempts of application of γ- or β-MnO_2 (i.e. nsutite or pyrolusite), characterized by chain-like tunnel structures, failed in lithiophorite synthesis. The presented data shed additional light on the genetic aspects of the problem regarding the place occupied by lithiophorite in a series of newly-formed oxyhydroxide minerals of manganese after the alumosilicate (spessartine) protore.

Cryptomelane formation. In a series of supergene alterations of silicate, carbonate, and manganese-bearing protore, cryptomelane occupies a definite place. In autochthonous, lens-like, laminated ore accumulations of the weathering crust profile it is formed, as a rule, after lithiphorite (during alteration of Mn-garnets) or after pyrolusite (during weathering of Mn-carbonates). In ores, formed with relatively long transportation of manganese (for example, of a cap) the cryptomelane is an essential mineral. The analysis of natural observations and experimental data allow one to think that the formation of cryptomelane may proceed during relatively mature or final stages of development of the weathering crust profile, when a pronounced oxidizing regime occurs and a sufficient amount of K ions are present in the system. The cryptomelane may be formed both during direct oxidation of Mn(II) ions by the scheme (Giovanoli, 1976, 1980):

$$[Mn(H_2O)]^{2+} + H_2O + K^+(Ba^{2+}) \rightarrow Me_x Mn_8 O_{16} + H^+,$$

where Me_x is ion of K in the case of cryptomelane or Ba^{2+} during hollandite formation, and through the intermediate products, in particular, γ-MnOOH (manganite).

Experimental works on the synthesis of the considered mineral illustrate the main features of the process dynamics (Hypolito *et al.*, 1982a, 1982b, 1984). These authors obtained synthetic cryptomelane as the result of oxidation of Mn(II) ions by oxygen at 25°C, initial pH less than 3, and gradual addition of KOH to the solution. The nucleation (nucleus formation) of cryptomelane takes place only with a local increase of pH. Similar phenomena may occur under natural conditions when in the neutral medium of infiltrating ground waters, the contact develops between the component-bearing solution and hydrolytically decomposing minerals which produce OH-ions. The experiments show that cryptomelane forming during the nucleation stage will coexist with manganite and nsutite, formed proportionately to the pH increase.

In experiments with solutions characterized by high pH, nsutite (γ-MnO) is the first mineral that is formed in media, containing K ions at pH decrease. However, later nsutite may be transformed to cryptomelane. During acid treatment of some synthetic oxyhydroxide minerals of manganese in the presence of K^+ ions the following transformations may take place: nsutite $\rightarrow$ cryptomelane; hausmannite $\rightarrow$ nsutite; manganite $\rightarrow$ pyrolusite. However, it is necessary to point out that under the same experimental conditions it was not possible to achieve direct transformations of manganite and pyrolusite into cryptomelane. The most probable mechanism of formation of the considered mineral may be realized under conditions of low values of pH and in the presence of K^+, when cryptomelane nucleation is initiated mainly by H_3O^+ ions. It is essential that cryptomelane crystallinity is increased with the incorporation of K^+ and Mn(IV) ions into its structure, but it is decreased with an increase of Mn(II), H_2O, and ions of Zn(II) (as one of heavy metals) in the system.

Thus, cryptomelane synthesis proceeds most actively in the interval of pH of 1.00 to 2.90, with a gradual addition of KON and the passing of pure O_2 bubbles through the solution (Hypolito *et al.*, 1982a, 1982b, 1984).

During consideration of problems of the genesis of supergene manganese oxyhydroxide minerals of the hollandite-cryptomelane group, it is necessary to keep in mind that rigid stoichiometry is not always typical for these compounds. On the contrary, they are characterized by relatively wide ranges of substitutions, in which a number of ions of heavy and alkaline metals and alkaline earths, as well as of manganese of different valence, participate. In works by Giovanoli *et al.* (Giovanoli, 1985; Giovanoli and Balmer, 1981) it is shown that tunnel cavities in the structure of these minerals are built from four double chains of octahedrons of MnO_6. The cross-section of similar tunnels is about 3 $\overset{\circ}{A}$, allowing penetration of cations of radius of about 1.5 $\overset{\circ}{A}$ (for example NH_4^+, H_3O^+, K^+, Ba^{2+}, and Pb^+) into them. The stability of such supergene minerals depends on how their structure is balanced, mainly by cations in tunnels, from the action of such factors as values of pH, Eh, oxygen contents, pressure and temperature, as well as from geological settings of their occurrence. A crucial role belongs to the chemistry of solutions, ground waters, their ionic composition, ionic strength, and presence of dissolved organic components. In comparison with layered manganates (δ-MnO_2; 7 $\overset{\circ}{A}$-MnO_2, and 10 $\overset{\circ}{A}$-MnO_2), the tunnel manganates of the hollandite-cryptomelane group are characterized by absence or extremely low ion exchange properties, since cations in the tunnel sites are rigidly incorporated in the mineral structure.

Formation stages of nsutite (γ-MnO_2) and other modifications of MnO_2.   During formation of manganese-bearing weathering crusts, an essential role belongs to acid solutions that participate both in early stages of manganese leaching from bedrocks and in a number of its consequent phase transformations. Experimental investigations by Giovanoli and his colleagues (Giovanoli, 1976, 1980; Giovanoli *et al.*, 1976a, b) have shown that in the region of acid pH, with normal values of partial pressure of O_2 (or generally at oxidizing values of Eh) and in the absence of ions of alkaline and alkaline-earthy metals, the Mn^{2+} is directly oxidized to γ-MnO_2 (nsutite). This product is almost completely transformed to β-MnO_2 (pyrolusite) during consequent recrystallization in an acid medium. Schematically, the process of rigid oxidation of Mn(II)–Mn(IV) was considered above in reaction (3).

In those cases when, in acid solutions, the above products of intermediate oxidation of manganese, represented by Mn_3O_4 or by modifications of MnOOH, have been formed, their consequent oxidation to Mn(IV) shown in reactions (2 and 4) takes place. It is experimentally established (Giovanoli *et al.*, 1969, 1976b) that the oxidation of γ-MnOOH (manganite) in a diluted solution of HNO_3, saturated with oxygen, leads to its topochemical transformation to β-MnO_2 (pyrolusite). Oxidation of less stable oxide Mn(III) groutite (α-MnOOH) in similar conditions depends on the degree of its dispersion: the finest isolated crystals of this compound are oxidized quickly to Mn_5O_8 (a variety of 7 Å-manganate) which is transformed to β-MnO_2 during consequent oxidation. It is essential that in the acid medium, as a rule, the reaction of disproportionation of Mn_3O_4, α-MnOOH, and γ-MnOOH to γ-MnO_2 and Mn^{2+} (see reaction (7) above), proceeds with consequent quick oxidation of Mn(II) to Mn(IV) (reaction (3)).

4.4. Conclusion: On the Geochemical Model

It is necessary to keep in mind that the data on the chemistry of manganese, and the results of experiments, only schematically reflect the general trend of processes proceeding in natural environments of weathering. Relatively simple and obvious features of manganese behavior in different regimes of oxidation were accounted with examples of simple model systems, in which the sorption interactions in diluted solutions were taken into consideration. Nevertheless, this data allows one to explain without contradictions, mineral relationships observed in natural objects. The above general sequence of oxidizing transformations of silicate and carbonate protores is described by the following series (Bouladon *et al.*, 1965; Grandin, 1978, 1985; Grandin and Perseil, 1977, 1983; Leclerc and Weber, 1980; Nahon *et al.*, 1982, 1983; Weber *et al.*, 1979):

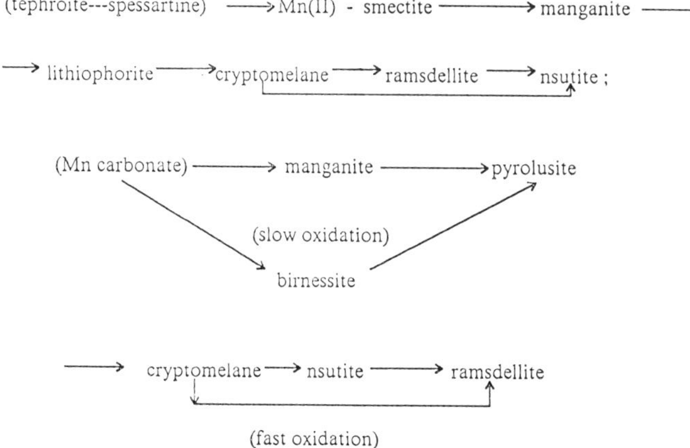

In these series of the mineral transformation sequences the geochemical trend of alterations is convincingly reflected in the development of the laterite weathering crust profile, from the neutral slightly oxidized environments to acid with normal regime of O_2 (Eh) of the medium and different participation of such components as Al^{3+}, K^+, H_3O^+, Ba^{2+}, etc. By the example of manganite formation (Nahon *et al.*, 1983) it is shown that chemical composition of this mineral in autochthonous occurrences varies noticeably depending on the initial matter type. The initial mineral-forming solutions (slightly alkaline with a soft oxidizing regime) are also enriched in Si, Ca, Mg and a number of transition metals, in addition to Mn, that are sorbed by the forming solid phase. The uniform and homogeneous distribution of these elements in the manganite structure, as shown with the data of electron microprobe scanning, points to this mechanism. Purer crystals of γ-MnOOH without admixtures are found in streaks filling in fractures, and voids, i.e. in aggregates, associated with a certain transportation of mineral-forming solutions.

In higher horizons of the weathering crusts during long, and intensive circulation of acid ground solutions with oxidizing values of Eh, accumulation of some Mn(IV) compounds occurs. For the minerals which form in this zone (lithiophorite, birnessite, cryptomelane, etc.) it is typical that their composition, as it has been shown above, reflects the geochemical features of primary rocks, protores (Perseil and Grandin, 1985). It is necessary to emphasize that sorption processes are the main mechanism of the given interaction. As a whole, the considered process may be described by the proceeding reactions of Mn(II) oxidation, in which an essential role belongs to sorption autocatalytic interactions (Stumm and Morgan, 1970):

$$Mn^{2+}_{(hydr.)} + O_{2(gas)} \underset{(slow)}{\rightleftharpoons} MnO_{2(solid)} \tag{1}$$

$$Mn^{2+}_{(hydr.)} + MnO_{2(solid)} \underset{(fast)}{\rightleftharpoons} Mn(II)\,MnO_{2(solid)} \tag{2}$$

$$Mn(II)\,MnO_{(solid)} + O_{2(gas)} \underset{(slow)}{\rightleftharpoons} 2MnO_{2(slow)}. \tag{3}$$

It is necessary to point out that in natural environments, where solutions and waters with relatively low to moderate concentrations of Mn and other transition metals are often found, the reactions (1) and (2) will most probably proceed. Their sorption and autocatalytic nature is shown by their pulsed and cyclic repetition. Here in the solution, together with the ion of $Mn^{2+}_{(hydr.)}$ with close functions in reaction (2) ions of heavy and alkaline metals and, to a lesser degree, alkaline earths may participate.

This proposition is especially valid for the formation of Co(III) and Ni(II)-bearing oxyhydroxide minerals of manganese, i.e. asbolanes in weathering crusts after ultrabasic rocks (Vitovskaya *et al.*, 1984). In spite of the fact that for manganese in similar weathering crusts the complete removal is characteristic, this metal is relatively enriched in the upper zones: in nontronite clays by one order of magnitude, in ochre by two orders of magnitude. The behavior of manganese and associated Co and Ni in the weathering crust profile may be illustrated by the example of laterite weathering crusts after ultrabasite complexes of Siberia and Bulong, Kalgoorlie region, West Australia (Elias *et al.*, 1981), where the maximum concentrations of these elements (MnO_2 to 61.35%; NiO to 14.77%; CoO to

10.87%) are observed in the zone of clays at the base of limonite (goethite) zone. The most common forms of manganese accumulations are nodules of different types and globules, the combination of both of which are predominated by early products: Ni, Co-asbolanes (Chukhrov *et al.*, 1980a, b), and to a lesser degree δ-MnO$_2$ and β-MnO$_2$. Later generations are represented by Ni and Co-asbolanes, minerals of the cryptomelane-hollandite group, romanchite, coronadite, rancieite, todorokite, and Ni-lithiphorite (Vitovskaya *et al.*, 1984). These authors pointed out that in the process of formation of the weathering crust profile after serpentinite, the repeated redeposition of manganese may take place. The composition of the minerals so formed reflects the medium chemistry.

Hence, in mineral transformations controled by the dissolution-deposition mechanism, the most important role belongs to processes of sorption accumulation of a number of metals (Ni, Co, Ba, K, etc.) reflecting the chemistry of both solutions and primary rocks. The essential part of this processes is the separation of Mn and Fe controlled by the regime of pH, Eh, chemistry of solutions, and kinetic parameters.

In the given context, the experimental data provides the grounds for the interpretation of the main facts observed in the natural environment, though it does not embrace the natural complexity of the whole phenomena.

Chapter 5

EVOLUTION OF MANGANESE ORE FORMATION IN THE EARTH'S HISTORY

5.1. General Classification of Manganese Deposits

At present on the basis of available information, the following general classification scheme of manganese ore deposits is proposed (Betekhtin, 1946; Rakhmanov, 1967; Roy, 1980, 1981; Strakhov *et al.*, 1968; Varentsov, 1962; Varentsov and Rakhmanov, 1974).

I. Weathering crusts:
1. Residual accumulations and products of their local redeposition (laterite type, deep leaching and redeposition). The manganese ores of this type are usually developed after initially manganese-poor accumulations in the zone of tropical weathering: deposits of West Africa (Gabon – Moanda, Ghana – Nsuta, etc.; Côte d'Ivoire – Ziemougoula), Australia (Groote Eylandt), Brazil (states: Amapa, Bahia, Moro do Urucum), etc. Manganese ores are composed of oxide minerals – cryptomelane, pyrolusite, manganite, lithiphorite, nsutite, and todorokite. Manganese ores are characterized by high quality (40.4 to 57.3% of Mn). Reserves of this type are rather considerable with many hundreds mln t of high grade manganese ores (Brazil's important new manganese province, 1978; Leclerc and Weber, 1980; Manganese, 1977; Nahon *et al.*, 1983; Slee, 1980; Varentsov, 1982).
2. Karst, infiltration accumulations. This type includes the considerable part of manganese ores of deposits of Kuruman and Postmasburg regions at the north of Cape Province of South Africa. Oxide manganese ores (braunite, bixbyite, hausmannite, and jacobsite) are localized in deposits filling in the cavities of paleokarst in the lower dolomite suite of the Lower Proterozoic Transvaal Supergroup. Manganese ores are characterized by high quality (more than 44% of Mn). The reserves constitute about 3 bln t calculated for manganese in the ore (Boardman, 1964; Roy, 1980; Ellison, 1983). Other deposits of smaller scales are known too.

II. Sedimentary (manganese ores accumulated in sedimentary basins):
1. Sedimentary (exogenic source of ore components: redeposited weathering crusts,

products of washing out of the source area; underwater leaching); the typical representatives of this type are the Lower Oligocene deposits of Ukraine (Nikopol'skoe, Bol'she-Tokmakskoe, etc.), Georgia (Chiaturskoe and others), a group of the Paleocene deposits of the eastern slope of the Northern Urals, etc. (Nikopol'ski manganese ore basin, 1964; Chiaturskoe manganese deposit, 1964). Ore-bearing scales are considerable: about 6% of the reserves of manganese ores of continents. The most commercial valuable ores are represented by oxide, oxidized ores composed of pyrolusite, manganite, and minerals from the romanechite (psilomelane)-cryptomelane group; carbonate ores (calcium rhodochrosite and manganocalcite) are also used in metallurgy;

2. Hydrothermal-sedimentary (endogenous source of ore components: hydrothermal solutions, exhalations, etc.); examples of such types are stratified deposits of manganese and iron-manganese ores in the Famennian marine siliceous sequence, Atasuiskii area of the Central Kazakhstan (Kalinin, 1963; Rozhnov *et al.*, 1967, 1980, 1984; Sapozhnikov *et al.*, 1961, 1963, 1967; Varentsov *et al.*, 1993). The association of iron-manganese and barium-lead-zinc mineralization is characteristic; the predicted reserves of more than 300 mln t of high grade phosphorus-free manganese ores (Rozhnov *et al.*, 1967) are the second highest in the U.S.S.R. after the Nikopol'skii–Chiaturskii basin. Braunite, braunite-hausmannite, and cryptomelane-coronadite-hollandite ores are distributed among manganese accumulations of this type; in oxidation zone: romanechite (psilomelane) and romanechite-vernadite varieties (Kalinin, 1963, 1963; Rozhnov *et al.*, 1980; Sapozhnikov *et al.*, 1961, 1963). In South Africa, the hydrothermal-sedimentary Kalahari deposit (reserves 7.5 bln t of the ore containing more than 30% of Mn), the Transvaal Supergroup, Lower Proterozoic has been studied and operated; the ores, associated with ferruginous-cherty (iron) formation and are represented essentially by braunite, to a lesser degree by kutnahorite, hausmannite, and other minerals (Beukes, 1983, 1989, 1993).

III. Volcanogenic:
 1. Hydrothermal;
 2. Contact-metasomatic. Manganese ores of the (1) and (2) types do not have essential commercial significance. However, in a number of cases they may be facies types in a series of volcanogenic-hydrothermal-sedimentary deposits of manganese; for example, vein bodies in a group of deposits of iron-manganese ores of Central Kazakhstan.

IV. Metamorphosed (regional, contact metamorphism of sedimentary and volcanogenic ore accumulations):
The characteristic representatives are deposits of India, composed of metamorphosed Precambrian sedimentary sequences partially enriched in the lateritization zone: deposits of the Sausar group of the manganese ore belt of Madhya Pradesh and Maharashtra States. The layers of oxide ores (braunite, bixbyite, hollandite, and jacobsite) are conformably alternated with manganese oxide-siliceous rocks (gondites), crystalline schists, and quartzites, altered to green schist-amphibolite stage. In rocks of the Khondalite Group the beds of oxide manganese ores are enclosed in the rock mass, metamorphosed to gran-

ulite facies (Andhra Pradesh and Orissa States). Similar types of deposits are known among the Precambrian sequences of the African and Brazilian Shields. These deposits are characterized by rather considerable reserves (hundreds mln t) (Dorr, 1972; Roy, 1980, 1981).

The limited frames of the work do not allow us to give a thorough description of the geochemistry of formation processes of Mn ores in the main types of paleobasins. The results of these studies are presented elsewhere and in our earlier publications.

5.2. On the Relationship of Formation of the World Ocean Metalliferous Sediments and Oxyhydroxide Nodules and Crusts

5.2.1. METALLIFEROUS SEDIMENTS

In the Mesozoic–Cenozoic sedimentary rock sequences of the World Ocean metalliferous sediments are widely developed, the typical hydrothermal accumulations of the axial zones. The hydrothermal solution, from which the components of these sediments precipitate, is deeply transformed seawater convectionally circulating within the magmatic/hydrothermal chamber and leaching heavy metals from enclosing basaltoid rocks. The thickness, dimension, and composition of such hydrothermal accumulations are associated with geologic and geophysical features of the active zone: distribution of geothermical anomalies, relation of mantle, magmatic masses (diapirs, chambers, etc.), structure of faults as the hydrothermal channels, spreading rate, specific features of circulations of fluids, and the presence of favorable conditions for the accumulation of hydrothermal sediments. Among the considered sediments, several groups are distinguished differing by composition, scale of mineralization, and formation settings (Embley and Jonasson, 1988; Leinen *et al.*, 1986; Levin, Varentsov *et al.*, 1987; Rona, 1983, 1987, 1988; Thompson, 1983; Varentsov, 1980; Varentsov *et al.*, 1983, 1984, 1985, 1989).

1. Sediments of sulphides of heavy metals associated with activity of relatively high temperature hydrothermal fluids (depressions of the Red Sea, accumulations of polymetallic sulphides in the region of the East Pacific Rise, in particular, in the zone of the Galapagos Rift, in the region of 21° N, the Mid-Atlantic Ridge, etc.). The representative example of development of metalliferous sediments of such a type is the region of the Galapagos Rift of the East Pacific Rise (Embley and Jonasson, 1988; Varentsov *et al.*, 1983). Depending on the combination of the above listed parameters, the hydrothermal sediments are represented by essentially massive sulphides, more rarely by oxyhydroxide ores, and to a lesser degree by todorokite (birnessite, celadonite), nontronite varieties. Different types of the postvolcanic mineralization are associated with the zone of large faults, along which basalt-andesite lavas were subjected to deep hydrothermal alteration, accompanied by removal Ca, K, Na, SiO_2, Mn, partially Fe, and accumulation and supply of Fe, S, Cu, and Zn. In this case the content of Si and Mg is changed depending on the degree of alteration and chloritization of the substrate. The temperature of hydrothermal solutions varies widely reaching 350°C. In their composition, from the data of $^{87}Sr/^{86}Sr$ ratios the seawater predominates over the 'juvenile' fluid in ratio of > 100 : 10. The

sulphide ores, developed on hydrothermal fields in the form of cone-like mounds 1 to 80 m high and up to 100 m in cross section, are localized near areas of discharge of high temperature hydrothermal solutions. The mineral composition of such ores is: quartz, amorphous silica, goethite, hematite, magnetite, sulphides of Fe, Cu, and Zn, covellite, digenite, atacamite, botallackite, and barite. In the deeper and central parts of deposits the massive sulphide ores are developed. In the altered effusives they are changed by streak-impregnated and stockwork-like ones. Within the modern continents, such deposits are similar to pyrite deposits in ophiolite zones of Cyprus, Oman, Newfoundland, etc. The characteristic feature of the considered ores is that in areas of development of relatively low temperature hydrothermal solutions (Varentsov *et al.*, 1983) hydrothermal mounds are formed. They are essentially composed of todorokite (birnessite)-nontronite (celadonite) that indicates a mainly temperature gradient control of separation of Mn, and Fe, as well as the heavy metals in the zone of circulation of hydrothermal fluids.

It is important to note that the deposits of metalliferous sediments in both volcanogenic and enclosing sediments may be found in tectonic settings, as a rule, indicative of both early and relatively mature stages of opening of ocean basins. The features of tectonic settings control the mineral composition and morphology of the deposits, except for cases of accumulation formation in sedimentary sequences with low rates of spreading and at areas of the development of underwater mountains. Thus, the considered type may include both stratiform, stockwork, and dispersed sulphides of high temperature nature and, to a lesser degree, stratiform laminated sulphates, silicates, carbonates, and oxyhydroxides that are relatively low temperature accumulations. However, a considerable part of the discussed deposits may be attributed to massive ores (Rona, 1988).

2. The accumulations of Mn–Fe oxyhydroxides, to a lesser degree, of sulphides of heavy metals occurring on altered basalts of the second seismic layer (basement), that are covered by sedimentary sequences. They also include Mn–Fe crust-like accumulations on the surface of basalts of rift valleys and fault zones and other sediments formed near hydrothermal vents. In those cases when these sediments are exposed to hydrogenetic activity of the bottom water they are characterized by high content of heavy metals and REE.

These products, as well as the above considered accumulations of massive sulphides, are usually associated with early stages of the ocean crust formation.

3. The relatively low temperature accumulations may be attributed to an independent type on a wide scale and massive development within the above considered sediments (see 1). Such metalliferous sediments compose the uplifts of the hydrothermal mounds and ridge types or they are accumulated in marginal sedimentary basins: depressions of the Red Sea, and in the Southwestern Pacific. Their early Jurassic equivalents are known in Mn-Fe deposits of Úrkút, Hungary, and adjacent regions of the Alpian zone (Varentsov *et al.*, 1988). The results of the study of the deep sea (DSDP) drilling data in the region of the Galapagos Rift (70th Leg of the R/V 'Glomar Challenger', Varentsov *et al.*, 1983) suggest that the given metalliferous sediments consist of green ferruginous clays (mixed-layered phase – nontronite-celadonite) and Mn-oxyhydroxides (todorokite, to a lesser degree birnessite). The significant feature of these accumulations is a sharp separation of Mn and Fe. Fe-clays were deposited at the initial stages, and Mn-oxyhydroxides were

deposited during the final stages. Considerable amounts of Mn were disseminated in the surrounding pelagic sediments and seawater.

4. The masses of Mn-Fe oxyhydroxides and associated components, accumulated in sedimentary depressions of axial and fault zones, for example, depressions adjacent with East Pacific Rise: Hess and Bauer, and small basins on flanks of the Mid-Atlantic Ridge (Varentsov, 1978). These sediments are characterized by a considerable dilution of Mn–Fe oxyhydroxide components by clastic, aeolian, and biogenic material.

5. Sediments accumulated at some distance from the axial zones and considerably diluted by biogenic and clastogenic material. The metalliferous and hydrothermal nature of ore components of such sediments may be determined after the corresponding geochemical recalculations allowing the removal of the diluting effect of biogenic and clastogenic components (Varentsov, 1980, 1981a, 1981b, 1983; Varentsov *et al.*, 1981a, 1981b). The hydrothermal activity of axial zones as a whole considerably influences the accumulation of heavy metals in sediments and concretions in the sedimentation history of the World Ocean. On the basis of analysis of accumulation rates of transition metals and their associations in the Cenozoic sediments of the central-eastern parts of the Pacific, it was established that at definite time intervals the dominating role belonged to the components of a hydrothermal nature derived from the East Pacific Rise (Leinen and Stakes, 1979; Leinen *et al.*, 1986).

5.2.2. MN–FE CRUSTS AND NODULES

The formation of polymetallic nodule and crust-like Mn–Fe ores is the characteristic feature of the most part of modern and ancient basins (Bezrnkov, 1976; Drittenbass, 1979; Exon, 1983; Glasby, 1977, 1978, 1986, 1988; Glasby *et al.*, 1982a, 1982b; Guillou, 1976; Jenkyns, 1977; Levin and Varentsov *et al.*, 1987; Roy, 1981, 1988; Roy *et al.*, 1982; Varentsov, 1975, 1980; Varentsov *et al.*, 1984, 1989). The nodules and crusts occur, as a rule, on the surface of the sediment–water interface in areas where the rates of their growth usually exceed the rates of sediment accumulations. The mineral composition of nodules and crusts is usually represented by tiny dispersed oxyhydroxides of Mn and Fe (Varentsov *et al.*, 1989). The nodules with high contents of Mn, Ni, Cu, Pb, and Zn are known in the World Ocean regions with optimum combination of a number of factors: (a) high biological productivity, (b) water depths exceeding the critical level of carbonate compensation (CCD), (c) regions of relatively active action of the bottom ocean currents. The largest deposits of such polymetallic nodules are developed in latitudinally elongated zones of the World Ocean, for example, in sub-equatorial regions, limited by Clarion and Clipperton fracture zones ($6°–20°$ N and $110°–180°$ W), where the radiolarian sediments overlie the Miocene and Paleogene sediments as a thin cover. The association of biogenic siliceous sediments with the nodules, characterized by anomalous enrichment, is of interest: Ni + Cu to 2%. As a rule, the cores of such nodules are composed of fragments of palagonitized basalts and hyaloclastites. Rather low rates of sedimentation (≤ 1 mm $\times 10^{-3}$ years) and pronounced erosion activity of bottom currents are observed. We emphasize that the region is adjacent to the western flank of the East Pacific Rise, characterized by tremendous hydrothermal activity (Leinen and Stakes, 1979; Leinen *et al.*, 1986).

It was found that nodules considerably enriched in Ni, Cu, and Mn are developed in areas of relative depressions of the bottom relief, where a noticeable role belongs to the oxidation diagenesis, whereas the high contents of Co are registered in the crust growths on seamounts (Exon, 1983; Varentsov *et al.*, 1989). In recent works it is reported that Mn–Fe oxyhydroxide nodules and crusts are developed as the result of hydrogenetic (hydrogenetic-diagenetic) or hydrothermal processes. The reality of the formation of Mn–Fe oxyhydroxide accumulations due to such an alternative supply of components is sufficiently obvious. However, in environments of the sedimentary zones the given sources are rarely manifested separately and independently: each of them may provide the definite contribution to the accumulation of the Mn–Fe oxyhydroxide deposits at different stages of their formation. The study of the mineralogy and geochemistry of the Mn–Fe oxyhydroxide growths on the substrate of seamounts of the Cape Verde plate, the Eastern Atlantic (Varentsov *et al.*, 1989), showed that the hydrothermal influence was noticeably manifested during the early stages of formation of Mn–Fe crusts, when the seamount (in the given case the Krylov Seamount) was located near the axial zone of the Mid-Atlantic Ridge. As this area of the lithospheric plate was moved off to the east, the role of hydrothermal processes was decreased, and in the accumulation of relatively young layers of the crust the participation of hydrogenetic factors was considerably greater than for more ancient accumulations.

The study of natural phenomena of formation of nodules, and crusts and the experimental data (Varentsov *et al.*, 1979) lead us to the conclusion that the oxyhydroxide phases of Mn, Fe, Ni, Co, and other metals are formed as the result of multistage chemisorption with autocatalytic oxidation interaction of active surfaces with component-bearing solutions (pore solutions, bottom seawater). A certain role during the early stages of accumulation may belong to microbiological phenomena, the essence of which probably differs mainly by kinetic parameters at the general trend of these processes (Varentsov *et al.*, 1979).

In the relatively vast literature discussing the role of hydrothermal processes in the World Ocean, there are extremely few evaluations of Mn amounts in metalliferous sediments and crusts: only in the work by Boström and Kunzendorf (1986) is a value of 5×10^{12} t of Mn mentioned, that is one order of magnitude higher than the amount of Mn in deep sea nodules – 10^{11} t (Wedepohl, 1980b). Irrespective of the absence of distinctly established ratios between amounts of Mn supplied by hydrothermal solutions to the ocean and accumulated in sediments, as well as the disseminated Mn in the seawater, it is necessary to take into account that a number of authors consider that 90% of Mn from the hydrothermal sources is buried in the form of metalliferous sediments (Bäcker and Fellerer, 1986; Boström and Kunzendorf, 1986; Bäcker and Lange, 1987; Scott, 1987). In this case, most calculations ignore the low temperature, halmyrolithic leaching of Mn from basalt rocks and clastic sediments on the ocean bottom, irrespective of the large scale of these processes and their rather significant contribution to the total balance of Mn in the ocean (Varentsov, 1971). Collier and Edmond (1984) also point to the hypothetical nature and rather approximate character of similar evaluations. According to these authors, there is almost a precise correspondence between the amounts of Mn introduced into the ocean and Mn consumed in the formation of deep sea nodules and sediments, and that Mn, delivered by hydrothermal sources to the ocean, constitutes about 90% of the total Mn introduced to the ocean. Lisitsyn *et al.* (1985) considered the role of a hydrothermal source to be significantly dominant. Taking into account all the complexity of biochemical transformations of manganese compounds, including the phenomena of multiple recycling (Landing and Bruland, 1987), the opinion may be formed that in so called hydrogenetic processes the initial nature of an element, in particular, Mn may loose its genetical roots. However, even approximated calculations of the balance of input and removal from the ocean, with the account of mechanisms of plate tectonics, indicate the leading role of hydrothermal and halmyrolithic processes.

5.3. Geochemical Evolution of Mn and Mn–Fe Ore Formation in Geological History

The analysis of the formation of Mn deposits in the history of the Earth's crust development allows one to focus particular attention to the basic concepts of the evolution of manganese ore process from the Archean to the modern time (Glasby, 1978, 1988; Levin, Varentsov, *et al.*, 1987; Roy, 1981, 1988; Varentsov, 1964; Varentsov *et al.*, 1984).

In the Archean (older than 2600 MA), the oxide manganese ores of the Iron Ore Group of India, rhodochrosite, kutnahorite accumulations of carboniferous of the Rio das Velhas Series, Brazil (2700 MA), and similar formations of other regions were accumulated in large, relatively shallow basins (Table 33).

In the Proterozoic (2600–570 MA), manganese was concentrated in carbonate-free, siliceous-carbonate, siliceous-ferruginous, and carbonate sediments, often altered to gondites, queluzites, eulysites, itabirites, jaspilites, and marbles, as well as in the sediments associated with the basic volcanites (see Table 33). The fixed paragenesis of Mn with different types of iron ore and carboniferous formations is noted. The ore formation took place in protogeosyncline basins on the Archean basement; it was known both in the zones similar to eugeosynclines and in zones close to miogeosynclines of the Phanerozoic. Reliable evidence for the manganese ore sediments accumulated in the large deep water basins is absent. The presented concepts did not change in the light of recently obtained data on tremendous accumulations of Mn ores in the Early Proterozoic Transvaal Supergroup (2500–2239 MA), Griqualand West Region, Kalahari, Cape Province, South Africa. Manganese deposits of this region are the lateral lithofacies, and in some sites alternate with iron ores and lavas, and they are substituted by dolomites up the section. Ore-bearing sediments were accumulated in the basin occurring on the Kaapvaal craton. Contribution of ore components proceeded mainly from hydrothermal sources associated with thick series of pillow-lavas. Manganese ores are represented by braunite, kutnahorite, and manganese calcite. In the region, over 13613 mln t of ore (1337 drill holes) with Mn content of 20% were explored. A 20 m thick bed is distinguished, containing 400 mln t of the ore with concentration of 38% of Mn (Beukes, 1983, 1989, 1993; Pouit, 1988). If we assume that 5026.3 mln t of Mn is concentrated in ores of the Kalahari Province (DeYoung *et al.*, 1984; Glasby, 1988), then this quantity constitutes 95.6% of Mn in ores of the Precambrian, or 78.8% of the World reserves of Mn ores on land from the Archean to Holocene (see Table 33). Nevertheless, a considerable part of Mn, delivered

Table 33. Distribution of Mn ore concentrations ($n \times 10^6$ t) in geological history for the main deposits on land (after Varentsov, 1964; Roy, 1981, 1988, 1989; De Young et al., 1984; Glasby, 1988).

No.	Deposit, region	Reserves ($n \times 10^6$ t)	World (%) reserves of Mn ores on land (Archean–Holocene)	World (%) reserves of Mn ores on land (PCm)	Stratigraphical unit	Absolute age ($n \times 10^6$ years)	Ore type, mineralogy	Lithology of enclosing sediments, rock association	Tectonic position, type of accumulation basis	Notes
1.	Deposits of Orissa State, India	12.4	0.19	0.24	Gangpur Group (Early Proterozoic)	846–946 sedimentation 1700–2000	Ghoriajor deposit; braunite, bixbyite, hollandite, jacobsite, hausmannite, vredenburgite, spessartine, rhodonite, Mn-pyroxene, Mn-amphiboles, Mn-micas	Micaceous schists		
					Khondalite Group (Late Archean-Early Proterozoic)	1600–2650	Mn-ores, weathering crusts: pyrolusite, cryptomelane, rarer manganite, 10 Å-manganate after braunite, hollandite, jacobsite, hausmannite, etc.	Rocks are metamorphosed by pregranulite facies: garnet-almandine, graphite granulites, etc. Ores occur conformably in enclosing rocks:		
					Iron-Ore Group (Early Archean)	2950–3200	Mn ores of weathering zone: pyrolusite, cryptomelane, manganite, 10 Å-manganate after braunite, etc.	Association: tuffogenic schists, silicites, dolomites	Platform (?)–internal shallow water shelf basin	

Table 33.　(Continued)

No.	Deposit, region	Reserves ($n \times 10^6$ t)	World (%) reserves of Mn ores on land (Archean–Holocene)	World (%) reserves of Mn ores on land (PCm)	Stratigraphical unit	Absolute age ($n \times 10^6$ years)	Ore type, mineralogy	Lithology of enclosing sediments, rock association	Tectonic position, type of accumulation basis	Notes
2.	Morro do Mina deposits, Brazil	1.4	0.12	0.03	Rio das Velhas Series (Late Archean)	> 2700	Rhodochrosite, manganocalcite, tephroite, spessartine, rhodonite, pyroxmangite and zone of weathering with Mn oxyhydroxides	Graphite phyllites, micaceous schists, quartzites; metamorphosed conglomerates, amphibolites, iron quartzites.	Geosyncline 0.02% basin (island-arc basin with exogenic sources (?) of Mn)	
3.	Anabarski massif, Siberia, Russia	–	–	–	Khapchanskaya Suite (Archean)	–	Mn-carbonates, Mn-silicates	Carboniferous Mn-bearing sediments, close to Rio das Velhas Series, Minas Gerais State, Brazil	–	
4.	East Africa	–	–	–	Lower part of the Turocka Series (Archean)				–	
5.	Serra do Novio deposit (Guinean Shield) Brazil	8.9	0.14	0.17	Amapa Series (Late Archean-Early Proterozoic)	1700–2700	Mn-carbonates: rhodochrosite with admixture of spessartine and rhodonite. In Mn silicate ores: spessartine with graphite, quartz, and Mn-amphibole	Amapa Series within Guinean Shield: (a) Lower Group of Jornal (amphibolites, crystalline schists, quartzites). (b) Upper Group: Serra do Novio (quartzites, biotite and graphite schists)	Miogeosyncline (?) back island-arc basin-(?)	0.14%

Table 33. (Continued)

No.	Deposit, region	Reserves ($n \times 10^6$ t)	World (%) reserves of Mn ores on land (Archean–Holocene)	World (%) reserves of Mn ores on land (PCm)	Stratigraphical unit	Absolute age ($n \times 10^6$ years)	Ore type, mineralogy	Lithology of enclosing sediments, rock association	Tectonic position, type of accumulation basis	Notes
	Azul deposit (Guinean Shield), Brazil	24.7	0.38	0.47						0.38%
6.	Ampaihi, Bekily deposits Madagascar Isl., Malagasy Republic	–	–	–	Graphit System (Early Proterozoic)	2420	Spessartine, rhodonite, tephroite; zone of weathering: Mn-oxyhydroxides	Crystalline schists (not volcanogenic)	–	
7.	Deposits of Andhra Pradesh State (India)	0.6	0.009	0.01	Penganga Beds (Proterozoic)		10 Å-manganate, birnessite, nsutite, ramsdellite, pyrolusite	Limestones, siliceous rocks, jaspers	Platform - internal, shallow, shelf basin	
					Khondalite Group, phase II (Middle Proterozoic)	1600	Conformable ore beds at different levels: braunite, hollandite, jacobsite, vredenburgite; weathering zone: Mn oxyhydroxides	Metasedimentary rocks of granulite facies of metamorphism: garnet, sillimanite, graphite granulites		0.01%
					Khondalite Group, phase I (Late Archean)	2650				20.01%

Table 33. (Continued)

No.	Deposit, region	Reserves ($n \times 10^6$ t)	World (%) reserves of Mn ores on land (Archean–Holocene)	World (%) reserves of Mn ores on land (PCm)	Stratigraphical unit	Absolute age ($n \times 10^6$ years)	Ore type, mineralogy	Lithology of enclosing sediments, rock association	Tectonic position, type of accumulation basis	Notes
8.	Nsuta deposit Ghana, West Africa	6.0	0.09	0.11	Birrimian Series (Early Proterozoic)	2200–2400	Mn carbonate, silicate ores: rhodochrosite, spessartine, rhodonite. In weathering zone: Mn oxyhydroxides	Lavas, tuffs of basic and intermediate composition, metamorphosed facies of green schists, sericite and graphite lutites and greywackes	Eugeosyncinal basin (inter-island-arc basin with endogenic source of Mn)	0.09%
9.	Deposit of Karanataka State, India	4.4	0.07	0.08	Dharwar Supergroup, Ranibennur-Dandeli Group (Early Proterozoic) Chitradurga Group (Early Proterozoic) Sandur Belt	2000–2400 2400	Syngenetic Mn oxide ores, conformably occurring in enclosing rocks: pyrolusite, cryptomelane, ramsdellite, manganite; with subordinate amounts of braunite, Fe-jacobsite	Phyllites, siliceous quartzites	Geosyncline (interisland-arc basin with exogenic source of Mn)	0.07%
10.	Kalahari deposit, West Griqualand, South Africa	5026.3	77.10	96.30	Transvaal Supergroup (Early Proterozoic)	2100–2500 (2239–2357)	Braunite, hausmannite, kutnahorite, manganocalcite	Sedimentary-diagenetic manganese-bearing sediments intercalated with ferruginous quartzites and are gradually replaced by dolomites up the section. Mn ores and Fe-quartzites are laterally substituted by basic lavas.	Transition from intracraton to marginal basin, hydrothermal-sedimentary genesis of ores	77.10%

Table 33. (Continued)

No.	Deposit, region	Reserves ($n \times 10^6$ t)	World (%) reserves of Mn ores on land (Archean–Holocene)	World (%) reserves of Mn ores on land (PCm)	Stratigraphical unit	Absolute age ($n \times 10^6$ years)	Ore type, mineralogy	Lithology of enclosing sediments, rock association	Tectonic position, type of accumulation basis	Notes
11.	Deposit of Tambao area, Burkina Faso (Upper Volta), and Mn deposit of Côte d'Ivoire, Ghana (West Africa	9.0	0.14	0.17	Birrimian System (Early Proterozoic)	2200–2400	Rhodochrosite, spessartine, rhodonite	Lavas and tuffs of basic and intermediate composition, metamorphosed to facies of green schists, sericite and graphite phyllites, greywackes. Thick weathering zone, laterite crust: Mn-oxyhydroxides	Geosyncline (interisland-arc basin with hydrothermal-sedimentary source, and accumulation of Mn)	0.14%
12.	Mn ore belt from Sao Joao del Rei deposit to Lafaiete deposit, Quadrilltero, Minas Gerais State, Brazil	–	–	–	Minas-Gerais Series (Middle-Early Proterozoic)	1350–2200	Rhodochrosite, managanocalcite, spessartine, rhodonite and other Mn silicates. In oxidation zone, weathering crust: Mn-oxyhydroxides	Ferruginous quartzites, dolomites; Mn ores occur conformably among phyllites, micaceous schists and amphibolites of Rio das Valhas Series, laterally replaced by carbonate facies of Fe quartzites	Miogeosyncline-platform (back arc, island-arc basin with distal hydrothermal-sedimentary source of Mn)	
13.	Deposits of Kisenge-Kamata belt, Zaire	–	–	–	Lukoshi complex	> 1845	Spessartine, braunite, bixbyite, manganite, rhodochrosite. In crust, oxidation zone: Mn-oxyhydroxides	Sericite-graphite schists, quartzites, stromatolite (Collenia) limestones with Mn carbonates		

Table 33.   (Continued)

No.	Deposit, region	Reserves ($n \times 10^6$ t)	World (%) reserves of Mn ores on land (Archean–Holocene)	World (%) reserves of Mn ores on land (PCm)	Stratigraphical unit	Absolute age ($n \times 10^6$ years)	Ore type, mineralogy	Lithology of enclosing sediments, rock association	Tectonic position, type of accumulation basis	Notes
14.	Deposits of Madhia Pradesh and Maharashtra States, India	22.3	0.34	0.43	Sausar Group (Late Proterozoic)	846–986	Braunite, bixbyite, hollandite, hausmannite, jacobsite, vredenburgite spessartine, rhodonite, Mn micas	Micaceous schists, carbonate-orthoquartzite rocks of different horizons	Miogeosyncine-platform (island-arc basin)	0.34
15.	Deposits of Gujarat and Rajasthan, India	> 1.2	> 0.02	> 0.02	Aravalli Supergroup=Chanpaner Group in Gujarat State (Early Proterozoic) Aravalli Supergroup and Champaner Group in Madhia Pradesh, Rajasthan, and Gudjarat States (Early Proterozoic)	1500–900 Sed. 2000 950–1500 Sed. 2000	Jhabua District Aravalli Supergroup: braunite, bixbyite, hollandite jacobsite, spessartine, rhodonite, Mn-pyroxene, Mn-amphiboles, Mn-micas; weathering zone. Panch Mahals District Champaner Group: rhodochrosite, spessartine, tephroite, rhodonite. Weathering zone.	Aravalli Supergroup: phyllites, quarizites Champaner Group: graphite schists		> 0.02%
16.	Moanda deposit, Gabon, West Africa	96.8	1.48	1.85	Francevillien Series (Early Proterozoic)	1740	Ca-rhodochrosite, manga-nodolomite. Thick zone of laterite Mn ores with wide set of Mn oxyhydroxide minerals	Black carbonaceous shales, sandstones, dolomites, ferruginous quartzites, phosphorites	Platform (internal shallow shelf basin)	1.48%

Table 33.   (Continued)

No.	Deposit, region	Reserves ($n \times 10^6$ t)	World (%) reserves of Mn ores on land (Archean–Holocene)	World (%) reserves of Mn ores on land (PCm)	Stratigraphical unit	Absolute age ($n \times 10^6$ years)	Ore type, mineralogy	Lithology of enclosing sediments, rock association	Tectonic position, type of accumulation basis	Notes
17.	Wafangzi deposit, China	0.7	0.01	0.01	Tiling Formation (Middle Proterozoic)	1250	Rhodochrosite, Fe-rhodochrosite, manganite	Association of silt, black shale, carbonate rocks: silts, schists, limestones	NE part of Djilu fault block on North-Chinese Platform (at the top of seamount of marine shelf basin)	0.01%
18.	Xiangtang deposit, China	> 4.5	> 0.07	> 0.09	Early Sinian sedimants (Late Proterozoic)	700–800	Rhodochrosite, kutnahorite, manganocalcite	Interglacial sediments of association of black high carbonaceous shales: carbonate-clayey sediments predominate	Yangze Platform (stagnated, semiclosed epicontinental depressional basin of moderate (shelf) depths	> 0.07
19.	Idikel', Tiouin, deposits, Ani Atlas, Morocco	–	–	–	(Middle Proterozoic)	900–1050	Braunite, jacobsite, hausmannite, manganocalcite, hollandite, Mn-dolomite, spessartine, rhodonite	Rhyolites, andesites, latites, trachytes, red tuffogenic schists, sandstones, conglomerates	Activized platform, aulakogene, shallow continental basin	–
20.	Otjosondu deposit, Namibia, Africa	–	–	–	Damara Super-group (Late-Middle Proterozoic)	620–1000	Braunite, jacobsite, hausmannite, bixbyite, vredenburgite, hollandite	Association of Mn ores with formation of iron quartzites: ferruginous quartzites, jaspilites, etc.	geosyncline (island arc basin with nonvolcanogenic source of Mn)	–

Table 33. (Continued)

No.	Deposit, region	Reserves ($n \times 10^6$ t)	World (%) reserves of Mn ores on land (Archean–Holocene)	World (%) reserves of Mn ores on land (PCm)	Stratigraphical unit	Absolute age ($n \times 10^6$ years)	Ore type, mineralogy	Lithology of enclosing sediments, rock association	Tectonic position, type of accumulation basis	Notes
21.	Deposits of areas of Malyi Khingan etc., Far East, Russia	0.2	0.003	0.004	Vendian-Lower Cambrian	540–630	Braunite, hematite, hausmannite, rhodochrosite, oligonite, magnetite, siderite, manganese silicates (rhodonite, etc.)	Siliceous-clayey schists, ferruginous quartzites, altered, metamorphosed rocks after tuffogenic sediments of basic composition	Activized blocks on Precambrian middle massifs (aulakogenes with hydrothermal-sedimentary sources of Mn (limited shallow water intercontinental basins)	
22.	Urandi deposit, Bahia State, Brazil	–	–	–	Late Proterozoic-Cambrian	520–790	Bixbyite, jacobsite, hausmannite, hollandite, Mn-dolomite, manganocalcite, spessartine, rhodonite	Mn ores are alternated with phyllites and crystalline schists	Geosyncline (island arc basin with nonvolcanogenic source of Mn)	
23.	Postmasburg deposit, South Africa	5.8	0.09	–	Early Proterozoic-modern time	0–2200	Braunite, bixbyite, paleokarst deposits of Mn	Products of leaching and redeposition from Mn-dolomites of Campbellrand Subgroup. In cherty dolomites braunite ores predominate: in dolomites without silicite-bixbyite varieties	Karst cavities in oxidation zone, formed since Proterozoic to modern time	0.09%

Table 33. (Continued)

No.	Deposit, region	Reserves ($n \times 10^6$ t)	World (%) reserves of Mn ores on land (Archean–Holocene)	World (%) reserves of Mn ores on land (PCm)	Stratigraphical unit	Absolute age ($n \times 10^6$ years)	Ore type, mineralogy	Lithology of enclosing sediments, rock association	Tectonic position, type of accumulation basis	Notes
24.	Moro do Urukum deposit, Brazil	27.8	0.42	2.10	Cambrian-Ordovician	438–590	Cryptomelane	Ferruginous quartzites, sandstones, conglomerates	Platform (epicontinental shallow basin with exogenic source of Mn)	0.42%
25.	Usinskoe deposit, Kuznetskii Alatau, Russia	28.0	0.43	2.15	Early Cambrian	540–590	Ca-rhodochrosite, manganocalcite, oxidation zone: Mn-oxyhydroxides	Limestones, dolomites, black limestone-shales	Eugeosyncline (island arc basin, shallow, mixed: hydrothermal and exogenic source of Mn)	0.43%
26.	Karazhal'skoe and other deposits, Central Kazakhstan	120.0	1.84	9.23	Famennian, Late Devonian	360–367	Braunite, hausmannite, goethite, manganocalcite and Mn limestones, Pb-Zn mineralization. In oxidation zone: vernadite, romanechite (psilomelane), cryptomelane, etc.	Red, variegated ferruginous-cherty limestones, argillites, jaspers, siliceous schists	Aulakogene (intercontinental shallow basin with hydrothermal source of Mn, Fe, accompanied by Pb, Zn sulphide mineralization).	1.84%
27.	Mn deposits of Leping area, China	3.0	0.05	0.23	Limestone Suite Huanglong (Middle Carboniferous)	333	Mn–Fe carbonates, Fe-oxyhydroxides	Limestone sediments of Huanglong Suite	Deposits of Mn–Fe ores in shallow marine sediments of platform intercontinental basin	0.05%

Table 33.   (Continued)

No.	Deposit, region	Reserves ($n \times 10^6$ t)	World (%) reserves of Mn ores on land (Archean–Holocene)	World (%) reserves of Mn ores on land (PCm)	Stratigraphical unit	Absolute age ($n \times 10^6$ years)	Ore type, mineralogy	Lithology of enclosing sediments, rock association	Tectonic position, type of accumulation basis	Notes
28.	Ulu-Telyakskoe deposit, Bashkirskoe Predural'e, Russia	1.4	0.02	0.11	Late Permian sediments (Ufimian stage)	258	Mn-limestones and marl. In oxidation zone: vernadite, romanechite	Limestones, marls, cherty-limestone and dolomite-anhydrite sediments	Platform shallow epicontinental evaporite basin with exogenic source of Mn	0.02%
29.	Mn deposit Zunyi area, China	> 2.0	> 0.03	0.15	Longtan Formation (Late Permian)	253	Rhodochrosite, manganocalcite, hausmannite, alabandite	Association of black, high carboniferous shales and argillites, in transgressive series	Carbonate platform sediments (inland basin of lacustrine-marine type, lagoonal-marine facies)	> 0.03
30.	Úrkút Mn deposit Bakoni Mountains, Hungary	44.0	0.67	3.38	Early Jurassic, Toarcian age	194–188	Rhodochrosite, and Fe-smectite, celadonite, manganite. Oxidation zone: pyrolusite	Siliceous limestones	Hydrothermal-sedimentary ores in limited riftogenous basin on carbonate platform (shelf interval of depths)	0.67%
31.	Molango Mn deposit, Mexico	523.6	8.03	40.27	Tamaan Formation, Kimmeridgian age Late Jurassic	156–150	Thin laminated Mn carbonates: kutnahorite, rhodochrosite	Taman Formation: intercalation of black, dark-grey shales and limestones	Platform, diagenetic-syngenetic mineralization in stagnated, oxygen-free basin of limited dimensions with depths of 100 to 300 m	8.03%

Table 33. (Continued)

No.	Deposit, region	Reserves ($n \times 10^6$ t)	World (%) reserves of Mn ores on land (Archean–Holocene)	World (%) reserves of Mn ores on land (PCm)	Stratigraphical unit	Absolute age ($n \times 10^6$ years)	Ore type, mineralogy	Lithology of enclosing sediments, rock association	Tectonic position, type of accumulation basis	Notes
32.	Groote Eylandt deposit Australia	152.9	2.34	11.76	Albian-Cenomanian sediments with superimposed Eocene Oligocene weathering crust	(113–97.5) (42–23)–0	Mn-carbonates (Mn-limestones, marls of Albian–Cenomanian age); in Eocene–Oligocene weathering crust cryptomelane, pyrolusite, etc.	Quartz sand, clays, pyrite-bearing Mn-marl	Albian–Cenomanian intracraton basin, in Eocene–Oligocene to the modern time laterite weathering crust with high grade Mn ores was formed	2.34%
33.	Imini and Tasdremt deposits, Marocco	0.7	0.01	0.05	Cenomanian–Turonian	97.5–88.5	Pyrolusite, romanechite (psilomelane), coronadite, hollandite, lithiphorite	Mn-oxide ores beds occur at several levels in dolomite sequence	Platform. Shallow basin: tide zone of transition from land to sea shallow water: superimposed karst layers of Mn oxide ores	0.01%
34.	Obrochishte deposit, Varna area, Bulgaria	5.0	0.08	0.38	Oligocene	38–25	Ca-rhodochrosite, manganocalcite, alabandite. In oxidation zone: pyrolusote, etc.	Diatomites, clays	Aulakogene, mobile, activized margin of platform	0.08%

Table 33. (Continued)

No.	Deposit, region	Reserves ($n \times 10^6$ t)	World (%) reserves of Mn ores on land (Archean–Holocene)	World (%) reserves of Mn ores on land (PCm)	Stratigraphical unit	Absolute age ($n \times 10^6$ years)	Ore type, mineralogy	Lithology of enclosing sediments, rock association	Tectonic position, type of accumulation basis	Notes
35.	Nikopol'skoe deposit, Ukraine	144.1	2.21	11.08	Early Oligocene	38	Pyrolusite, romanechite (psilomelane), manganite, cryptomelane, 10 Å-manganite, etc. Ca-rhodochrosite, manganocalcite. Zone of oxidation: MN-oxyhydroxides	Orthoquartzite, glauconite sands, clays	Slope of Ukrainian crystalline shield (coastal shallow water-littoral of vast sea basin; exogenic source of Mn)	2.21%
36.	Bol'she-Tokmakskoe deposit, Ukraine	203.5	3.12	15.65	Lower Oligocene	38	Ca-rhodochrosite, manganocalcite, pyrolusite, romanechite (psilomelane), manganite, 10 Å-manganate, etc.	Orthoquartzite, glauconite sands, clays	Slope of Ukrainian crystalline shield (coastal shallow water-littoral of vast sea basin; exogenic source of Mn)	3.12%
37.	Chiaturskoe deposit, Georgia	38.0	0.58	2.92	Lower Oligocene	38	Ca-rhodochrosite, manganocalcite, pyrolusite, manganite, etc.	Sandstones quartz-polymictic, glauconite diatomites, clays, illite-montmorillonite	Median massif of folded region (geosyncline) back island arc basin	0.58%
38.	Deposits of West Transvaal, South Africa	3.7	0.06	0.28	Weathering crust formed since Proterozoic to present time	2000–0	Oxides, hydroxides of Mn		Zone of weathering, karst formation, redeposition after Lower Proterozoic rocks, predominantly Mn dolomites etc.	0.06%

Table 33. (Continued)

No. Deposit, region	Reserves ($n \times 10^6$ t)	World (%) reserves of Mn ores on land (Archean–Holocene)	World (%) reserves of Mn ores on land (PCm)	Stratigraphical unit	Absolute age ($n \times 10^9$ years)	Ore type, mineralogy	Lithology of enclosing sediments, rock association	Tectonic position, type of accumulation basis	Notes
World reserves of Mn ores on land Total ($n \times 10^6$ t)	6519.4	100.0	100.0						
	1300.2	19.94	–	Phanerozoic	0.510–0.000				
Distribution:	550.0	8.44	42.31	Cenozoic	0.065–0.000				
land Mn ores in	560.3	8.71	43.70	Mesozoic	0.248–0.065				
main geochrono-	181.9	2.79	13.99	Paleozoic	0.59–0.248				
logical units	5219.2	80.06	–	Precambrian	0.59–4.5				
(era, subera)	27.5	0.42	0.53	Middle and Late Proterozoic	1.6–0.9				
	5143.7	78.90	9855	Early Proterozoic	2.5–1.6				
	48.0	0.74	0.92	Archean–Early Proterozoic (unstratified layers)	4.5–1.6				
World Ocean Mn–Fe nodules (i.e. 10^{11} t Mn)	10^5			Eocene–Holocene	54.9–0 m.y.	MN-Fe-oxyhydroxides	Siliceous, clayey, rarer carbonate sediments, often borders of erosion hiatuses	Regions of World Ocean with low sedimentation rates, as a rule below CCD	Wedepohl, 1980
World Ocean 5×10^6 Mn–Fe metalliferous (i.e. 5×10^{12} t Mn) sediments				Middle Jurassic–Holocene	181.0–0 m.y.	Mn–Fe oxyhydroxides, sulphides of poly-metals and Fe, Fe-smectites, etc.	In different proportions are mixed with siliceous, clayey, carbonate sediments	Characteristic accumulations of axial zones of oceans and seas, as well as faults, and of seamounts and intermantle fractures	Bostrom, K., and Kunzendorf, H., 1986

by hydrothermal solutions to this Early Proterozoic basin of sedimentation, was bonded to a limited degree to low valence (braunite) oxides and diagenetic carbonate ores, and was dispersed in the water mass because the oxygen content in the atmosphere of that period constituted approximately one hundredth of the modern value.

In the Paleozoic (570–230 MA), manganese deposits of volcanogenic-sedimentary (basaltoid) formations were mainly formed. These deposits were accumulated in eugeosyncline rift basins in environments of relatively shallow depressions (Magnitogorski synclinorium) with rather active underwater volcanism. The Famennian hydrothermal-sedimentary ores of Central Kazakhstan, the associations of Usinskoe and Ulutelyakskoe types were accumulated in comparatively shallow basins (see Table 33).

In the Mesozoic (230–65 MA), apart from rather small manganese deposits of limestone-dolomite formation of North Africa, the formation of nodules and metalliferous sediments in pelagic environments of the Early Jurassic–Cretaceous is of interest (marginal seas of Tethys, Alpine-Mediterranean region, Timor Isl, Reti, Indonesia).

It is important to emphasize that in the Middle Jurassic the break-up of the supercontinent took place, leading to the initiation of formation of the modern oceans.

Cenozoic (65.0 MA). At the end of the Paleogene at the northern periphery of the Paratethys, under the conditions of the marine littoral-shelf, the largest manganese deposits were formed (Varna region of Bulgaria, the south of Ukraine, Western Georgia, Mangyshlak Peninsula). However, irrespective of the commercial significance of these deposits, the total reserves of Mn in them (390.6 mln t of Mn) are rather small in geological history (6.1% from the amount of Mn in ores on the land; see Table 33).

The beginning of the formation of deep sea iron-manganese nodules is mainly associated with the boundary of the Paleogene and Neogene, when as a result of the opening of the Drake Passage, the massive invasion of Antarctic waters into more northern regions took place. The development of the World Ocean, manifested in its widening and deepening and accompanied by gigantic endogenous activity of the axial zones, led to manganese ore accumulation unknown in scales within the Earth's history: 10^{11} t of Mn in the form of nodules and crusts and 5×10^{12} t of Mn as metalliferous sediments (Boström and Kunzendorf, 1986; Wedepohl, 1980b).

BIBLIOGRAPHY

Alexander, L.T. and Cady, J. (1962) Genesis and hardening of laterite. U.S. Depart. Agriculture, Technic. Bull., No. 1282, 1–90.

Anderson, B.J., Jenne, E.A. and Chao, T.T. (1973) The sorption of silver by poorly crystallized manganese oxides. *Geochim. Cosmochim. Acta* **37**(3), 611–622.

Andreev, S.I., Kazmin, Yu.B. and Egiazarov, B.H. *et al.* (1984) In: *Iron-Manganese Concretions of the World Ocean*, Nedra, Moscow, pp. 18–61.

Anikeeva, L.I., Andreev, S.I. and Kazmin, Yu.B. *et al.* (1984) In: *Iron-Manganese Concretions of the World Ocean*, Nedra, Moscow, pp. 62–104.

Appelbaum, J.A. and Hamann, D.R. (1980) Chemisorption on semiconductor surfaces. In: Smith, J.R. (Ed.), *Theory of Chemisorption*, Springer-Verlag, Berlin, Heidelberg, New York, pp. 43–68.

Arrhenius, G. (1963) Pelagic sediments. In: Hill, M.N. (Ed.), *The Sea, Vol. 3*, Interscience, New York, pp. 655–727.

Arrhenius, G.O.S. and Bonnatti, E. (1965) Neptunism and volcanism in the ocean. In: Sears, M. (Ed.), *Progress in Oceanography, Vol. 3*, Pergamon Press, New York, pp. 7–22.

Arrhenius, G.O. and Tsai, A.G. (1981) *Structure, Phase Transformation and Pre-Biotic Catalysis in Marine Manganese Minerals*, Scripps Inst. Oceanography, La Jolla, CA, U.S.A., Vol. 81-28, pp. 1–19.

Bäcker, H. and Fellerer, R. (1986) Marine mineral exploration examples. In: Kunzondorf, H. (Ed.), *Marine Mineral Exploration*, Elsevier, Amsterdam, pp. 191–244.

Backer, H. and Lange, J. (1987) Recent hydrothermal metal accumulation, products and conditions of formation. In: Teleki, P.G., Dobson, M.R., Moore, J.R. and Stackelberg, U. (Eds.), *Marine Minerals. Advances in Research and Resource Assesment*, D. Reidel, Dordrecht, pp. 317–337.

Balashov, Yu.A. (1976) *Geochemistry of Rare Earth Elements*, Nauka, Moscow, 268 pp.

Balistrieri, L.S. and Murray, J.W. (1982a) The surface chemistry of δ-MnO$_2$ in major ion seawater. *Geochim. Cosmochim. Acta* **46**, 1041–1052.

Balistrieri, L.S. and Murray, J.W. (1982b) The adsorption of Cu, Pb, Zn and Cd on goethite from major ion seawater. *Geochim. Cosmochim. Acta* **46**, 1253–1265.

Balitskii, D.K. and Kornev, T.Ya. (1984) Manganese ore mineralization of the Precambrian series of the Eniseiskii Ridge. In: *Manganese Ore Mineralization on the USSR Territory*, Nauka, Moscow, pp. 82–89.

Banerdji, P.K. (1981) Lateritization process: What we know about them? *Nature and Resources* **17**(3), 21–25.

Banerdji, P.K. (1984) About the modern state of knowledge on the problem of ore formation in the lateritization process. In: *XXVII International Geological Congress. Reports*, Nauka, Moscow, Vol. 12, pp. 106–115.

Barcelos, J.P. and Bücni, J. (1986) Mina de minério de ferro-manganes de Miguel Congo, Minas Gerais. (Capítulo VI). In: *Principas Depósitos Minerais do Brasil, Vol. II, Ferro e Metais da Indústria do Aço*, Coordinador-geral: Schobbenhaus, C. and Coelho, C.E.S., Brasilia. Ministerio das Minas é Energia, pp. 86–95.

Basilio, J.A. and Brondi, M.A. (1986) Distrito Manganesífero da Região de Licínio de Almeida, Bahia, (Capítulo XV). In: *Principas Depositos Minerais do Brasil, Vol. II, Ferro e Metais da Industria do Aço*, Coordenador-geral: Schobbenhaus, C. and Coelho, C.E.S., Brasilia, Ministerio das Minas e Energia, pp. 177–185.

Baskov, E.A. (1976) *Paleohydrogeological Analysis during Metallogenic Investigations*, Nedra, Leningrad, 200 pp.

Baskov, E.A. (1983) *Foundations of Paleohydrogeology of Ore Deposits*, Nedra, Leningrad, 263 pp.

Baskov, E.A. and Pavlov, D.I. (1985) Subsurface waters and mineralization processes. In: *Subsurface Waters and the Evolution of Lithosphere. Proc. All-Union Conference*, Nauka, Moscow, Vol. 1, pp. 5–19.

Basu, A., Blanchar, D.P. and Brannon, J.G. (1982) Rare earth elements in the sedimentary cycle: A pilot study of the first leg. *Sedimentology* **29**, 737–742.

Baturin, G.N. (1986) *Geochemistry of Iron-Manganese Concretions of the Ocean*, Nauka, Moscow, 328 pp.

Baynton, W.U. (1984) Cosmochemistry of the rare earth elements: Meteorite studies. In: Henderson, P. (Ed.), *Rare Earth Element Geochemistry* (Developments in Geochemistry, Vol. 2), Elsevier, Amsterdam, Oxford etc., pp. 63–114.

Beadle, C.W. and Burges, A.A. (1953) Further note on laterites. *Austral. J. Sci.* **15**(5), 170–171.

Beaudoin, B., Lesavre, A. and Pélissonnier, H. (1976) Action des eaux superficielles dans le gisement de manganèse d'Imini (Maroc). *Bull. Soc. Geol. France (7)* **XVIII**(1), 95–100.

Beevers, J.R. (1966) A chemical investigation into the role of sorption processes in ore genesis. Commonwealth of Australia. Dept. of National Development. Bureau of Mineral Resources. Geology and Geophysics. Report No. 106, 63 pp.

Benjamin, M.M. and Leckie, J.O. (1981a) Multiple-site adsorption of Cd, Cu, Zn, and Pb on amorphous iron oxyhydroxide. *Jour. Coll. Inter. Sci.* **79**, 209–221.

Benjamin, M.M. and Leckie, J.O. (1981b) Competitive adsorption of Cd, Cu, Zn, and Pb on amorphous iron oxyhydroxide. *Jour. Coll. Inter. Sci.* **83**(2), 410–419.

Berner, R.A. (1981) A new classification of sedimentary environments. *Jour. Sed. Petrology* **51**, 359–365.

Berry, J.E. and Berglin, C.I.K. (1968) Mineral processing studies on Groote Eylandt manganese ores. *Trans. Inst. Engrs. Austr. Mech. Chem. MC4*, pp. 144–148.

Betekhtin, A.G. (1946) *Commercial Manganese Ores of the USSR*, Publ. Acad. Sci. of the USSR, Moscow–Leningrad, 315 pp.

Betekhtin, A.G. (Ed.) (1964a) *Chiaturskoe Manganese Deposit*, Nedra, Moscow, 244 pp.

Betekhtin, A.G. (Ed.) (1964b) *Nikopol'skii Manganese Ore Basin*, Nedra, Moscow, 535 pp.

Beukes, N.J. (1973) Precambrian iron-formation in Southern Africa. *Econ. Geol.* **68**, 960–1004.

Beukes, N.J. (1983) Palaeoenvironmental setting of iron formations in the depositional basin of the Transvaal Supergroup, South Africa. In: Trendall, A.F. and Morris, R.C. (Eds.), *Iron Formation: Facts and Problems*, Elsevier, Amsterdam etc., pp. 131–209.

Beukes, N.J. (1984a) Iron-formation and the evolution of iron and manganese ore deposits. Transvaal Supergroup, South Africa. In: *XXVII International Geological Congress*, Moscow, 4–14 August, 1984, Abstracts, Vol. 2, section 04-05. Nauka, Moscow, p. 20.

Beukes, N.J. (1984b) Sedimentology of the Proterophytic Kalahari manganese deposit, Transvaal Supergroup, South Africa. In: *XXVII International Geological Congress*, Moscow, 4–14 August, 1984, Abstracts, Vol. VI, section 12. Nauka, Moscow , pp. 32–33.

Beukes, N.J. (1984c) Sedimentology of the Kuruman and Griquatown iron-formations, Transvaal Supergroup, Griqualand West, South Africa. *Precambrian Research* **24**, 47–84.

Beukes, N.J. (1986) The Transvaal Sequence in Griqualand West. Mineral deposits of Southern Africa. In: Anhaeusser, C.R. and Maske, S. (Eds.), Geol. Soc. S. Afr., Johannesburg, Vols. I and II, pp. 819–828.

Beukes, N.J. (1988) Depositional setting of Early Proterozoic manganese deposits of the Transvaal Supergroup, Griqualand West, South Africa. In: *IAS International Symposium on Sedimentology Related to Mineral Deposits*, Beijing, China, July 30–August 4, 1988, Abstracts, p. 6.

Beukes, N.J. (1989) Sedimentological and geochemical relationship between carbonate, Iron Formation and manganese deposits in the Early Proterozoic Transvaal Supergroup, Griqualand West, South Africa. In: *XXVIII International Geological Congress*, Washington, DC, U.S. A., July 9–19, 1989, Abstracts, Vol. 1, pp. 1–143.

Beukes, N.J. (1993) A review of manganese deposits associated with the Early Proterozoic Transvaal Supergroup in Northern Cape Province, South Africa. In: *XVI International Colloquium on Africa Geology*, September 14–16, 1993, Eznlwini, Swaziland, Mbabane, Swaziland, Extended abstracts, Vol. 1. pp. 37–38.

Beukes, N.J., Gutzmer, J. and Kleyenstüber, A.S.E. (1993) Iron and manganese deposits of the Transvaal Supergroup in Griqualand West, Excursion Guide, South Africa Field Workshop, September, 1993, IGCP Project 318: Genesis and correlation of marine polymetallic oxide. Rand Afrikaans University and Mintek, Johannesburg, R.S.A., 223 pp.

Bezrukov, P.L. (Ed.) (1970a) *The Pacific Ocean. Vol. VI, book I. Sedimentation in the Pacific*, Nauka, Moscow, 427 pp.

Bezrukov, P.L. (Ed.) (1970b) *The Pacific Ocean. Vol. VI, book II. Sedimentation in the Pacific*, Nauka, Moscow, 419 pp.

Bezrukov, P.L. (Ed.) (1976) *Iron-Manganese Concretions of the Pacific*, Nauka, Moscow, 301 pp.

Black, R. (1967) Sur l'ordonnance des chaines métamorphiques en Afrigue Occidentale. *Chr. Mines.* **364**, 225-238.

Blokh, A.M. (1969) *Structure of Water and Geological Processes*, Nedra, Moscow, 216 pp.

Blokh, A.M. and Kochenov, A.V. (1964) Admixture elements in the bone phosphate of fossil fish. *Geology of REE Deposits, Issue 24*, Nedra, Moscow, 83 pp.

Boardman, L.G. (1940) The geology of the manganese deposits on Aucampsrust, Postmasburg. *Trans. Geol. Soc. S. Afr.* **43**, 27–36.

Boardman, L.G. (1961) Manganese in the Union of South Africa. *Commonwealth Mining Metall. Congr. 7th* **1**, 201–216.

Boardman, L.G. (1964) Further geological data on the Postmasburg and Kuruman manganese ore deposits, Northern Cape Province. In: Haughton, S.H. (Ed.), *The Geology of Some Ore Deposits in Southern Africa, 2*, Geol. Soc. S. Afr., Johannesburg, pp. 415–439.

Bobrovnik, D.P. and Khmelevskii, V.A. (1966) Major features of mineralogy and geochemistry of the Burshtyn-skoe manganese deposit. In: *Problems of Mineralogy of Sedimentary Associations, book 7*, Publ. L'vovskii University, L'vov, pp. 97–114.

Bobrovnik, D.P. and Khmelevskii, V.A. (1968) Tortonian sedimentary manganese-carbonate rocks of the south-western margin of the Russian Platform and the conditions of their formation. *Lithology and Mineral Resources* **I**, 117–125.

Bockris, J.O.M., Devanathan, M.A.V. and Müller, K. On the structure of charged interfaces. *Proc. Roy. Soc. A.* **274** (M. 63), 55.

Boeglin, J.-L., Melfi, A.J., Nahon, D. and Tardy, Y. (1980) Solutions solides caractérisant plusieurs générations de rhodochrosites dans les protores des districts de Conselheiro Lafaiete, au Brésil et de Ziemougoula en Côte-d'Ivoire. *C. R. de l'Acad. Sci., Paris, Série D* **291**, 13–15.

Boiko, T.F., Sotskov, Yu.P and Maeva, M.M. (1986) Behaviour of lanthanides and Y in the process of rock weathering. *Lithology and Mineral Resources* **3**, 12–26.

Bolton, B.R., Berents, H.W. and Frakes, L.A. (1990) Groote Eylandt manganese deposits. In: Hugnes, F.E. (Ed.), *Geology of the mineral deposits and Papua New Guinea*, Monograph No. 14, The Australian Institute of Mining and Metallurgy, pp. 1575–1579.

Bolton, B.R. and Frakes, L.A. (1982) A shallow-water marine manganese deposit: Groote Eylandt, Australia. In: *International Association on the Genesis of Ore Deposits (IAGOD), VI Symposium*, Tbilisi, September 6–12, 1982. Collected abstracts, Tbilisi, 1982, p. 287.

Bolton, B.R. and Frakes, L.A. (1984) On the origin of manganese giants: a preliminary comparative investigation of the Chiatura (USSR) and Groote Eylandt (Australia) deposits. In: *XXVII International Geological Congress*, Moscow, 4–14 August, 1984, Abstracts, Vol. VI (Section 12), Nauka, Moscow, p. 46.

Bolton, B.R. and Frakes, L.A. (1985) Geology and genesis of manganese oolite, Chiatura, Georgia, USSR. *Geol. Soc. Am. Bull.* **96**, 1398–1406.

Bolton, B.R., Frakes, L.A. and Cook, J.N. (1988) Petrograpphy and origin of inversed graded manganese pisolite from Groote Eyland, Australia. *Ore Geol. Rev.* **4**, 47–69.

Bolton, B.R., McHugh, L. and Frakes, L.A. (1984) The geology and geochemistry of secondary manganese deposits on Groote Eylandt, Australia. In: *XXVII International Geological Congress*, Moscow, August 4–14, 1984. Abstracts, Vol. VI (Section 12), Nauka, Moscow, pp. 46–47.

Bonhomme, M. (1962) Contribution a l'étude géochronologique de la plateforme ouest africaine. These Ann. Fac. Sci. Clermond Ferrand (Géol. et Min.).

Bonhomme, M.G. and Bertran-Sarfati, J. (1982) Correlation of Proterozoic sediments of Western and Central Africa and South America based upon radiochronological and palaeontological data. *Precambrian Research* **18**, 171–194.

Bonhomme, M.G., Cordani, U.G., Kawashita, K., Macedo, M.-H. and Filho, A.T. (1982) Radiochronological age and correlation of Proterozoic sediments in Brazil. *Precambrian Research* **18**, 103–118.

Bonhomme, M.G., Gauthier-Lafaye, F. and Weber, F. (1982) An example of Lower Proterozoic sediments: The Francevillian in Gabon. *Precambrian Research* **18**, 87–102.

Boström, K. (1973) The origin and fate of ferromanganese an active ridge sediments. Acta Univ. Stockholmiensis. *Stockholm Contributions in Geology* **27**(2), 149–243.

Bostrom, K. and Kunzendorf, H. (1986) Marine hard mineral resources. In: Kunzendorf, H. (Ed.), *Marine Mineral Exploration*, Elsevier, Amsterdam, pp. 21–53.

Bottke, H. (1969) Die Eisenmanganese der Grube Dr. Geier bei Bingen/Rhein als Verwitterungsbildungen des Mangans vom Typ Lindener Mark. *Mineral. Deposita* **4**(4), 355–367.

Bouladon, J. and Graciansky, P.-Ch., de (1985) Le minéralisations dites de converture (plomb, zinc, cuivre, uranium, barytine, fluorine . . .) de Trias an Pliocène en France. Leurs realtion area les phénomènes connexes de l'ouverture de la Téthys et de l'Atlantique Nord. *Chron. Rech. Miniere* **53**, No. 480, 17–33.

Bouladon, J. and Jouravsky, G. (1956) Le gîtes de manganese du Maroc (suivi d'une description des gisements du Précambrien III). In: *XX Congress Geologico International Symposium Sorbe Yacimientos de Manganeso*, Vol. II, Africa, Mexico, pp. 217–248.

Bouladon, J., Weber, F., Veysset, C. and Favre-Mercuret, R. (1965) Sur la situation géologique et la type métallogénique du gisement de manganese de Moanda, près de Franceville (République Gabonaise). *Bull. Serv. Carte Géol. Als. Lorr., Strassbourg* **18**(4), 253–276.

Boyle, E.A., Edmond, J.M. and Sholkovitz, E.R. (1977) The mechanism of iron removal in estuaries. *Geochim. Cosmochim. Acta* **41**, 1313–1324.

Brazil's important new manganese province. (1978) *Mining Mag.* **138**(3), 181.

Bricker, O.P. (1965) Some stability relations in the system $Mn-O_2-H_2O$ at 25° and one atmosphere total pressure. *Amer. Mineral.* **50**, 1296–1354.

Bros, R., Stille, P., Gauthier-Lafaye, F., Weber, F. and Clauer, N. (1992) Sm–Nm isotopic dating of Proterozoic clay mineral: An example from the Francevillian sedimentary series, Gabon. *Earth Planet. Sci. Lett.* **113**, 207–218.

Bruce, T. and Benkovich, S. (1970) *Mechanisms of Bioorganic Reactions*, Mir Publishers, Moscow.

Bruevich, S.M. (1946) Average chemical composition of oceanic water, and "normal" seawater according to the recent data. *Probl. Arktiki* **4**.

Bruland, K.W. (1983) Trace elements in seawater. In: Riley, J.P. and Chester, R. (Eds.), *Chemical Oceanography*, Academic Press, London etc., Vol. 8. pp. 157–220.

Bugel'skii, Yu.Yu. (1979) *Ore-Bearing Weathering Crusts of Humid Tropics*, Nauka, Moscow, 286 pp.

Bugel'skii, Yu.Yu. (1980) Hydrogeochemistry of the weathering crust. In: *Problems of the Theory of the Weathering Crust and Exogenic Deposits*, Nauka, Moscow, pp. 154–161.

Bugel'skii, Yu.Yu., Vitovskaya, I.V., Nikitina, A.P. and Slukin, A.D. (1983) Regularities of mineralization formation in the weathering crust. In: *Problems of Petrology, Mineralogy, and Ore Genesis*, Nauka, Moscow, pp. 56–60.

Burkov, V.V. and Podporina, E.K. (1967) Rare earths in the crust of weathering of granitoids. *Dokl. AN SSSR* **177**(3), 691–694.

Burns, R.G. (1976) The uptake of cobalt into ferromanganese nodules, soils and synthetic manganese (IV) oxides. *Geochim. Cosmochim. Acta* **40**, 95–102.

Burns, R.G. and Burns, V.M. (1975) Mechanism for nucleation and growth of manganese nodules. *Nature, London* **255**, No. 5504, 130–131.

Burns, R.G. and Burns, V.M. (1979) Manganese oxides. In: Burns, R.G. (Ed.), *Marine Minerals, Short Course Notes, November 1979, 6*, Mineralogical Society of America, Washington, DC, U.S.A., pp. 1–46.

Bursill, L.A. and Grzinic, G. Incommensurate superlattice ordering in the hollandites $Ba_x\ Ti_{8-x}\ Mg_x\ O_{16}$ and $Ba_x\ Ti_{8-2x}\ Ga_{2x}O_{16}$. *Acta Cryst., B* **36**, 2902–2913.

Buser, W., Graf, P. and Feitknecht, W. (1954) Beitrag zur Kennttnis der Mangan (II) – Manganite und des δ-MnO_2. *Helvetica Chim. Acta* **37**, 2322–2333.

Button, A. (1976) Transvaal and Hamersley basins – Review of basin development and mineral deposits. *Min. Sci. Engineering* **8**, 262–293.

Button, A. and Tyler, N. (1981) The character and economic significance of Precambrian palaeoweathering and erosion surfaces in Southern Africa. *Econ. Geol.*, 75th Anniv. Vol., 676–699.

Calvert, S.E. and Price, N.B. (1977) Shallow water, continental margin and lacustrine nodules: distribution and geochemistry. In: Glasby, G.P. (Ed.), *Marine Manganese Deposits*, Elsevier, Amsterdam, pp. 45–86.

Cannon, W.F. and Force, E.R., (1983) Potential for high-grade shallow marine manganese deposits in North America. In: Shanks, W.C., III (Ed.), *Cameron Volume of Unconvential Mineral Deposits*, Am. Inst. Min. Metal. Eng. Soc. Mining Engineers, New York, N.Y., pp. 175–190.

Chen, C.C., Golden, D.C. and Dixon, J.B. (1986) Transformation of synthetic birnessite to cryptomelane: An electron microscopy study. *Clays and Clay Minerals* **34**, 565–571.

Choi, Hunsoo and Kim, Soo Jin (1992) Chemistry and dehydration behavior of (Ca, Mg)-buserite from the Junggun Mine, Korea. *Journ. of the Mineralogical Society of Korea*, 102–108.

Chowdhury, M.K.R., Venkatsh, V., Anankatsh, V., Anandalwar, M.A. and Paul, D.K. (1965) Recent concepts on the origin of Indian laterites. *Proc. Nat. Acad. Sci.*, 31, Pt. A, No. 6, 547–558.

Chuiko, V.T. and Gavrilyuk, A.I. (1967) Study of adsorption (coprecipitation) of microamounts of nickel by iron hydroxide in various electrolytes. *Visnik Lviv. Derzhav. Un-tu, Ser. Khimichna, (USSR)* 9, 57–66.

Chuiko, V.T. and Gavrilyuk, A.I. (1969) Sorption (coprecipitation) of microamounts of copper by iron hydroxides in the presence of macroamounts of different electrolytes. *Visnik Lviv. Derzhav. Un-tu, Ser. Khimichna, (USSR)* 11, 49–54.

Chukhrov, F.V., Drits, V.A. and Gorshkov, A.I. (1987a) About structural transformations of manganese oxides of oceanic Fe–Mn concretions. *Proc. Acad. Sci. USSR, Geol. Series*, No. I, 3–14.

Chukhrov, F.V., Drits, V.A., Gorshkov, A.I., Sakharov, B.A. and Dikov, Yu.P. (1987b) Structural models of vernadite. *Proc. Acad. Sci. USSR, Geol. Series*, No. 12, 3–16.

Chukhrov, F.V., Gorshkov, A.I., Rudnitskaya, E.S., Berezovskaya, V.V. and Sivtsov, A.V. (1978) About vernadite. *Proc. Acad. Sci. USSR, Geol. Series*, No. 6, 5–19.

Chukhrov, F.V., Gorshkov, A.I., Rudnitskaya, E.S. and Sivtsov, A.V. (1978) To the characteristic of birnessite. *Proc. Acad. Sci. USSR, Geol. Series*, No. 9, 67–76.

Chukhrov, F.V., Gorshkov, A.I., Vitovskaya, I.V., Drits, V.A., Sivtsov, A.V. and Rudnitskaya, E.S. (1980a) Crystallochemical nature of Co-Ni asbolane. *Proc. Acad. Sci. USSR, Geol. Series*, No. 6, 73–81.

Chukhrov, F.V., Gorshkov, A.I., Vitovskaya, I.V., Drits, V.A., Sivtsov, A.V. and Dikov, Yu.P. (1980b) About crystallochemical nature of Ni-asbolane. *Proc. Acad. Sci. USSR, Geol. Series*, No. 9, 108–120.

Chukhrov, F.V., Gorshkov, A.I., Tyuryukanov, A.N., Berezovskaya, V.V. and Sivtsov, A.V. (1980c) To geochemistry and mineralogy of manganese and iron in young products of supergenesis. *Proc. Acad. Sci. USSR, Geol. Series*, No. 7, 5–24.

Chukhrov, F.V., Gorshkov, A.I., Drits, V.A., Sivtsov, A.V. and Dikov, Yu.P. (1982) New structural variety of asbolane. *Proc. Acad. Sci. USSR, Geol. Series*, No. 6, 69–77.

Chukhrov, F.V., Gorshkov, A.I., Drits, V.A., Fin'ko, V.I., Sivtsov, A.V., Dikov, Yu.P. and Sakharov, B.A. (1983a) Structural disordered asbolanes with tetrahedral coordination of manganese. *Proc. Acad. Sci. USSR, Geol. Series*, No. 12, 85–95.

Chukhrov, F.V., Gorshkov, A.I., Drits, V.A., Shterenberg, L.E., Sivtsov, A.V. and Sakharov, B.A. (1983b) Mixed-layered asbolane-buserite minerals and asbolanes in oceanic iron-manganese concretions. *Proc. Acad. Sci. USSR, Geol. Series*, No. 5, 91–100.

Chukhrov, F.V., Shterenberg, L.E., Gorshkov, A.I., Drits, V.A., Sivtsov, A.V., Sakharov, B.A., Aleksandrova, V.A. and Rudnitskaya, E.S. (1983c) On the nature of 10Å manganese mineral of Fe–Mn oceanic nodules. *Lithology and Mineral Resources (USSR)* 3, 33–41.

Chukhrov, F.V., Gorshkov, A.I., Berezovskaya, V.V. and Sivtsov, A.V. (1984) To mineralogy of lateritic weathering crusts. *Proc. Acad. Sci. USSR, Geol. Series*, No. 7, 108–125.

Chukhrov, F.V., Gorshkov, A.I. and Drits, V.A. (1989) *Supergene Manganese Oxides*, Nauka, Moscow, 208 pp.

Chukhrov, F.V., Shanin, L.L. and Ermilova, L.P. (1965) About possibility of determination of absolute age of potassium-bearing manganese minerals. *Proc. Acad. Sci. USSR, Geol. Series*, No. 2, 3–6.

Chukhrov, F.V., Zbyagin, B.B., Rudnitskaya, E.S. and Ermilova, L.P. (1966) About nature and genesis of halloysite. *Proc. Acad. Sci. USSR, Geol. Series*, No. 5, 3–20.

Chukhrov, F.V., Zvyagin, B.B., Gorshkov, A.I. and Ermilova, L.P. (1975) Delta-hydroxides of iron. In: *Supergene Iron Oxides in Geological Processes*, Nauka, Moscow, pp. 70–84.

Chukhrov, F.V., Zvyagin, B.B., Gorshkov, A.I., Ermilova, L.P., Korovushkin, V.V., Rudnitskaya, E.S. and Yakubovskaya, N.Yu. (1976) Ferroxyhite – A new modification of FeOOH. *Proc. Acad. Sci. USSR, Geol. Series*, No. 5, 5–24.

Churaev, N.V. (1973) Influence of surface forces on the motion of liquids in porous media. In: *Advances of Colloidal Chemistry*, Nauka, Moscow, pp. 79–86.

Cloud, P. (1972) A working model for the primitive Earth. *Amer. Journ. Sci.* 272, 537–548.

Cloud, P. (1973) Paleoecological significance of the banded iron-formation. *Econ. Geol.* 68, 1135–1143.

Cloud, P.E. (1976) Beginnings of biospheric evolution and their biochemical consequences. *Paleobiology* 2, 351–387.

Cloud, P. (1980) Early biogeochemical systems. In: *Biogeochemistry of Ancient and Modern Environment. B*, Springer-Verlag, Berlin.

Cloud, P. (1983) Bended iron-formation – Agardualist's dilemma. In: Trendall, A.F. and Morris, R.C. (Eds.), *Iron-Formation: Facts and Problems*, Developments in Precambrian Geology, Vol. 6, Elsevier, Amsterdam, pp. 401–416.

Coelho, C.E.S. and Rodrigues, O.B. (1986) Jazida de Manganês do Azul, Serra dos Carajás, Pará, (Capítulo XII). In: *Principas Depósitos Minerais do Brasil, Vol. II, Ferro e Metais da Indústria do Aço*, Coordinador-geral: Schobbenhaus, C. and Coelho, C.E.S., Brasilia, Ministerio das Minas e Energia, pp. 145–152.

Coffman, J.S. and Palencia, C.M. (1984) Manganese availability – Market economy countries. Bureau of Mines, Washington, DC, Government Print. Office, IC., 8978, 26 pp.

Cole, W.F., Wadsley, A.D. and Walkley, A. (1947) An X-ray diffraction study of MnO_2. *Electrochem. Soc. Trans.* **92**, 1–22.

Collier, R. and Edmond, J. (1984) The trace element geochemistry of marine biogenic particulate matter. *Prog. Oceanogr.* **13**, 113–199.

Cornell, R.M., Giovanoli, R. and Schindler, P.W. (1987) Effect of silicate species on the transformations of ferrihydrite into goethite and hematite in alkaline media. *Clays and Clay Minerals* **35**, 21–28.

Corliss, J.B., Lyle, M. and Dymond, J. (1978) The chemistry of hydrothermal mounds near the Galapagos Rift. *Earth Planet. Sci. Lett.* **40**, 12–24.

Crerar, D.A., Cormic, R.K. and Barnes, H.L. (1980) Geochemistry of manganese: An overview. In: *Geology and Geochemistry of Manganese. Vol. 1. General Problems, Mineralogy, Geochemistry, Methods*, Akadémiai Kiadó, Budapest, pp. 293–334.

Crerar, D.A., Fischer, A.G. and Plaza, C.L. (1980) Metallogenium and biogenic deposition of manganese from Precambrian to Recent time. In: *Geology and Geochemistry of Manganese. Vol. 3. Manganese on the Bottom of Recent Basins*, Akadémiai Kiadó, Budapest, pp. 285–303.

Crerar, D.A., Knox, G.W. and Means, J.L. (1979) Biogeochemistry of bog iron in the New Jersey Pine Barrens. *Chem. Geol.* **24**, 111–135.

Crittenden, M.D., Cuttitta, F., Rose, H.J., Jr. and Fleischer, M. (1962) Studies on manganese oxide minerals IV. Thaliuym in some manganese oxides. *Amer. Mineral.* **47**, 1461–1467.

Cronan, D.S. (1972) Regional geochemistry of ferromanganese nodules in the World Ocean. In: Horn, D.R. (Ed.), *Ferromanganese Deposits on the Ocean Floor*, Lamont-Doherty Geological Observatory, National Science Foundation, NY, Washington, DC, pp. 19–30.

Cronan, D.S. (1980) *Underwater Minerals*, Academic Press, London, 362 pp.

Cseh-Nemeth, J. and Grasselly, Gy. (1966) Data on the geology and mineralogy of the manganese ore deposit of Úrkút II. *Acta Miner. Petrogr. Szeged* **17**, 89–114.

Cseh-Nemeth, J., Grasselly, Gy., Nemecz, E. and Szabo, Z. (1971) Jurassic manganese ores of Hungary. Proc. IMA-IAGOD Meeting'70 IAGOD, *Soc. Mining Geol. Japan* **3** (Special Issue), 461–465.

Cseh-Nemeth, J, Konda, J, Szabo, Z. and Grasselly, Gy. (1980) Sedimentary manganese deposits of Hungary. In: *Geology and Geochemistry of Manganese, Vol. II, Manganese Deposits on Continents*, Akadémiai Kiadó, Budapest, pp. 199–221.

Cullers, R.L. Chandhuri, S., Kiibane, N. and Koch, R. (1979) Rare earths in size fractions and sedimentary rocks of Pennsylvanian-Permian age from the mid-continent of the U.S.A. *Geochim. Cosmochim. Acta* **37**, 1499–1512.

Davis, J.C. (1973) *Statistics and Data Analysis in Geology*, J.W. Wiley & Sons, New York.

Davis, J.A., James, R.O. and Leckie. J.O (1978) Surface ionization and complexation at the oxide/water interface. I. Computation of electrical double layer properties in simple electrolytes. *Journal Coll. Inter. Sci.* **63**(3), 480–499.

Davis, J.A. and Leckie, J.O. (1978) Surface ionization and complexation at the oxide/water interface. II. Surface properties of amorphous iron oxyhydroxide and adsorption of metal ions. *Journal of Coll. Inter. Sci.* **67**(1), 90–107.

De Andrade, F.G., Nakashima, J. and Podestá, P.R. (1986) Depósito de Manganés da Serra de Buritirama, Pará, (Capítulo XIII). In: *Principas Depósitos Minerais do Brasil, Vol. II, Ferro e Metais da Industria do Aço*, Coordenador-geral: Schobbenhaus, C. and Coelho, C.E.S., Brasilia Ministerio das Minas e Energia, pp. 153–166.

De Baar, H.J.W., Bacon, M.P. and Brewer, P.G. (1985a) Rare earth elements in the Pacific and Atlantic oceans. *Geochim. Cosmochim. Acta* **49**, 1943–1959.

De Baar, H.J.W., Brewer, P.G. and Bacon, M.P. (1985b) Anomalies in rare earth distributions in seawater: Gd and Tb. *Geochim. Cosmochim. Acta* **49**, 1961–1969.

Dement'ev, V.S. (1978) Dependence between content of iron and microquantities of metals in ground waters of Kendyktaslary. *Izvest. Akad. Nauk Kaz. SSR. Ser. Geol.* **6**, 60–61.

Demina, L.L. (1982) *Forms of Migration of Heavy Metals in the Ocean (at the Early Stages of Oceanic Sedimentation)*, Nauka, Moscow, 119 pp.

Demina, L.L., Gordeev, V.V. and Fomina, L.L. (1978) Forms of iron, manganese, zinc, and copper in river water and their changes in the zone of mixing the river waters with seawater (on the example of the rivers of the basins of the Black, Caspian, and Azov Seas). *Geokhimia (USSR)* **8**, 1211–1229.

Deryagin, B.V. (1983) The main tasks of the studies in surface forces. In: *Surface Forces and Boundary Layers of Liquids*, Nauka, Moscow, pp. 3–12.

Desai, M.V. and Ganguly, A.K. (1970) *Interaction of Trace Elements with Organic Constituents in the Marine Environments*, Bhabha Atomic Research Centre, Bombay, India, 102 pp.

Desai, M.V.M. and Ganguly, A.K. (1980) Organo-metallic interactions of manganese and other heavy metals in the marine environment. In: *Geology and Geochemistry of Manganese. Vol. 1*, Akadémiai Kiadó, Budapest, pp. 389–410.

De Villiers, J. (1956) The manganese deposits of the Union of South Africa. In: *XX Congresso Geologico International. Symposium Sobre Yacimentos de Manganese*, Tomo II, Africa, Mexico, pp. 39–72.

De Villiers, J. (1960) *The Manganese Deposits of the Union of South Africa. Handbook 2*, compiled by John de Villiers from reports by Boardman, L.G., De Villiers, J.E., De Villiers, John, Enslin, J.F., Visser, D.J.L., Geological Survey, Dept. of Mines, The Government Printer, Pretoria, 280 pp.

De Villiers, J.E. (1983) The manganese deposits of Griqualand West, South Africa: Some mineralogical aspects. *Econ. Geol.* **78**, 1108–1118.

De Villiers, P.R. (1967) New stratigraphic correlation and interpretation of the geological structure of the Postmasburg-Sishen Area, *Geol. Survey South Africa, Annals*, 39–42.

De Villiers, P.R. (1973) The geology and mineralogy of the Kalahari manganese-field north of Sishen, Cape Province. South Africa Geol. Survey, Dept. Mines. Memoir 59, 84 pp.

DeYoung, J.H. Jr, Sutphin, D.M. and Cannon, W.P. (1984) International strategic minerals inventory summary report – Manganese. U.S. Geol. Surv. Circ., 930-A, 22 pp.

Divina, T.A. (1976) Manganese-bearing capacity of sedimentary rocks of the Siberian Platform. *Proc. of SNIIGGIMS, 240*, Novosibirsk, pp. 56–67.

Dobrosmyslova, I.G., Koriakova, M.D. and Saenko, G.N. (1979) Seasonal migration of metals between water and microphyts of Sea of Japan. *Interaction between Water and Living Matter, Intern. Symp. Odessa*, Vol. II, Nauka, Moscow, p. 159.

Doncoisne, P.L. (1985) World situation of manganese resources and reserves. *Minerals and Society* **9**(3), 353–369.

Dorr, J.V.N., II (1972) Ferrugenous and associated manganese formations of Brazil. In: *Geology and Genesis of the Precambrian Ferrugenous-Siliceous and Manganese Formations of the World. Proc. International Symposium*, Naukova Dumka, Kiev, pp. 103–111.

Dorr, J.V.N., II, Varentsov, I.M. and Grasselly Gy., (1971–1972) Report on the business meeting of the Working Group on Manganese Formations (now Commission on Manganese) August 24, 1972, Montreal, Canada. *Acta Mineral.-Petrograph., Szeged, Hungary*, **20**, 388–391.

Drittenbass, W. (1979) Sedimentologie and Geochimie von Kisen-Mangan führenden Knellen und Krusten in Jura der Trento-Zone (östliche Sudalpen Norditalien). *Ecologae Geologicae Helvetiae, Basel* **72/2**, 313–345.

Drost-Hansen, W. (1969) Structure of water near solid interface. *Industr. and Engineer. Chemistry* **61**(11), 10–47.

Drost-Hansen, W. (1977) Effects of vicinal water on colloidal stability and sedimentation processes. *Journ. Colloid Interface Science* **58**(2), 251–262.

Dublyanskii, V.N. (1971) Calcite sinter barriers (buttes) of karst cavities of the Mountaneous Crimea. In: *Caves*, No. 10–11, Permskii University, Perm', pp. 57–65.

Dublyanskii, V.N. and Polkanov, Yu.A. (1974) Composition of water chemogenous and mechanical sediments of karst cavities of the Mountaneous Crimea. In: *Caves*, No. 14–15, Permski University, Perm', pp. 32–38.

Duddy, I. (1981) Redistribution and fractionation of rare earth and other elements in a weathering profile. *Chem. Geol.* **30**, 363–381.

Duse, R.A., Qwinn, J.G., Olney, Ch.E., Piotrowicz, S.R., Ray, B.J. and Wade, T.L. (1972) Enrichment of heavy metals and organic compounds in the surface microlayer of Narragansett Bay Rhode Island. *Science* **176**, 161–163.

Duursma, E.K. (1968) Organic chelation of Co^{60} and Zn^{65} by leucine in relation to sorption by sediments. In: *Symposium on Organic Matter in Natural Waters*, Alaska, pp. 387–397.

Dzotsenidze, G.S. (1965) *Influence of Volcanism on Formation of Sediments*, Nedra, Moscow, 155 pp.

Dzotsenidze, G.S. (1965) About the genesis of the Chiaturskoe manganese deposit. *Lithology and Mineral Resources* **1**, 3–18.

Dzotsenidze, G.S. (1969) *Role of Volcanism in the Formation of Sedimentary Rocks and Ores*, 2nd edition revised and added, Nedra, Moscow, 344 pp.

Edington, D.N. and Callender, E. (1970) Minor element geochemistry of Lake Michigan ferromanganese nodules. *Earth, Planet. Sci. Lett.* **8**, 97–100.

Efremov, I.F., Voronina, L.A. and Samigullina, G.V. (1980) Polymorphism of boundary liquid layers and the problem of the surface lyophilization. *Chemistry and Technology of Water* **2**(6), 525–532.

Ehrlich, H.L. (1972) The role of microbes in manganese nodule genesis and degradation. In: Horn, D.R. (Ed.), *Ferromanganese Deposits on the Ocean Floor*, Conference, Arden House, Harriman, NY, January 20–22, 1972, Lamont-Doherty Geological Observatory, Columbia University and IDOE-NSF, pp. 63–70.

Elgabaly, M.M. (1950) Mechanism of zinc fixation by colloidal clays. *Soil Sci.* **69**, 167.

Elias, M., Donaldson, E.M. and Giorgetta, N. (1981) Geology, mineralogy and chemistry of lateritic nickel-cobalt deposits near Kalgoorlie, Western Australia. *Econ. Geol.* **76**, 1775–1783.

Ellison, T.D. (1983) Manganese. *Mining Journ. Annual Rev.* 67–69.

Embley, R.W. and Jonasson, I.R. (1988) Submersible investigation of an extinct hydrothermal system on the Galapagos Ridge: Sulphide mounds, stockwork zone and differentiated lavas. *Canad. Mineral.* **26**(3), 517–539.

Erdey-Gruz, Tibor (1974) *Transport Phenomena in Aqueous Solutions*, Akadémiai Kiadó, Budapest, 512 pp.

Ergin, Yu.V. (1983) *Magnetic Properties and Structure of Electrolyte Solutions*, Nauka, Moscow, 183 pp.

Ergin, Yu.V. and Kostrova, L.I. (1969) Study of hydration of ions in aqueous solutions by the magnetochemical method. *Zhurn. Strukt. Khim. (USSR)* **10**, 971–978.

Eschenbrenner, V. and Grandidn, G. (1970) La séquence de cuirasses et ses difféenciation entre Agnibilékrou et Diebougou (Haute-Volta). *Cah. ORSOM, Ser. Géol.* **2**, 205–246.

Eswaran, H. and Raghu Mohan, N.G. (1973) The microfabric of petrolinthite, *Soil Sci. Soc. Am. Proc.* **37**, 79–82.

Exon, N.P. (1983) Manganese nodules deposits in the central Pacific ocean and their variation with latitude. *Mar. Mining* **4**(1), 79–107.

Feitknecht, W. and Marti, W. (1945) Über die Oxydation von $Mn(OH)_2$ mit molekularem Sauerstoff. *Helv. Chim. Acta* **28**, 129–148, 149–156.

Ferguson, J. and Bubela, B. (1974) The concentration of Cu(II), Pb(II) and Zn(II) from aqueous solution by particulate algal matter. *Chem. Geol.* **13**, 163–186.

Fermor, L.L. (1906) Manganese in India. *Mining Geol. Inst. India, Trans.* **1**, 69–131.

Fermor, L.L. (1909) The manganese ore deposits of India. *Memor. Geol. Survey of India* **37** 610 pp.

Fischer, A.G. (1984) The two Phanerozoic supercycles. In: Berggren, W.A. and van Couvering, J.A. (Eds.), *Catastrophes and Earth history: The New Uniformitarianism*, Princeton University Press, Princeton, NJ, pp. 129–150.

Fleet, A.J. (1983) Hydrothermal and hydrogenous ferromanganese deposits: Do they form a continuum? The rare earth element evidence. In: Rona, P.A., Boström, K., Laubrier, L. and Smith, K.L. (Eds.) *Hydrothermal Processes on Seafloor Spreading Centres*, Plenum, New York, pp. 535–555.

Fleischer, M. (1960) Studies of manganese oxide minerals. III. Psylomelane. *Amer. Mineral.* **45**, 176–187.

Fleischer, M. (1964) Manganese oxide minerals. VIII. Hollandite. In: *Advancing Frontiers in Geology and Geophysics*, Hyderrabad, Osmania University Press, pp. 221–232.

Fleischer, M. and Faust, G. (1963) Studies of manganese oxide minerals. VII. Lithiophorite. *Schweiz. Mineralogische und Petrographische Mitteilungen* **43/1**, 197–216.

Fleischer, M. and Richmond, W.E. (1943) The manganese oxides: A preliminary report. *Econ. Geol.* **38**, 269–286.

Fleischer, M., Richmond, W.E. and Evans, H.T., Jr. (1962) Studies of the manganese oxides. V. Ramsdellite, MnO_2, an orthombic dimorph of pyrolysite. *Amer. Mineral.* **47**, 47–58.

Force, E.R. and Cannon, W.F. (1988) Depositional model for shallow-marine manganese deposits around black-shale basins. *Econ. Geol.* **83**, 93–117.

Forsh-Menshutkina, T.B. (1973) Hydrochemistry of the Onega Lake. General mineralization and basic ion composition of water. In: *Hydrogeochemistry of the Onega Lake and Its Tributaries*, Nauka, Leningrad, pp. 132–175.

Frakes, L.A. and Bolton, B.R. (1982) Depositional environments of sedimentary manganese. In: *International Association on the Genesis of Ore Deposits (IAGOD), VI Symposium*, Tbilisi, September 6–12, 1982,

Collected abstracts, Tbilisi, p. 290.

Frakes, L.A. and Bolton, B.R. (1984a) The impact of climate on the geochemistry of fine grained marine rocks in intra-cratonic basins. In: *XXVII International Geological Congress*, Moscow, August 4–14, 1984, Abstracts, Vol. 2 (Sections C.04, C.05), Nauka, Moscow, p. 55.

Frakes, L.A. and Bolton, B.R. (1984b) Origin of manganese giants: sea-level change and anoxic-oxic history. *Geology* **12**, 83–86.

Frakes, L.A. and Bolton, B.R. (1992) Effects of ocean chemistry, sea level and climate on the formation of primary sedimentary manganese ore deposits. *Econ. Geol.* **82**(2), 1207–1217.

Frank, H.S. (1965) The structure of water. *Fed. Proc.* **24**, No. 2, pt. II, 1–11.

Frank, H.S. (1972) Structural models. In: Franks, F.N. (Ed.), *Water*, Plenum Press, New York, Vol. 2, pp. 515–544.

Frank, H.S. and Wen, W.Y. (1957) III. Ion-solvent interaction in agueons solutions: A suggested picture of water structure. *Discuss. Faraday Soc.* **24**, 133–140.

Frazer, F.W. and Belcher, Ch.B. (1975) Mineralogical studies of the Groote Eylandt manganese ore deposits. *Proc. Austral. Inst. Min. Metall.*, No. 254. 29–36.

Frenzel, G. (1980) The manganese ore minerals. In: *Geology and Geochemistry of Manganese, Vol. 1. General Problems*, Akadémiai Kiadó, Budapest, pp. 25–158.

Frost, D.C., Ishitani, A. and McDowell, C.A. (1972) X-ray photoelectron study of transition metals oxides. *Molecul. Physic.* **24**(1), 861.

Gabano J.P., Etienne, P. and Laurent, J.F. (1965) Etude des properties de surface du bioxide de manganese. *J. Electrochmica Acta* **10**, 1947–963.

Gardner, L.R. (1974) Organic versus inorganic tracemetal complexes in sulphidic marine water – Some speculative calculations based on available stability constants. *Geochim. Cosmochim. Acta* **38**, 1297–1302.

Garrels, R.M. (1960) *Mineral Equilibria at Low Temperature and Pressure*, Harper Brothers, New York, 254 pp.

Garrels, R.M. (1982) Geochemical equlibrium in sedimentary systems. *Estudioa Geol.* **38**, 289–294.

Garrels, R.M. and Christ, C.L. (1965) *Solutions, Minerals and Equilibria*, Harper and Row, New York, XIII, 450 pp.

Garrett, W.D. (1968) Organic chemistry of natural sea surface film. in: *Symposium on Organic Matter in Natural Waters*, Alaska, pp. 469–477.

Gaudefroy, C. (1960) Caractères distinctifs de la pyrolusite ex manganite. *Notes Serv. Géol. Maroc.* **19**, 77–86.

Gauthier-Lafaye, F. and Weber, F. (1989) The Francevillian (Lower Proterozoic) uranium ore deposits of Gabon. *Econ. Geol.* **84**, 2267–2285.

Gavrilyuk, A.I., Chuiko, V.T. and Golovaty, R.N. (1970) Sorption of microamounts of copper and nickel by iron hydroxide from solutions of sodium chloride in the presence of some complexing agents. In: Chuiko, V.T. (Ed.), *Complexing, Molecular Interaction and Co-Precipitation in Some System*. Dnepropetr. Gos. Un-t, Dnepropetrovsk, pp. 212–216.

Giovanoli, R. (1976) Vom Hexaquo-Mangan zum Mangansediment. Reaktionesseqenzen feinteiliger fester Manganoxidhydroxide, *Chimia* **30**, 102–103.

Giovanoli, R. (1980) On natural and synthetic manganese nodules. In: *Geology and Geochemistry of Manganese, Vol. 1*, Akadémiai Kiadó, Budapest, pp. 159–202.

Giovanoli, R. (1985) Layer structures and tunnel structures in manganates. *Chem. Erde* **44**(3), 227–244.

Giovanoli, R. and Balmer, B. (1981) A new synthesis of hollandite. A possibility for immobilization of nuclear waste. *Chimia* **35**(2), 53–54.

Giovanoli, R., Brütsch, R., Stadelmann, W. and Bürki, P. (1979) Use of specific surface measurements for investigation of solid state reactions. In: Gregg, S.J. and Sing, K.S.W. (Eds.), *Characterisation of Porous Solids, Proc. Symp.*, Neuchâtel, July 9–12, 1978, Society of Chemistry, London, pp. 313–320.

Giovanoli, R., Bühler, H. and Sokolowska, K. (1973) Synthetic lithiphorite: electron microscopy and X-ray diffraction. *Jour. Microscopie* **18**, 271–284.

Giovanoli, R., Bürki, P. and Giuffredi, M. (1975) Layer structured manganese oxide-hydroxides. IV: The buserite group: Structure stabilization by transition elements. *Chimia* **29**(12), 517–520.

Giovanoli, R., Feitknecht, W. and Georges, P. (1976a) Homogene Klimbildung und Keimwachstrum von γ-MnO. *Chimia* **30**(5), 268–269.

Giovanoli, R., Feitknecht, W., Maurer, R. and Häni, H. (1976b) Über die Reactionen von Mn_3O_4 mit Säuren. *Chimia* **30**, 307–309.

Giovanoli, R. and Leuenberger, U. (1969) Über die Oxydation von Manganoxidhydroxid. *Helvetica Chemica*

Acta **52**, fasc. 8, No. 233, 2333–2347.

Glasby, G.P. (Ed.) (1977) *Marine Manganese Deposits*, Elsevier Oceanograph. Ser., Vol. 15, Elsevier, Amsterdam, 523 pp.

Glasby, G.P. (1978) Deep-sea manganese nodules in the stratigraphic record: evidence from DSDP cores. *Mar. Geol.* **28**, 51–64.

Glasby, G.P. (1988) Manganese deposition through geological time: Dominance of the post-Eocene deep-sea environment. *Ore Geol. Rev.* **4**, 135–144.

Glasby, G.P., Stroffers, P., Sioulas, A., Thijssen, T. and Friedrich, G. (1982) Manganese nodule formation in the Pacific Ocean: A general theory. *Geo-Marine Letters* **2**, 47–53.

Glasby, G.P. and Thijssen, T. (1982) Control of the mineralogy and composition of marine manganese nodules by the supply of divalent transition metal ions. *N. Jb. Miner. Abh.* **145**(3) 291–307.

Glemser, O, Gattow, G. and Meisiek, H. (1961) Über Manganoxyde VII. Darstellung und Eigensehaften von Braunsteinen, 1 (Die δ-Gruppe der Braunsteine). *Zeitschr. Anorg. Allg. Chemie* **309**, 1–36.

Goldberg, E.D. (1954) Marine geochemistry. I. Chemical scavengers of the sea. *Jour. Geol.* **62**(3), 249–265.

Goldberg, E.D. (1961) Chemistry in the oceans. In: Sears, M. (Ed.), *Oceanography*, Amer. Assoc. Advanc. Sci. Public., Washington, DC, No. 67, pp. 583–597.

Goldberg, E.D. and Arrhenius, G.O.S. (1958) Chemistry of Pacific pelagic sediments. *Geochim. Cosmochim. Acta* **13**, 153–212.

Goldberg, E.D., Koide, M., Schmitt, E.A. and Smitt, R.H. (1963) Rare earth distributions in the marine environment. *Journ. Geophys. Res.* **68**, 4209–4217.

Goldschmidt, V.M. (1937) The principles of distribution of chemical elements in minerals and rocks. *J. Chem. Soc.* 655–672.

Golovko, V.A. (1976) Geochemical features of conditions of accumulation of iron and manganese in the Riphean basin of the eastern Prisayan'e. *Proc. of Higher Institutions. Geology and Exploration*, **9**, 95–97.

Golovko, V.A. (1984) Geological-geochemical conditions of formation of supergene deposits of manganese and their practical significance. In: *Manganese Ore Formation on the USSR Territory*, Nauka, Moscow, pp. 35–45.

Golovko, V.A., Mstislavskkii, M.M, Nasedkina, V.H., Sherman, M.L., Mkrtych'yan, A.K., Ustalov, V.V. and Storozhenko, A.A. (1982) Manganese-bearing capacity of the Precambrian of the Eniseiskii Ridge. In: *Geology and Geochemistry of Manganese*, Nauka, Moscow, pp. 94–104.

Golovko, V.A. and Nasedkina, V.H. (1982) Composition and genesis of manganese ores of the Porozhinskoe deposit (Eniseiskii Ridge). In: *Geology and Geochemistry of Manganese*, Nauka, Moscow, pp. 104–109.

Goncharov, V.V., Romanova, I.I., Samoilov, O.Ya. and Yashkichev, V.I. (1967) A quantitative characteristic of near hydration of certain ions in diluted water solutions. I. The energy of activation of self-diffusion. *Zhurn. Strukt. Khim. (USSR)* **8**, 613–617.

Gordeev, V.V. and Lisitsyn, A.P. (1979) Microelements. In: *Oceanology. Chemistry of the Ocean. Vol. I. Chemistry of the Ocean Waters*, Nauka, Moscow, pp. 337–375.

Gorham, E. and Swaine, D.J. (1956) The influence of oxidizing and reducing conditions upon the distribution of some elements in lake sediments. *Limnol., Oceanogr.* **10**, 268–279.

Grahame, D.C. (1947) The electrical double layer and the theory of electrocapillarity. *Chem. Rev.* **41**, 441.

Grahame, D.C., Roth, M.A. and Cummings, J.I. (1952) The differential capacity and the electrical double layer. The role of the anion. *Journ. Amer. Chem. Soc.* **74**, 4422.

Grandin, G. and Perseil, E.A. (1977) Le gisement de manganèse de Mokta (Côte d'Ivoire). Transformations mineralogiques des minerais par action meteorique. *Bull. Soc. Géol.*, (7) **XIX**(2), 309–317.

Grandin, G. and Perseil, E.A. (1983) Les minéralosations manganésifères volcano-sedimentaires du Blafo-Guéto (Côte d'Ivoire). Paragenesis-alteration climatique. *Mineralium Deposita* **18**, 99–111.

Grasselly, Gy. and Cseh-Nemeth, J. (1961) Data on the geology and mineralogy of the manganese ore deposit of Úrkút I. *Acta Mineral. Petrograph., Szeged, Hungary* **14**, 3–25.

Grasselly, Gy. and Varentsov, I.M. (1974) Report on the Business Meeting of the Commission on Manganese, 4th Symposium of the IAGOD, Varna, Bulgaria, September 25, 1974. *Acta Mineral. Petrograph., Szeged, Hungary* **21**(2), 313–316.

Grasselly, Gy. and Varentsov, I.M. (1975) Letters the Commission on Manganese (IAGOD); Information on the IGCP Project: Genesis of Manganese Ore Deposits. *Acta Mineral. Petrograph., Szeged, Hungary* **22**(1), 189–192.

Greenslate, J.L., Frazer, J.Z. and Arrhenius, G. (1973) Origin and deposition of selected transition elements in the seabed. In: Morgenstern, M. (Ed.), *The Origin and Distribution of Manganese Nodules in the Pacific and Prospects for Exploration*, Symposium, Honolulu, Hawaii, July 23–25, 1973, Valdivia Manganese Exploration Group, University of Hawaii and IDOE-NSF, pp. 43–69.

Gribov, E.M., Varand, E.L. and Pasashnikova, G.K. (1984) Conditions of formation of manganese ores of the Porozhinskoe deposit. In: *Manganese Ore Formation on the USSR Territory*, Nauka, Moscow, pp. 95–101.

Grobbelaar, W.S. and Beukes, N.J. (1986) The Bishop and Glosam manganese mines and the Beeshoek Iron Ore Mine of the Postmasburg area. In: Anhaeusser, C.R. and Maske, S. (Eds.), *Mineral Deposits of Southern Africa*, Geol. Soc. S. Afr., Johannesburg, I and II, pp. 957–961.

Grutter, A. and Buser, W. (1952) Untersuchungen an Mangansedimnten. *Chimia* **11**, 132–133.

Guillou, J.-J. (1976) Role possible du mecanisme tranagressif dans la genese des gites marines de faret de manganese un example ordovician (Sierradde Caurel, Lugo, Espagne). *C. R. Acad. Sci., Paris, Série D* **282**, 2021–2024.

Gümbel, C.W. (1861) *Geognostische Bercheibung des Bayerischen Alpengebirges und seines Vorlandes. 1,* Perthes, Gotha, 950 pp.

Gümbel, C.W. (1878) Über die im stille Ocean auf dem Meeresgrunde vorkommenden Manganknollen. *Sitsungsber. d. bauer Akad. d. Wiss. Math. Phys. Cl.* **2**, 189–209.

Gümbel, C.W., v. (1888a) Die mineralogisch-geologische Beschaffenheit der auf Forschungreise S.M.S. "Gaselle" gesammelton Meeresgrung-Ablagerungen. *Die Forschungsreise S.M.S. "Gazelle" in den Jahren 1874 bis 1876. 2. Teil. Physik und Chemie*, Hydrographisches Amt der Admiralität, Berlin, pp. 69–116.

Gümbel, K.W., von (1888b) *Grundzüge der Geologie*, Th. Fischer, Kassel, 1063 pp.

Gümbel, W.V. (1889) Geologische Bemerkungen über die warmen Quellen von Gastein und ihre Ungebung. *Sitzber, d. bayer, Akad. d. Wiss. Math.-phys. Cl.* **19**, 341–408.

Gümbel, C.W., V. (1891) Geologische Bemerkungen die Thermen von Bormino und das Ortlergebirge. *Sitzber, d. bayer. Akad. d. Wiss. Math-phys. Cl.* **21**, 79–120.

Halbach, P., Friedrich, G. and von Stackelberg, U. (Eds.) (1988) *The Manganese Nodule Belt of the Pacific Ocean: Geological Environment, Nodule Formation and Mining Aspects*, Ferdinand Enke Verlag, Stuttgart, 254 pp.

Haralyi, N.L.E. and Walde, D.H.G. (1986) Os minérios de Ferro e Manganês da Região de Urucum, Corumbá, Mato Grosso do Sul, (Capítulo XI). In: *Principas Depósitos Minerais do Brasil, Vol. II, Ferro e Metais da Indústria do Aço*, Coordinador-geral: Schobbenhaus, C. and Coelho, C.E.S., Brasilia, Ministerio das Minas e Energia, pp. 127–144.

Harman, H.H. (1967) *Modern Factor Analysis*, University of Chicago Press, Chicago, 474 pp.

Haskin, M.A. and Haskin, L.A. (1966) Rare earth in European shales: A redetermination. *Science* **154**, 307–309.

Haskin, L.A., Wildeman, T.R., Frey, F.A., Collins, K.A., Keedy, C.R. and Haskin, M.A. (1966) Rare earth in sediments. *Journ. of Geophys. Res.* **71**(24), 6091–6105.

Healy, T.W., Herring, A.P. and Fuerstenau, D.W. (1966) The effect of crystal structure and the surface properties of a series of manganese oxides. *Journ. Interface Sci.* **21**, 435–444.

Hein, J.R. and Bolton, B.R. (1993) Composition and origin of the Moanda manganese deposit, Gabon. In: *16th International Colloquium on Africa Geology*, Eznlwini-Swaziland, 14–16 September, 1993, Extended Abstracts, Geological Survey and Mines, Mbabane, Swaziland, 1, pp. 150–152.

Hein, J.R., Bolton, B.R., Nziengui, P., McKirdy, D. and Frakes, L. (1989) Chemical, isotopic and lithologic associations within the Moanda manganese deposit, Gabon. In: *Abst. 28th International Geological Congress*, July 9–19, Washington, DC., pp. 2–46.

Heisenberg, W. (1987) Des Naturgesetz und Struktur der Materie. In: *Schritte über Grenzen*, Progress, Moscow, pp. 107–122.

Heisenberg, W. (1987) Sprache und Wirklichkeit in der modern Physik, In: *Schritte über Grenzen*, Progress, Moscow, pp. 208–225.

Hem, J.D. (1963) Chemical equilibria and rates of manganese oxidation. Chemistry of manganese in natural water. Geol. Survey Water-Supply Paper. 1667-A, U.S. Government Printing Office, Washington, 1–64.

Hem, J.D. (1965) Reduction and complexing of manganese by gallic acids. U.S. Geol. Surv., Water-Supply, Paper 1667 D, 1–28.

Hem, J.D. (1972) Chemical factors that influence the availability of iron and manganese in aqueous systems. *Geol. Soc. America Spec. Paper* **140**, 17–24.

Hem, J.D. (1981) Rates of manganese oxidation in aqueous systems. *Geochim. Cosmochim. Acta* **45**, 1369–1374.

Hem, J.D. and Lind. C.J. (1983) Nonequilibrium models for predicting forms of precipitated manganese oxides.

Geochim. Cosmochim. Acta **47**, 2037–2046.

Hewett, D.F. (1964) Veins of hypogene manganese oxide minerals in the Southwestern United States. *Econ. Geol.* **59**(8), 1429–1472.

Hewett, D.F., Corwall, H.R. and Erd, R.C. (1968) Hypogene veins of gibbsite, pyrolusite and lithiphorite in Nye County, Nevada. *Econ. Geol.* **63**(4), 360–371.

Hewett, D.F. and Fleischer, M. (1960) Deposits of manganese oxides. *Econ. Geol.* **55**, 1–55.

Hewett, D.F., Fleischer, M. and Conklin, N. (1963) Deposits of the manganese oxides supplement. *Econ. Geol.* **58**, 1–51.

Hodgson, J.F. (1960) Cobalt reactions with montmorillonite. *Proc. Soil Sci. Soc. Amer.* **24**, 165.

Holland, H.D. and Beukes, N.J. (1990) A paleoweathering profile from Griqualand West, South Africa: Evidence for a dramatic rise in atmospheric oxygen between 2.2 and 1.9 B.Y. B.P. *Amer. Journ. of Sci.* **290-A**, 1–34.

Holtrop, J.F. (1965) The manganese deposits of the Guiana shield. *Econ. Geol.* **60**(6), 1185–1212.

Horn, D.R., Horn, B.M. and Delach, M.N. (1973) Copper and nickel content of ocean ferromanganese deposits and their relation to properties of the substrate. In: Morgenstein, M. (Ed.), *The Origin and Distribution of Manganese Nodules in the Pacific and Prospects for Exploration*, Symposium, Honolulu, Hawaii, 23–25 July, 1973, Valdivia Manganese Exploration Group, University of Hawaii and IDOE-NSF, pp. 77–84.

Horne, R. (1969) *Marine Chemistry. The Structure of Water and the Chemistry of the Hydrosphere.* John Wiley & Sons, New York, 568 pp.

Hutchinson, D.E. (1957) *Treatise on Limnology. Geography, Physics, and Chemistry. Vol. I*, Chapman and Hall, New York, Wiley, London, 1015 pp.

Hypólito, R., Giovanoli, R. and Valarelli, J.V. (1982a) Obtençao de criptomelana a partir de $Mn^{2+}1$ de outros mínerais sinteticos de manganos. *An. Acad. Brasil Sciénc.* **54**(4), 713–720.

Hypólito, R., Giovanoli, R., Valarelli, J.V. and Sonoki, N.T. (1982b) Sintese de criptomelana. *Boletin I.G. Instituto de Geociências, USP* **13**, 1–13.

Hypólito, R., Valarelli, J.V., Giovanoli, R. and Netto, S.M. (1984) Gibbs free energy of formation of synthetic cryptomelane. *Chimia* **38**(12), 98–99.

Idzikowski, S. (1971) Adsorption of inorganic ions on α-Fe_2O_3 from mixtures of strong electrolytes. Part I. Adsorption of silver, cupric and aluminium ions as a function of indifferent salt concentration. *Roczniki Chemii (Ann. Soc. Chim. Polonorum)* **45**, 1139–1149.

Idzikowski, S. (1972a) Adsorption of inorganic ions on α-Fe_2O_3 from mixtures of strong electrolytes. Part II. Relationship between the increase in adsorption and concentration of the mixture's components. *Roczniki Chemii (Ann. Soc. Chim. Polonorum)* **46**, 167–177.

Idzikowski, S. (1972b) Adsorption of inorganic ions on α-Fe_2O_3 from mixtures of strong electrolytes. Part III. Adsorption of cupric and aluminium chlorides in presence of an indifferent electrolyte. *Roczniki Chemii (Ann. Soc. Chim. Polonorum)* **46**, 1009–1016.

Idzikowski, S. (1973) Adsorption of inorganic ions on α-Fe_2O_3 from mixtures of strong electrolytes. Part IV. Adsorption of silver, cupric and aluminium sulphate in presence of an indifferent electrolyte. *Roczniki Chemii (Ann. Soc. Chim. Polonorum)* **47**, 231–238.

Indurm, M. and Senior, B.R. (1978) Palaeomagnetic ages of Late Cretaceous and Tertiary weathered profiles in the Eromanga Basin, Queensland. *Palaeogeogr., Palaeoclimatol., Palaeoecolog.* **24**(4), 263–277.

Indurm, M. and Schmidt, P.W. (1984) Progress of IGCP Projects: No. 129. Lateritization processes. Geological correlation, No. 12, Paris, IGCP, IUGS, UNESCO, pp. 30–37.

Ingri, J. and Ponter, Ch. (1987) Rare earth abundance patterns in ferromanganese concretions from the Gulf of Bothnia and the Barents Sea. *Geochim. Cosmochim. Acta* **51**, 155–161.

Ionov, V.I., Mazitov, R.K. and Samoilov, O.Ya. (1969) Dependence of near hydration of a salted-out cation on the hydration of the salting-out agent cations according to NMR data. *Zhurn. Strukt. Khim. (USSR)* **10**, 407–410.

Ivanenkov, V.N. (1979) The main salt composition of the oceanic water. In: *Chemistry of the Ocean. Vol. I. Chemistry of the Waters of the Ocean*, Nauka, Moscow, pp. 43–47.

Ivanenkov, V.N., Chernyakova, A.M., Gusarova, A.N. and Sapozhnikov, V.V. (1979) Typization of the waters and chemo-oceanographic regionalization of the World Ocean. In: *Chemistry of the Ocean. Vol. I. Chemistry of the Waters of the Ocean*, Nauka, Moscow, pp. 75–84.

James, R.O. and Healy, T.W. (1972) Adsorption of hydrolyzable metal ions of the oxide-water interface. *Journ. Coll. Inter. Sci.* **40**(1), 42–81.

James, R.O., Stiglich, P.J. and Healy, T.W. (1975) Analysis of models of adsorption of metal ions at oxide-water interfaces. *Faraday Discuss., Chem. Soc.* **59**, 142–156.

Jencks, W.P. (1969) *Catalysis in Chemistry and Enzymology*, McGraw-Hill Book Company, New York.

Jenkyns, H.C. (1977) Fossil nodules. In: Glasby G.P. (Ed.), *Marine Manganese Deposits*, Elsevier, Amsterdam, pp. 85–108.

Jones, B. (1992) Manganese precipitates in the karst terrain of Gayman, British West Indies. *Can. Jour. Earth Sci.* **29**, 1125–1139.

Jones, T.S. (1991) Manganese, 1990. Annual Report. U.S. Department of the Interior, Bureau of Mines, Washington, D.C., U.S. Government Print. Office, 16 pp.

Jones, T.S. (1992) Manganese, 1991. Annual Report, U.S. Department of the Interior. Bureau of Mines, Washington, D.C., U.S. Government Print. Office. 23 pp.

Jouravsky, G. and Pouit, G. (1965) Les minéralisations manganésifères de la région de Bou Arfa. Colloque sur les gisements stratiformes de plomb, zinc et manganese du Maroc. *Notes et Mém. Serv. Géol. Maroc.*, No. 181, 113–124.

Kaeding, L., Brockamp, O. and Harder, H. (1983) Submarin-hydrothermale Entstehung der sedimentären Mangan-Lagerstätte Úrkút/Ungarn. *Chem. Geol.* **40**, 251–268.

Kalinin, V.V. (1963) *Iron-Manganese Ores of the Karazhal Deposit*, Publ. Acad. Sci. USSR, Moscow, 124 pp.

Kauzmann, W. (1976) Pressure effects on water and the validity of theories of water behaviour. In: *Colloques Internationaux du C.N.R.S., No. 246. L'Eau et les Systemes Biologiques*, pp. 63–71.

Kavitskii, M.L., Mkrtych'yan, A.K., Storozhenko, A.A. and Ustalov, V.V. (1980) Porozhinskoe deposit of manganese. *Exploration and Preservation of Mineral Resources*, No. 3, 13–16.

Kheraskov, N.P. (1951) Geology and genesis of the eastern Bashkirian manganese deposits. In: *Problems of Lithology and Stratigraphy of the USSR*, In memory of Acad. A.D. Archangel'skii. Publ. Acad. Sci. USSR, Moscow, pp. 328–348.

Khmelevskii, V.A. (1968) Geologic structure and genetic type of the Burshtynskoe manganese deposit. *Geol. Coll. L'vov. Geol. Society*, No. II, 108–115.

Khmelevskii, V.A. and Gurzhii. D.V. (1975) About distribution of manganese in the Burshtynskoe deposit. In: *Geology and Geochemistry of Combustible Minerals*, No. 42. Naukova Dumka, Kiev, pp. 73–83.

Kholodov, V.N. (1983) Life and works of Acad. N.M. Strakhov. In: Strakhov, N.M. (Ed.), *Selected works. General Problems of Geology, Lithology and Geochemistry*, Nauka, Moscow, pp. 6–28.

Kholodov, V.N. (1985) Introduction of the translation editor. In: Roy, S., *Manganese Deposits*, Mir, Moscow, pp. 5–9.

Kim, S.J. (1974) Mechanisms of formation of supergene manganese oxide ores at the Janggun Mine, Korea. *Neues Jahrb. für Mineral.*, Vol. 8, Stuttgart, pp. 371–384.

Kim, Soo Jin (1977) Janggunite, A new manganese hydroxide mineral from the Janggun mine, Bonghwa, Korea. *Mineral. Mag.* **41**, 519–523.

Kim, Soo Jin (1980) Genesis of manganese carbonate and oxide deposits at the Janggun mine, South Korea. In: *Geology and Geochemistry of Manganese. Vol. II. Manganese Deposits on Continents*, Akadémiai Kiadó, Budapest, pp. 297–317.

Kim, Soo Jin, (1990) Crystal chemistry of hexagonal 7Å phyllomanganate minerals. *Journ. Miner. Soc. Korea* **3**, 34–43.

Kim, Soo Jin (1991) New characterization of takanelite. *Amer. Mineral.* **76**, 1426–1430.

Kim, Soo Jin (1993) Chemical and structural variations in rancieite-takanelite solid solution series. *N. Jb. Miner. Mh*, H. 5 (Mai), 233–240.

Klein, C. and Beukes, N.J. (1989) Geochemistry and sedimentology of a facies transition from limestone to iron-formation deposition in the Early Proterozoic Transvaal Supergroup, South Africa. *Econ. Geol.* **84**(7), 1733–1774.

Kleyenstüber, A.S.E. (1984) The mineralogy of the manganese-bearing Hotazel Formation of the Proterozoic Transvaal sequence in Griqualand West, South Africa. *Trans. Geol. Soc. South Africa* **87**, 257–272.

Kleyenstüber, A.S.E. (1988) The correlation of the mineralogy with the sedimentary cycles of the Proterozoic manganese-bearing Hotazel Formation in South Africa. In: *IAS International Symposium on Sedimentology Related to Mineral Deposits*, Abstracts, July 30–August 4, 1988. Beijing, China, pp. 107–108.

Klinkhammer, G.P., Eldefield, H. and Hudson, A. (1983) Rare earth elements in seawater near hydrothermal vents. *Nature* **305**, 185–188.

Kornev, T.Ya. (1972) Diabase-albitophyric formation and some problems of Precambrian stratigraphy of the Eniseiskii Ridge. *Proc. SNIIGGIMS*, Novosibirsk, 122, pp. 63–72.

Koshy, E., Desai, M.V. and Ganguly, A.K. (1969) Studies on organo-metallic interactions in the marine environment. *Current Sci., India* **38**, 555–558.

Koski, R.A., Hein, J.R and Bouse, R.M. (1989) Stratigraphy and geochemical characteristics of Tertiary sediment-hosted Mn-oxide deposits in the Southwestern United States: Implications for genesis. In: *28th Intern. Geol. Congr.*, Abstracts, Vol. 2, Washington, DC, U.S.A., pp. 215–216.

Kosygin, Yu.A. and Kulish, E.A. (Eds.) (1984) *Main Types of Ore Formations. Terminological Manual*, Nauka, Moscow, 316 pp.

Krainov, S.R. (1973) *Geochemistry of Rare Elements in Subsurface Waters*, Nedra, Moscow, 296 pp.

Krainov, S.R., Ryzhenko, B.I., Lyal'ko, V.I., Dobrovol'skii E.V. and Solomin, G.A. (1985) Computer modelling of interactions in the water-rock system at different stages of lithogenesis. In: *Subsurface Waters and Evolution of Lithoshere. Proc. All-Union Conference, I*, Nauka, Moscow, pp. 91–105.

Krauskopf, K.B. (1956) Factors controlling the concentration of thirteen rare metals in sea water. *Geochim. Cosmochim. Acta* **9**, 1–32B.

Krauskopf, K.B. (1957) Separation of manganese from iron in sedimentary processes. *Geochim. Cosmochim. Acta* **12**, 61–84.

Krestov, G.A. (1973) *Thermodynamics of Ion Processes in Solutions*, Khimiya (USSR), Leningrad, 303 pp.

Krestov, G.A. and Abrosimov, V.K. (1967) The influence of temperature on negative hydration of ions. *Zhurn. Strukt. Khim. (USSR)* **8**, 822–826.

Landing, W.M. and Bruland, K.W. (1980) Manganese in the North Pacific. *Earth and Planet. Sci. Lett.* **49**, 45–56.

Landing, W.M. and Bruland, K.W. (1987) The contrasting biogeochemistry of iron and manganese in the Pacific Ocean. *Geochim. Cosmochim. Acta* **51**, 29–43.

Leclerc, J. and Weber, F. (1980) Geology and genesis of the Moanda manganese deposits, Republic of Gabon. In: *Geology and Geochemistry of Manganese. Vol. II. Manganese Deposits on Continents*, Publishing House of Hungarian Academy of Sciences, Budapest, pp. 89–109.

Leinen, M., Rea, D.K. and Anderson, R.N. *et al.* (1986) Initial Reports, DSDP, Vol. 92, Washington (U.S. Government Printing Office), 617 pp.

Leinen, M. and Stakes, D. (1979) Metal accumulation rates in the Central Equstorial Pacific during Cenozoic time. *Geol. Soc. Amer. Bull. Part 1*, **90**, 357–375.

Levin, L.E., Varentsov, I.M., Baskakova, D.K. and Virta, A.M. (1967) Mineralization of the sedimentary cover of the Pacific and its relation with spreading and volcanism. *Bull. Mosc. Society of Nature Invest., Geol. Division* **62**(4), 3–17.

Linnik, P.N. and Nabivanets, B.I. (1977) Determination of different forms of ions of metals in natural waters. *Hydrobiol. Journal* **1**, 103–111.

Lisitsyn, A.P. and Bogdanov Yu.A. (Eds.) (1986) *Metalliferous Sediments of the Red Sea*, Nauka, Moscow, 288 pp.

Lisitsyn, A.P., Bogdanov, Yu.A., Murdmaa, I.O. *et al.* (1976) Metalliferous sediments and their genesis. In: *Geologic-Geophysical Investigations of the Southeastern Part of the Pacific*, Nauka, Moscow, pp. 289–379.

Lisitsyn, A.P., Gordeev, V.V., Demina, L.L. and Lukashin, V.N. (1985) Marine geochemistry of manganese. *Proc. Acad. Sci. USSR, Geol. Series*, No. 3. 3–29.

Loganathan, P. and Burau, R.G. (1973) Sorption of heavy metals by hydrous manganese oxide. *Geochim. Cosmochim. Acta* **37**, 1277–1293.

Lopez-Eyzaguirre, C. and Bisque, E. (1975) Study of the weathering of basic, intermediate and acidic rocks under tropical humid conditions. *Quarterly of the Colorado School of Mines* **70**(1), 1–59.

Lyashchenko, A.K. (1968) Localization of ions and hydrate complexes in the structure of aqueous solution. *Zhurn. Struct. Khim. (USSR)* **9**, 781–787.

Lyashchenko, A.K. (1972) Structural model of the water solutions of electrolytes on the density data. In: *Physical Chemistry of Solutions*, Nauka, Moscow, pp. 5–12.

Lyashchenko, A.K. (1973) Problems of the structure of aqueous solutions of electrolytes. I. Aqueous solution of an electrolyte as a structural system. *Izv. AN SSSR. Ser. Chim. (USSR)* **2**, 287–293.

Lyklema, J. (1968) The structure of the electrical double layer on porous surfaces. *Journ. Electro-Anal. Chem.* **18**, 341–348.

Lyklema, J. (1971) The electrical double layer on oxides. *Croatica Chemica Acta* **43**, 249–260.

Mackenzie, F.T. and Wollast, R. (1977) Thermodynamic and kinetic controls of global chemical cycles of the elements. In: Stumm, W. (Ed.), *Global Chemical Cycles and Their Alterations by Man. Dahlem Konferenzen*,

Springer-Verlag, Berlin, pp. 45–59.

Makatova, I.N. and Imanalieva, S.V. (1977) Influence of electrolytes on sorption of copper ions by iron hydroxide. In: *Fertilizers and Inorganic Materials from Kazakhstan Minerals. Proc. Inst. of Chem. Sci. of the Kaz. Acad. Sci.* **44**, 104–106.

Manceau, A. and Combes, J.M. (1988) Structure of Mn and Fe oxides and oxyhydroxides: A topological approach by EXAFS: Physics Chemistry Minerals, 15. 283–295.

Manceau, A., Llorca, S. and Calas, G. (1987) Crystal chemistry of cobalt and nickel in lithiophorite and asbolan from New Caledonia. *Geochim. Cosmochim. Acta* **51**, 105–113.

Manceau, A., Gorshkov, A.I. and Drits, V.A. (1992) Structural chemistry of Mn, Fe, Co, and Ni in manganese hydrous oxides: Part I. Information from XANES spectroscopy. *Amer. Mineral.* **77**, 1133–1143.

Manceau, A., Gorshkov, A.I. and Drits, V.A. (1992) Structural chemistry of Mn, Fe, Co, and Ni in manganese hydrous oxides. Part II. Information from EXAFS spectroscopy and electron and X-ray diffraction. *Amer. Mineral.* **77**, 1144–1157.

Manganese (World) (1977) *Miner. Trade Notes* **74**(11), 14–16.

Manganese (1990) *Mining Magazine* **162**(1), 36–40.

Mank, V.V., Ovcharenko, F.D. and Malyarenko, A.V. (1983) Radiospectroscopic study of the structure of boundary water layers. In: *Surface Forces and Boundary Layers of Liquids*, Nauka, Moscow, pp. 126–131.

Marchenko, Z. (1968) *Spectrophotometric Identification of Elements*, Wydawnictwa-Naukowo-Techniczne, Warszawa, 501 pp.

Matveeva, L.A., Vasil'eva, L.A., Neklyudova, Ye.A. and Rozhdestevenskaya, Z.S. (1978) Effect of soil organic acids on behaviour of aluminium in diluted solutions. *Weathered Crust* **16**, pp. 212–229.

Maynard, J.B. (1983) *Geochemistry of Sedimentary Ore Deposits*, Springer-Verlag, New York, 305 pp.

McIntosh, J.L., Farag, J.S. and Slee, K.J. (1975) Groote Eylandt manganese deposits. In: Knight, C.L. (Ed.), *Economic Geology of Australia and Papua New Guinea. 1. Metals*, Aust. Inst. Min. Metall. Monograph, Vol. 5, pp. 815–821.

McKelvey, V.E., Wright, N.A. and Bowen, R.W. (1983) Analysis of the world distribution of metal-rich subsea manganese nodules. Geol. Surv. Circular 886. U.S. Geol. Sur., Alexandria, VA, 55 pp.

McKenzie, R.M. (1970) The reaction of cobalt with manganese dioxide minerals. *Australian Journ. Soil Res.* **8**, 97–106.

McKenzie, R.M. (1971) The synthesis of birnessite, cryptomelane and some other oxides and hydroxides of manganese. *Miner. Magazine* **38**, 493–502.

McKenzie, R.M. (1972) The sorption of some metals by the lower oxides of manganese. *Geoderma* **8**, 29–35.

McKenzie, R.M. (1980) The adsorption of lead and other heavy metals on oxides of manganese and iron. *Australian Journ. Soil Res.* **18**, 61–73.

McMichael, B. (1989) Manganese: EMD – The power in the market. *Ind. Miner (London)*, No. 260, 42–59.

Melfi, A.J. and Pedro, G. (1973) Étude sur l'altération expérimentale de silicates manganésiféres et la formation exogène des gisments de manganèse. *Groupe Fr. Argiles, Bull.* **26**, 91–105.

Melfi, A.J., Valarelli, J.V., Pedro, G., Levi, F. and Hypolito, R. (1973) Étude sur la formation d'oxydes et d'hydroxydes de manganèse par altération expérimentale de silicates manganésifires. *C.R. Acad. Sci. Paris, Série D.*, **277**, 5–8.

Menard, H.W. and Frazer, J.Z. (1978) Manganese nodules on the sea floor: Inverse correlation between grade and abundance. *Science* **199**(3), 969–971.

Mero, J.L. (1962) Ocean-floor manganese nodules. *Econ. Geol.* **37** 747–767.

Mero, J.L. (1965) *The Mineral Resources of the Sea*, Elseiver, Amsterdam, 312 pp.

Meyers, P.A. and Quinn, I.G. (1974) Organic matter on clay minerals and marine sediments – Effect on adsorption of dissolved copper, phosphate and lipids from saline solutions. *Chem. Geol.* **13**, 63–68.

Michard, A., Albaréde, F., Michard, G., Minster, J.F. and Charlow, J.L. (1983) Rare-earth elements and uranium in high-temperature solutions from East Pacific Rise hydrothermal vent field (13°N). *Nature* **303**, No. 5920, 795–797.

Michard, G. (1969) Dépôt de traces de manganèse par oxidation. *C. R. Acad. Sci., Paris, Série D*, **269**, 1811–1814.

Migdisov, A.A., Gradusov, B.P., Bredanova, N.V., Bezrogova, E.V., Savel'ev, B.V. and Smirnova, O.N. (1983) Major and minor elements in hydrothermal and pelagic sediments of the Galapagos mounds area, Leg 70, Deep Sea Drilling Project. In: Honnores, J. and Von Hersen, R.P. *et al.*, *Init. Repts. DSDP, Vol. 70*, Washington, U.S. Government Printing Office, pp. 277–295.

Mikhailov, I.G. and Syrnikov, Yu.P. (1960) On the influence of ions on the structure of water. *Zhurn. of Struct.*

Chem. **1**, 12–27.

Minachev, Kh.M., Antoshin, G.V. and Shkiro, E.S. (1975) Study of the state of nickel in multicomponent zeolite catalyst by X-ray photoelectron spectroscopy. *Metallofizika Kiev*, **60**, 55–59.

Minaeva, L.G., Kizilshtein, L.Ya. and Besedin, V.B. (1974) Some of the biochemical aspects of concentration of organic compounds by means of coacervate droplets of hydrate of ferrous iron. Doklady AN SSSR, 215, 6. 1470–1472.

Mishchenko, K.P. and Poltaratskii, G.M. (1976) *Thermodynamics and Structure of Aqueous and Non-Aqueous Solutions of Electrolytes*, Khimiya, Leningrad, 328 pp.

Mkrtych'yan, A.K., Savan'yak, Yu.V., Ustalov, V.V. and Tsykin, R.A. (1982) Manganese-bearing formations of the Eniseiskii Ridge. In: *Geology and Geochemistry of Manganese*, Nauka, Moscow, pp. 89–94.

Mkrtych'yan, A.K., Tsykin, R.A. and Ustalov, V.V. (1979) Manganese-bearing sediments and manganese ores of the eastern part of the Altai-Sayanskaya Mountainoous region and Eniseiskii Ridge. *Geology of Ore Deposits*, No. 5, 105–112.

Mkrtych'yan, A.K., Tsykin, R.A. and Savan'yak, Yu.V. (1980) Manganese-bearing capacity of the Eniseiskii Ridge. In: *New Data on Manganese Deposits of the USSR*, Nauka, Moscow, pp. 205–210.

Morgan, J.J. and Stumm, W. (1964) Colloid chemical properties of manganese dioxide. *J. Colloid Interface Sci.* **19**, 347–359.

Morgan, J.J. and Stumm, W. (1965) The role of multivalent metal oxides in limnological transformations as exemplified by iron and manganese. In: Jaag, O. (Ed.), *Advances in Water Pollution Researches. Proc. Second Intern. Conf.*, Tokyo, August 1964, 1, A, Pergamon Press, New York, pp. 103–131.

Mosevich, V.A. (1964) Variability of the ion composition of surface waters of the Karel'skii isthmus. In: *Lakes of the Karel'skii Isthmus. Limnology and Technology of Investigations*, Nauka, Moscow, Leningrad, pp. 3–15.

Mstislavskii, M.M. (1981) Paleotectonical features of localization of the Oligocene manganese ore deposits of the USSR South. *Repts. Acad. Sci. USSR* **260**(5), 1207–1212.

Mstislavskii, M.M. (1985) If there are "classic sedimentary" deposits of manganese of the Chiaturskoe type in nature. *Geology of Ore Deposits* **27**(6), 3–16.

Mstislavskii, M.M. (1988) New view on the formation of manganese deposits. *Soviet Geology*, No. 6, 116–122.

Mstislavskii, M.M. and Potkonen, N.I. (1990) Porozhinskoe manganese deposit in the Eniseiskii Ridge. *Geology of Ore Deposits*, No. 3, 82–95.

Murdmaa, I.O. and Skornyakova, N.S. (Eds.) (1986) *Iron-Manganese Concretions of the Central Part of the Pacific*, Nauka, Moscow, 344 pp.

Murray, J. (1876) Preliminary report on speciments of the sea bottom. *Proc. Roy. Soc. London* **24**, 471–532.

Murray, J. and Irvine, R. (1895) On the chemical changes with take place in the composition of the sea water associated with blue muds on the floor of the ocean. *Trans, Roy. Soc. Edinburgh* **37**, 481–508.

Murray, J. and Renard, A.F. (1884) On the microscopic characters of volcanic ashes and cosmic dust, and their distribution in the deep sea deposits. *Proc. Roy. Soc. Edinburgh* **12**, 474–495.

Murray, J. and Renard, A.F. (1891) *Deep-Sea Deposits*, Rep. Sci. Results Explor. Voyage Challenger, 525 pp.

Murray, J.W. (1974) The surface chemistry of hydrous manganese dioxide. *J. Colloid Interface Sci.* **46**, 357–371.

Murray, J.W. (1975a) The interaction of cobalt with hydrous manganese dioxide. *Geochim. Cosmochim. Acta* **39**, 635–647.

Murray, J.W. (1975b) The interaction of metal ions at the manganese dioxide-solution interface. *Geochim. Cosmochim. Acta* **39**, 1531–1543.

Murray, J.W. and Brewer, P.G. (1977) Mechanisms of removal of manganese, iron and other trace metals from seawater. In: Glasby, G.P. (Ed.), *Marine Manganese Deposits*, Elsevier, Amsterdam, pp. 291–325.

Murray. J.W. and Dillard, J.G. (1979) The oxidation of cobalt(II) adsorbed on manganese dioxide. *Geochim. Cosmochim. Acta* **43**, 781–787.

Murray, J.W., Dillard, J.G, Giovanoli, R., Moers, H. and Stumm, W. (1985) Oxidation of Mn(II): initial mineralogy, oxidation state and ageing. *Geochim. Cosmochim. Acta* **49**, 463–470.

Musatov D.I., Ustalov, V.V., Kachevskii, L.K., Blagodatskii, A.V., Storozhenko, A.A. and Leonov, O.I. (1980) Manganese mineralization in the Precambrian of the Eniseiskii Ridge. In: *New Data on Manganese Deposits of the USSR*, Nauka, Moscow, pp. 200–205.

Musinu, A., Paschina, G., Piccaluga, G. and Mangini, M. (1982) The sulphate ion in aqueous solution: An X-ray diffraction study of $ZnSO_4$ solution. *Journ. Appl. Crystal.* **15**, 621–625.

Nahon, D., Beavais, A., Boeglin, J.-L., Ducloux. and Nziengui-Mapangou, P. (1983) Manganite formation in the first stage of the lateritic manganese ores in Africa. *Chem. Geol.* **40**, 25–42.

Nahon, D., Colin, F. and Tardy, Y. (1982) Formation and distribution of Mg, Fe, Mn-smectites in the first stages of the lateritic weathering of forsterite and tephroite. *Clay Minerals* **17**, 339–348.

Nahon, D., Janot, C., Karpoff, A.M., Paquet, H. and Tardy, Y. (1977) Mineralogy, petrography and structures of iron crusts (ferricretes) developed on sandstones in western part of Senegal. *Geoderma* **19**, 263–277.

Nance, W.B. and Tayler, S.R. (1976) Rare earth patterns and crustal evolution, 1: Australian post-Archean sedimentary rocks. *Geochim. Cosmochim. Acta* **40**, 1539–1551.

National Minerals Advisory Board (1981) Manganese Reserves and Resources of the World and their Industrial Implications. Natl. Acad. Sci., Washington, D.C., NMAB-374, 334 pp.

Nemoshkalenko, V.V., Aleshin, V.G. (1976) *Crystal Electron Microscopy*, Naukova Dumka Publishers.

Nesbitt, H.W. (1979) Mobility and fractionation of rare earth elements during weathering of a granodiorite. *Nature* **279**, 206–210.

Nevolina, N.A., Samoilov, O.Ya and Seifer, A.L. (1969) Dependence of near hydration of cations Na^+, K^+, and NH_3^+ on pressure. *Zhurn. Struct. Khim. (USSR)* **10**, 203–207.

Nissenbaum, A. and Kaplan. I.R. (1972) Chemical and isotopic evidence for the in situ origin of marine humic substances. *Limnol. Oceaongr.* **17**, 570–582.

Nissenbaum, A. and Swaine, D.J. (1976) Organic matter-metal interaction in recent sediments: the role of humic substances. *Geochim. Cosmochim. Acta* **40**, 809–816.

Nosova, T.A. and Samoilov, O.Ya. (1964) Dependence of dehydration and hydration on hydration of salted-out ion. *Zhurn. Struct. Khim. (USSR)* **5**, 363–370.

Novikov, A.I. (1972) On the chemistry of coprecipitation of small amounts of elements with hydrated oxides. In: *Coprecipitation with Hydrated Oxides*, Izd-vo Tadzhik. Gos. Un-ta, Dushanbe, 1, pp. 5–42.

Novikov, A.I. and Goncharova, L.K. (1972) Coprecipitation of Pb with Fe(III) and Zr hydroxides. In: *Coprecipitation with Hydrated Oxides*, Izd-vo Tadzhik. Gos. Un-ta, Dushanbe, 1, pp. 69–90.

Novikov, A.I. and Knyazev, N.A. (1977) On the chemistry of coprecipitation of small amounts of elements with hydroxides. II. Coprecipitation of zinc with iron hydroxide in conditions of complexation. In: *Coprecipitation with Hydroxides*, Izd-vo Tadzhik. Un-ta, Dushanbe, 2, pp. 5–21.

Novikov, A.I., Shaffert, A.A. and Yakubova, N.S. (1977) Study of dependence beyween hydrolysis of 3-d elements and their sorption by precipitation of hydroxides. In: *Coprecipitation with Hydroxides*, Izd-vo Tadzhik. Un-ta, Dushanbe, 2, pp. 81–86.

Ostroumov, E.A. (1956) Distribution of titanium in the sediments of the Sea of Okhotsk. *Geochimiya* **1**, 90–95.

Ostwald, J. (1975) Mineralogy of manganese oxides from Groote Eylandt. *Mineral. Deposita* **10**, 1–12.

Ostwald, J. (1980) Aspects of the mineralogy, petrology and genesis of the Groote Eyland manganese ores. In: *Geology and Geochemistry of Manganese. Vol. II. Manganese Deposits on Continents*, Akadémiai Kiadó, Budapest, pp. 149–181.

Ostwald, J. (1981a) Evidence for biochemical origin of the Groote Eylandt manganese ores. *Econ. Geol.* **76**(3), 556–567.

Ostwald, J. (1981b) Some observations on the mineralogy and genesis of chalcophanite. *Mineral. Mag.* **44**, pp. 109–111.

Ostwald, J. (1982) Characterisation of Groote Eyland manganese minerals. *BHP Technical Bull. (Australia)* **26**(1), 52–55.

Ostawld, J. (1984) The influence of clay mineralogy on the crystallisation of the tetravalent manganese layer-lattice minerals. *Neues Jahrb. Mineralogie Monash.* **1**, 9–16.

Ostwald, J. (1987a) Hybrid-structured lithiophorite. *Austral. Mineral.* **2**, 15–31.

Ostwald, J. (1987b) Manganese carbonates in the Groote Eylandt sedimentary manganese deposit. *Neues Jahrb. Mineralogie Abh.* **157**, 265–301.

Ostwald, J. (1988) Mineralogy of the Groote Eylandt manganese oxides: A review. *Ore Geol. Rev.* **4**, 3–45.

Ostwald, J. (1992) Genesis and paragenesis of the tetravalent manganese oxides of the Australian continent. *Econ. Geol.* **87**, 1237–1252.

Ostwald, J. and Bolton, B.R. (1990) Diagenetic braunite in sedimentary rocks of the Proterozoic Manganese Group, Western Australia. *Ore Geology Rev.* **5**, 315–323.

Ostwald, J. and Bolton, B.R. (1992) Glauconite formation as a factor in sedimentary manganese deposit genesis. *Econ. Geol.* **87**, 1336–1344.

Parc, S., Nahon, D., Tardy, Y. and Vieillard, P. (1989) Estimated solubility products and fields of/and stability for cryptomelane, nsutite, birnessite and lithiophorite based on natural lateritic weathering sequences. *Amer. Mineral.* **74**, 466–475.

Park, C.F. (1956) On the origin of manganese. 20th International Geologic. *Congress. Sympos. Manganese*, Mexico, 1, pp. 75–98.

Parsons, T.R. and Seki, H. (1968) Importance and general implications of organic matter in aquatic environments. In: *Symposium on Organic Matter in Natural Waters*, Alaska, pp. 1–27.

Pasashnikova, G.K. (1984) Mokhovoe manganese deposit of the Porozhinskoe ore area. In: *Manganese Ore Formation on the USSR Territory*, Nauka, Moscow, pp. 101–109.

Patterson, S.H. (1971) Investigations of ferruginous bauxite and other mineral resources on Kausi and a reconnaissance of ferruginous bauxite deposits on Main Hawaii. U.S. Geolog. Surv., Prof. Paper No. 656, 71 pp.

Perseil, E.A. and Bouladon, J. (1971) Microstructures des oxydes de manganèse à la base du gisement de Moanda et leur signification génétigue. *C.R. Acad. Sci. Paris* **273**, 278–279.

Perseil, E.A. and Giovanoli, R. (1983) Contribution à la connaissance des constituants des nodules polymétalliques à partir des donneés fournies par l'étude des produits de synthèse, ainsi que des oxydes dans les gisements terrestres. *Bull. Mus. Nant. Hist. Nat., Paris*, 4e ser., 5, sect. C, No. 2, 163–190.

Perseil, E.A. and Grandin, G. (1978) Évolution minéralogique du manganèse dans trois gisements d'Afrique de l'Ouest: Mokta, Tambao, Nsuta. *Mineral. Deposita, (Berl.)* **13**, 295–311.

Perseil, E.A. and Grandin, G. (1985) Altération supergéne des protores à grenats manganésifères dans quelques Gisements d'Afrique de l'Ouest. *Mineral. Deposita* **20**(3), 211–219.

Perseil, E.A. and Pinet, M. (1976) Contribution à la connaissance des romannéchites et des cryptomélanes-coronadites-hollandites. Trait essentiel et paragèneses. *Contrib. to Mineral. and Petrol.* **55**, 191–204.

Picard, G.L. and Felbeck, G.T., Jr. (1976) The complexation of iron by marine humic acid. *Geochim. Cosmochim. Acta* **40**, 1347–1350.

Pillai, T.V., Desai, M.V., Mathew, E., Ganapathy, S. and Ganguly, A.K. (1971) Organic materials in the marine environment and the associated metallic elements. *Current Sci., India* **40**, 75–81.

Piper, D.Z. (1974) Rare earth elements in the sedimentary cycle: A summary. *Chem. Geol.* **14**, 285–304.

Piper, D.Z. and Williamson, M.E. (1977) Composition of Pacific ocean ferromanganese nodules. *Mar. Geol.* **23**(4), 285–303.

Plumb, K.A., Derrick, G.M. and Wilson, I.H. (1980) Precambrian geology of the McArthur River-Mount Isa region, Northern Australia. In: Henderson, R.A. and Stephenson, P.J. (Eds.), *The Geology and Geophysics of the Northern Australia*, Geological Soc. Australia, Queensland Div., pp. 71–88.

Podporina, E.K. and Burkov, V.V. (1977) Rare earth elements in the process of weathering of biotite pyroxenites. *Lithology and Mineral Resources*, No. 4, 39–53.

Popov, N.I., Fedorov, K.N. and Orlov, V.M. (1979) *Seawater* (Reference manual), Nauka, Moscow, 327 pp.

Portyannikov, D.I. and Ageenko, N.F. (1979) Sungaiskoe manganese occurreence (Salairskii Ridge). In: *Geology and Minerals of the Altaiskii Region*, Proc. Conf., Barnaul, p. 33.

Posselt, H.S., Anderson, F.J. and Weber. W.J., Jr. (1968) Cation sorption on colloidal manganese dioxide. *Environmental Science and Technology* **2**(12), 1087–1093.

Potkonen, N.I. (1985) The first finding of the manganosite and hansmannite in Porozhinskoe deposit of manganese (Eniseiskii Ridge). *Doklady AN SSSR* **285**(2), 431–434.

Potter, P.W., II (1970) *Computer Modelling in Low Temperature Geochemistry. Review of Geophysics and Space Physics*, American Geophysical Union, Vol. 17, No. 4, pp. 850–880.

Potter, R.M. and Rossman, G.R. (1979) The tetravalent manganese oxides: Identification, hydration and structural relationships by infrared spectroscopy. *Amer. Mineral.* **64**, 1199–1218.

Poit, G. (1964) Les gites de manganese marocains encaissés dans les formations carbonatées: elements pour une synthèse. *Chronique Mines Rech. Min.*, No. 3397, 371–380.

Poit, G. (1976) La concentration de manganese de l'Imini (Maroc). Peut-elle être d'origine karstique? *C. R. Sommaire des Sciences Bull. de la Société Géologique de France* **5**, 227–229.

Poit, G. (1988) Quelle est la plus grosse concentration metallique a la surface emergee du globe? *Chron. Rech. Miniere* **58**, No. 492, 55–56.

Pracejus, B. (1990) Groote Eylandt manganese norm – A new application of mineral normalization techniques on supergene alteration products. In: Parnell, J, Lianjun, Ye. and Ghangming, Chen (Eds.), *Sediment-Hosted Mineral Deposits*, Spec. Publ. Int. Ass. Sediment, Blackwell, Oxford, 11, pp. 3–16.

Pracejus, B. and Bolton, B.R. (1992a) Geochemistry of supergene manganese oxide deposits, Groote Eylandt, Australia. *Econ. Geol.* **87**, 1310–1335.

Pracejus, B. and Bolton, B.R. (1992b) Interdependence of Mn, Fe and clay mineral formation on Groote Eylandt,

Australia: A model for modern and ancient weathering environments. In: Skinner, H.G.W. and Fitzpatrick (Eds.), *Biomineralization: Processes of Iron and Manganese – Modern and Ancient Environments*, Catena Supplement 21, Cremlingen, pp. 371–397.

Pracejus, B., Bolton, B.R. and Frakes, L.A. (1986) Electrochemical process in Mn-ores from Groote Eylandt (N.T. Australia): Significance for the generation of supergene Mn-deposits. *12th Int. Sed. Congress*, Abstract, Canberra, pp. 247–248.

Pracejus, B., Bolton, B.R. and Frakes, L.A. (1988a) Mineral normalization for supergene Mn-oxides and associated rocks on ores from Groote Eylandt, N.T. Australia. *IAS Symposium on Sedimentology Related to Mineral Deposits*, Abstract, Beijing, China, pp. 202.

Pracejus, B., Bolton, B.R. and Frakes, L.A. (1988b) Nature and development of supergene manganese deposits, Groote Eylandt, Northern Territory, Australia. *Ore Geol. Rev.* **4**, 71–98.

Pracejus, B., Bolton, B.R., Frakes, L.A. and Abbot, U. (1990a) Rare earth element geochemistry of supergene manganese deposits from Groote Eylandt, Northern Territory, Australia. *Ore Geol. Rev.* **5**, 293–314.

Pracejus, B., Varga, R., Madgwick, J, Frakes., L.A. and Bolton, B.R. (1990b) Effect of mineral composition on microbiological leaching of manganese oxides. *Chem. Geol.* **88**(1–2), 143–149.

Proceedings of the First International Seminar on Lateritization Processes (1981) Trivandrum, December, 1979. Oxford and IBH Publishing Co., New Delhi, 150 pp.

Progress of IGCP Projects: No. 129 (1984) Lateritization processes. Geological correlation, No. 12. Paris, pp. 30–37.

Pronina, N.V. and Varentsov, I.M. (1973) On the special features of nickel and cobalt sorption from sea water by natural ferric and manganese hydrous oxides. *Dokl. Acad. Nauk, SSSR* **210**(4), 944–947.

Pronina, N.V., Varentsov, I.M., Spektorova, L.V., Spektorov, K.S. and Ovsyannikova, M.N. (1973) Study of the sorption of nickel and cobalt (biogenic forms) from sea water by natural ferric and manganese hydrous oxide. *Geokhimiya* **6**, 876–886.

Putilina, V.S. and Varentsov, I.M. (1980) Interaction between organic matter and heavy metals in the waters of recent basins – A review of the current state of the problem. *Chem. Erde* **39**, 398–310.

Putilina, V.S. and Varentsov, I.M. (1982) Iron in waters of the Northern Atlantic. In: *Geology and Geochemistry of Manganese*, Nauka, Moscow, pp. 244–252.

Putilina, V.S. and Varentsov, I.M. (1983) Experimental study of the processes of ferromanganese ore formation in recent basin: sorption of iron (II) by manganese dioxide from seawater in the presence of an organic complexing agent. *Chem. Erde* **42**, 321–337.

Putilina, V.S. and Varentsov, I.M. (1984) About the role of organic component in the process of absorption of iron (II) by manganese dioxide from seawater (experimental data to the model of formation of iron-manganese ores in modern basins). *Geochemistry*, No. 9, 1352–1363.

Rakhmanov, V.P. (1967) *Manganese Ores. Achievements in Studies of the Major Sedimentary Mineral Resources*, Nauka, Moscow, pp. 80–100.

Ramdohr, P. and Frenzel, G. (1956) Die Manganerze. *Internat. Geol. Cong.*, 20th session, Symposium del Manganeso, Mexico City, 1, pp. 19–73.

Ramsdell, L.S. (1932) An X-ray study of psylomelane and wad. *Amer. Mineral.* **17**, 143–149.

Rashid, M.A. (1972a) Role of humic acid of marine origin and their different molecular weight fractions in complexing di- and trivalent metals. *Soil Sci.* **111**, 298–306.

Rashid, M.A. (1972b) Role of quinone groups in solubility and complexing of metals in sediments and soils. *Chem. Geol.* **9**, 241–247.

Rashid, M.A. (1972c) Amino acids associated with marine sediments and humic compounds and their role in solubility and complexing of metals. *Intern. Geol. Congress*, 24th Session, Section 10, Geochemistry. Montreal, pp. 346–353.

Rashid, M.A. (1974) Absortion of metals on sedimentary and peat humic acids. *Chem. Geol.* **13**, 115–123.

Report of the Conference on Scientific Ocean Drilling (1982), November 16-18, 1981, Sponsored by the Joint Oceanographic Institutions for Deep Earth Sampling (JOIDES), J.O.I. Inc. Washington, D.C, 112 pp.

Reuter, J.H. and Perdue, E.M. (1977) Importance of heavy metal-organic matter in natural waters. *Geochim. Cosmochim. Acta* **41**, 325–334.

Reyna, J.G. (Ed.) (1956a) *Symposium Sobre Yacimientos de Manganeso. XX Congreso Geologico International. I, El Manganesoen General*, Mexico, 155 pp.

Reyna, J.G. (Ed.) (1956b) *Symposium Sobre Yacimientos de Manganeso. XX Congreso Geologico Internacional. II, Africa*, Mexico, 300 pp.

Reyna, J.G. (Ed.) (1956c) *Symposium Sobre Yacimientos de Manganeso. XX Congreso Geologico Internacional. III. America*, Mexico, 438 pp.

Reyna, J.G. (Ed.) (1956d) *Symposium Sobre Yacimientos de Manganeso. XX Congreso Geologico Internacional. IV. Asia y Oceania*, Mexico, 336 pp.

Reyna, J.G. (Ed.) (1956e) *Symposium Sobre Yacimientos de Manganeso. XX Congreso Geologico Internacional. V. Europe*, Mexico, 376 pp.

Richmond, W.E. and Fleischer, M. (1942) Cryptomelane, a new name for the commonest of the "Psilomelane" minerals. *Amer. Mineral.* **27**, 607–610.

Richmond, W.E., Fleischer, M. and Morse, M.E. (1969) Studies on manganese oxide minerals. IX. Rancieite. *Bull. Soc. Fr. Mineral. Cristallogr.* **92**, 191–195.

Rocci, G. (1965) Essai d'interprétation de mesures geochronologiques. La structure de l'Ouest Africain. *Coll. Intern. CNRS. Géochronologique Absolue*, Nancy, 151, pp. 257–273.

Rode, E.Ya. (1952) *Oxygen Compounds of Manganese (Artificial Compounds, Minerals, and Ores)*, Publ. Acad. Sci. Moscow, USSR, 398 pp.

Rodrigues, O.B., Kosuki, R. and Coelho, Filho, A. (1986) Distrito Manganesífero de Serra do Navio, Amapa, (Capítulo XIV). In: *Principas Depositos Minerais do Brasil, Vol. II, Ferro e Metais da Industria do Aço*, Coordenador-geral: Schobbenhaus, C., and Coelho, C.E.S.., Brasilia, Ministerio das Minas e Energia, pp. 167–175.

Romankevich, Ye.A. (1977) *Geochemistry of Organic Matter in the Ocean*, Nauka, Moscow, 256 pp.

Rona, P.A. (1987) Oceanic ridge crest processes. Reviews of Geophysics, 25, No. 5 (U.S. National Report to International Union of Geodesy and Geophysics 1983-1986), 1089–1114.

Rona, P.A. (1988) Hydrothermal mineralization at oceanic ridges. *Canad. Mineral.* **26**, 431–465.

Rona, P., Boström, K., Laubier, L. and Smith, K.L., Jr. (Eds.) (1983) Hydrothermal processes at seafloor spreading centers. Proc. NATO Adv. Res. Inst. Dept. Earth Sci. Cambridge University, 5-8 April 1982–NATO Conf. Series 4, 12, New York and London, Plenum Press, 796 pp.

Ronov, A.B. (Ed.) (1980) *Geochemistry of Elements-Hydrolysates*, Nauka, Moscow, 239 pp.

Ronov, A.B., Balashov, Yu.A. and Migdisov, A.A. (1967) Geochemistry of rare earth elements in sedimentary cycle. *Geochemistry*, No. 1, 3–19.

Rosencwaig, A., Wertheim, G.K. and Guggenheim, H.J. (1971) Origins of satellites on inner-shell photoelectron spectra. *Phys. Rev. Lett.* **27**(8), 479.

Roy, Supriya (1965) Comparative study of the metamorphosed manganese protores of the world – The problem of the nomenclature of the gondites and kodurites. *Econ. Geology* **60**, 1238–1260.

Roy, Supriya (1968) Mineralogy of the different genetic types of manganese deposits. *Econ. Geol.* **63**, 760–786.

Roy, Supriya (1969) Classification of manganese deposits. *Acta Mineral.-Petrograph., Acta Universitatis Szegediensis, Szeged, Hungaria*, **XIX**(1), 67–83.

Roy, Supriya (1972) Metamorphism of sedimentary manganese deposits. *Acta Miner. Petrogr., Szeged* **20**, 313–324.

Roy, Supriya (1973) Genetic stidies on Precambrian manganese formation of India with particular reference to the effect of metamorphism. Genesis of Precambrian Iron Manganese Deposits, UNESCO, Paris, Earth Science, 229–242.

Roy, Supriya (1973) Mineralogy of Indian manganese ores. *Indian Min. and Engineer. Journ.* **12**(6), 55–60.

Roy, Supriya (1980) Genesis of sedimentary manganese formations processes and products in Recent and older geological ages. In: *Geology and Geochemistry of Manganese. Manganese Deposits on Continents*, Akadémiai Kiadó, Budapest, 2, pp. 13–44.

Roy, Supriya (1980) Manganese ore deposits of India. In: *Geology and Geochemistry of Manganese. Manganese Deposits on Continents*, Akadémiai Kiadó, Budapest, 2, pp. 237–264.

Roy, Supriya (1981) *Manganese Deposits*, Academic Press, London, 458 pp.

Roy, S. (1988) Manganese metallogenesis: A review. *Ore Geol. Rev.* **4**(1–2), 155–170.

Roy, S., Bandopadhyay, P.S. and Base, P.K. (1982) Geology and genesis of the manganese deposits of the Penganga Beds, Adilabad District, Andhra Pradesh, India. I: Collected Abstracts, Intern. Ass. on the Genesis of Ore Deposits (IAGOD), VI Symposium, Tbilisi, September 6-12, 1982, Tbilisi, pp. 305–306.

Rozhnov, A.A. (1967) About geologic and genetical features of manganese mineralization of the western part of the Dzhail'minskaya through and the place of manganese mineralization in a series of occurrences of iron and polymetals of the area. In: *Manganese Deposits of the USSR*, Nauka, Moscow, pp. 311–324.

Rozhnov, A.A., Buzmakov, E.I., Negovora, N.M. and Sereda, V.Ya. (1984) New data on manganese-bearing capacity of the Atasuiskii area (Central Kazakhstan). In: *Manganese Ore Formation on the USSR Territory*, Nauka, Moscow, pp. 123–131.

Rozhnov, A.A., Buzmakov, E.I. and Sereda, V.Ya. (1980) New data on the geological structure of iron-manganese deposits of the Atasuiskii area (Central Kazakhstan). In: *New Data on Manganese Deposits of the USSR*, Nauka, Moscow, pp. 158–170.

Ruetschi, P. and Giovanoli, R. (1982) The behaviour of MnO_2 in strongly acidic solution. *Journ. of Apllied Electrochemistry* **12**, 109–114.

Ruppert, H. (1980) Fixation of metals on hydrous manganese and iron oxides phases in marine Mn–Fe nodules and sediments. *Chem. Erde* **39**, 97–132.

Saenko, G.N., Koryakova, M.D., Makienko, V.D. and Dobrosmylova, I.G. (1975) Concentration of polyvalent metals by algae of the Sea of Japan. *Use of Inorganic Resources of Ocean Water* **1**, 96.

Saenko, G.N., Radkevich, R.O. and Belcheva, N.N. (1977) *Marine Geology and Geological Structure of Nourishment Areas of the Seas of Japan and Okhotsk*, Far East Dept. USSR Ac. Sci., Vladivostok, pp. 138–163.

Safronova, A.A. (1985) About the role of water in the formation of carbonate powder. In: *Subsurface Waters and the Evolution of Lithospere. Proc. All-Union Conf., II*, Nauka, Moscow, pp. 132–134.

Samoilov, O.Ya. (1957) *Structure of Water Solutions of Electrolytes and Hydration of Ions*, Publ. Acad. Sci. USSR, Moscow, 185 pp.

Samoilov, O.Ya. (1957) A new approach to the study of hydration of ions in aqueous solutions. *Discuss. Faraday Soc.* **14**, 141–146.

Samoilov, O.Ya. (1959) Coordination number and translation of particles in aqueous solutions of electrolytes. *Dokl. AN SSSR (USSR)* **126**, 330–333.

Samoilov, O.Ya. (1966a) To the theory of salting out from aqueous solutions. I. General problems. *Zhurn. Strukt. Khim.* **7**, 15–23.

Samoilov, O.Ya. (1966b) To the theory of salting out from aqueous solutions. II. Dependence of dehydration and hydration on hydration of salted out cation. *Zhurn. Struk. Khim.* **7**, 175–178.

Samoilov, O.Ya. (1967) General questions of the theory of hydration of ions in aqueous solutions. In: *Condition and Role of Water in Biological Objects*, Nauka, Moscow, pp. 31–41.

Samoilov, O.Ya. (1972) Residence time of ionic hydration. In: Horne, R.A. (Ed.), *Water and Aqueous Solutions*, NSF, New York, pp. 597–611.

Samoilov, O.Ya. (1980) The phenomenon of the increase in the mobility of water molecules in aqueous solutions of electrolytes. Discovery of the USSR. In: *Bull. of Inventions, Discoveries, Trademarks*, diploma No. 230.

Samoilov, O.Ya. and Tikhomirov (1960) Salting out and exchange of the nearest to ions water molecules in aqueous solutions. *Radiokhimiya (USSR)* **2**, 183–191.

Sapozhnikov, D.G. (1963) *Karadzhal'skoe Iron-Manganese Deposit*, Publ. Acad. Sci. USSR, 196 pp.

Sapozhnikov, D.G. (1967a) Some geochemical conditions of formation of manganese deposits. In: *Manganese Deposits of the USSR*, Nauka, Moscow, pp. 11–33.

Sapozhnikov, D.G. (Ed.) (1967b) *Manganese Deposits of the USSR, Proc. Meet. on Major Genetical Types and Geochemistry of Manganese Deposits in the USSR*, Nauka, Moscow, 460 pp.

Sapozhnikov, D.G. (Ed.) (1980) *New Data on Manganese Deposits of the USSR*, Nauka, Moscow, 243 pp.

Sapozhnikov, D.G. (Ed.) (1984) *Manganese Ore Formation on the USSR Territory*, Nauka, Moscow, 293 pp.

Sapozhnikov, D.G. and Vitovskaya, I.V. (Eds.) (1987) *Exogenic Ore Formation (Al, Ni, Mn)*, Nauka, Moscow, 249 pp.

Savan'yak, Yu.V., Lomaev, V.G. and Khokhlov, A.P. (1984) Manganese-bearing formations of the southern part of the Eniseiskii Ridge. In: *Manganese Ore Formation on the USSR Territory*, Nauka, Moscow, pp. 79–82.

Schellmann, W. (1978) Behaviour of nicked, cobalt and chromium in ferruginous lateritic nickel ores. *Bull. Bur. Rech. Géol. et Minier. (deuxième série)*, Sect. II, 3, 275–282.

Schmidt, P.W., Currey, D.T. and Ollier, C.D. (1976) Sub-basaltic weathering, damsites, palaeomagnetsim and the age of laterization. *Jour. Geol. Soc. Austral.* **23**(4), 367–370.

Schmidt, P.W. and Embleton, B.J.J. (1976) Palaeomagnetic result from sediments of the Perth Basin, Western Australia and their bearing on the timing of regional laterization. *Palaeogeogr., Palaeoclimatol., Palaeoecolog.* **19**(4), 257–273.

Schmidt, P.W., Prasad Vanka and Ramam, P.K. (1983) Magnetic ages of some Indian Laterites. *Palaeogeogr., Palaeoclimatol., Palaeoecolog.* **44**(3–4), 185–202.

Schnitzer, M. and Hansen, E.N. (1970) Organo-metallic intereactions in soils. *Soil Sci.* **109**, 333 –340.

Schopf, J.W. (1981) The Precambrian development of an oxygenic atmosphere. In: Armstrong, F.C. (Ed.), *Genesis of Uranium- and Gold-Bearing Precambrian Quartz-Pebble Conglomerates*, U.S. Geol. Survey. Prof. Paper, No. 1161, pp. B1–B11.

Schoph, J.W. (Ed.) (1983) *Earth's Earlest Biosphere. Its Origin and Evolution*, Princeton University Press, Princeton, NJ, 543 pp.

Schsanning M., Nass K., Egeberg R.K., Bomé F. (1988) Cycling of manganese in the permanently anoxic Drammensfjord. *Mar. Chem.* **23**, 365–387.

Schufle, J.A., Chin-Tsung Huang and Drost-Hansen, W. (1976) Temperature dependence of surface conductance and a model of vicinal (interfacial) water. *Journ. of Colloid. and Interface Sci.* **54**(2), 184–202.

Schweisfurth, R., Eleftheriadis, D., Gundlach, H., Jacobs, M. and Jung, W. (1978) Microbiology of the precipitation of manganese. In: Krumbein, W. (Ed.), *Environmental Biogeochemistry and Geomicrobiology*, Ann Arbor Science Publication Inc., Ann Arbor, MI, U.S.A., pp. 923–928.

Scott S.D. (1987) Seafloor polymetallic sulfides: Scientific curiosities or mines of future? In: Teleki, P.G., Dobson, M.R., Moore J.R. and von Stackelberg U. (Eds.), *Marine Minerals Advances in Research and Resource Assessment*, D. Reidel, Dordrecht, pp. 277–300.

Senyavin, M.M. (1959) Elements of ion exchange theory and ion exchange chromatography. In: *Ion Exchange and Its Application*, Nauka, Moscow, pp. 84–126.

Sevastianov, V.F., Volkov, I.I. (1967) Redistribution of chemical elements in the oxidized layer of the sediments in the Black Sea and formation of ferromanganese nodules. In: *Proceedings of the Institute of Oceanology of the USSR Acad. Sci.*, Nauka, Moscow, 83, pp. 135–152.

Shackleton N.J. (1986) Paleogene stable isotope events. *Palaeogr., Palaeoclimtol., Palaeoecol.* **40**, 183–198.

Shatskii, N.S. (1954) On manganese-bearing formations and manganese metallogeny. *Proc. Acad. Sci. USSR, Geol. Series*, No. 4, 3–37.

Shatskii, N.S. (1960) Parageneses of sedimentary and volcanogenic rocks and formations. *Proc. Acad. Sci. USSR, Geol. Series*, No. 5, 3–23.

Shatskii, N.S. (1963) On geologic formations. In: *Selected Works, 3*, Nauka, Moscow, pp. 16–51.

Shcheka, V.A. and Zakutin, V.P. (1985) Bed zoning of redox states of subsurface waters and its influence on water migration of iron and manganese. In: *Subsurface Waters and Evolution of Lithosphere. Proc. All-Union Conf., II*, Nauka, Moscow, pp. 163–166.

Shcherbina, V.V. (1959) Geochemical foundations of separation of rare earth elements. In: *Geology of Deposits of Rare Elements*, Issue 3, Gosgeoltekhizdat, Moscow, pp. 44–59.

Shnyukov, E.F., Beloded, R.M. and Tsemko, V.P. (1974) *Mineral Resources of the World Ocean*, Naukova Dumka, Kiev, 205 pp.

Shnyukov, E.F., Beloded, R.M. and Tsemko, V.P. (1979) *Mineral Resources of the World Ocean*, Naukova Dumka, Kiev, 254 pp.

Shnyukov, E.F. and Mitropol'skii, A.Yu. (1987) *Metallogenic Investigations in Seas and Oceans*, VINITI, Ore Deposits Series, Vol. 17, Moscow, 146 pp.

Shnyukov, E.F. and Orlovskii, G.N. (1979) *Iron-Manganese Concretions of the Indian Ocean*, Inst. Geol. Sci., Kiev, 60 pp.

Shnyukov, E.F., Starostenko, V.I., Shcherbakov, I.B., Belevtsev, R.Ya. *et al.* (1984) *Geology and Metallogeny of the Northern and Equatorial Parts of the Indian Ocean*, Naukova Dumka, Kiev, 168 pp.

Shvartsev, S.L. (1978) *Hydrogeochemistry of Supergene Zone*, Nedra, Moscow, 287 pp.

Shvartsev, S.L. (1980) Occurrence in stages of weathering as the result of change of hydrogeochemical conditions of environment. In: *Problems of the Theory of Formation of Weathering Crust and Exogenic Deposits*, Nauka, Moscow, pp. 161–168.

Shvartsev, S.L. (1985) Physicochemical and geological evolution of the water-rock system. In: *Subsurface Waters and Evolution of Lithosphere. Proc. All-Union Conf., I*, Nauka, Moscow, pp. 253–266.

Sipalo-Žuljević, J. and Wolf, R.H.H. (1973) Sorption of lanthanum(III), cobalt(II) and iodide ions at trace concentrations of ferric hydroxide. *Microchimica Acta (Wien)*, 315–320.

Skinner, H.C.W. and Fitzpatrick, R.W. (Eds.) (1992) *Biomineralization Processes of Iron and Manganese*, Catena supplement 21, Lowerence, KS, U.S.A., Catena Verlag, 400 pp.

Slee, K.J. (1980) Geology and origin of the Groote Eylandt manganese oxide deposits, Australia. In: *Geology and Geochemistry of Manganese. Vol. II. Manganese Deposits on Continents*, Akadémiai Kiadó, Budapest, pp. 125–148.

Smirnov, V.I. (Ed.) (1979) *Metalliferous Sediments of the Southeastern Part of the Pacific*, Nauka, Moscow,

280 pp.

Smith, J.R. (Ed.) (1980) *Theory of Chemisorption*, Springer-Verlag, Berlin, New York, 240 pp.

Smith, W.C. and Gebert, H.W. (1969) Manganese at Groot Eylandt, Australia. In: Jones, M. (Ed.), *Mining and Petroleum Geology*, Ninth Commonwealth Mining and Metallurgy Congress and the Institution of Mining and Metallurgy, London, pp. 585–604.

Sokolov, D.S. (1962) *Basic Conditions of Karst Development*, Gosgeoltekhizdat, Moscow, 332 pp.

Sönge, P.G. (1977) Timing aspects of the manganese deposits of the Northern Cape Province (South Africa). In: Klemm, D.D. and Schneider, H.-J. (Eds.), *Time- and Strate-Bound Ore Deposits*, Springer-Verlag, Berlin, pp. 115–122.

Sorem, R.K. and Cameron, E.N. (1960) Manganese oxides and associated minerals of the Nsuta manganese deposits, Ghana, West Africa. *Econ. Geol.* **55**, 278–310.

Sorem, R.K. and Fewkes, R.H. (1979) *Manganese Nodules. Research Data and Methods of Investigation*, IFI/Plenum Data Co., New York, 723 pp.

South African Committee for Stratigraphy (SACS) (1980). Lithostratigraphy of the Republic of South Africa, South West Africa/Namibia and the Republics of Bophuthatswana, Transkei and Venda. *Handbk. Geol. Surv. S. Afr.*, 8, 690 pp.

Stepanov, V.I. (1979) Regularities in the distribution salinity. In: *Oceanology. Chemistry of the Ocean. Vol. 1. Chemistry of the Waters of the Ocean*, Nauka, Moscow, pp. 58–75.

Stevenson, F.J. and Goh, K.M. (1971) Infra-red spectra of humic acids and related substances. *Geochim. Cosmochim. Acta* **35**, 471–473.

Strakhov, N.M. (1947) *Iron Ore Facies and Their Analogs in the Earth's History. Proc. GIN Acad. Sci. USSR*, **73**, Publ. USSR Acad. Sci., Moscow, 267 pp.

Strakhov, N.M. (1960a) *Foundations of the Theory of Lithogenesis. Vol. I. Types of Lithogenesis and Their Distribution on the Earth's Surface*, Publ. Acad. Sci. USSR, Moscow, 212 pp.

Strakhov, N.M. (1960b) *Foundations of the Theory of Lithogenesis. Vol. II. Regularities of Composition and Distribution of Humid Sediments*, Publ. Acad. Sci. USSR, Moscow, 574 pp.

Strakhov, N.M. (1962a) *Foundations of the Theory of Lithogenesis*, Vol. III. Publ. Acad. Sci. USSR, Moscow, 530 pp.

Strakhov, N.M. (1962b) *To the Understanding of Lithogenesis of Volcanogenic-Sedimentary Type. Proc. Acad. Sci. USSR, Geol. Series*, No. 5. 3–17.

Strakhov, N.M. (1963) *Types of Lithogenesis and Their Evolution in the Earth's History*, Gosgeolizdat, Moscow, 535 pp.

Strakhov, N.M. (1964) On problems and some results of studies of geochemistry of the Paleogene manganese ore basin of the USSR South. *Lithology and Mineral Resources*, No. 1, 3–10.

Strakhov, N.M. (1965a) To the understanding of underwater volcanogenic-sedimentary rock formation (state of knowledge and problem). In: *Volcanogenic-Sedimentary Formations and Mineral Resources*, Nauka, Moscow, pp. 11–23.

Strakhov, N.M. (1965b) Types of manganese accumulations in modern water reservoirs and their significance for the understanding of manganese ore process. *Lithology and Mineral Resources*, No. 4, 18–49.

Strakhov, N.M. (1968) *Problems of Sedimentary Manganese Ore Process. VIIth Vernadskii Reading*, Nauka, Moscow, 30 pp.

Strakhov, N.M. (1974a) On the exhalation at mid-oceanic ridges as sources of ore elements in ocean sediments. *Lithology and Mineral Resources*, No. 3, 20–37.

Strakhov, N.M. (1974b) Localization of ore nodules of iron and manganese in the Pacific and its genetic sence. *Lithology and Mineral Resources*, No. 5, 3–17.

Strakhov, N.M. (1976a) *Problems of Geochemistry of the Modern Oceanic Lithogenesis*, Nauka, Moscow, 300 pp.

Strakhov, N.M. (1976b) Conditions of formation of concretion iron-manganese ores in modern water reservoirs. *Lithology and Mineral Resources*, No. I, 3–19.

Strakhov, N.M., Varentsov, I.M., Kalinenko, V.V., Tikhomirova, E.S. and Shterenberg, L.E. (1967) On the understanding of mechanism of manganese ore process (with the example of the Oligocene ores of the USSR South). In: *Manganese Deposits of the USSR*, Nauka, Moscow, pp. 34–56.

Strakhov, N.M., Shterenberg, L.E., Kalinenko, V.V. and Tikhomirova, E.O. (1968) *Geochemistry of the Sedimentary Manganese Ore Process. Proc. Geol. Inst. Acad. Sci. USSR*, Vol. 185, Nauka, Moscow, 495 pp.

Strauss, C.A. (1964) The iron ore deposits in the Sishen area, Cape Province, In: Haughton, S.H. (Ed.), *The*

Geology of Some Ore Deposits in Southern Africa, 2, Geol. Soc. S. Afr., pp. 393–403.

Stuermer, D.N. and Harvey, G.R. (1974) Humic substances from sea water. *Nature* **250**, 480–481.

Stumm, W. and Giovanoli, R. (1976) On the nature particulate manganese in simulated lake waters. *Chimia* **30**, 423–425.

Stumm, W., Hohl, H. and Dalang, F. (1976) Interaction of metal ions with hydrous oxide surfaces. *Croatica Chim. Acta (CCACAA)* **48**(4), 491–504.

Stumm, W., Huang, C.P. and Jenkins, S. (1970) Specific chemical interaction affecting the stability of dispersed systems. *Croatica Chem. Acta.* **42**, 223–245.

Stumm, W. and Morgan, J.J. (1970) *Aquatic Chemistry*, Wiley Interscience, New York, 2nd ed., 583 pp.

Stunzhas, P.A. (1979) Structure of water and properties of water and water solutions. In: *Oceanology. Chemistry of Ocean.* Vol. I. *Chemistry of Ocean Water*, Nauka, Moscow, pp. 11–42.

Sugimura, Y., Suzuki, Y. and Miyake Y. (1978) Chemical forms of minor metallic elements in the oceans. *Journ. Oceanogr. Soc. Japan* **34**, 93–96.

Sully, A.H., *Manganese*, Butterworth Sci. Publ., London.

Swallow, K.C., Hume, D.H. and Morel, F.M.M. (1980) Sorption of copper and lead by hydrous ferric oxide. *Environ. Sci. Techn.* **14**, 1326–1331.

Szabo, Z. and Grasselly, Gy. (1980) Genesis of manganese oxide ores in Úrkút, Hungary. In: *Geology and Geochemsitry of Manganese, Vol. II. Manganese Deposits on Continents*, Akadémiai Kiadó, Budapest, pp. 223–236.

Szabo, Z., Grasselly, Gy. and Cseh-Nemeth, J. (1981) Some conceptual questions regarding the origin of manganese in the Úrkút deposit, Hungary. *Chem. Geol.* **34**, 19–29.

Takematsu, N. (1979) Sorption of transition metals on mangenese and iron oxides and silicate minerals. *Journ. Oceanogr. Soc. Japan* **35**, 36–42.

Tarasevich, Yu.I. (1983) On the structure of boundary water layers in mineral dispersions. In: *Surface Forces and Boundary Layers of Liquids*, Nauka, Moscow, pp. 147–151.

Taylor, R.M., McKenzie, R.M. and Norrish, K. (1964) The mineralogy and chemistry of manganese in some Australian soils. *Austral. Journ. Soil Res.* **2**, 235–248.

Taylor, S.R. and McLennan, S.M. (1981) The composition and evolution of the continental crust: Rare earth element evidence from sedimentary rock. *Philos. Trans. R. Soc. London, Ser. A* **301**, 381–399.

Taylor, S.R. and MacLennan, S.M. (1985) *The Continental Crust: Its Composition and Evolution*, Blackwel, London, 312 pp.

Teleki, P.G., Dobson, M.R., Moore, J.R. and von Stackelberg, U. (Eds.) (1987) *Marine Minerals. Advances in Research and Resource Assessment. Proceedings of the NATO Advanced Research Workshop on Marine Minerals: Resource Assessment Strategies/1985: University of Wales* (NATO ASI Series, Series C: Mathematical and Physical Sciences, Vol. 194), D. Reidel, Dordrecht, Boston, 588 pp.

Tewari, P.H., Campbell, A.B. and Lee, W. (1972) Adsorption of Co^{2+} by oxides from aqueous solution. *Canad. Journ. Chem.* **50**, 1642–1648.

Thein, J. (1990) Paleogeography and geochemistry of the "Cenomano-Turonian" formations in the manganese district of Imini (Morocco) and their relations to ore deposition. *Ore Geol. Rev.* **5**, 257–291.

Thomas, J.M. and Thomas, W.J. (1967) *Introduction to the Principles of Heterogenous Catalysis*, Academic Press, London, New York.

Thompson G. (1983) Hydrothermal fluxes in the ocean. In: Riley J.P. and Chester R. (Eds.), *Chemical Oceanography*, Academic Press, London, 8, pp. 271–357.

Thorstenson, D.C. (1970) Equilibrium distribution of small organic molecules in natural waters. *Geochim. Cosmochim. Acta* **34**, 745–770.

Thrailkill, J. (1968) Chemical and hydrologic factors in the excavation of limestone caves. *Geol. Soc. of Amer. Bull.* **79**, 19–46.

Tkach, B.I. (1974) On the role of organic matter in concentration of mercury. *Geokhimiya* **9**, 1410–1412.

Tsykin, R.A. (1974) Weathering crust and karst formations of the Mazul'skoe limestone deposit. In: *Weathering Crust*, No. 14, Nauka, Moscow, pp. 199–207.

Tsykin, R.A. (1977) On mineralogy of caves of carbonate karst of the Southern Siberia. In: *Minerals and Rocks of the Krasnoyarskii Region*, Krasnoyarsk, pp. 51–59.

Tsykin, R.A. (1980) On the conditions of formation of covered karst. *Geology and Geophysics*, No. 8. 52–58.

Tsykin, R.A. (1980) Metasomatic rocks and ores of the supergene zone. In: *Problems of the Theory of Formation of Weathering Crust and Exogenic Deposits*, Nauka, Moscow, pp. 142–153.

Tsykin, R.A. (1985) *Sediments and Mineral Resources of Karst*, Nauka, Novosibirsk, 165 pp.

Tsykin, R.A. and Kostenenko, L.P. (1984) Mesozoic-Cainozoic sediments of the Porozhinskoe manganese deposit. In: *Manganese Ore Formation on the USSR Territory*, Nauka, Moscow, pp. 109–116.

Turekian, K.K. (1968) Deep sea deposition of barium, cobalt and silver. *Geochim. Cosmochim. Acta* **32**, 603–612.

Turekian, K.K. (1977) The fate of metals in the oceans. *Geochim. Cosmochim. Acta* **41**, 1139–1144.

Turekina, K.K. and Wedepohl, K.H. (1961) Distribution of the elements in some major units of the Earth's crust. *Bull. Geol. Soc. Am.* **72**(2), 175–192.

Turner, S. and Buseck, P.R. (1979) Manganese oxide tunnel structures and their integrowths. *Science* **203**, 456–458.

Udodov, P.A., Solodovnikov, R.S., Shvartseva, N.M. and Rychkova, E.S. (1975) On relation between antimony and organic matter of humus type in natural waters and water solutions. *Izvest. Tomsk Polytechn. Inst.* **297**, 68–73.

U.S. Congressional Office of Technology Assessment. Strategic Materials (1985) Technologies to Reduce U.S. Import Vulnerability, May, OTA–ITE–248, 409 pp.

Ustalov, V.V. (1982) Structures, formations and manganese-bearing capacity of the Vorogovskii trough (Eniseiskii Ridge). Author's Abstr. Diss. Cand. Geol. Min. Sci. MGU, Moscow.

Van der Weijden, C.H. (1976) Some geochemical controls of Ni and Co concentrations in marine ferromanganese deposits. *Chem. Geol.* **18**, 65–80.

Van Schalkwyk, J.F. and Beukes, N.J. (1986) The Sishen Iron Ore Deposit, Griqualand West. In: Anhaeusser, C.R. and Maske, S. (Eds.), *Mineral Deposits of Southern Afirca*, Geol. Soc. S. Afr., Johannesburg I and II, pp. 931–956.

Varadan, G.N. (1979) International seminar on laterization process (IGCP Project-129). Excursion IV-A. Hyderabad.

Varand, E.L. and Andreev, O.V. (1984) Porozhinskii area and its perspectives. In: *Manganese Ore Formation on the USSR Territory*, Nauka, Moscow, pp. 89–95.

Varentsov, I.M. (1962) On the major manganese-bearing formations. In: *Sedimentary Ores of Iron and Manganese. Proc. Geol. Inst. of Acad. Sci. USSR*, Vol. 70, Publ. Acad. Sci. USSR, Moscow, pp. 119–173.

Varentsov, I.M. (1964) *Sedimentary Manganese Ores*, Elsevier, Amsterdam, 119 pp.

Varentsov, I.M. (1971) On the leaching of manganese in the course of interaction of basic volcanic materials with seawater. *Soc. Min. Geol. Japan* (Spec. Issue), 466–473.

Varentsov, I.M. (1972) On the main aspects of formation of iron-manganese ores in modern basins. In: *Mineral Deposits. International Geological Congress*, XXIV session, Reports of the Soviet Geologists, Nauka, Moscow. pp. 158–173.

Varentsov, I.M. (1975) Geochemical aspects of formation of iron-manganese ores in modern shelf seas. In: *Problems of Lithology and Geochemistry of Sedimentary Rocks and Ores*, Nauka, Moscow, pp. 150–165.

Varentsov, I.M. (1976) Geochemistry of transition metals in the process of formation of iron-manganese ores in modern basins. In: *Mineral Deposits. International Geological Congress*, XXIV session, Reports of the Soviet Geologists, Nauka, Moscow, pp. 79–96.

Varentsov, I.M. (1978) The geochemistry of heavy metals in Upper Cenozoic sediments near the crest of the Mid-Atlantic Ridge, Latitude 23° N, drilled on DSDP Leg 45. In: Melson, W.G. and Rabinowitz, P.D. *et al.*, *Initial Reports of the Deep Sea Drilling Projects, 45*, Washington, D.C. (U.S. Government Printing Office), pp. 349–377.

Varentsov, I.M. (1980a) Metalliferons sediments of the Northern Atlantic (geochemistry, features of formation). In: *Marine Geology, Sedimentology, Sedimentary Petrography, and Geology of the Ocean. International Geological Congress*, XXVI session, Reports of the Soviet Geologists, Nedra, Leningrad, pp. 29–42.

Varentsov, I.M. (1980b) Foreword. In: Varentsov I.M. and Grasselly, Gy. (Eds.), *Geology and Geochemistry of Manganese. Vol. 1, General Problems: Mineralogy, Geochemistry, Methods*, Akadémiai Kiadó, Budapest. Joint edition with E. Schweizerbartsche Verlagsbuchhandeung (Nägele u. Obermiller), Stuttgart, pp. 13–19.

Varentsov, I.M. (1980c) Geochemistry of transition matals in processes of ferromanganese ore formations in Recent basins. In: Varentsov I.M., Grasselly Gy. (Eds.), *Geology and Geochemistry of Manganese. Vol. 1*, Akadémiai Kiadó, Budapest, pp. 367–387.

Varentsov, I.M. (1981a) Geochemical history of post-Jurassic sedimentation in the Central Northwestern Pacific, Southern Hess Rise, Deep Sea Drilling Project Site 465. In: Thide, J. and Vallier, T.L. *et al.*, *Initial Reports of the Deep Sea Drilling Project, 62*, Washington (U.S. Government Printing Office), pp. 819–832.

Varentsov I.M. (1981b) Geochemical history of post-Jurassic sedimentation in the Central Northwestern Pacific,

Southern Hess Rise, Deep Sea Drilling Project Site 466. In: Thiede J. and Vallier T.L. *et al.*, *Initial Reports of the Deep Sea Drilling Project, 62*, Washington (U.S. Government Printing Office), pp. 833–845.

Varentsov, I.M. (1982) Groote Eylandt manganese deposit, Australia. In: *Geology and Geochemistry of Manganese*, Nauka, Moscow, pp. 66–83.

Varentsov, I.M. (1982) Groote Eylandt manganese oxide deposits, Australia (I. General Characteristics). *Chem. Erde* **41**, 157–173.

Varentsov, I.M. (1983a) Trace element geochemical history of Late Mesozoic sedimentation in the Southwest Atlantic, Falkland Plateau, Site 511. In: Ludwig, W.J. and Krasheninnikov, V.A. *et al.*, *Initial Reports of the Deep Sea Drilling Project, 71*, Washington (U.S. Government Printing Office), pp. 391–407.

Varentsov, I.M. (1983b) Geochemical history of post-Middle Jurassic sedimentation in the Southwestern Atlantic, Deep Sea Drilling Project Leg 71: Ba, Sr and major components. In: Ludiwg, W.J. and Krashenninnikov, V.A. *et al.*, *Initial Reports of the Deep Sea Drilling Project, 71*, Washington (U.S. Government Printing Office), pp. 423–442.

Varentsov, I.M. (1984) Lithology, mineralogy, and geochemistry of the sedimentary cover of the Southwestern Atlantic (Leg 71). Geochemical history of the post Middle Jurassic sedimentation: Ba, Sr, and main components. In: *Problems of the Lithology of the World Ocean. Mineralogy and Geochemistry of the Atlantic Ocean*, Nauka, Moscow, pp. 152–185.

Varentsov, I.M. (1985) Mineralogy and geochemistry of hydrothermal sediments of the Galapagos rift zone (Leg 70). Geochemical associations of the main components, Ba, Sr as indicators of sedimentary processes: hydrothermal sediments of the Galapagos rift. In: *Problems of Lithology of the World Ocean. Mineralogy and Geochemistry of the Pacific*, Nauka, Moscow, pp. 202–213.

Varentsov, I.M. (1989) On the major problems in understanding of manganese deposts genesis. In: Kholodov, V.N. (Ed.), *Lithogenesis and Ore Formation (Criteria of Separation of Exogenic and Endogenic Processes)*, Nauka, Moscow, pp. 151–157.

Varentsov, I.M. (1989) To the lithologic-geochemical model of formation of manganese ores in the supergene zone. In: Timofeev, P.P. (Ed.), *Sedimentary Cover of the Earth in Space and Time: Sedimentolithogenesis. International Geological Congress*, XXVIII Session (Washington, July, 1989), Reports of the Soviet Geologists, Nauka, Moscow, pp. 114–120.

Varentsov, I.M. (1993) Distribution of rare earth elements in Mn–Fe oxyhydroxide crusts from Bezymiannya Seamount, Atlantic: Geochemical history of deposition. *Chem. Erde* **53**, 133–157.

Varentsov, I.M., Bakova, N.V., Dikov, Yu.P., Gendler, T.S. and Giovanoli, R. (1979a) On the model of Mn, Fe, Ni, Co ore formation in Recent Basins: Experiments with synthesis of Me-hydroxide phases on Mn_3O_4. *Mineralium Deposita* **14**, 281–296.

Varentsov, I.M., Bakova, N.V., Dikov, Yu.P., Gendler, T.S. and Giovanoli, R. (1979b) Synthesis of Mn, Fe, Ni, Co oxide-hydroxide phases on manganese oxides: on a model for transition metal ore formation in Recent basins. *Acta Mineral.-Petrograph., Szeged* **XXIV/I**, 63–90.

Varentsov, I.M., Bakova, N.V., Dikov, Yu.P. and Gendler, T.S. (1981) On the model of formation of ores of Mn, Fe, Ni, and Co in modern basins: experiments on synthesis of hydroxide phases of these metals on Mn hydroxides. In: *Lithology at the New Stage of Evolution of Geological Knowledge*, Nauka, Moscow, pp. 270–302.

Varentsov, I.M., Bazilevskaya, E.S., Belova, I.V. and Semenova, M.G. (1967) Features of distribution of Ni, Co, Cu, V, and Cr in ores and enclosing sediments of the South-Ukrainian manganese ore basin. In: *Manganese Deposits of the USSR*, Nauka, Moscow, pp. 179–198.

Varentsov, I.M. and Blazhchishin, A.I. (1976) Iron-manganese concretions. In: *Geology of the Baltic Sea*, Vilnyus, Mokslas, pp. 307–348.

Varentsov, I.M., Dikov, Yu.P. and Bakova, N.V. (1978) On the model of formation of Fe–Mn ores in modern basins (experiments on synthesis of oxide phases of Mn, Fe, Ni, and Co on iron hidroxides). *Geochemistry* **8**, 1198–1210.

Varentsov, I.M., Dikov, Yu.P. and Bakova, N.V. (1980) On the study of Fe–Mn ore formation in Recent basins: Experiments on the synthesis of Mn, Fe, Ni, Co hydroxide pases on iron hydroxides (γ-FeOOH). In: Varentsov, I.M. and Grasselly, Gy. (Eds.), *Geology and Geochemistry of Manganese. Vol. 3., Manganese on the Bottom of Recent Basins*, Akadémiai Kiadó, Budapest. Joint edition with Schweizerbartsche Verlagsburchhandlung (Nägele u Obermiller), Stuttgart, pp. 91–104.

Varentsov, I.M., Drits, V.A., Gorshkov, A.I., Sivtsov, A.V. and Sakharov, B.A. (1989) Formation processes of Mn-Fe crusts in the Atlantic: mineralogy, geochemistry of main and trace elements, Krylov Seamount.

In: Kholodov, V.N. (Ed.), *Genesis of Sediments and Fundamental Problems of Lithology*, Nauka, Moscow, pp. 59–78.

Varentsov, I.M., Drits, V.A. and Gorshkov, A.I. (1991), Rare earth element indicators of Mn–Fe oxzhydroxide crust formation on Krylov Seamount, Eastern Atlantic. *Mar. Geol.* **96**, 71–84.

Varentsov, I.M. and Golovin, D.I. (1987) Groote Eylandt manganese deposit, North Australia: K–Ar age of cryptomelane ores and aspects of genesis. *Reports Acad. Sci. USSR* **294**(1), 203–207.

Varentsov, I.M. and Grasselly, Gy. (1969) Provisional Programme of the Working Group on Manganese Formations Sponsored by the International Association on the Genesis of ore Deposits (IAGOD). *Acta Mineral.-Petrograph., Szeged, Hungary* **19**(1), 101–104.

Varentsov, I.M. and Grasselly, Gy. (1970) Report on the statutory meeting of the Working Group on Manganese Formations held on August 31, 1970, in Kyoto during the IMA-IAGOD Meeting' 70. *Acta Mineral.-Petrograph., Szeged, Hungary* **19**(2), 209–216.

Varentsov, I.M. and Grasselly, Gy. (Eds.) (1980a) *Geology and Geochemistry of Manganese. Vol. 1. General Problems: Mineralogy, Geochemistry, Methods*, Akadémiai Kiadó, Budapest. Joint edition with E. Schweizerbartsche Verlagabuchhandlung (Nägele u. Obermiller), Stuttgart, 463 pp.

Varentsov, I.M. and Grasselly, Gy. (Eds.) (1980b) *Geology and Geochemistry of Manganese. Vol. 2. Manganese Deposits on Continents*, Akadémiai Kiadó, Budapest. Joint edition with E. Schweizerbartsche Verlagsbuchhandlung (Nägele u. Obermiller), Stuttgart, 513 pp.

Varentsov, I.M. and Grasselly, Gy. (Eds.) (1980c) *Geology and Geochmeistry of Manganese. Vol. 3. Manganese on the Bottom of Recent Basins*, Akadémiai Kiadó, Budapest. Joint edition with E. Schweizerbartsche Verlagsbuchhandlung (Nägele u. Obermiller), Stuttgart, 357 pp.

Varentsov, I.M., Grasselly, Gy. and Szabo, Z. (1988) Ore-formation in the Early Jurassic Basin of Central Europe: Aspects of mineralogy, geochemistry and genesis of the Úrkút Manganese Deposit, Hungary. *Chem. Erde* **48**, 257–304.

Varentsov, I.M. and Pronina, N.V. (1973) On the study of mechanisms of iron-manganese ore formation in recent basins; the experimental data on nickel and cobalt. *Miner. Deposita* **8**(2), 161–178.

Varentsov, I.M., Pronina N.V. (1976) Experimental study of ferromanganese ores formation in recent basins: Data on iron, manganese, nickel and cobalt. In: *25th International Geological Congress*, Sydney, 3, p. 799.

Varentsov, I.M. and Rakhmanov, V.P. (1974) Manganese deposits. In: *Ore Deposits of the USSR*, Nedra, Moscow, I, pp. 109–167.

Varentsov, I.M. and Rakhmanov, V.P. (1980) Manganese deposits of the USSR (A Review). In: *Geology and Geochemistry of Manganese. Vol. II. Manganese Deposits on Continents*, Akadémiai Kiadó, Budapest, pp. 319–391.

Varentsov, I.M., Rakhmanov, V.P., Gurvich, E.M. and Grasselly, Gy. (1984) Genetical aspects of formation of manganese deposits in geological history of the Earth's crust. In: *XXVII International Geological Congress, Sect. 12*, Reports, 12, Nauka, Moscow, pp. 149–156.

Varentsov, I.M., Sakharov, B.A., Rateev, M.A. and Choporov D.Ya. (1981) Geochemical history of post-Jurassic sedimentation in the Central Northwestern Pacific Northern Hese Rise, Deep Sea Drilling Project Site 464. In: Thiede J. and Vallier T.L. *et al., Initial Reports of the Deep Sea Drilling Project, 62*, Washington (U.S. Government Printing Office), pp. 805–818.

Varentsov, I.M., Sakharov, B.A., Drits, V.A., Tsipursky, S.I., Choporov, D.Ya. and Aleksandrova, V.A. (1983) Hydrothermal deposits of the Galapagos Rift Zone, Leg 70: Mineralogy and geochemistry of major components. In: Honnorez, J. and Von Herzen, R.P. *et al., Initial Reports of the Deep Sea Drilling Project, 70*, Washington (U.S. Government Printing Office), pp. 235–268.

Varentsov, I.M. and Stepanets, M.I. (1970) Experiments on modeling of leaching processes of manganese by seawater from volcanic materials of the basic compositions. *Reports Acad. Sci. USSR* **190**(3), 679–682.

Varentsov, I.M., Timofeev, P.P. and Rateev, M.A. (1981) Geochemical history of post-Jurassic sedimentation in the Central Northwestern Pacific, Western Mid-Pacific Mountains, Deep Sea Drilling Project Site 463. In: Thiede, J. and Vallier T.L. *et al., Initial Report of the Deep Sea Drilling Project, 62*, pp. 785–804.

Varentsov, I.M., Veimarn, A.B., Rozhnov, A.A., Shchibrik, V.I. and Sokolova, A.L. (1993) Geochemical model of formation of manganese ores of the Famennian riftogenic basin of Kazakhstan (main components, rare earths, trace elements). *Lithology and Mineral Resources* **3**, 56–79.

Varentsov, I.M., Zaitseva, L.V. and Putilina, V.S. (1985) Experimental investigations of main ions of seawater in the process of uptake of Cu(II) by manganese hydroxides – On the geochemistry of formation of polymetallic concretion ores in modern basins. *Geochemistry*, No. 5. 710–722.

Varentsov, I. M., Zaitseva, L.V. and Putilina, V.S. (1985) On the Geochemistry of Nodular Polymetallic Ore Formation in Recent Basins: Experimental Data on the Role of Major Ions of Seawater in the Process of Copper Sorption by Manganese Hydroxides. *Chem. Erde* **44**, 193–225.

Varshal, G.M., Kosheleva, I.Ya., Sirotkina, I.S., Veliukhanova, T.K., Intskirveli, L.N. and Zaluiukina, N.S. (1979) Study of organic substances of surface water and their interactions with ions of metals. *Geokhimiya* **4**, 598–607.

Varshal, G.M. and Senyavin, M.M. (1964) The process of distributive chromatography of a mixture of rare-earth elements on paper in the light of the structural theory of salting out. *Zhurn, Strukt. Khim. (USSR)* **5**, 681–690.

Vasconcelos, P.M., Becker, T.A., Renne, P.R. and Brimhall, G.H. (1992) Age and Duration of Weathering by $^{40}K-^{40}Ar$ and $^{40}Ar/^{39}Ar$ Analysis of Potassium-Manganese Oxides. *Science* **258**, 451.

Vdovenko, V.M., Gurikov, Yu.V. and Legin, E.K. Thermodynamics of two-structure water model. III. Temperature dependence of thermodynamic properties of ice-like and disordered structures. *Zhurn. Struct. Khim.* **8**, 403–407.

Vernadskii, V.I. (1954a) Manganese geochemistry in relation with the concept on mineral resources (report at conference, April, 1935). In: *Selected Works, 1*, Publ. Acad. Sci. USSR, Moscow, pp. 528–542.

Vernadskii, V.I. (1954b) Assays of geochemistry. Third assay. Geospheres. Manganese history. Energy of geospheres. In: *Selected Works, 1*, Publ. Acad. Sci. USSR, Moscow, pp. 61–97.

Vernadskii, V.I. (1960) Oceanography and geochemistry. In: *Selected Works, 5*, Publ. Acad. Sci. USSR, Moscow, pp. 231–288.

Vernadskii, V.I. (1988) The scientific thought as a planetary phenomenon. In: Vernadskii, V.I. *Phylosophical Thoughts of the Naturalist*, Nauka, Moscow, pp. 19–208.

Vernadskii, V.I. (1981) On the scientific World outlook. In: Vernadskii, V.I. (Ed.), *Phylosophical Thought as a Planetary Phenomenon*, Nauka, Moscow, pp. 191–234.

Vincienne, H. (1956) Observations geologiques sur quelques, cîtes Marocains de manganese singénétique. In: *XX Congreso Geologico International Symposium Sorbe Yacimientos de Manganeso, II*, Africa, Mexico, pp. 249–269.

Vinogradov, A.P. (1962) Mean content of chemical elements in main types of igneous rocks of the Earth's crust. *Geochemistry* **7**, 555–571.

Vinogradova, E.N., Prokhorova, G.V. and Sevastyanova, T.N. (1968) Polarographic determination of Ni and Co in soil and natural waters. *Vestn. Moskovsk. Universiteta* **5**, 74–75.

Visser, D.J.L. (1954) Deposits of manganese ore on Rooinekke and neighbouring farms, Distract Hay. *Trans. Geol. Soc. S. Afr.* **57**, 61–75.

Vitovskaya, I.V. (Ed.) (1985) *Volcanogenic-Sedimentary and Hydrothermal Manganese Deposits (Central Kazakhstan, Malyi Caucases, and Eniseiskii Rdge)*, Nauka, Moscow, 198 pp.

Vitovskaya, I.V., Kapustkin, G.R. and Sivtsov, A.V. (1984) Distribution and forms of occurrence of manganese in weathering crusts of serpentinites. In: *Manganese Ore Formation in the USSR Territory*, Nauka, Moscow, pp. 45–56.

Vladimirov, A.I. and Gorovoi, L.Ya. (1972) Some problems of stratigraphy of the Precambrian sediments of the Priangarian part of the Eniseiskii Ridge in relation with their syngenetic manganese- and phosphate-bearing capacity. *SNIIGGIMS*, No. 122, Novosibirsk, 73–78.

Volkhin, V.V., Sokolova, T.S. and Shulga, E.A. (1979) The role of liquid phase in sorption processes with inorganic sorbents. *Zhurnal Prikladnoy Chimii* **52**(3), 529–532.

Volkov, I.I. (1979) Iron-manganese concretions. In: *Chemistry of Ocean: Geochemistry of Bottom Sediments. Oceanology,* 2, Nauka, Moscow, pp. 414–467.

Volkov, I.I., Shterenberg, L.E. and Fomina, L.S. (1980) Iron- manganese concretions. In: *Geochemistry of Diagenesis of the Pacific Sediments (Transocean Profile)*, Nauka, Moscow, pp. 169–223.

Vologodskii, G.P. (1975) *Karst of the Irkutskii Amphitheater*, Nauka, Moscow, 124 pp.

Wadsley, A.D. (1950) Synthesis of some hydrated manganese minerals. *Amer. Mineral.* **35**, 485–499.

Wadsley, A.D. and Walkley, A. (1951) The sturcture and reactivity of the oxides of manganese. *Review Pure and Applied Chem.* **1**, 203–213.

Weber, F. (1968) Une série Precambrienne du Gabon: Le Francevillien. Sédimentologie, géochimie, relations avec les gîtes minéraux associés. Strasburg. *Mém. Serv. Carte Geol. Alsace-Lorraine* **28**, 328 pp.

Weber, F., Leclerc, J. and Millot, G. (1979) Epigénies manganésufeères successives dans le gisement de Moanda (Gabon). *Sci. Géol., Bull.* **32**, 147–164.

Wedepohl, K.H. (1980a) Geochemical behaviour of manganese. In: Varentsov, I.M. and Grasselly (Eds.),

Geology and Geochemistry of Manganese, Publishing House of Hungarian Academy of Sciences, Budapest, 1, pp. 334–351.

Wedepohl K.H. (1980b) Potential sources for manganese oxide precipitation in the oceans. In: Varentsov, I.M. and Grasselly G. (Eds.), *Geology and Geochemistry of Manganese*, Publishing House of Hungarian Academy of Sciences, Budapest, 3, pp. 13–22.

Wershaw, R.L. (1976) Organic chemistry of lead in natural waters systems. Geol. Surv. Prof. Paper 957, 13–16.

White, W.B. (1984) Rate processes: chemical kinetic and karst land form development. In: La Fleur, R.G. (Ed.), *Groundwater as a Geomorphic Agent*, (Binghamton Symposium in Geomorphology, International Series, No. 13), Allen and Unwin Inc., Boston, pp. 227–248.

Williams, D. (1975) *The Metals of Life*, Mir, Moscow, 236 pp.

Wolfram, T. and Ellialtioglu, S. (1980) Concepts of surface states and chemisorption on d-baund perovskites, In: Smith, J.R. (Ed.), *Theory of Chemisorption*, Springer-Verlag, Berlin, pp. 149–180.

Wood, W.W. (1985) Origin of caves and other solution openings in the unsaturated (vadose) zone of carbonate rocks: A model for CO_2 generation. *Geology* 13, 822–824.

Wrosley, T.R., Nance, R.D. and Moody, J.B. (1986) Tectonic cycles and the history of the earth's biogeochemical and paleoceanographic record. *Paleoceanography* 1, 233–263.

Yanchuk, E.A. and Khmelevskii, V.A. (1976) Minerals of oxidized manganese ores of the Burshtynskoe deposit. *Proc. L'vov. Gos. Univers. (L'vov. Geol. Society), L'vov*, 2, No. 30. 79–84.

Yanchuk, E.A. and Khmelevskii, V.A. (1981) Mineralogic- geochemical features of the oxidation process of calcium- manganese carbonate ores (on the example of the Burshtynskoe deposit). *Proc. L'vov. Gos. Univers. (L'vov. Geol. Society), L'vov* 35, 2, 38–43.

Yanchuk, E.A. and Khmelevskii, V.A. (1982) Manganese in weathering crusts of carbonate rocks of the south-western margin of the Eastern-European Platform. In: *Geology and Geochemistry of Manganese*, Nauka, Moscow, pp. 205–209.

Yanchuk, E.A., Khmelevskii, V.A. and Panchenko N.A. (1986) Mineralogy and geochemistry of oxidized manganese carbonate ores (Nikopol'skoe and Burshtynskoe deposits of the Ukr.SSR). In: *Conditions of Formation of Ore Deposits. Proc. VI LAGOD Symposium*, Tbilisi, September 6–12, 1982, Vol. 2, Nauka, Moscow pp. 757–764.

Yashvili, L.P and Gukasyan, R.Kh. (1973) About application of cryptomelane for K–Ar dating of manganese ores of the Sevkar-Sarigyukhskoe deposit (Arm. SSR). *Repts. Acad. Sci. USSR* 212(1), 185–188.

Yastremskii, P.S. and Samoilov, O.Ya. (1963) Stabilization of the structure of aqueous solutions by molecules of nonelectrolyte and the dielectric permeability. *Zhurn. of Struct. Chim.* 4, 844–849.

Yatsimirsky K.B., Kostromina I.A. and Scheka Z.A. *et al.* (1966) *The Chemistry of Complex Compounds of Rare Earth Elements*, Naukova Dumka, Kiev, 493 pp.

Zaitseva,L.V. and Varentsov, I.M. (1988) On the model of formation of polymetallic nodular ores in modern basins: experimental data on the role of major ions of seawater (chloride systems) in the sorption process of Cu(II) on birnessite ($7\overset{\circ}{A}$–MnO_2). *Geochemistry*, No. 6, 879–891.

Zaitseva, L.V. and Varentsov, I.M. (1990) Influence of major ions of seawater (chloride systems) on Cu(II) sorption by manganese oxyhydroxides: use of experimental data in a model of polymetallic ore formation in Recent basins. *Chem. Erde* 50, 255–268.

Zwicker, W.K., Meijer, W.O.J.G. and Jaffe, H.W. (1962) Nsutite a widespread manganes oxide mineral. *Amer. Mineral.* 47, 246–266.

INDEX

acid leaching processes, 274, 284, 285

active site, 228, 249, 252, 259, 260, 261, 263, 273

adsorption, 228, 251, 252, 256–258, 260, 262, 263, 269, 271

adsorption of minor, transition elements, 219–281

Africa, 22, 109, 290, 295

Africa North, 188, 189, 302

Africa South (see South Africa)

Africa West, 23, 57, 107, 109, 110, 116, 118–120, 135, 214–218, 282, 292–294

alabandite, 289, 299

Alambaiskaya Suite, of the Lower Cambrian, Salair Ridge, Siberia, Russia, 207

alkalinity, 211, 241

alletanite, 209

Alpine-Mediterranean region, Mn deposits of, 5, 285, 302

aluminium, 41, 71, 74–76, 83, 94, 95, 97, 100, 101, 114, 122, 125, 129, 142, 144, 174, 184, 185, 196, 207, 269, 270, 276, 280

Amapa Series (Late Archean–Early Proterozoic), 290

Amapa, state of Brazil, Mn deposits of, 117, 282

America, 216

amesite, 180

amino acids, 265, 267

amorphous silica, 285

Amphaihi, Bekily, Mn deposits, Madagascar Island, 291

amphibole, 133

amphibole, Mn, 117, 289, 294

Anabarskii Massif, Siberia, Russia, Mn mineralization of, 290

Andhra Pradesh, state of India, Mn deposits of, 284, 291

Antarctic water (bottom), 302

Anti-Atlas, region, manganese deposits of, North Africa, 188, 295

antimony, 154, 155, 186, 270

apatite, 180, 200

Aravalli Supergroup (Early Proterozoic), 294

Arnhem Land, North Australia, 23, 25

artesian basin, 132, 133

Asbesheuwels Subgroup (iron formations), South Africa, 170, 176, 183

asbolane, 10, 88, 140, 141, 146, 151, 168, 169, 280, 281

asbolane-busserite I, mixed-layered minerals, 10

associations (assemblages) geochemical, 94–102

associations of rocks, 289–301

associations, paragenetic of manganese deposits, 94

atacamite, 285

Atlantic, 10, 120, 287

Australia, 22–24, 26, 74, 77, 80, 83, 84, 86, 87, 89, 90, 92, 93, 95, 96, 104, 107, 109, 110, 135, 280, 282, 299

authigenic, 59, 117, 206, 214

authigenic formation of oxyhydroxide minerals . . . , 22, 152, 158, 159

autigenic minerals, 59, 206

autocatalytic accumulation, of transitional metals and REE, 22, 117, 148, 168, 214, 280

autocatalytic oxidation, 117, 168, 169, 210, 212, 219, 225–227, 240, 242, 243, 272, 280, 287

Azul, Mn deposit (Guinean Shield), Brazil, 117, 291

bacteria, 4, 22, 214

Bahia, state of Brazil, Mn deposits of, 117, 282

Baikalides, Siberia, Mn deposits of, Russia, 198

Bakoni Mountains, Mn deposits of, Hungary, 20, 298

Baltic Sea, 83, 88, 90, 91, 93

barite, 180, 285

barium, 41, 45, 47, 49, 51, 53, 58, 74–76, 94, 95, 99, 102, 122, 125, 142–145, 151, 152, 280, 281, 283

basalt, 170, 284–287

Bauer Basin, Pacific, Mn–Fee metalliferous sediments, 286

biotite, 133

birnessite, viii, 10, 36, 39, 43, 51, 57, 59, 71, 72, 99, 100, 105, 108, 117, 124, 127, 197, 210, 217, 228, 240–250, 263, 271–273, 275, 279, 280, 284, 285, 291

Birrimian Series (System), volcanogenic-sedimentary, 130, 214, 292, 293

Bishop Mn deposit, Postmasburg region, vi, 174, 176, 177, 181

bixbyite, 103, 109, 173, 174, 176–180, 182, 183, 282, 283, 289, 293–296

Black Reef Formation, South Africa, 170

black shale, 110, 111, 113, 116, 117, 216, 294, 295, 297, 298
Bol'she-Tokmanskoe (Bol'shoi Tokmak) Mn deposit, Ukraine, 7, 283, 300
boron, 77–79, 83, 88, 89, 178, 183
botallackite, 285
Bouarfa Mn deposit, Morocco, 195
boulder-lump(y), manganese ore variety, the Groote Eylandt deposit, 32, 35, 37, 42–46, 52, 54, 71
boulder mangangese ores . . . , 32, 42, 114, 140, 168
boundary water layers, 255–257
braunite, 109, 119, 123, 126, 127, 173, 178–180, 182, 183, 191, 209, 276, 282, 283, 288, 289, 291–297, 302
Brazil, 109, 110, 116, 178, 216, 282, 288, 290, 291, 293, 296, 297
Buffalo Springs Group, South Africa, 170
Burshtynskoe Mn deposit, Western Ukraine, 20, 197, 198
buserite, 10, 140, 141, 146, 168, 197, 209, 210, 228, 275
buserite I–buserite II, the mixed-layered minerals, 10
buserite – "defective lithiophorite", the mixed layered minerals, 10
Bushweld Intrusive Complex, South Africa, 183

cadmium, 270
calcite, 26, 190, 191, 205, 206, 209, 210
calcium, 41, 74–76, 95, 100, 113, 114, 122, 125, 129, 142–145, 185, 190, 191, 196, 205–207, 217, 268, 271, 272, 280, 284
Cambrian, 24, 198, 201, 207, 296, 297
Campbellrand Subgroup, 170, 173, 175–178, 180, 181, 183, 184, 296
cap, manganese (cuirass), vi, vii, 118, 120, 124–127, 130, 131, 217, 218, 277
Cape Province, Mn deposits of, South Africa, 171, 172, 175, 282, 288
Cape Verde Plate, Atlantic, Fe–Mn crusts of, 287
carbon, 113
carbonate compensation depth (CCD), 286, 301
carbonates, manganese, 27, 28, 38, 57, 70, 72, 94, 97, 99, 106, 108–110, 113, 116, 120, 122, 124, 127, 128, 131, 136, 147, 149, 168, 277, 279, 283, 290, 292, 293, 297, 299
Carboniferous, 183
catagenesis, 132, 133
catalytic oxidation, 227, 274
celedonite, 136, 147, 148, 155, 158, 284, 285, 298
Central Kazakhstan, the Famennian hydrothermal-sedimentary ores of, 11, 20, 182, 283, 297, 302

cerium, 44, 45, 47, 49, 51, 53, 71, 80–82, 91, 92, 94, 95, 97, 101, 116, 143, 145, 154–160, 163, 187, 268
Challenger, H.M.S., 5
Champaner Group (Early Proterozoic), 294
Chapskaya Series, the Vendian, Siberia, 198
chemisorption, 117, 168, 210, 212, 219, 225, 227, 241, 242, 251, 258–260, 272, 287
Chiatura (Chiaturskoe) manganese deposit, Georgia, 3, 6, 7, 14, 105, 135, 283, 300
Chitradurga Group (Early Proterozoic), 292
chlorine, 258
chlorite, 200, 207, 214
chloritization, 284
chromium, 77–79, 129, 142–145, 167
Clarion and Clipperton fracture zones, deposits of Mn–Fe nodules of, 286
cobalt, 9, 77–79, 83, 85, 87, 122, 125, 143, 145, 150, 151, 184, 186, 190, 191, 219, 220, 223–229, 231, 234, 235, 237, 239–241, 260, 267, 269, 272, 273, 280, 281, 287
complex compounds, with organic matter, of metals, 264–274
complex-forming component, 264–274
concretions of manganese oxyhydroxides . . . , 32, 33, 37, 39, 43, 46, 48–51, 54, 55, 57, 60, 63–65, 69, 71, 73, 74, 77, 80, 83, 85–87, 89–93, 107, 121, 125–127, 177, 206, 209, 210, 273
concretions, digital, Groote Eylandt deposit, 39, 74, 77, 80, 83, 84–86, 88–93
copper, 9, 22, 44, 77–79, 83, 95, 99, 129, 131, 143, 145, 151, 184, 186, 240–243, 264, 267–270, 272, 273, 284, 285, 287
coronadite, 54, 103, 183, 189–191, 281, 283, 299
covellite, 285
Crimea Mountains, Mn karst caves of, 205
crusts, infiltrated accumulations of manganese oxy-hydroxides, 107, 115
crusts, oxyhydrodixe, of the World Ocean, viii, 5, 8, 9, 284–287, 302
cryptomelane, 10, 35, 36, 38–40, 42, 43, 45–54, 56–65, 67–72, 83, 84, 94, 96–100, 102–104, 106, 108, 110, 112, 115, 117, 118, 121–131, 140, 141, 146–152, 159, 168, 179, 182, 183, 189–191, 193–195, 206, 207, 209, 210, 215–218, 242, 277–283, 289, 292, 297, 299, 300
crystal chemistrym 10, 152, 243, 253, 264
Csárdahegy area, of the Úrkút deposit, 135, 139, 142, 144, 146, 148–151, 154–158, 160–166
cuirass, cap, manganese, 115, 124–131, 217
Cyprus, Mn mineralization of, 285

Damara Supergroup (Late–Middle Proterozoic), 295
deep-sea deposits, 284–288

Deep Sea Drilling Project (DSDP), 8, 13, 285
deep sea environment, 302
Devonian, 20, 207, 297
Dharwar Supergroup, 4, 292
diagenesis, 132, 201, 287
diagenesis, in sediments, 201, 287
diagenesis, in ores, 201
diagenetic processes, 8, 72, 136, 287
diaspore, 180, 183
dickite, 38
digenite, 285
dispersed sulphides, 285
Djilu fault block, North-Chinese Platform, 295
dolomite, 110, 111, 133, 137, 170, 172, 173, 175,
      177, 178, 180, 181, 185–191, 193, 194, 196–
      204, 207–209, 282, 289, 292–294, 296, 298,
      299
Drake Passage, 302
Dr. Guier Mine, near Bingen city, region of Rhine
      River, Germany, 207
Durnovskoe Mn deposit, Siberia, 201

East European Platform, Mn deposit of, 196
East Pacific Rise, region of accumulations poly-
      metallic sulphides and metalliferous sedi-
      ments, 284, 286
electrical double layer, structural features of, 252,
      252
electrolyte aqueous solutions, structural aspects of
      . . . , viii, 251, 253
electron diffraction pattern, 67, 69
electron microdiffraction, 10, 13, 59, 141, 275
electron microscopy, 38, 61, 230, 275
electron paramagnetic resonance spectroscopy
      (EPR), 256, 257
electrostatic aspects, 256
envelope grain, manganese oxyhydroxide, Groote
      Eylandt deposit, 39
Enisei Ridge, manganese deposits (Central Siberia),
      vii, 20, 197–199
Eocene, 104–106, 108, 110, 136, 137, 139, 207, 299,
      301
ephesite (Na-margarite), 180
epigenesis, 132, 133, 167, 201
epitaxial growth, 221, 224, 227, 230, 242
epitaxy, 221
eulysite, 288
Europe, 207
europium, 80–83, 91, 93, 154–158, 161, 162, 166,
      184, 187
evolution, of manganese ore formation in the Earth
      history, viii, ix, 7, 282, 288
experimental studies, 219–281, 287

factor analysis, 94, 95

factor, groups (clusters assemblages), 94, 95, 97–
      101
factor loadings, 94, 95
factor scores, 96, 99, 101
Fairfield-dolomites, of the Campbellrand Subgroup,
      South Africa, 178
feitknechtitem 275
feldspar, 103, 191
ferralite profile, 130
ferralite soils, 126, 127, 130
ferricrete, massive Groote Eylandt deposit, 40
ferruginous pisolites, 37
Francevillian Series (Early Proterozoic), 109, 110,
      115, 294
fulvo-compounds, 21, 264, 265, 268, 269

Galapagos Rift, the region of metalliferous sedi-
      ments, Pacific, 136, 284, 285
gallium, 143, 145
Gamagara Formation, South Africa, 172, 174–176,
      179, 181, 182
Gamohaan Formation, of the Cambellrand Subgroup,
      South Africa, 184
Gangpur Group (Early Proterozoic), manganese de-
      posits of, India, 289
garnet, Mn, vii, 121, 122, 124–129, 131, 215, 216,
      277, 289, 291
general classification of manganese deposits, 282–
      284
germanium, 264
Germany, 207
Ghaap Group, South Africa, 170, 176
Ghoriajor Mn deposit of, India, 289
gibbsite, 70, 112, 115, 277
glanconite, 25, 26, 38, 70, 300
glauconite-smectite, mixed-layered phase, 70
glanconization, 70
globule, of manganese oxyhydroxide, the Groote Ey-
      landt deposit, 42, 46, 48, 49, 54
Glomar Challenger, R.V., 8, 285
Glosam, Mn deposit, Postmasburg region, vi, 174,
      176, 177, 181
glycerol treatment method, 60
glycol treatment method, 60
godolinium, 80–82
goethite, 32, 33, 36, 39, 40, 42, 43, 49–51, 53, 54,
      56, 57, 59, 61–63, 66, 67, 71, 83, 84, 104,
      112–115, 122, 123, 136, 140, 141, 146, 149,
      150, 153, 168, 189, 200, 202, 207, 208, 251,
      252, 264, 281, 285, 297
gold, 267
gondite (i.e. spessartine) association, and rock, 4,
      283, 288
Graphit System (Early Proterozoic), 291

greenstone series, of manganese volcanogenic-sedimentary formations, 6, 201

Griqualand West Region, Mn deposits of, South Africa, vi, 14, 170, 171, 175, 176, 180, 288, 292

Griquatown fault zone, Mn deposits of, South Africa, 171

Griquatown Iron-Formation, South Africa, 176, 183, 188

Groote Eylandt Group (Beds) (Layers), Precambrian, 24, 25, 27, 30, 32, 45

Groote Eylandt Mn deposits, North Australia, v, 23, 28–31, 33–35, 39, 41, 44, 45, 47, 49, 51, 53, 54, 58–66, 68–70, 72, 74, 77, 80, 83–93, 95–98, 100–102, 104–107, 109, 135, 195, 282, 299

ground water, 114, 116, 130, 132, 274, 276–278

groutite, 114, 117, 123, 217, 279

Guinier-De Wolff camera, 230, 237, 238, 242

Gujarat, state, manganese deposits of, India, 294

hafnium, 154, 155, 186

halloysite, 38, 50, 52, 57, 67, 68, 70, 94, 97

halmyrolysis, 5, 287

halmyrolytic leaching of Mn, 5, 287, 288

hardground, manganese oxyhydroxide ores, the Groote Eylandt deposit, 39, 72, 75, 78, 81, 83–93

Hartley lava, of the Olifantshoek Group, South Africa, 181

hausmannite, 103, 117, 119, 172, 178, 182, 188, 201, 238, 240, 275, 276, 282, 283, 287, 289, 292, 295–298

Helmholtz plane, 258

hematite, 33, 39, 40, 49, 51, 109, 112, 140, 146, 172, 173, 176, 178–181, 188, 189, 200, 207, 251, 252, 285, 296

Hess Basin, Pacific, Mn–Fe metalliferous sediments of, 286

Heynskop Formation, South Africa, 172

High Atlas region, Mn deposits of, North Africa, 188

hollandite, 42, 48, 54, 103, 189, 191, 207, 242, 278, 281, 283, 291, 294–296, 299

Hotazel Formation, South Africa, 170, 172, 188

Huanglong Suite (Middle Carboniferous), China, 297

humic acids, 21, 22, 214, 267

humus, 214, 264, 267

humus compounds, 264, 265, 267, 269

Hungary, 20, 134–139, 142, 144, 146, 147, 154, 155, 285, 298

hydration of irons (see also iron hydration), 243, 251, 253, 255

hydrogen, 241

hydrogenic activity of the bottom water, 4, 285–287

hydrogoethite, 140, 141, 146, 147, 149–153, 159, 168

hydrohausmannite, 117, 241

hydrohematite, 207

hydrolysis, 22, 123, 129, 159, 168, 214, 228, 229, 240, 248, 259, 265

hydrolytic reaction, 123, 128, 242, 259, 260, 264, 273, 274, 277

hydromica, 38, 43, 50, 52, 59, 66, 94, 97, 108, 140, 141, 146–150, 152, 153, 168, 196, 200, 202

hydrothermal fields, 285

hydrothermal sediments, 135, 153, 283, 284

hydrothermal solution, 133, 136, 283–285, 287, 302

hydrothermal vent, 285

hypotheses of manganese ore formation, 5

Idikel manganese deposit, Morocco, 295

illite, 113, 114, 205, 206, 216, 300

Imini Mn deposit, Morocco, vii, 188–190, 194, 195, 299

Imini-Tasdremt type of manganese formations, 196

India, 22, 107, 288, 289, 291, 292, 294

infiltration Mn accumulations, 20, 282

infrared spectra, 38, 66, 135, 140, 269

inter-element relationships, 94–102, 163–167

International Geological Congress, 11, 12, 14, 23

iodine, 258

ion-exchange, sorption, 226–228, 239–242, 245, 251, 258, 259

ion hydration (see hydration of ions)

ion hydration, in electrolyte aqueous solutions, viii, 253

Irkutsk Amphitheater, Mn karst deposits of, Russia, 206

iron, 1, 3, 41, 45, 47, 49, 51, 53, 54, 71, 74–76, 83, 85, 86, 94, 95, 97, 100, 109, 110, 116, 122, 125, 128, 129, 142–145, 150, 170, 177, 178, 185, 196, 202, 207, 208, 211, 217, 219–268, 270–273, 275, 283–285, 287, 288

iron-manganese ratio, 71, 83, 85, 86

Iron Ore Group, of India, 288, 289

itabirite, 288

jacobsite, 172, 173, 178, 182, 282, 283, 289, 291, 292, 294–296

Janggun Mn mining area, South Korea, 208

Janggun Suite, Cambrian, South Korea, 208

janggunite, 208, 209

jaspilite, 288, 295

Jhabua District, manganese deposits of, India, 294

Jurassic, 5, 104, 134–137, 169, 203, 205, 285, 298, 301, 302

K–Ar age of Mn oxyhydroxide (cryptomelane ores, Groote Eylandt), v, 102–104, 106, 108, 110

Kaapvaal craton, Mn deposits of, South Africa, 170, 171, 288

Kalahari Mn deposits, South Africa, 11, 283, 292

Kalahari Province, Mn deposits of, South Africa, 288

kaolinite, clay, 30–35, 38, 40, 42, 43, 46–54, 58, 59, 66–70, 72, 74, 77, 83, 84, 94, 97, 108, 110, 112–114, 120, 125, 126, 140, 141, 146, 148, 152, 168, 200, 202, 206

Kapstewel Mn deposit, Postmasburg region, South Africa, vi, 176, 178

Karagasskaya Suite, Prisayan'e region, Siberia, Russia, 206

Karanataka State, India, Mn deposits of, 292

Karazhal'skoe and other Mn deposits, Central Kazakhstan, 9, 182, 297

Karoo Supergroup, South Africa, 177

karst, vi, 114, 132–136, 141, 150, 152, 173, 176, 177, 180, 192, 201, 203, 282

karst caves, modern, manganese mineralization in, vii, 139, 167, 205, 206

karst, manganese ores in, vi, 132, 134, 138, 140, 159, 167, 169, 176, 188, 192, 194, 195, 197, 202–204, 207, 282, 300

Khapchaskaya Suite (Archean) (Siberia), Russia, 290

Khondalite Group, India, 283, 289, 291

kinetics of oxidation, 213, 227, 241, 281

Kisenge-Kamata belt, Mn deposits of, Zair, 293

Klipfonteinheuwel Formation of the Campbell Subgroup, South Africa, 181

Klippan Formation, of the Campbellrand Subgroup, South Africa, 181

Koegas Subgroup, South Africa, 170, 172

Korannaberg fold belt, Mn deposits of, South Africa, 170, 171

Korea, 208, 209

Kosovskaya Suite of the Upper Tortonian, Ukraine, 196

Krasnoyarsk region, Mn karst deposits of, Russia, 206

Krest cave, Krasnoyarsk region, Mn karst deposits of, Russia, 206

Krylov Seamount, Mn–Fe oxyhydroxide crusts of, 287

Kuruman iron-formation, South Africa, 175, 176, 178, 183, 184, 188

Kuruman region, Mn deposits of, Cape Province, South Africa, 282

kutnahorite, 209, 283, 288, 292, 295, 298

Kuznetskii Alatau Ridge, Mn deposits of, Siberia, Russia, 297

Labinskoe Mn deposit, Russia, 7

Lafaiete, Mn deposit, Quidrilltero, Minas Gerais State, Brazil, 293

Lafayette region, Mn deposits of, Brazil, 216

Langmuir adsorption equation, 228

lanthanium, 44–47, 51, 53, 80–83, 88, 90, 94, 95, 100, 101, 143, 145, 154–159, 161–164, 187

laterite, crust, 20, 21, 43, 45, 48, 50, 52, 54, 58, 59, 70–72, 88, 99, 105–107, 110, 117, 118, 128, 216, 217, 279, 293

laterite, (Fe), 27, 30, 31, 34, 35, 40, 43, 45–47, 51, 52, 54, 85, 88

laterite manganese deposits, 20, 22, 48, 58, 70, 75, 78, 81, 83, 85, 116, 214, 294

laterite, (Mn), 33, 40, 42, 43, 46, 47, 49, 52–54, 91, 280, 282, 293, 299

lateritization, deep supergene alterations, vi, 27, 58, 59, 70, 71, 85, 87, 91, 106–110, 115, 283

lattice, crystal, 103, 104

layers of boundary water, 255–263

leaching, 20, 40, 109, 113, 115, 117, 118, 123, 129, 131, 132, 134, 167, 176, 177, 180–183, 188, 194, 195, 210, 274, 278, 282–284

lead, 77–79, 143, 145, 190, 191, 269, 270, 272, 283, 297

lepidocrocite (γ-FeOOH), 219–221, 224–226, 241

Leping area, Mn deposits of, China, 297

Liassic, 136–141, 147, 149–151, 167, 168, 195

limonite, 27, 139, 206, 207, 281

limonite pisolite, of Groote Eylandt Mn deposit, 32

lithiophorite, 11, 27, 36, 38, 39, 42, 43, 46–50, 52–54, 57–59, 70, 72, 99, 100, 102, 104, 105, 108–110, 112, 115, 121–125, 127–129, 131, 141, 168, 179, 180, 183, 189, 207, 215, 216, 276, 277, 279–282, 299

lithium, 122, 125, 183, 257, 276, 277

Lohatlha deposit, Postmasburg region, vi, 173, 176, 177, 181

Longtan Formation (Late Permian), 298

Lower Cretaceous carbonate manganese bearing rocks . . . Groote Eylandt deposit, 25

Lukoshi Complex (Early Proterozoic), 293

lumpy manganese ores, 32, 37, 52, 57, 71, 110, 114

lutetium, 154–158, 187

Madagascar, 291

Madhya Pradesh, State of India, Mn deposits of . . . , 283, 294

magnesium, 1, 41, 74–76, 94, 95, 97, 99–101, 113, 114, 122, 125, 129, 142, 144, 185, 190, 191, 196, 207, 217, 243, 250, 253, 263, 268–272, 280, 284

magnetic measurements, 106, 253

magnetite, 182, 184, 285, 296

Magnitogorskii Synclinorium, Mn deposits of, 302

Maharashtra, State of India, Mn deposits of, 283, 294

major ions, of the background electrolyte solution, the effect on sorption of the heavy metals, viii, 242, 261

Makganyene Diamictite, South Africa, 170

Malyi Khingan, Mn deposits of, Far East, Russia, 296

manganate, (δ), (2.4 Å), 197, 241, 278, 281

manganate, (7 Å), 128, 197, 240–242, 263, 271, 272, 277–279

manganate, (10 Å), 7, 13, 36, 38, 42, 53, 71, 72, 88, 105, 108, 130, 140, 141, 146–152, 159, 168, 179, 197, 209, 210, 241, 278, 289, 291, 292, 300

manganese-iron ratio, 71, 83, 85, 86, 177

managanese dioxide: α-MnO_2, 54, 128, 130

manganese oxide: β-MnO_2, 105, 130, 277, 281

manganese oxide: γ-MnO_2, 128, 120, 277

manganese oxide: δ-MnO_2 (see 2.4 Å-MnO_2)

manganite, vii, 7, 43, 46, 52, 57, 63, 72, 94, 97, 104, 105, 108, 111, 113, 114, 116, 117, 121, 122, 125, 127, 128, 130, 131, 136, 148, 168, 180, 194, 200–203, 208–210, 215–217, 275, 277–280, 282, 283, 289, 293, 295–298, 300

manganocalcite, 7, 38, 39, 196, 197, 200, 214, 283, 288, 290, 292, 293, 295, 297, 298–300

Manganore Iron Formation, South Africa, 176, 178

manganosiderite, 200

manganosite, 191, 201

mangcrete, cemented (oolitic-pisolitic), Groote Eylandt deposit, 37, 39, 73

mangcrete, laminar, Groote Eylandt deposit, 39, 73, 75, 78, 81, 83, 85–93

mangcrete, massive, Groote Eylandt deposit, 39, 73, 75, 78, 81, 83, 85–93

Mangyshlak Mn deposit, 7, 302

manjiroite, 179

Mapedi Formation, South Africa, 172, 175, 181

Maremane dome, South Africa, 172, 173, 175, 176, 179, 181

martite, 189

massive sulphide ores, 285

mercury, 258

Mesozoic, 111, 119, 177, 206, 207, 284, 301

metal-humus complexes, 265–271

metalliferous sediments of the World Ocean, vii, 5, 8, 9, 11, 13, 136, 153, 184, 284, 285, 287, 301

metamorphism, 132, 283, 291

Mexico, 298

mica, 70, 94, 103, 136, 140, 289, 294

mica-smectite clay, 66, 67

microorganism, 17, 213, 240

microscopic features, 55, 56, 59–63, 113

Mid-Atlantic Ridge, the regions of metalliferous sediments, 284, 286, 287

Mikheevskaya depression, Enisei Ridge, Siberia, Russia, 198

Minas-Gerais Series (Middle–Early Proterozoic), Brazil, 293

Minas Gerais State, Mn deposits of, Brazil, 117, 290, 293

Miocene, 105–107, 110, 136, 196, 286

Moanda deposit, Gabon, West Africa, v, 109, 110, 112, 115–117, 135, 195, 216, 282, 294

model, general geochemical, viii, xi, xii, 16, 213, 214, 279

model geochemical, of the karst Mn ore formation, vii, 210

model of manganese ore formation in supergene zone, vii, 213, 214

model, partial (particular) geochemical, xi, xii, 16, 22, 213, 214

modeling, experimental, of main links of manganese ore formation, 219–274

Mokhovyi (Mokhovoi) area of the Porozohinskoe Mn deposit, Siberia, Russia, 199, 202, 203

Mokta, Mn deposits of, Côte d'Ivoire, vi, 118, 120, 123–131, 214

Molongo, Mn deposits of, Mexico, 298

molybdenum, 77–79, 264, 267

monjiroite, 54

montmorillonite, 40, 196, 200, 300

Mooidraai Formation, 170

Moro do Mina, Mn deposits of, Brazil, 290

Moro do Urucum, State of Brazil, Mn deposits of, 282, 297

Mössbauer spectra, 235, 236, 242

Mössbauer spectroscopy, 230, 242

Mullaman Beds, Lower–Middle Cretaceous, 24, 25, 27, 30, 45

muscovite, 183, 200, 208

Newfoundland, Mn mineralization of, 285

nickel, 9, 22, 77–79, 83, 122, 125, 129, 142–145, 151, 184, 186, 219, 220, 222, 223, 225–232, 234, 235, 239–241, 267, 269, 272, 273, 280, 281, 287

Nikolaevskoe Mn deposit, Prisayan'e region, Siberia, Russia, 207

Nikopol' (Nikopol'skoe) manganese deposit, 3, 7, 135, 283, 300

niobium, 143, 145, 186

nodule chemistry, 83

nodule, manganese oxyhydroxide ore variety, the Groote Eylandt deposit, 32–35, 42, 48, 49, 51, 53, 54, 56, 71

nodules, oxyhydroxide of the World Ocean, viii, 5, 8, 9, 11, 13, 83, 88, 90, 260, 284, 286, 287,

301, 302
nonspecific sorption, 258–260
nontronite, 38, 40, 136, 280, 284, 285
North-Urals Mn deposits, 6
Nsuta, Mn deposits, Ghana, vi, 116–118, 120–122, 127, 131, 135, 282, 292
nsutite, 11, 38, 39, 46, 48, 50, 52, 58, 59, 108, 110, 112, 114, 115, 117, 121, 123–127, 130, 131, 209, 215–218, 277–279, 282, 291
nucleation, 277, 278

Obrochishte Mn deposit, Varna area, Bulgaria, 299
Olifantshoek Group, South Africa, 170–173, 175, 177, 181
Oligocene, 3, 6, 7, 20, 104–106, 108, 110, 135, 207, 283, 299, 300
oligonite, 296
Ol'khon Island, Baikal Lake, Siberia, Russia, 206
Oman, 285
Ongeluk Formation, South Africa, 170, 181
Ongeluk/Hekpoort Lavas of the Transvaal Sequence, South Africa, 181
oolite, manganese ores, Groote Eylandt deposit, 31, 33, 34, 37, 39, 40, 42, 43, 46–52, 55, 61, 68, 69, 72–74, 76, 77, 79, 80, 82–93, 105, 110
ophiolite zones, 285
Ordovician, 188, 297
ore chocalade, manganese and iron oxyhydroxides, Groote Eylandt deposit, 40, 74, 77, 80
organic matter, viii, 21, 113, 114, 133, 134, 211–214, 217, 264–281
organic matter, the role of . . . in the sorption process of transition metals, viii, 264–274
Orissa, state of India, 284, 289
Otjosondu Mn deposit, Namibia, Africa, 295
oxides of iron, 27, 39
oxides of manganese, 27, 39, 121
oxygen, 10, 106, 259, 262, 269, 274, 276–279, 302
oxyhydroxides of iron, 10, 27, 31, 33, 40, 42, 43, 46, 47, 54, 67, 94, 97, 101, 109, 115, 133, 136, 140, 193, 242, 243, 252, 258, 285, 286, 297, 301
oxyhydroxides of manganese, 10, 13, 27, 30, 31, 33–36, 38, 42, 43, 46–49, 51, 54, 59, 60, 67, 68, 72, 97, 99, 100, 101, 103, 104, 106, 109–111, 114, 115, 117, 123, 124, 126, 133, 135, 136, 138, 139, 177, 181, 193, 194, 200, 202, 205, 206, 208, 209, 228, 242, 243, 261, 274, 277, 285, 286, 290–293, 297, 301

Pacific, 10, 136, 285, 286
palaeomagnetism, 106
Palaeophytic, South Africa, 181
palagonite, 286
palagonitized basalts and hyaloclastites, 286

Paleogene, 286, 302
paleomagnetic determinations, of deep weathering and lateritization . . . , 106–108
paleokarst, vi, 170, 196
Panch Mahals District, Mn deposits of, India, 294
paragenesis, mineral, 22, 103, 127, 214
paragenetic, assemblages of components, v, 94, 102
parent rock, 109
partridgeite, 178, 180
pearls, Mn cave, in Groote Eylandt deposit, 39
Penchenginskaya Suite, Precambrian (Middle Protevozoic), Siberia, Russia, 197, 198
Penganga Beds (Proterozoic), 291
Permian, 188, 298
phosphate, 94, 97, 153, 158, 159, 164
phosphorous, 41, 74–76, 94, 95, 97, 101, 102, 142, 177, 185, 207, 283
phosphorite, 294
physical sorption, 251
pisolite manganese oxyhydroxides, 27, 33, 34, 37, 39, 40, 42, 43, 46–48, 51, 55, 56, 71, 85–87, 107, 108, 110, 112, 126
pisolite, manganese ore, 31–33, 37, 40, 42, 44, 47, 49, 50, 55, 68, 69, 72–74, 76–84, 88–93, 105, 106, 114, 115, 126
plagioclase, Ca, Na, 133, 196
Pleistocene, 3, 203
Pod'emskaya Suite, the Vendian, Siberia, Russia, 198, 199, 201, 202
polianite (pyrolusite), 114, 115, 179, 180
polymetallic nodules and crust-like Mn–Fe ores, 9, 286–288
Porozhinskaya grabensyncline, Eniseiskii Ridge, Siberia, Russia, 198
Porozhinskoe Mn deposit, Eniseiskii Ridge, Siberia, 20, 198–200, 204
porphyry series, of manganese volcanogenic-sedimentary formations, 6, 201
Postmasburg Group, South Africa, 170, 172
Postmasburg manganese deposits, Griqualand West, Cape Province, South Africa, vi, 20, 133, 135, 170, 173, 175, 176, 178–183, 195, 296
potassium, 41, 45, 47, 49, 51, 53, 54, 71, 74–76, 95, 99, 100–106, 114, 122, 125, 129, 142–145, 185, 190, 191, 196, 225, 226, 253, 261, 271, 278, 280–282, 284
Potoskuiskaya Suite, the Riphean, Siberia, Russia, 197, 198
Precambrian, 109–111, 113, 119, 120, 173, 182, 188, 191, 197, 214, 283, 284, 288, 296, 301
precipitation, 174
Predcarpathian marginal trough, Mn deposits of, 196
Prieniseiskaya eugeosynclinal zone, Siberia, Russia, 198

Priol'khon'e region, Mn mineralization of, Siberia, Russian, 206
Proterophytic, South Africa, 170, 181
protore, alteration of, (formation of the primary residual ores), vi, 21, 22, 109, 117–120, 122, 124, 127, 131, 214, 276
protore, carbonate, alteration of, vii, 22, 117, 121, 123, 128, 131, 133, 141, 216, 218, 276, 277, 279
protore, silicate, alteration of . . . , vii, 22, 121, 123–125, 214, 218, 276, 277, 279
psilomelane (see romanechite)
pyrite, 25, 26, 110, 111, 113, 116, 134, 169, 185–187, 196, 200, 216, 285, 299
pyrolusite, 1, 3, 7, 35, 36, 38–40, 42, 43, 45–54, 56–59, 66–68, 71, 72, 83, 84, 96, 98, 104, 105, 108–113, 116, 117, 121, 124–128, 130, 131, 140, 141, 146–153, 159, 168, 172, 177, 179, 180, 182, 183, 189, 190, 193–195, 198–200, 202, 203, 206–208, 210, 217, 218, 277–279, 282, 283, 288, 291, 292, 298–300
pyrolusite, radial ore variety, Groote Eylandt deposit, 39, 76, 79, 82, 83, 86–90, 92, 93
pyroxene, 133, 289, 294
pyroxmangite, 117, 290

quartz, 25, 32, 37, 38, 40, 42, 43, 46, 50, 52, 54, 56, 57, 59, 60, 63, 66, 72, 83, 84, 94, 103, 104, 110, 112, 113, 123, 126, 128, 129, 140, 141, 146–153, 159, 168, 170, 178, 182, 190, 191, 200, 208, 209, 216, 230, 252, 256, 263, 285, 290, 299
queluzite, 288

Rajastan State, Mn deposits of, India, 294
ramsdellite, 11, 50, 55, 121, 123, 124–128, 130, 131, 179, 215–218, 279, 291, 292
rancieite, 11, 197, 200, 202, 281
Ranibennur-Dandeli Group (Early Proterozoic), 292
rare earth elements, vi, 80–83, 88, 90–93, 135, 152–167, 169, 184, 187, 268, 285
Recent, 219
Red Sea, depressions of, accumulations of poly-metallic sulphides and oxyhydroxide sediments, 136, 282, 285
redox potential, 101, 102, 133, 210, 211
Reivilo Formation, South Africa, 176, 181
residual accumulations, soil, 107
Reti Island, Indonesia, Mn deposits of, 302
rhodochrosite, (Ca), 7, 26, 36, 111, 113, 115–117, 119, 121–124, 127, 136, 141, 147–149, 151, 153, 155, 156, 158, 168, 196–198, 200–203, 208–211, 216, 217, 283, 288, 290, 292–299
rhodonite, 117, 119, 198, 209, 289, 290–296
rift, 136, 185

Rio das Velhas Series, Brazil, 288, 290, 293
Riphean, Siberia, Russia, 197, 198
romanechite (psilomelane), 3, 36, 38, 43, 48, 50, 52, 53, 54, 57–60, 72, 83, 84, 99, 100, 102, 105, 108, 110, 172, 174, 177, 179, 180, 182, 183, 189–191, 193–195, 199, 200, 202, 203, 205, 206, 281, 283, 297–300
Rooinekke Iron-Formation Member, South Africa, 172
Russia, 197, 290, 295–298
rutile, 251, 252

Salair Ridge, Mn deposits of, Siberia, Russia, 207
salting out, salting in, in electrolyte aqueous solutions, viii, 253, 255
samarium, 80–83, 91, 93, 154–158, 161, 164–166, 184, 187
sand, Mn ore-bearing, Groote Eylandt deposit, 30–32, 39, 40, 71, 76, 79, 82
Sandur Belt, manganese deposits of, India, 292
Sao Joao del Rei Mn deposit, Quadrilltero, Minas Gerais State, Brazil, 293
saponite, 215
saprolite, 20, 107
Sausar Group, (Late Proterozoic), manganese deposits of, India, 294
scandium, 74–79, 154, 155, 186, 267
scanning electron microscope, 61–65, 221, 224, 227, 229, 230, 232, 241
Schmidsdrif Subgroup, South Africa, 170
seawater, viii, 220, 227, 228, 243–247, 249, 261, 270–272, 284, 285, 287
secondary ores, manganese oxyhydroxide, the Groote Eylandt deposit, 39
serpentine, 117, 281
Serra do Novio Mn deposit (Guinean Shield), Brazil, 117, 290
siderite, 182, 184–187, 200, 211, 296
siliceous ore, manganese oxyhydroxide, the Groote Eylandt deposit, 32, 37, 39, 40, 72, 76, 82–87, 89, 90, 92, 93
silicon, 41, 71, 94, 95, 97, 100–102, 114, 122, 125, 129, 142, 144, 185, 190, 191, 196, 207, 258, 276, 280, 284
siltstone, manganiferrous, calcarious, the Groote Eylandt deposit , 25, 26, 38, 40
silver, 251, 267
Singaiskoe occurrence, of Mn ore, Salair Ridge region, Siberia, Russia, 207
Sishen Fe deposit, South Africa, 175, 181, 182, 188
Sishen Shale Member, of the Gamagara Formation, South Africa, 133, 176, 177
smectite, vii, 25, 38, 43, 52, 70, 94, 97, 136, 140, 141, 146, 147, 168, 215, 279, 298, 301

sodium, 41, 95, 98–100, 114, 122, 125, 129, 142, 144, 185, 205, 243–250, 253, 257, 261, 263, 264, 270–272, 284
soil, Mn ore-bearing, Groote Eylandt deposit, 30, 31, 35, 45, 47, 53
soil, residual, 30, 31, 35, 45, 47, 53
sorbed layer, 221–224, 226, 233–240, 248, 264
sorption, viii, 169, 213, 219, 226, 230–233, 237, 240, 242–249, 251, 252, 255, 256, 258, 260, 261, 271, 273, 274, 279, 281
sorption processes, 22, 214, 242–248, 258, 271–274, 280
sorption, relationship between various kinds of . . . , viii, 242, 247, 251, 258–260
sorption synthesis, vii, viii, 219, 227, 240–242, 281
Sosnovskaya Suite (Riphean), Siberia, Russia, 197, 198
South Africa, 11, 14, 20, 133, 135, 170, 185–187, 282, 283, 288, 292, 300
South-Ukranian Mn deposits, 6, 20, 195, 300
Southern Siberia, Mn karst deposits of, 206, 207
Soviet Union (see: USSR)
specific sorption, 251, 258, 259
spessartine, 4, 119, 128, 129, 131, 214, 216, 276, 277, 279, 289–296
spherulite (manganiferous), 37, 39, 40, 76, 79, 82, 83
stalactite, Mn-bearing, 206
Stern layer, 251, 258, 261, 263
stockwork sulphides, 285
stratiform sulphides, 284, 285
stromatolite, 172, 173
strontium, 44, 45, 47, 49, 51, 53, 74–76, 95, 98–101, 142–145, 186, 268
structure of acqueous solutions of electrolytes, viii, 242, 251, 253
structure of water, near the solution/solid phase interface, viii, 251, 253, 255–264
submarine volcanism, 284–286
Sukhopitskaya Series (the Riphean), Siberia, Russia, 197
sulphur, 142, 144, 185, 206, 229, 237, 253, 261, 266, 267, 271
synthesis, sorption, of the oxyhydroxide phases of Mn, Fe, Ni, Co and Cu on birnessite, 240–242
synthesis, sorption, of the oxyhydroxide phases of Mn, Fe, Ni, and Co, on iron oxyhydroxides 219–227
synthesis, sorption, of the oxyhydroxide phases of Mn, Fe, Ni, Co and Cu on manganese oxyhydroxides (hausmannite), 219, 226–240

takanelite, 210
takanelite-rancieite, 210

Taman Formation (Kimmeridgean age, Late Jurassic), 298
Tambao Mn deposit, Burkino Faso, v, 52, 118, 120, 122, 124, 126–128, 131, 214, 293
tantalum, 154, 155, 186
Tasdremt Mn deposit, Morocco, 190, 191, 299
Teiskaya Series, Precambrian (Middle Protecozoic), Siberia, Russia, 197, 296, 297
tephroite, vii, 119, 214–216, 279, 290, 291, 294
terbium, 154, 155, 187
Tertiary, 105–107, 109, 111, 120, 207
Tethyan region, 302
Tethys, marginal seas of, Mn deposits of, 136, 302
thulium, 154, 155
Tiling Formation (Middle Proterozoic), 295
Timor Island, Mn deposits of, 302
tin, 251
Tiouine Mn deposit, Morocco, 190, 191, 295
titanium, 41, 74–76, 83, 85, 94, 95, 97, 100, 101, 122, 125, 129, 142–145, 150, 185, 196, 202, 251, 267, 270
todorokite, 10, 36, 38–40, 42, 47, 48, 53, 54, 57–59, 66–69, 71, 72, 83, 84, 99, 100, 102, 105, 110, 130, 141, 146, 148, 168, 281, 282, 284, 285
Tokminskaya Suite, the Riphean, Russia, 197, 198
transmission electron microscope, 66–68, 230, 241, 242
Transvaal Supergroup (Sequence), Early Proterozoic, 14, 170–173, 181, 184–187, 282, 283, 288, 292
Triassic, 137, 188
Turocka Series (Archean), East Africa, 290
Tungusikskaya Series, the Riphean, Siberia, Russia, 197

Ukraine, vi, 20, 109, 196, 293, 300
Ulu-Telyakskii subtype of manganese formations, 196
Ulu-Telyakskoe Mn deposit, Bashkirskoe Predural'e, Russia, 298, 302
United States of America, (U.S.A.), 11
Urandi Mn deposit, Brazil, 296
uranium, 77–79, 154, 155, 186, 264
Úrkút manganese deposit, Hungary, vi, 20, 134–139, 142, 144, 149–151, 154–158, 160–166, 285, 298
Usinskoe, Siberia, Russia, 297, 302
U.S.S.R., 283

vadose manganese minerals, 3
vanadium, 77–79, 143, 145, 186, 264, 267
Vendian, Siberia, Russia, 198–204, 296
Verbovetskii Beds, of the Upper Tortonian, Ukraine, 196

342 *Index*

vernadite, 10, 39, 59, 76, 97, 200, 202, 203, 241, 283, 297, 298
vicinal water, 253–264
vicinal water layer, 253–264
viscosity, of electrolyte solution, e.g. seawater, 261–264
volcanogenic sediments, 7, 201, 207, 214, 283, 285
Vorogovskii Orogenic Trough, Siberia, Russia, 198–200, 291
vredenburgite, 289, 291, 294, 295

wad, 3, 4, 37, 76, 83, 84, 174, 189, 205, 206, 209
Wafangzi ferromanganese deposit, China, 295
water, structural properties near the solution/solid phase interphase, 253–264
weathering crust, v, 20, 50, 59, 70, 118, 128, 133, 141, 152, 168, 202, 203, 206, 207, 212–219, 278, 282, 289, 299
weathering crust, profile, 21, 50, 88, 91, 106, 107, 112, 116, 117, 121, 128, 131, 214–217, 274, 277, 281
weathering crust, tropical, v, 2-, 37, 50, 52, 59, 67, 70, 72, 118, 282, 291, 299
West Transvaal, Mn deposits of, South Africa, 300
Western Ukraine, manganese deposits of, 20, 196
Wolhaarkop Breccia, South Africa, 175, 176, 178
Wolkenberg Group, South Africa, 170
World Ocean Mn–Fe metalliferous sediments reserves, vii, 284, 287, 301

World Ocean Mn–Fe nodules reserves, viii, 286, 287, 301
world reserves of Mn ores on land, 288, 289–302
Wryburg Formation, South Africa, 170

X-ray diffraction, 10, 11, 58–60, 66, 141, 197, 206, 209, 224–226, 230, 237, 242, 275
X-ray diffractometry, 38, 44, 59, 66, 135, 140, 146, 148
X-ray energy dispersion analysis, 10
X-ray photoelectron spectroscopy, 10, 221, 226, 230, 242, 275
Xiangtan Mn deposit, China, 295

Yangze Platform, China, Mn deposits of, 295
ytterbium, 80–82, 90, 153, 154–158, 160, 162, 163, 167, 187, 268
yttrium, 44, 45, 47, 49, 51, 53, 95, 99, 101, 143, 145, 186, 268

Ziemougoula Mn deposits, Côte d'Ivoire, 214, 216, 282
zinc, 77–79, 131, 142–145, 184, 186, 251, 267, 269, 278, 283–285, 297
zirconium, 44, 71, 77–79, 83, 85, 87–89, 143, 145, 184, 186
Zunyi area, Mn deposits of, China, 298

Solid Earth Sciences Library

Publications:

1. E.I. Galperin: *The Polarization Method of Seismic Exploration.* 1983
   ISBN 90-277-1555-6

2. S. Hastenrath: *The Glaciers of Equatorial East Africa.* 1984
   ISBN 90-277-1572-6

3. R.K. Verma: *Gravity Field, Seismicity and Tectonics of the Indian Peninsula and the Himalayas.* 1985
   ISBN 90-277-1864-4

4. R. Haenel, L. Rybach and L. Stegena (eds.): *Handbook of Terrestrial Heat-Flow Density Determination.* With Guidelines and Recommendations of the International Heat-Flow Commission. 1988
   ISBN 90-277-2589-6

5. E. Roth and B. Poty (eds.): *Nuclear Methods of Dating.* 1989
   ISBN 0-7923-0188-9

6. G. Wagner and P. Van den haute: *Fission-Track Dating.* 1992
   ISBN 0-7923-1624-X

7. J. Lima-de-Faria: *Structural Mineralogy.* An Introduction. 1994
   ISBN 0-7923-2821-3

8. I.M. Varentsov: *Manganese Ores of Supergene Zone: Geochemistry of Formation.* 1996
   ISBN 0-7923-3906-1

Kluwer Academic Publishers – Dordrecht / Boston / London